7. Ring Theory

8. Topology

An italicized page number refers to a term defined in an exercise.

The Structure of Proof

with Logic and Set Theory

Michael L. O'Leary

Concordia University
River Forest, Illinois

PRENTICE HALL
Upper Saddle River, New Jersey 07458

Library of Congress Cataloging-in-Publication Data

O'Leary, Michael L.
The structure of proof : with logic & set theory / Michael L. O'Leary.
p. cm.
Includes bibliographical references and index.
ISBN 0-13-019077-2
1. Proof theory. I. Title.

QA9.54 .O45 2002
511.3–dc21

2001040035

Acquisition Editor: George Lobell
Editor-in-Chief: Sally Yagan
Vice President/Director of Production and Manufacturing: David W. Riccardi
Executive Managing Editor: Kathleen Schiaparelli
Senior Managing Editor: Linda Mihatov Behrens
Production Editor: Bob Walters
Manufacturing Buyer: Alan Fischer
Manufacturing Manager: Trudy Pisciotti
Marketing Manager: Angela Battle
Marketing Assistant: Vince Jansen
Director of Marketing: John Tweeddale
Editorial Assistant: Melanie VanBenthuysen
Art Director: Jayne Conte
Cover Design: Kiwi Design

Printed in the United States of America

10 9 8 7 6 5 4 3 2 1

ISBN: 0-13-019077-2

Pearson Education LTD., *London*
Pearson Education Australia PTY, Limited, *Sydney*
Pearson Education, Singapore, Pte. Ltd.
Pearson Education North Asia Ltd., *Hong Kong*
Pearson Education Canada, Ltd., *Toronto*
Pearson Educaciûn de Mexico S.A. de C.V.
Pearson Education - Japan, *Tokyo*
Pearson Education Malaysia, Pte. Ltd.

for William F. Phillips,
my grandfather

Contents

Preface

There are many approaches that one can take with regard to a course dedicated to teaching proof writing. Some prefer to teach the mathematics and the structure of the proofs simultaneously. Others choose to teach the methods of proof and then apply those methods to various topics. This is the strategy of this text. Here we boldly jump into a discussion of logic and examine some details with the belief that if the details are not understood well, then the application of those details will suffer later. Moreover, it is the understanding of how logic interacts with mathematics that empowers the student to have the courage and confidence to tackle greater problems in courses such as *Abstract Algebra* or *Topology*. How we accomplish this is outlined below.

TEXT OUTLINE

This text is designed for a one semester course on the fundamentals of proof writing for students with a modest calculus background. The text is divided into three parts: **Logical Foundations**, **Main Topics**, and **Coming Attractions**.

Part I: Logical Foundations. Since it is a requirement for any proof, the text begins with an introduction to mathematical logic. This part begins by studying sentences with and without variables and concludes by writing basic paragraph-style proofs.

1. **Propositional Logic.** In this chapter we translate propositions using logical symbols and translate the symbols into English. Connectives and truth tables are covered, and formal two-column proofs are introduced. These proofs require students to carefully follow the rules of logic and serve as a model for paragraph proofs.
2. **Predicates and Proofs.** Here we cover basic sets, quantification, and negations of quantifiers. Two-column proofs are written using propositional forms with quantifiers. Strategies that are covered include Direct and Indirect Proof, Biconditional Proof, and Proof by

Cases. A transition to writing paragraph-style proofs is included throughout.

Part II: Main Topics. The logic covered in the first part can be viewed as an advanced organizer for writing proofs. This is best seen in the first application of Part II: set theory. This is a natural first choice, for set theory is just one step from logic. Sets are found throughout the part as we study induction, well-ordered sets, congruence classes, relations, equivalence classes, and functions.

3. **Set Theory.** The basic set operations as well as inclusion and equality are covered. Two sections are devoted to families of sets and operations with them.
4. **Mathematical Induction.** Various forms of mathematical induction are studied. Applications include combinatorics, recursion, and the Well-Ordering Principle. Under recursion we look at the Fibonacci sequence and (just for fun!) the golden ratio makes an appearance. (This is the inspiration for the cover graphic.)
5. **Number Theory.** We take a look at the axioms of number theory and discuss topics such as divisibility, greatest common divisors, primes, and congruences.
6. **Relations and Functions.** As a prelude to our look at functions, relations are examined. The main example is the equivalence relation. Our look at functions includes the notions of well-defined, domain, range, one-to-one, onto, image, pre-image, and cardinality.

Part III: Coming Attractions. The topics seen in this last part will become familiar to the student of mathematics. Even so, logic and sets are not forgotten. We see these subjects at work when we study rings and then move from the discrete to the continuous and study topology.

7. **Ring Theory.** The study of ring theory is viewed as a generalization of the study of the integers. We encounter integral domains, fields, subrings, ideals, factors, homomorphisms, and polynomials.
8. **Topology.** The text closes with a look at the various spaces and subsets that are important to topology. Topics include metric spaces, normed spaces, open and closed sets, isometries, and limits.

As this overview suggests, the text contains more topics than can be covered during a semester. This provides greater flexibility for the instructor, and it gives the student a one-stop reference on the basics of undergraduate mathematics and the proof structures needed for success. This will benefit mathematics students as they take their upper level courses in algebra and analysis. It will also benefit mathematics education students who must teach many of the fundamental concepts.

COURSE STRUCTURES

The following flowchart presents the basic chapter dependencies of the text. Section numbers within parentheses indicate a section that is not necessary for the next topic.

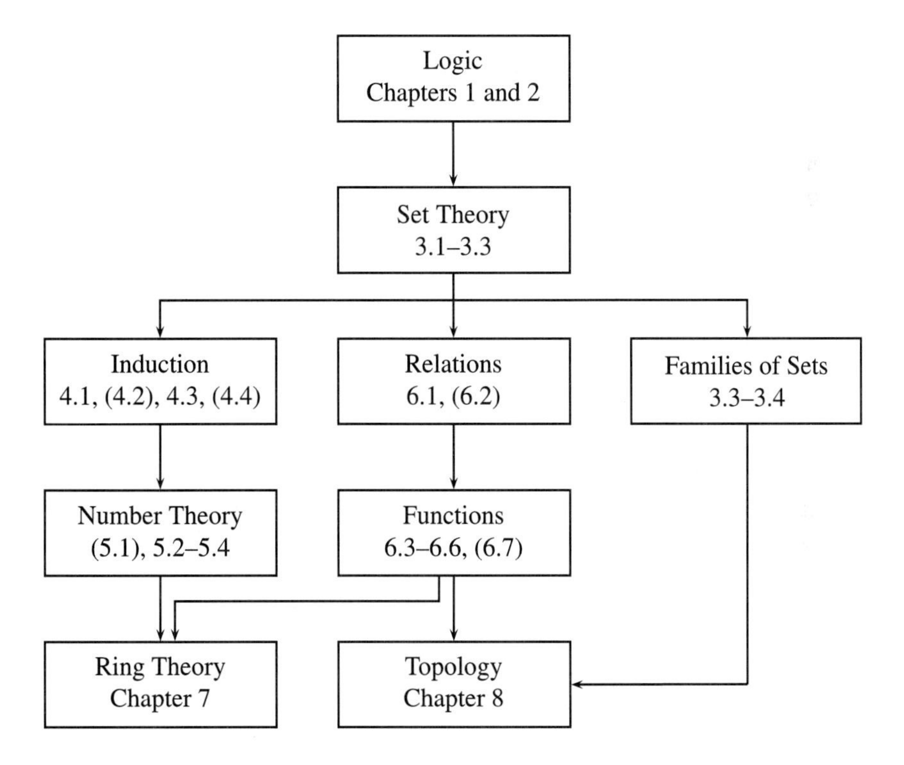

Notes concerning the dependencies:

1. Section 2.4 introduces two-column proofs with quantifiers and covers universal and existential paragraph proofs. This section can be used without the two-column proofs.
2. Section 2.6 covers various proof methods, the most important of which is proof of biconditionals. If time warrants, this section can be skipped, and the methods introduced as needed.
3. Section 4.2 serves to provide applications of mathematical induction to topics in combinatorics. This section is not referenced elsewhere in the text.

4. Section 4.4 includes the Well-Ordering Principle, which is used in the proofs of the Division Algorithm (5.2.2), Theorem 5.2.6, the fact that every ideal of $\mathbb{Z}$ is principal in Section 7.2, and the Polynomial Division Algorithm (7.5.3). If Section 4.4 is skipped, these results can be used without working through their proofs.
5. Section 5.1 covers the number theory axioms. Although they are mentioned occasionally throughout chapter 5, this section can be omitted.
6. A few examples in Chapters 6 and 7 use congruence classes.
7. Section 6.2 is referenced in Section 7.3 when it is noted that the set of cosets forms a partition for a ring.
8. Families of sets make occasional appearances in Chapters 6 and 7 but they are more crucial to Sections 8.2 and 8.3.
9. Section 6.7 on cardinalities can be safely omitted.

The outline of the text was designed with two course strategies in mind:

1. *A course which has Parts I and II as its focus.* This provides the student with a good introduction to mathematical logic with applications. The goal should be to cover most or all of Part I and choose desired topics from Part II. These topics should include the chapters on set theory (3) and mathematical induction (4). Topics from Part III can be used if time permits or as projects for advanced students.
2. *A course which has Parts II and III as its focus.* This may be appropriate for a class with students who are already strong in logic. Here, Part I can serve as a reference. The goal would be to cover most of Part II and selected topics from Part III.

EXERCISES

Of course, exercises play a crucial role in the text. Each section concludes with a set of problems to solve. These **Exercises** range from basic tests of understanding to the practice of writing proofs using the topics just discussed. Certain exercises have solutions given in the back. Each chapter in Parts II and III also concludes with an exercise set. These **Chapter Exercises** are intended to be questions that reflect a higher level of difficulty than the section exercises, combine topics from the entire chapter or possibly earlier in the text, or introduce new ideas inspired by the topics in the chapter.

SOLUTIONS

Solutions are provided to selected exercises indicated by boldface. If only a part of an exercise has a solution, then the letter of the part is in boldface, else the exercise number is in boldface. The solutions take the form of complete answers, partial solutions, or hints.

DEFINITIONS, THEOREMS, AND PROOFS

The layout of the text is designed to emphasize proof writing. Main definitions and theorems are set apart with lines from the rest of the text. This includes the proofs of theorems which are given in detail. The proofs are within the theorem area and indented. The idea is to allow the student to quickly focus on the result and its demonstration. Furthermore, the theorems and main definitions are numbered sequentially by section so that they can be found easily.

EXAMPLES

Examples are used extensively throughout the text. These include computations or short results that illustrate a theorem or, in some cases, are theorems themselves. Key examples are denoted by the word **Example** and are indented from the rest of the text.

DIAGRAMS

Diagrams are used when appropriate. These not only display traditional topics such as function and relations, but they also serve to illustrate the more "abstract" notions of logic. The diagrams are often completely boxed as figures.

HISTORICAL NOTES

When a mathematician is noted in the text, usually with regard to a famous theorem, a footnote is included giving a brief description of the mathematician including dates and cities of birth and death.

APPENDICES

There are three appendices that serve as references for the student. The first one, **Logic Summary**, provides a listing of the logic results cited in Part I. The next appendix, **Summation Notation**, provides a quick reference for the *sigma* summation notation. These ideas are used with the generalized set operations, mathematical induction, and polynomials. The last appendix, the **Greek Alphabet**, lists all upper and lowercase Greek letters and their names.

INSTRUCTOR'S MANUAL

An instructor's solution manual is available. To receive a copy, contact the mathematics editor at Prentice Hall.

TECHNICAL NOTES

This document was typeset by the author using LaTeX. The PSTricks macro package was used for the diagrams. Other packages include: amsfonts, amsmath, amssymb, eucal, fancyheadings, graphics, ifthen, longtable, mathpi, mathtime, multicol, multind, times, and ulem. The document was prepared using MiKTeX 1.20 and compiled to PostScript using DVIPS 5.86.

ACKNOWLEDGMENTS

Thanks are due to George Lobell, the mathematics editor at Prentice Hall, and Bob Walters, my production editor, for their help and patience throughout various stages of the project. I wish to also thank the following reviewers for their helpful comments:

Gregory Budzban	*Southern Illinois University, Carbondale*
John J. Buoni	*Youngstown State University*
Maurice Burke	*Montana State University*
Evan Fisher	*Lafayette College*
Iraj Kalantari	*Western Illinois University*
Steen Pedersen	*Wright State University*
Stephen Spielberg	*University of Toledo*
Joel Zinn	*Texas A & M University*

Thanks also go to Kenneth Mangels and Manfred Boos for their reading of the manuscript and suggestions. I also appreciate the patience of my students over the last two years who had to suffer through early editions (some very early!) of the manuscript. I especially want to mention Sean Bowen, Rebecca Goude, Hector Hernandez, Alan Huebner, Jeremiah Johnson, and Nikol Ziegelbein who provided valuable student perspectives.

On a personal note, I would like to express my gratitude to my parents for their support and understanding throughout the years (especially since I moved to this "far away" land), to my brother for providing chauffeur services at Christmas and Angel games during the summer, to my grandfather who introduced me to computers, to David Elfman who taught me about science and programming, to Paul Eklof, Kenneth Mangels, and Robert Meyer who taught me a lot of mathematics, and to Barb Sebela who taught me to like Chicago.

Michael L. O'Leary
Concordia University
River Forest, Illinois

I Logical Foundations

In this first part we will study the basic components of proofs: propositions and predicates. We will learn about truth tables and write two-column logic proofs. These will provide the structure for our ultimate goal of writing paragraph proofs.

1 Propositional Logic

When we calculate an average, solve an equation, or differentiate a function, we are *working with* mathematics. We believe that our steps are legitimate, and we trust the result. But why should we? How do we know, for instance, that the product rule for derivatives gives the correct answer? The reason for our trust is that these results have been proven. They have been demonstrated to be true either by us or by someone else. When one proves sentences about numbers, equations, or functions, then one is *doing* mathematics. This is the subject of this book. We do not want to simply work with math, but we want to do it. We want to write proofs. To begin this process, we must look at the sentences that form the proofs.

1.1. PROPOSITIONS

Here is an example of a mathematical sentence:

if a polygon has three sides, then it is a triangle.

Since this sentence is true, we say that it has a ***truth value*** of **true**. The truth value of the next sentence is **false**:

there is a largest integer.

In mathematics we are only interested in statements that can be either true or false. This leads us to our first definition.

1.1.1. Definition

A ***proposition*** is a sentence that is either true or false, but not both. Propositions are also known as ***statements***.

Although no proposition can be true and false at the same time, there are many sentences that are neither true nor false.

Example. These sentences are not propositions.

1. *Oh no!*—Is this true or false? Neither! It is simply an exclamation.
2. *Study hard.*—This is a command. It is not stating that we do study hard, but it is giving an order.
3. *This sentence is false.*—If it is true, then it must be false. Conversely, if it is false, then the sentence is true. Hence, it cannot be a proposition, for both cases are impossible.

Sometimes sentences can be neither true nor false based on factors such as imprecision or poor structure.

Example. Take the sentence:

he is sad.

Is this true or false? It is impossible to know because we do not know to whom *he* refers. In this sentence the word *he* is acting like a variable as in $x + 2 = 5$. Since the value of x is unknown, it cannot be determined whether $x + 2 = 5$ is true or false. However, if we know the value of x, we could make a determination. Likewise, if we knew to which person *he* referred, then the sentence would be a proposition. (Be careful! This does not mean that we would necessarily know its truth value. We would simply know that it had one.)

Example. Random sequences of words like

big the runs fox,

or nonsense as

the sky runs faster than pink,

are not propositions. The first is not a sentence, and how are we to make sense of the second?

Let us now examine some basic types of propositions. Consider the following statements:

(1) *the base angles of an isosceles triangle are congruent,*

and

(2) *a square has no right angles.*

Using the words *and* and *or*, we may create two new propositions:

the base angles of an isosceles triangle are congruent,
and a square has no right angles;

and

the base angles of an isosceles triangle are congruent,
or a square has no right angles.

We know that the first proposition is false because proposition (2) is false. We also know that the second proposition is true because (1) is true. This leads to the following:

1.1.2. Definition

- When a proposition is formed by combining two propositions with the word *and*, it is called a ***conjunction***, and the two propositions are called ***conjuncts***. A conjunction is true when both of its conjuncts are true and false otherwise.
- When a proposition is formed by combining two propositions with the word *or*, it is called a ***disjunction***, and the two propositions are called ***disjuncts***. A disjunction is true when at least one of its disjuncts is true and false otherwise.

Example. The conjunction,

$3 + 6 = 9$, *and all even integers are divisible by two,*

is true, but

all integers are rational, and 4 *is odd*

is false because the second conjunct is false.

There are many variations on the conjunction. To illustrate, consider the sentence: *Nero fiddled, and Rome burned.* This could be written as:

Nero fiddled while Rome burned.
Nero fiddled, but Rome burned.
Nero fiddled plus Rome burned.
Nero fiddled, yet Rome burned.
Nero fiddled; however, Rome burned.

Each of these have their own nuances. However, they are all conjunctions. This is because each sentence asserts that both clauses are true.

Example. The disjunction

$3 + 7 = 9$, *or all even integers are divisible by three,*

is false since both disjuncts are false. On the other hand,

$3 + 7 = 9$, *or circles are round*

is true.

We must remember that only one disjunct needs to be true for the entire disjunction to be true. For this reason, the logical disjunction is sometimes called an ***inclusive or***. If an ***exclusive or*** is needed, make it explicit in the translation. For example, take the proposition

the number is greater than five,
or the number is less than five.

To make it exclusive, write:

the number is greater than five,
or the number is less than five, but not both.

Both *and* and *or* are called ***connectives*** because they connect propositions together to make new ones. Although *and* and *or* require two propositions, there is a connective that needs only one. It is the word *not*.

1.1.3. Definition

A ***negation*** of a proposition is a statement that has the opposite meaning of the original proposition. Therefore, the truth value of a negation is the opposite of the original proposition's truth value.

For example, the negation of $3 + 8 = 5$ is $3 + 8$ *is not equal to* 5 or, more simply, $3 + 8 \neq 5$. In this case we say that the proposition $3 + 8 = 5$ has been ***negated***.

Example. Connecting *not* with the proposition *a square has no right angles* yields

it is not the case that a square has no right angles.

A translation without the *not* is

it is false that a square has no right angles.

However, both of these are usually written as

a square has right angles.

A common type of mathematical proposition is exemplified by the following sentence. It states that the second proposition is true when the first one is true:

if the figure is a square,
then the figure is a rectangle.

Two propositions have been combined using the words *if* and *then.* We will consider this pair a connective. The if-then structure connects the two propositions. The first is called the ***hypothesis***, and the second is called the ***conclusion***.

1.1.4. Definition

A proposition of the form

if *hypothesis*, then *conclusion*

is called a ***conditional***. It is also known as an ***implication*** or an ***if-then statement***.

Example. In the above conditional, the hypothesis is

the figure is a square,

and the conclusion is

the figure is a rectangle.

Determining the truth value of a conditional is an interesting problem. The matter is straightforward when the hypothesis is true. In this case a conditional is true when the conclusion is true and false when the conclusion is false. For example, *if* $4+5=9$, *then* $2+3=8$ is false. If the hypothesis is false, then we cannot draw any conclusion from the conditional. Since we do not want to say that it is false under these circumstances, the conditional is said to be true. For instance, both *if* $2+3=8$, *then* $4+5=9$ and *if* $2+3=8$, *then* $4+5=19$ are true.

Since conditionals are so common, it is not surprising that they can be written in many ways.

Example. Here are more ways that *if the figure is a square, then the figure is a rectangle* can be written.

- The word *implies* is sometimes used in place of *if-then.* With this our example conditional becomes

  the figure is a square
  implies that the figure is a rectangle.

This usage is not as common as *if-then*, but as a connective, it does resemble *and* and *or* more closely.

- Another way to write this is to use *only if*:

  the figure is a square
  only if the figure is a rectangle.

  The *only if* phrase means that the proposition after *only if* must be true in order for the first proposition to be true.
- There are times when *then* does not accompany *if*, as when the conclusion precedes the hypothesis:

  the figure is a rectangle if the figure is a square.

  This is simply the original proposition. Other times it is simply left out as in

  if the figure is a square, the figure is a rectangle.
- Other possibilities include:

  the figure is a rectangle when the figure is a square;
  the figure is a square yields that the figure is a rectangle;
  the figure is a square gives that the figure is a rectangle.

Example. Sometimes a sentence that is an implication is not written in a way to make this obvious. For instance,

squares have four sides

is an implication. We have two options for the translation:

if the shape is a square, then it has four sides,

or

if the shape has four sides, then it is a square.

To identify the if-then structure, find the subject of the sentence. Since a sentence states something about its subject, if we have identified the subject, then we know the hypothesis. For this reason, the first translation is the correct one.

When studying an implication, we sometimes need to investigate the different ways that its hypothesis and conclusion relate to each other. For example, take the conditional,

$(*)$ *if it is raining, then I have my umbrella.*

To determine its truth value, we can examine the statement,

if I do not have my umbrella, then it is not raining.

1.1.5. Figure

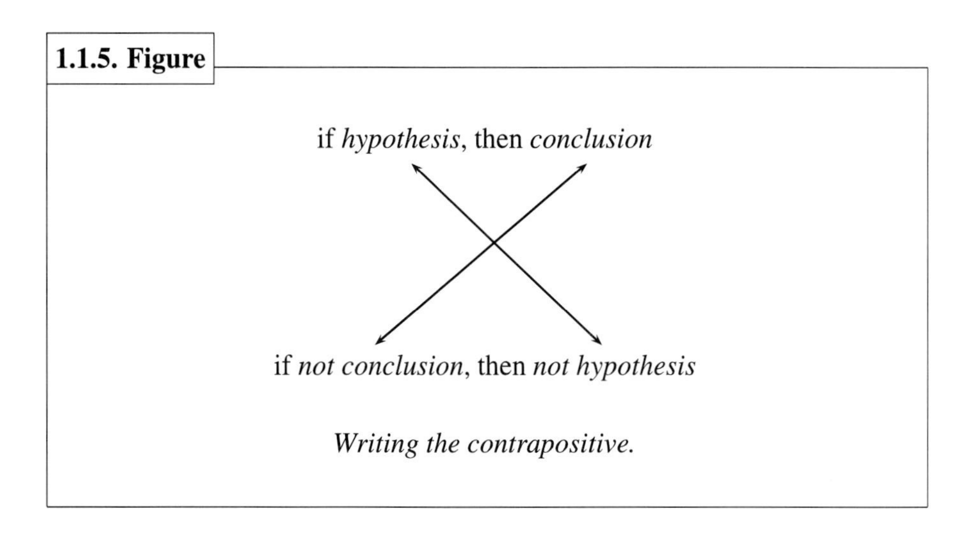

Writing the contrapositive.

This sentence is called the ***contrapositive*** of the original conditional. It is formed by negating both the hypothesis and the conclusion and then switching places. (See Figure 1.1.5.) In Section 1.4 we will prove that a conditional and its contrapositive are true at exactly the same times. In our example, if we know that it cannot be raining when I do not have the umbrella, then the original implication $(*)$ must be true.

The ***converse***, on the other hand, is different from the contrapositive in two important ways. First, it is formed by simply switching the hypothesis and the conclusion. (See Figure 1.1.6.) Second, it is possible that an implication has a different truth value than its converse. For instance, the converse of $(*)$ is

if I have my umbrella, then it is raining.

It could be true that whenever it is raining, I have my umbrella. However, it may be the case that I use my umbrella when it is sunny for some shade, which would mean that the converse is false.

Example. We need to carefully consider one possible implication translation: the use of *necessary* and *sufficient*. The word *necessary* means "needed" or "required," and *sufficient* means "adequate" or "enough." Thus, the sentence

it is raining is sufficient for me to have my umbrella

translates our original implication $(*)$. In other words, the fact that it is raining is enough for us to know that I have my umbrella. Likewise,

having my umbrella is necessary for it to be raining

1.1.6. Figure

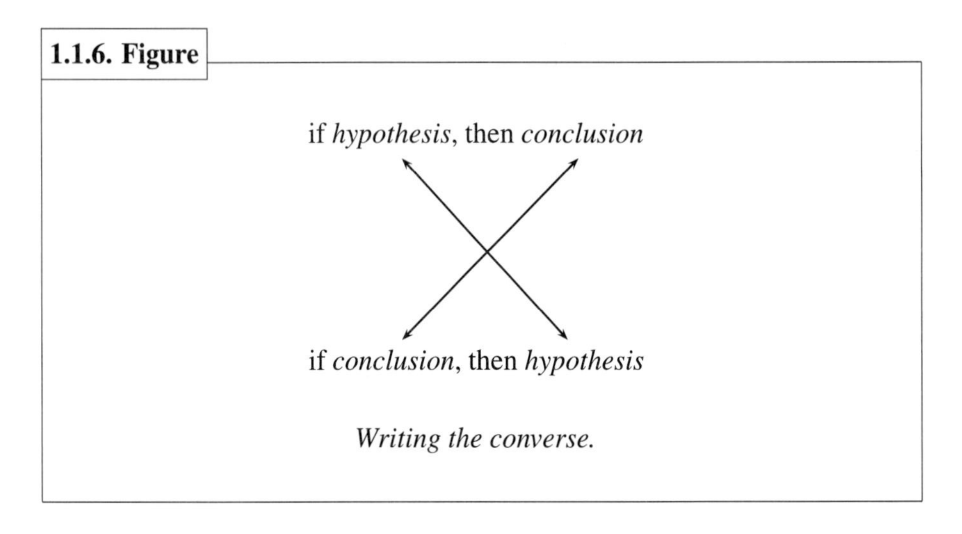

Writing the converse.

is satisfactory, for it means that having my umbrella is a condition that must bc truc whcn it is raining. Remember that the hypothesis is sufficient for the conclusion and the conclusion is necessary for the hypothesis:

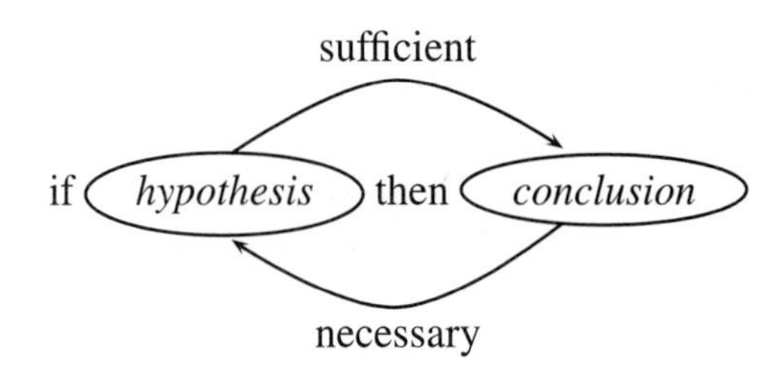

Suppose we know that both (∗) and its converse are true. We may then write:

if I have my umbrella, then it is raining,
and if it is raining, then I have my umbrella.

This has an unnecessary redundancy. To remedy this, notice that the first conditional can be rewritten as

it is raining if I have my umbrella

and the second conditional can be formulated as

it is raining only if I have my umbrella.

We may then substitute the original conjunction with the more compact

it is raining if and only if I have my umbrella

or the equivalent

I have my umbrella if and only if it is raining.

This is an example of our last proposition type.

1.1.7. Definition

A proposition of the form

proposition 1 if and only if *proposition 2*

is called a ***biconditional***.

The idea is that if it is true, then both propositions (those found on either side of the *if and only if*) are true at exactly the same times. They are either both true or both false. The *if and only if* phrase is our last connective.

Example. The proposition

a polynomial is a quadratic
if and only if it has at most two roots

can be rewritten as

a polynomial being quadratic
is necessary and sufficient for it having at most two roots.

Example. A ***definition*** is a particular type of biconditional that gives meaning to a word. For example, to define a parallelogram we may use the sentence

a parallelogram is a quadrilateral
that has parallel opposite sides.

This is shorthand for the biconditional

a quadrilateral is a parallelogram
if and only if its opposite sides are parallel.

Typically, the first half of the biconditional contains the word that is being defined, and the second half has the conditions that must be satisfied for an object to be designated by the term. Despite definitions being biconditionals, it is traditional to write a definition only as an if-then with the biconditional being understood. For example,

a quadrilateral is a parallelogram
if its opposite sides are parallel.

EXERCISES

1. Give two examples of sentences that are propositions from each of the following subjects:
   (a) calculus
   (b) algebra
   (c) literature
   (d) baseball
2. Identify each of the following as either a proposition or not a proposition. If it is a proposition, state whether it is true or false. If it is not a proposition, explain why.
   (a) *Hello.*
   (b) *Baseball players swing bats.*
   (c) $3+8=18$
   (d) $3+x=18$
   (e) *Define the term so that it is meaningful.*
   (f) *A triangle is a three sided polygon.*
   (g) *He will run for president.*
   (h) *This statement is true.*
   (i) *This statement is not true.*
3. Give the truth value of each of the listed propositions.
   (a) *A system of equations always has a solution, or a quadratic equation always has a real solution.*
   (b) *It is false that every polynomial in one variable has a derivative.*
   (c) *Vertical lines have no slope, and lines through the origin have a positive y-intercept.*
   (d) *Every integer is even, or every even number is an integer.*
   (e) *If every parabola intersects the x-axis, then a hyperbola has only one vertex.*
   (f) *The sine function is periodic if and only if every exponential function is always nonnegative.*
   (g) *Compounding interest is advantageous for savers, but it is a negative for those who borrow money.*
   (h) *It is not the case that $2+4 \neq 6$.*
   (i) *The distance between two points is always positive if every line segment is horizontal.*
   (j) *The derivative of a constant function is zero is necessary for the product rule being true.*
   (k) *The derivative of the sine function being cosine is sufficient for the derivative of the cosine function being sine.*
   (l) *Any real number is negative or positive, but not both.*

4. For each sentence fill in the blank using as many of the words *and*, *or*, *if*, and *if and only if* as possible to make the sentence true.
   (a) *Triangles have 3 sides* _______ $3 + 5 = 6$.
   (b) $3 + 5 = 6$ _______ *triangles have 3 sides.*
   (c) *Ten is the largest integer* _______ *zero is the smallest.*
   (d) *The derivative of a function that is a horizontal line is zero* _______ *increasing functions have positive slope.*
5. Determine whether each of the following propositions is a conjunction:
   (a) *The derivative is continuous, but the sum is negative.*
   (b) *The derivative is continuous implies the sum is negative.*
   (c) *The derivative is continuous if and only if the sum is negative.*
   (d) *The derivative is continuous when the sum is negative.*
   (e) *The derivative is continuous, yet the sum is negative.*
   (f) *The derivative is continuous is necessary and sufficient for the sum being negative.*
6. Identify the hypothesis and the conclusion for each of the following conditionals.
   (a) *If the triangle has two congruent sides, then it is isosceles.*
   (b) *The polynomial has at most two roots if it is a quadratic.*
   (c) *The data is widely spread only if the standard deviation is large.*
   (d) *The function being constant implies that its derivative is zero.*
   (e) *The system of equations is consistent is necessary for it to have a solution.*
   (f) *A function is even is sufficient for its square to be even.*
7. For each conditional in the previous problem:
   (a) Write its contrapositive and converse.
   (b) Write its inverse, where the ***inverse*** is the contrapositive of the converse.
8. Rewrite the implication

   if the integer is divisible by four, then it is even

   using the following words:
   (a) *implies*
   (b) *if* but not using *then* (do this two different ways)
   (c) *only if*
   (d) *necessary*
   (e) *sufficient*
9. Write each of the following using the words *if-then* or *if and only if*.
   (a) *Pentagons have five sides.*
   (b) *A triangle is a three-sided polygon.*
   (c) *Quadratics have at most two solutions.*
   (d) *All multiples of nine are divisible by three.*

   (e) *A necessary and sufficient condition for a polynomial to be a quadratic is that its degree is two.*
   (f) *The derivative of a constant is zero.*
   (g) *Lines with positive slope increase from left to right.*
   (h) *Polynomials always have derivatives.*
10. Explain why a proposition cannot be both true and false at the same time.
11. As noted in the section, the sentence *he is sad* can be considered a proposition if the person to whom *he* refers is identified. This means that the words on a page are not what makes a proposition. It must be something else. With this in mind, what exactly is a proposition, and how are the words on the page related to it?

1.2. PROPOSITIONAL FORMS

Consider what is involved in solving equations in algebra. Given an equation, we follow memorized rules to manipulate the symbols to find the value of the variable. Although we could explain why the steps are legitimate, we typically do not. The goal is to quickly answer the question and be confident about the answer. We will follow a similar strategy when we work to determine the truth value of propositions. We will use variables to represent propositions. We will then use rules that will allow us derive new propositions from given ones. Although we will understand the steps, we wish to arrive at a point when all of this becomes automatic, just like with algebra. The system that we will set up is called the ***propositional calculus***.

Let capital letters like P, Q, or R represent propositions. These are called ***propositional variables***. To assign a proposition to a variable, use the := symbol instead of a regular equal sign. We do this because sometimes the proposition includes an equal sign and two would be confusing.

Example. To assign the proposition

$$\textit{for all real numbers } x \textit{ and } y,\ x + y = y + x$$

to the propositional variable P, write

$$P := \textit{for all real numbers } x \textit{ and } y,\ x + y = y + x.$$

If a proposition includes connectives, then we will use various symbols to represent them. These symbols are listed in the next table.

connective	symbol
and	$\wedge$
or	$\vee$
not	$\sim$
implies	$\Rightarrow$
if and only if	$\Leftrightarrow$

The notation $P \Leftrightarrow Q$ is a combination of $P \Rightarrow Q$ and $P \Leftarrow Q$, the latter meaning "P if Q."

Example. To see how this works, make the assignments—

$P :=$ *it is raining,*
$Q :=$ *I have my umbrella,*
$R :=$ *I am struck by lightning.*

The following symbols represent the indicated propositions:

$P \wedge Q$	*It is raining, and I have my umbrella.*
$\sim R$	*I am not struck by lightning.*
$\sim P \vee \sim Q$	*It is not raining, or I do not have my umbrella.*
$Q \Rightarrow R$	*If I have my umbrella, then I am struck by lightning.*
$R \Leftrightarrow P$	*I am struck by lightning if and only if it is raining.*

These symbols are not propositions, but they represent propositions like expressions represent numbers in algebra. The symbols are instead called ***propositional forms***. Designate them with lower case letters, like p, q, or r, and use $:=$ for assignments.

Example. Make the following assignments:

$$p := R \Leftrightarrow (P \wedge Q),$$
$$q := (R \Leftrightarrow P) \wedge Q.$$

Since p and q are propositional forms, we may create new forms. The propositional form $p \wedge q$ is

$$[R \Leftrightarrow (P \wedge Q)] \wedge [(R \Leftrightarrow P) \wedge Q],$$

and $\sim q \Rightarrow p$ is

$$\sim[(R \Leftrightarrow P) \wedge Q] \Rightarrow [R \Leftrightarrow (P \wedge Q)].$$

Using the above assignments, what is the meaning of $R \Leftrightarrow P \wedge Q$? The possible interpretations are

I am struck by lightning if and only if
it is both raining and I have my umbrella

and

I am struck by lightning if and only if it is raining,
and I have my umbrella.

One way to eliminate the ambiguity is to introduce parentheses. The propositional form $R \Leftrightarrow (P \wedge Q)$ represents the first interpretation while $(R \Leftrightarrow P) \wedge Q$ is the second. However, as in algebra, another way to remove ambiguity is to introduce an ***order of operations***.

1.2.1. Order of Operations

To interpret a propositional form, read from left to right and use the following precedence:

1. propositional forms within parentheses (innermost first),
2. negations,
3. conjunctions,
4. disjunctions,
5. conditionals,
6. biconditionals.

Despite this order, we will sometimes use parentheses for clarification, so we will often write $P \Rightarrow (Q \wedge R)$ for $P \Rightarrow Q \wedge R$ or $(P \Rightarrow Q) \Rightarrow R$ in place of $P \Rightarrow Q \Rightarrow R$.

Example. To write the propositional form $\sim P \wedge Q \vee R$ with parentheses, we begin by interpreting $\sim P$. According to the order of operations, the conjunction is next, so we evaluate $(\sim P) \wedge Q$. This is followed by the disjunction, and we have $[(\sim P) \wedge Q] \vee R$.

Example. To interpret $P \wedge Q \vee R$ correctly, use the order of operations. We discover that it has the same meaning as $(P \wedge Q) \vee R$, but how is this distinguished from $P \wedge (Q \vee R)$ in English? Parentheses are not appropriate, for they are not used this way in sentences. Instead use *either-or*. Then using the assignments on page 15, $(P \wedge Q) \vee R$ can be translated as

either it is raining and I have my umbrella,
or I am struck by lightning.

Notice how *either-or* works as a set of parentheses. We can use this to translate $P \wedge (Q \vee R)$:

it is raining, and either I have my umbrella
or I am struck by lightning.

Be careful to note that the *either-or* phrasing is logically inclusive. For instance, some colleges require their students to take either logic or mathematics. This choice is meant to be exclusive in the sense that only one is needed for graduation. However, it is not logically exclusive. A student may take logic to satisfy the requirement yet still take a math class.

Example. Let us interpret $\sim(P \wedge Q)$. We may try translating this as *not P and Q*, but this represents $\sim P \wedge Q$ according to the order of operations. To handle a proposition such as $\sim(P \wedge Q)$, use a phrase like *it is not the case* or *it is false*. Therefore, $\sim(P \wedge Q)$ becomes

it is not the case that P and Q

or

it is false that P and Q.

To exemplify, make the assignments:

$P :=$ *the robin is bobbing along,*
$Q :=$ *the warbler is singing.*

Then,

the robin is not bobbing along, and the warbler is singing

is a translation of $\sim P \wedge Q$. On the other hand, $\sim(P \wedge Q)$ can be

it is not the case that robin is bobbing along
and the warbler is singing.

To interpret $\sim P \wedge \sim Q$, use *neither-nor*:

neither is the robin bobbing along,
nor is the warbler singing.

Every propositional form represents a proposition that is either true or false depending upon the situation and the structure of the form. We can examine these possible truth values by constructing a ***truth table***. We will begin by giving the five basic ones. Let p and q be propositional forms. The truth table for a negation is:

p	$\sim p$
T	F
F	T

On the left of the vertical line are all possible truth values for p. Use T for **true** and F for **false**. On the right are the results of those substitutions. Namely, when p is true, $\sim p$ is false, and when p is false, $\sim p$ is true. Similarly, the truth tables for conjunctions and disjunctions are:

p	q	$p \wedge q$
T	T	T
T	F	F
F	T	F
F	F	F

p	q	$p \vee q$
T	T	T
T	F	T
F	T	T
F	F	F

Since an implication is false only when its hypothesis is true and conclusion is false and the biconditional is true exactly when the truth values of p and q are the same, the truth tables for a conditional and a biconditional are as follows:

p	q	$p \Rightarrow q$
T	T	T
T	F	F
F	T	T
F	F	T

p	q	$p \Leftrightarrow q$
T	T	T
T	F	F
F	T	F
F	F	T

These truth tables along with the order of operations will allow us to find truth tables for any propositional form.

Example. To write the truth table of $P \Rightarrow (Q \wedge \sim P)$, list the propositional variables and build the propositional form using the order of operations:

$$P,\ Q,\ \sim P,\ Q \wedge \sim P,\ P \Rightarrow (Q \wedge \sim P).$$

On the left of the vertical line place columns for P and Q. Then fill in all possible truth value combinations for these two propositional variables. On the other side put a column for the other entries in the above sequence. Then moving from left to right, use the appropriate columns to find truth values for each of the pieces of our propositional form. The result is the truth table of $P \Rightarrow (Q \wedge \sim P)$:

P	Q	$\sim P$	$Q \wedge \sim P$	$P \Rightarrow (Q \wedge \sim P)$
T	T	F	F	F
T	F	F	F	F
F	T	T	T	T
F	F	T	F	T

The table says that when P is **true** and Q is **false**, $P \Rightarrow (Q \wedge {\sim}P)$ is **false**. The lines show how columns are formed from earlier columns using the connectives.

Example. We will use a truth table to find the truth value of

if the derivative is zero and the second derivative is positive, then the function has a relative minimum

when the derivative is zero, but neither is the second derivative positive nor does the function have a relative minimum. First define:

$P :=$ *the derivative is zero,*

$Q :=$ *the second derivative is positive,*

$R :=$ *the function has a relative minimum.*

So, P is true, but Q and R are false. The proposition is represented by $(P \wedge Q) \Rightarrow R$, and its truth table is:

P	Q	R	$P \wedge Q$	$(P \wedge Q) \Rightarrow R$
T	T	T	T	T
T	T	F	T	F
T	F	T	F	T
T	**F**	**F**	**F**	**T**
F	T	T	F	T
F	T	F	F	T
F	F	T	F	T
F	F	F	F	T

The fourth line shows that the proposition is true. Notice that we could have answered the question by simply writing one line from the truth table:

P	Q	R	$P \wedge Q$	$(P \wedge Q) \Rightarrow R$
T	F	F	F	T

To make clear the truth value pattern, note that if there are n variables, the number of rows is twice the number of rows for $n - 1$ variables. To see this, start with one propositional variable. Such a truth table has only two rows. Add a variable, we get four rows. The pattern is obtained by writing the one variable case twice. The first time it has a T written in front of each row. The second copy has an F. To obtain the pattern for three variables, copy the two-variable pattern twice. (See Figure 1.2.2.) To generalize, if there are n variables, there will be 2^n rows.

We close the section with two important types of propositional forms.

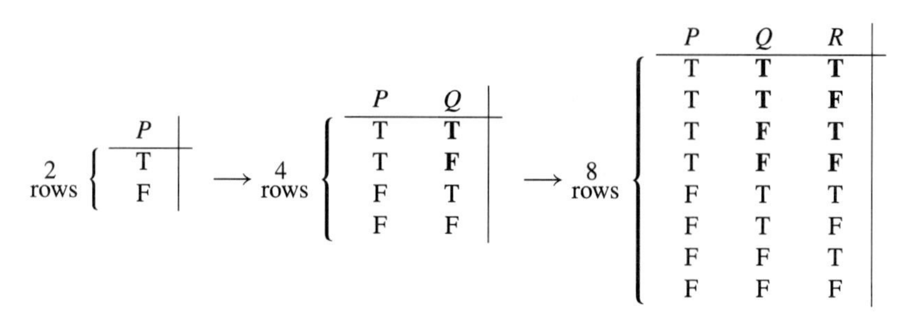

Truth table patterns.

Example. Both $P \vee \sim P$ and $P \Rightarrow P$ share an important property. Their columns in their truth tables are all **true**. For example, the truth table of $P \vee \sim P$ is:

P	$\sim P$	$P \vee \sim P$
T	F	T
F	T	T

However, the columns for $P \wedge \sim P$ and $P \Leftrightarrow \sim P$ are all **false**. To check the first one, examine its truth table:

P	$\sim P$	$P \wedge \sim P$
T	F	F
F	T	F

Based on these two examples, we make the next definition.

1.2.3. Definition

A propositional form p is a ***tautology*** if each entry of the column for p in its truth table is **true**. If every entry in its column is **false**, then it is a ***contradiction***.

EXERCISES

1. Write the typical pattern of Ts and Fs for the truth table of a propositional form with:
   (a) 4 variables. (b) 5 variables.

2. Define:

$$P := \textit{the angle sum of a triangle is 180},$$
$$Q := 3 + 7 = 10,$$
$$R := \textit{the sine function is continuous}.$$

Translate the following propositional forms into English:

(a) $P \vee Q$
(b) $P \wedge Q$
(c) $P \wedge \sim Q$
(d) $Q \vee \sim R$
(e) $Q \Leftrightarrow \sim R$
(f) $R \Rightarrow Q$
(g) $(P \vee R) \Rightarrow \sim Q$
(h) $Q \Leftrightarrow (R \wedge \sim Q)$
(i) $\sim(P \wedge Q)$
(j) $\sim(P \vee Q)$
(k) $P \vee Q \wedge R$
(l) $P \wedge Q \vee R$

3. Write the following sentences as propositional forms using the variables P, Q, and R as defined in the previous problem.
   (a) *The sine function is continuous, and* $3 + 7 = 10$.
   (b) *The angle sum of a triangle is 180, or the angle sum of a triangle is 180.*
   (c) *If* $3 + 7 = 10$, *then the sine function is not continuous.*
   (d) *The angle sum of a triangle is 180 if and only if the sine function is continuous.*
   (e) *The sine function is continuous if and only if* $3 + 7 = 10$ *implies that the angle sum of a triangle is not 180.*
   (f) *It is not the case that* $3 + 7 \neq 10$.
4. Define R to be *the determinant is nonzero* and S as *the matrix is invertible*. Translate $R \Rightarrow S$ once using the word *sufficient* and once using *necessary*.
5. Assign

$$P := \textit{the integer is even},$$
$$Q := \textit{the integer is divisible by } 2.$$

Translate $P \Leftrightarrow Q$ once using *if and only if* and once with *necessary and sufficient*.

6. Write the truth table for each of the given propositional forms.

(a) $\sim P \Rightarrow P$
(b) $P \Rightarrow \sim Q$
(c) $(P \vee Q) \wedge \sim(P \wedge Q)$
(d) $(P \Rightarrow Q) \vee (Q \Leftrightarrow P)$
(e) $P \wedge Q \vee R$
(f) $P \vee Q \Rightarrow R$
(g) $P \Rightarrow (Q \wedge \sim[R \vee P])$
(h) $P \Rightarrow Q \Leftrightarrow R \Rightarrow S$
(i) $P \vee (\sim Q \Leftrightarrow R) \wedge Q$
(j) $\sim P \vee Q \wedge P \Rightarrow Q \vee \sim S$

7. Let P and Q be propositions. Write the propositional form for the exclusive or, *P or Q but not both*, and find its truth table.

8. Check the truth value of these propositions using truth tables.
   (a) *If* $2 + 3 = 7$, *then* $5 - 9 \neq 0$.
   (b) *If some functions have a derivative implies that a square is round, then everyone loves mathematics.*
   (c) *Either snow is hot or two is even implies that three is even.*
   (d) *Every even integer is divisible by* 4 *if and only if either* 7 *divides* 21 *or* 9 *divides* 12.
   (e) *This is a math book, and if sine is an increasing function, then cosine is a decreasing one.*
9. Give examples of the following:
   (a) a conjunction that is a tautology
   (b) a disjunction that is a contradiction
   (c) a negation that is a tautology
   (d) an implication that is a contradiction
   (e) a biconditional that is a tautology
   (f) a biconditional that is a contradiction
   (g) a propositional form with two variables that is a tautology
   (h) a propositional form with two variables that is a contradiction

---

1.3. RULES OF INFERENCE

We are about ready to start writing some basic proofs. For this an introduction to the rules of logic is in order. There are two types. The first is the subject of this section.

Take a collection of propositional forms, $h_1, h_2, \ldots, h_n$, and call them ***hypotheses*** or ***premises***. For any propositional form p, if p is true when $h_1, h_2, \ldots, h_n$ are all true, then we say that p ***logically follows*** from these hypotheses and call p a ***theorem*** or a ***conclusion*** of $h_1, h_2, \ldots, h_n$. We will denote this relationship using a symbol called a ***turnstile*** ($\vdash$):

$$h_1, h_2, \ldots, h_n \vdash p.$$

This is called a ***rule of inference*** or sometimes a ***valid argument***. It can be read as "$h_1, h_2, \ldots, h_n$ yield p" or "give p."

Example. For propositions P and Q,

$$P, P \Rightarrow Q \vdash Q$$

because whenever we have P and $P \Rightarrow Q$, we must also have Q.

Example. Since $P \wedge Q$ is always true if both P and Q are true, we may write

$$P, Q \vdash P \wedge Q.$$

Furthermore, if we only know that P is true, we can still write

$$P \vdash P \vee Q.$$

We will use two methods to demonstrate rules of inference. The first involves truth tables. Suppose $h_1, h_2, \ldots, h_n \vdash p$. When one of the hypotheses is false, we cannot draw a conclusion about the truth value of p. However, if they are all true, then p must hold. This means that either $h_1 \wedge h_2 \wedge \cdots \wedge h_n$ is false or p true. Hence, we may define $\vdash$ using an implication where the hypothesis is $h_1 \wedge h_2 \wedge \cdots \wedge h_n$ and the conclusion is p:

1.3.1. Definition

Let $h_1, h_2, \ldots, h_n$, and p be propositional forms. Then,

$$h_1, h_2, \ldots, h_n \vdash p$$
if and only if
$(h_1 \wedge h_2 \wedge \cdots \wedge h_n) \Rightarrow p$ is a tautology.

Example. The following truth table shows $(P \vee Q) \Rightarrow Q, P \vdash Q$ since it demonstrates that $([(P \vee Q) \Rightarrow Q] \wedge P) \Rightarrow Q$ is a tautology.

P	Q	$P \vee Q$	$(P \vee Q) \Rightarrow Q$	$[(P \vee Q) \Rightarrow Q] \wedge P$	$([(P \vee Q) \Rightarrow Q] \wedge P) \Rightarrow Q$
T	T	T	T	T	T
T	F	T	F	F	T
F	T	T	T	F	T
F	F	F	T	F	T

If there is a time when p is false even though $h_1, h_2, \ldots, h_n$ are all true, then p does not logically follow from $h_1, h_2, \ldots, h_n$. The notation for this is

$$h_1, h_2, \ldots, h_n \nvdash p,$$

and it is called an ***invalid argument***. In other words, $h_1, h_2, \ldots, h_n \vdash p$ is false. To prove this, we must show that

$$(h_1 \wedge h_2 \wedge \cdots \wedge h_n) \Rightarrow p$$

is not a tautology.

Example. To see that $(P \wedge Q) \Rightarrow Q,\ P \not\models Q$, examine the truth table:

P	Q	$P \wedge Q$	$(P \wedge Q) \Rightarrow Q$	$[(P \wedge Q) \Rightarrow Q] \wedge P$	$([(P \wedge Q) \Rightarrow Q] \wedge P) \Rightarrow Q$
T	T	T	T	T	T
T	F	F	T	T	F
F	T	F	T	F	T
F	F	F	T	F	T

Since the last column is not all **true**, Q does not logically follow from $(P \wedge Q) \Rightarrow Q$ and P.

Notice that F appears for $([(P \wedge Q) \Rightarrow Q] \wedge P) \Rightarrow Q$ on a line when the hypotheses, $(P \wedge Q) \Rightarrow Q$ and P, are both true yet the conclusion Q is false. Because of this, we may shorten the procedure for showing an argument invalid by simply finding an assignment for all of the propositional variables that simultaneously makes all of the hypotheses true yet the conclusion false.

Example. Show that $(P \wedge Q) \Rightarrow Q,\ P \not\models Q$ by checking the appropriate line and columns from its truth table:

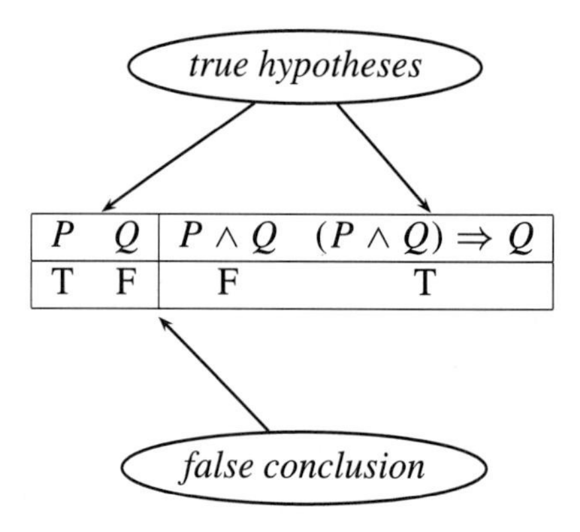

Example. To prove $(P \wedge Q) \Rightarrow R \not\models (P \Rightarrow R)$, we must find a substitution for P, Q, and R that make the hypothesis $(P \wedge Q) \Rightarrow R$ true yet the conclusion $P \Rightarrow R$ false. To make $P \Rightarrow R$ false, we need P to be true and R to be false. Therefore, use:

P	Q	R	$(P \wedge Q) \Rightarrow R$	$P \Rightarrow R$
T	F	F	T	F

Now suppose we want to show this familiar result from high school geometry:

opposite angles in a parallelogram are congruent.

In other words,

if $\square ABCD$ is a parallelogram, then $\angle B \cong \angle D$,

which translates to:

Given: $\square ABCD$ is a parallelogram,
Prove: $\angle B \cong \angle D$.

We did not use truth tables. Instead, we drew a diagram,

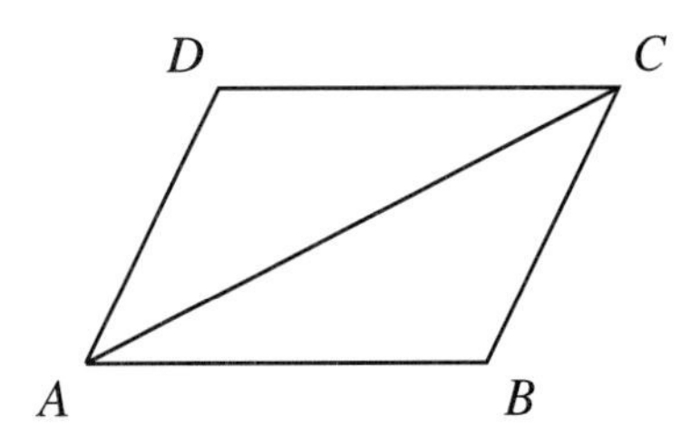

and wrote a proof:

Proof.

1.	$\square ABCD$ is a parallelogram	Given
2.	Have line $\overleftrightarrow{AC}$	Postulate
3.	$\angle DAC \cong \angle ACB$	Alternate interior angles
4.	$\angle CAB \cong \angle ACD$	Alternate interior angles
5.	$\overline{AC} \cong \overline{AC}$	Reflexive
6.	$\triangle ACD \cong \triangle CAB$	ASA
7.	$\angle B \cong \angle D$	Corresponding parts ■

Look closely at its structure. The proof is a sequence of propositions starting with the givens and ending with the desired conclusion. The middle propositions follow from earlier ones by rules of logic.

Let us generalize. Take a propositional form p and let $h_1, h_2, \ldots, h_k$ be hypotheses. A ***proof*** of p from $h_1, h_2, \ldots, h_k$ is a finite sequence of propositional forms,

$$p_1, p_2, \ldots, p_n,$$

such that $p_n = p$ and every proposition is either a hypothesis (h_i for some i) or follows by a rule of logic. This is the second method that we will use to prove $h_1, h_2, \ldots, h_k \vdash p$. Note, however, that if there is one mistake in the logic, then the sequence is not a proof. Although professors will often give partial credit for "proofs" that have some "minor" mistakes, this does not mean that a "proof" with only one mistake is any closer to being a proof than one with many mistakes. That one mistake may be insurmountable.

This definition is the motivation for calling p a theorem of $h_1, h_2, \ldots, h_k$, for with regard to propositions, a ***theorem*** is a statement that is proven. Other types of theorems are lemmas and corollaries. A ***lemma*** is usually a technical theorem used to prove another theorem, and a ***corollary*** is a theorem that is quickly proven from another theorem. On the other hand, a statement that is assumed to be true is called an ***axiom***. (In geometry, axioms are usually called ***postulates***.) The hypotheses used to prove a theorem will either be axioms or other theorems.

At this point to write proofs we will use only the following rules of inference.

1.3.2. Theorem

Let p, q, r, and s be propositional forms.

Modus Ponens* ***[MP]***
$p \Rightarrow q,\ p \vdash q$

Modus Tolens† ***[MT]***
$p \Rightarrow q,\ \sim q \vdash \sim p$

Constructive Dilemma [CD]
$(p \Rightarrow q) \wedge (r \Rightarrow s),$
$p \vee r \vdash q \vee s$

Simplification [Simp]
$p \wedge q \vdash p$

Conjunction [Conj]
$p,\ q \vdash p \wedge q$

Disjunctive Syllogism [DS]
$p \vee q,\ \sim p \vdash q$

Addition [Add]
$p \vdash p \vee q$

Destructive Dilemma [DD]
$(p \Rightarrow q) \wedge (r \Rightarrow s),$
$\sim q \vee \sim s \vdash \sim p \vee \sim r$

Transitivity [Trans]
$p \Rightarrow q,\ q \Rightarrow r \vdash p \Rightarrow r$

(Note: Transitivity is also known as the ***Hypothetical Syllogism***‡ or the ***Chain Rule***.)

To prove each of these rules of inference, appeal to the appropriate truth table.

*In Latin *modus* means "standard" or "measure," and *ponens* means "putting." Hence, this rule allows us to deduce the conclusion when we "put down" the hypothesis.

†In Latin *tolens* means "taking away," so *modus tolens* is the rule that allows us to "take away" the hypothesis when the negation of the conclusion is true.

‡A *syllogism* is a type of argument that involves two premises and a conclusion.

Example. To show that the Disjunctive Syllogism is true, we must show that

$$[(P \vee Q) \wedge \sim P] \Rightarrow Q$$

is a tautology. This is seen by its truth table.

P	Q	$P \vee Q$	$\sim P$	$(P \vee Q) \wedge \sim P$	$[(P \vee Q) \wedge \sim P] \Rightarrow Q$
T	T	T	F	F	T
T	F	T	F	F	T
F	T	T	T	T	T
F	F	F	T	F	T

To use the listed rules of inference, match the pattern on the left of the turnstile and conclude the pattern on the right.

Example. Each argument is shown valid by the indicated rule:

1. $(P \wedge Q) \Rightarrow \sim R,\ P \wedge Q \vdash \sim R$ *Modus Ponens*
2. $P \vdash P \vee (Q \wedge R)$ *Addition*
3. $\sim\sim P,\ (Q \vee R) \Rightarrow \sim P \vdash \sim(Q \vee R)$ *Modus Tolens*

When appealing to these rules, we will follow them exactly as stated.

Example. These are invalid uses of the rules of inference, although each are valid arguments.

1. $P \vdash Q \vee P$ *"Addition"*
   Given a proposition P, we may use Addition to conclude a disjunction in which P is the first disjunct, not the second.
2. $P \vee Q,\ \sim Q \vdash P$ *"Disjunctive Syllogism"*
   To apply the Disjunctive Syllogism, the first disjunct must be negated, not the second.

Example. A check of its truth table shows

$$(P \wedge Q) \Rightarrow R \nvdash P \Rightarrow R,$$

although it appears that it is an application of Simplification. The problem is that this rule of inference is "applied" to only a portion of a hypothesis. This is incorrect. A rule of inference must use an entire hypothesis because we are deriving sentences from other sentences.

We are now ready to write some proofs. Initially they will be two-column in format with each line being numbered. In the first column will be a sequence of propositions forms. It will begin with the hypotheses and end with the conclusion. In the right-hand column will be the reasons that allowed us to include each proposition. Initially, the only reasons that we will use are:

- *Given* (for hypotheses), and
- rules of inference.

A rule of inference is cited by giving the line numbers used as the hypotheses followed by the abbreviation for the rule. Furthermore, all our proofs will begin with the word *Proof* and close with an end of proof symbol (we will use the box, ■). This is a matter of style. Thus, a proof of

$$(P \vee Q) \Rightarrow (Q \wedge R),\ P \vdash Q$$

can be written as:

Proof.

1.	$(P \vee Q) \Rightarrow (Q \wedge R)$	Given
2.	P	Given
	$\langle$Show $Q\rangle$	
3.	$P \vee Q$	2 Add
4.	$Q \wedge R$	1, 3 MP
5.	Q	4 Simp ■

Notice that after the last hypothesis what is needed to be shown is written with angled parentheses. This is a ***comment*** and is used as a reminder of the goal of the proof. It helps us to follow the structure of the proof but is technically not part of it. We will always use them in our two-column proofs.

The following three examples show how to use various rules of inference in proofs.

Example. Show $P \Rightarrow Q,\ Q \Rightarrow R,\ S \vee {\sim}R,\ {\sim}S \vdash {\sim}P$ is valid.

Proof.

1.	$P \Rightarrow Q$	Given
2.	$Q \Rightarrow R$	Given
3.	$S \vee {\sim}R$	Given
4.	${\sim}S$	Given
	$\langle$Show ${\sim}P\rangle$	
5.	$P \Rightarrow R$	1, 2 Trans
6.	${\sim}R$	3, 4 DS
7.	${\sim}P$	5, 6 MT ■

Example. Prove $P \Rightarrow Q,\ (P \Rightarrow Q) \Rightarrow (T \Rightarrow S),\ P \vee T,\ \sim Q \vdash S$.

Proof.

1.	$P \Rightarrow Q$	Given
2.	$(P \Rightarrow Q) \Rightarrow (T \Rightarrow S)$	Given
3.	$P \vee T$	Given
4.	$\sim Q$	Given
	$\langle$Show $S\rangle$	
5.	$T \Rightarrow S$	1, 2 MP
6.	$(P \Rightarrow Q) \wedge (T \Rightarrow S)$	1, 5 Conj
7.	$Q \vee S$	3, 6 CD
8.	S	4, 7 DS ■

Example. Prove that $P \Rightarrow Q,\ Q \Rightarrow R,\ \sim R \vdash \sim Q \vee \sim P$.

Proof.

1.	$P \Rightarrow Q$	Given
2.	$Q \Rightarrow R$	Given
3.	$\sim R$	Given
	$\langle$Show $\sim Q \vee \sim P\rangle$	
4.	$(Q \Rightarrow R) \wedge (P \Rightarrow Q)$	1, 2 Conj
5.	$\sim R \vee \sim Q$	3 Add
6.	$\sim Q \vee \sim P$	4, 5 DD ■

EXERCISES

1. Use truth tables to prove the rules of inference found in Theorem 1.3.2 that were not proven in the section.
2. Use truth tables to show that the following are valid.
   (a) $\sim P \vee Q,\ \sim Q \vdash \sim P$
   (b) $\sim(P \wedge Q),\ P \vdash \sim Q$
   (c) $P \Rightarrow Q,\ P \vdash Q \vee R$
   (d) $P \Rightarrow Q,\ Q \Rightarrow R,\ P \vdash R$
   (e) $P \vee (Q \wedge R),\ \sim P \vdash R$
3. Show the following by giving only one line of a truth table.
   (a) $\sim(P \wedge Q) \nvdash \sim P$
   (b) $(P \wedge Q) \Rightarrow R,\ P \nvdash R$
   (c) $P \Rightarrow (Q \vee R),\ P \nvdash Q$

(d) $(P \wedge Q) \Rightarrow R \nvdash Q \Rightarrow R$
(e) $(P \Rightarrow Q) \vee (R \Rightarrow S),\ P \vee R \nvdash Q \vee S$
(f) $\sim(P \wedge Q) \vee R,\ (P \wedge Q) \vee S \nvdash R \wedge S$
(g) $P \vee R,\ Q \vee S,\ R \Leftrightarrow S \nvdash R \wedge S$

4. Identify the rule of inference used to justify the following:
(a) $(P \Rightarrow Q) \Rightarrow P,\ P \Rightarrow Q \vdash (P \Rightarrow Q) \Rightarrow Q$
(b) $P,\ Q \vee R \vdash P \wedge (Q \vee R)$
(c) $P \vdash P \vee (R \Leftrightarrow \sim P \wedge \sim[Q \Rightarrow S])$
(d) $P,\ P \Rightarrow (Q \Leftrightarrow S) \vdash Q \Leftrightarrow S$
(e) $(P \vee Q) \vee Q,\ [(P \vee Q) \Rightarrow Q] \wedge [(Q \Rightarrow (S \wedge T)] \vdash Q \vee (S \wedge T)$
(f) $P \vee (Q \vee S),\ \sim P \vdash Q \vee S$
(g) $P \Rightarrow \sim Q,\ \sim\sim Q \vdash \sim P$
(h) $(P \Rightarrow Q) \wedge (Q \Rightarrow R),\ \sim Q \vee \sim R \vdash \sim P \vee \sim Q$
(i) $(P \Rightarrow Q) \wedge (Q \Rightarrow R) \vdash P \Rightarrow Q$

5. Fill in the blanks to make the arguments valid.
(a) $P,$ ______ $\vdash Q$
(b) $P \Rightarrow Q,\ \sim Q \vdash$ ______
(c) $P \vee Q,$ ______ $\vdash Q$
(d) $(P \wedge Q) \Rightarrow R,$ ______ $\vdash R$
(e) *if a number is divisible by 4, then it is divisible by 2; the number is not divisible by 2* $\vdash$ ______
(f) ______, *the value fell outside the interval* $\vdash$ *the hypothesis should be rejected*
(g) *the function is increasing or decreasing; it is not increasing* $\vdash$ ______
(h) *if the integer is divisible by 8, then it is divisible by 4;* ______ $\vdash$ *if the integer is divisible by 8, then it is divisible by 2*

6. Find all errors in the following "proofs."
(a) "$Q \wedge S,\ (P \Rightarrow \sim Q) \wedge (R \Rightarrow \sim S) \vdash \sim R$"

Attempted Proof.

1.	$Q \wedge S$	Given
2.	$(P \Rightarrow \sim Q) \wedge (R \Rightarrow \sim S)$	Given
	$\langle$Show $\sim R\rangle$	
3.	Q	1 Simp
4.	S	1 Simp
5.	$Q \vee S$	3, 4 Add
6.	$\sim P \wedge \sim R$	2, 5 DD
7.	$\sim R$	6 Simp ■

(b) "$P,\ R \Rightarrow (P \vee Q) \vdash R$"

Attempted Proof.

1.	P	Given
2.	$R \Rightarrow (P \vee Q)$	Given
	⟨Show R⟩	
3.	$P \vee Q$	1 Add
4.	R	1, 3 MT ■

(c) "$P \vee Q,\ P \Rightarrow S \vdash S$"

Attempted Proof.

1.	$P \vee Q$	Given
2.	$P \Rightarrow S$	Given
	⟨Show S⟩	
3.	$(P \vee Q) \Rightarrow S$	2 Add
4.	S	1, 3 MP ■

(d) "$(P \Rightarrow Q) \Rightarrow R,\ P,\ Q \vdash R$"

Attempted Proof.

1.	$(P \Rightarrow Q) \Rightarrow R$	Given
2.	P	Given
3.	Q	Given
	⟨Show R⟩	
4.	$Q \Rightarrow R$	1, 2 MP
5.	R	3, 4 MP ■

7. Arrange each group of lines into a proof for the given argument and then supply the appropriate reasons.

(a) $P \Rightarrow Q,\ R \Rightarrow S,\ P \vdash Q \vee S$

- P
- $Q \vee S$
- $R \Rightarrow S$
- $(P \Rightarrow Q) \wedge (R \Rightarrow S)$
- $P \vee R$
- $P \Rightarrow Q$

(b) $P \Rightarrow Q,\ Q \Rightarrow R,\ P \vdash R \vee Q$

- P
- $P \Rightarrow R$
- $P \Rightarrow Q$
- R
- $Q \Rightarrow R$
- $R \vee Q$

(c) $(P \Rightarrow Q) \vee (Q \Rightarrow R),\ \sim(P \Rightarrow Q),\ \sim R,\ Q \vee S \vdash S$

- $\sim(P \Rightarrow Q)$
- S
- $(P \Rightarrow Q) \vee (Q \Rightarrow R)$
- $Q \Rightarrow R$
- $\sim R$
- $\sim Q$
- $Q \vee S$

(d) $(P \vee Q) \wedge R,\ (Q \vee S) \Rightarrow T,\ \sim P \vdash \sim P \wedge T$

- $P \vee Q$
- Q
- $Q \vee S$
- T
- $(P \vee Q) \wedge R$
- $(Q \vee S) \Rightarrow T$
- $\sim P$
- $\sim P \wedge T$

8. Give proofs for the following.
   (a) $P \Rightarrow Q,\ P \vee (R \Rightarrow S),\ \sim Q \vdash R \Rightarrow S$
   (b) $P \Rightarrow Q,\ Q \Rightarrow R,\ \sim R \vdash \sim P$
   (c) $P \Rightarrow Q,\ R \Rightarrow S,\ \sim Q \vee \sim S \vdash \sim P \vee \sim R$
   (d) $[P \Rightarrow (Q \Rightarrow R)] \wedge [Q \Rightarrow (R \Rightarrow P)],\ P \vee Q,$
   $\sim(Q \Rightarrow R),\ \sim P \vdash \sim R$
   (e) $(P \Rightarrow Q) \Rightarrow (R \Rightarrow S),\ S \Rightarrow T,\ P \Rightarrow Q,\ R \vdash T$
   (f) $P \Rightarrow (Q \wedge R),\ (Q \vee S) \Rightarrow (T \wedge U),\ P \vdash T$
   (g) $P \Rightarrow (Q \wedge R),\ \sim(Q \wedge R),\ (Q \wedge R) \vee (\sim P \Rightarrow S) \vdash S$
   (h) $(P \vee Q) \Rightarrow (\sim R \wedge \sim S),\ Q \Rightarrow R,\ P \vdash \sim Q$
   (i) $N \Rightarrow P,\ [(P \vee Q) \vee R] \Rightarrow (S \vee T),\ (S \vee T) \Rightarrow T,\ N \vdash T$
   (j) $P \Rightarrow Q,\ Q \Rightarrow R,\ R \Rightarrow S,\ S \Rightarrow T,\ P \vee R,\ \sim R \vdash T$
   (k) $(P \vee Q) \Rightarrow (R \vee S),\ (R \Rightarrow T) \wedge (S \Rightarrow U),\ P,\ \sim T \vdash U$
   (l) $P \Rightarrow Q,\ Q \Rightarrow R,\ R \Rightarrow S,\ (P \vee Q) \wedge (R \vee S) \vdash Q \vee S$
   (m) $[(P \vee \sim Q) \vee R] \Rightarrow (S \Rightarrow P),\ (P \vee \sim Q) \Rightarrow (P \Rightarrow R),$
   $P \vdash S \Rightarrow R$
9. Consider the following argument:

   If this argument is an instance of Modus Ponens, *then it is invalid.*
   This argument is an instance of Modus Ponens.
   Therefore, this argument is invalid.

   The argument seems to conclude that itself is invalid, yet it appears like a legitimate application of *Modus Ponens*. Is it valid? Explain.

1.4. RULES OF REPLACEMENT

In the study of mathematics and logic, we are interested in the truth values of propositions. It will often be important to be able to substitute one proposition for another if we are sure that their truth values will coincide. This brings us to the second type of logic rule.

1.4.1. Definition

We say that any two propositional forms p and q are ***(logically) equivalent*** if their columns in their truth tables are the same. The notation for this is $p \equiv q$.

Example. Any implication is equivalent to its contrapositive but not equivalent to its converse. Their truth table shows this.

P	Q	$P \Rightarrow Q$	$\sim Q$	$\sim P$	$\sim Q \Rightarrow \sim P$	$Q \Rightarrow P$
T	T	T	F	F	T	T
T	F	F	T	F	F	T
F	T	T	F	T	T	F
F	F	T	T	T	T	T

The first and fourth columns on the right show $P \Rightarrow Q \equiv \sim Q \Rightarrow \sim P$. The first and last show $P \Rightarrow Q \not\equiv Q \Rightarrow P$.

Example. All tautologies are logically equivalent. The same can be said for contradictions.

Sometimes a logical equivalence is called a ***rule of replacement***. To understand the terminology, suppose $P \equiv Q$ and P appears in a propositional form. If P is replaced by Q in the form, then the new propositional form is equivalent to the original.

Example. Because $P \Rightarrow Q \equiv \sim P \vee Q$, we may substitute one for the other at any time. For example,

$$Q \wedge (P \Rightarrow Q) \equiv Q \wedge (\sim P \vee Q).$$

To see this, examine the truth table:

P	Q	$P \Rightarrow Q$	$\boldsymbol{Q \wedge (P \Rightarrow Q)}$	$\sim P$	$\sim P \vee Q$	$\boldsymbol{Q \wedge (\sim P \vee Q)}$
T	T	T	**T**	F	T	**T**
T	F	F	**F**	F	F	**F**
F	T	T	**T**	T	T	**T**
F	F	T	**F**	T	T	**F**

Example. Consider $R \wedge (P \Rightarrow Q)$. Since $P \Rightarrow Q \equiv \sim Q \Rightarrow \sim P$, we may replace $P \Rightarrow Q$ by $\sim Q \Rightarrow \sim P$. This gives $R \wedge (\sim Q \Rightarrow \sim P)$. It is left as an exercise to show

$$R \wedge (P \Rightarrow Q) \equiv R \wedge (\sim Q \Rightarrow \sim P).$$

If we limit our proofs to the rules of inference, we will quickly realize that there will be little of interest that we can prove. We would not be able to write proofs for such clearly valid arguments as $P \vdash Q \vee P$ or $P \vee Q,\ {\sim}Q \vdash P$. To fix this, we will add certain rules of replacement to our list of logic rules.

1.4.2. Theorem

Let p, q, and r be propositional forms.

Commutative Laws [Com]
$p \wedge q \equiv q \wedge p$
$p \vee q \equiv q \vee p$

Associative Laws [Assoc]
$(p \wedge q) \wedge r \equiv p \wedge (q \wedge r)$
$(p \vee q) \vee r \equiv p \vee (q \vee r)$

Distributive Laws [Distr]
$p \wedge (q \vee r) \equiv (p \wedge q) \vee (p \wedge r)$
$p \vee (q \wedge r) \equiv (p \vee q) \wedge (p \vee r)$

Contrapositive Law [Contra]
$p \Rightarrow q \equiv {\sim}q \Rightarrow {\sim}p$

Double Negation [DN]
$p \equiv {\sim}{\sim}p$

De Morgan's Laws [DeM]*
${\sim}(p \wedge q) \equiv {\sim}p \vee {\sim}q$
${\sim}(p \vee q) \equiv {\sim}p \wedge {\sim}q$

Tautology [Taut]
$p \wedge p \equiv p$
$p \vee p \equiv p$

Material Equivalence [Equiv]
$p \Leftrightarrow q \equiv (p \Rightarrow q) \wedge (q \Rightarrow p)$
$p \Leftrightarrow q \equiv (p \wedge q) \vee ({\sim}p \wedge {\sim}q)$

Material Implication [Impl]
$p \Rightarrow q \equiv {\sim}p \vee q$

Exportation [Exp]
$(p \wedge q) \Rightarrow r \equiv p \Rightarrow (q \Rightarrow r)$

To prove each of these rules, use a truth table.

Example. To see that the first De Morgan's Law is true, examine the truth table:

p	q	$p \wedge q$	$\boldsymbol{{\sim}(p \wedge q)}$	${\sim}p$	${\sim}q$	$\boldsymbol{{\sim}p \vee {\sim}q}$
T	T	T	**F**	F	F	**F**
T	F	F	**T**	F	T	**T**
F	T	F	**T**	T	F	**T**
F	F	F	**T**	T	T	**T**

*Augustus De Morgan (Madura, India, 1806 – London, England, 1871): De Morgan's mathematical work included early contributions to abstract algebra, mathematical logic, and the study of relations.

The next truth table demonstrates a Material Equivalence:

p	q	$\boldsymbol{p \Leftrightarrow q}$	$p \Rightarrow q$	$q \Rightarrow p$	$\boldsymbol{p \Rightarrow q \wedge q \Rightarrow p}$
T	T	**T**	T	T	**T**
T	F	**F**	F	T	**F**
F	T	**F**	T	F	**F**
F	F	**T**	T	T	**T**

As with the rules of inference, the rules of replacement must be used exactly as stated. This includes times when it seems unnecessary or a step may appear obvious.

Example. An example of this is the use of the Associative Law. It is common practice to move parentheses freely when solving equations. Although we will treat parentheses in propositional forms this way in the near future, equivalences such as

$$(P \wedge Q) \wedge (R \wedge S) \equiv [P \wedge (Q \wedge R)] \wedge S$$

must be carefully demonstrated. Fortunately, our example only requires two applications of Associativity. Using the boxes as a guide, notice that

$$\boxed{(P \wedge Q)} \wedge (\boxed{R} \wedge \boxed{S})$$

is of the same form as the right-hand side of the Associative Law. Hence, we may regroup and obtain:

$$[\,\boxed{(P \wedge Q)} \wedge \boxed{R}\,] \wedge \boxed{S}.$$

One more application yields the result. We may, therefore, write a sequence of equivalences:

$$(P \wedge Q) \wedge (R \wedge S) \equiv [(P \wedge Q) \wedge R] \wedge S \equiv [P \wedge (Q \wedge R)] \wedge S.$$

This example has illustrated the justification for writing $H_1 \wedge H_2 \wedge H_3$ for $(H_1 \wedge H_2) \wedge H_3$ or $H_1 \wedge (H_2 \wedge H_3)$.

Rules of replacement are used in proofs like rules of inference. The rule's abbreviation is noted along with the lines used.

Example. It appears that it will be a problem to show

$$P \wedge Q,\ R \wedge S \vdash \sim\!R \Rightarrow P,$$

for how are we to proceed from hypotheses that are conjunctions to a conclusion that is an implication? In this case, try working backwards from the conclusion by using Material Implication. It will allow us to introduce the conditional into the proof.

Proof.

1.	$P \wedge Q$	Given
2.	$R \wedge S$	Given
	$\langle$Show $\sim R \Rightarrow P\rangle$	
	$\langle$Show $R \vee P\rangle$	
3.	R	2 Simp
4.	$R \vee P$	3 Add
5.	$\sim\sim R \vee P$	4 DN
6.	$\sim R \Rightarrow P$	5 Impl ■

Notice the introduction of the second comment. It is intended to be a translation of the first. Although it did not follow directly from the first, the intermediate step is included in the proof. Further notice that the first hypothesis was not used. *This happens very rarely!* Proofs that do not use all of their hypotheses should be double checked carefully, for there is probably a mistake! What we have actually proven is

$$R \wedge S \vdash \sim R \Rightarrow P.$$

The next example uses a number of replacement rules.

Example. Show $P \Rightarrow Q,\ R \Rightarrow Q \vdash (P \vee R) \Rightarrow Q$.

Proof.

1.	$P \Rightarrow Q$	Given
2.	$R \Rightarrow Q$	Given
	$\langle$Show $(P \vee R) \Rightarrow Q\rangle$	
3.	$(P \Rightarrow Q) \wedge (R \Rightarrow Q)$	1, 2 Conj
4.	$(\sim P \vee Q) \wedge (\sim R \vee Q)$	3 Impl
5.	$(Q \vee \sim P) \wedge (Q \vee \sim R)$	4 Com
6.	$Q \vee (\sim P \wedge \sim R)$	5 Dist
7.	$(\sim P \wedge \sim R) \vee Q$	6 Com
8.	$\sim(P \vee R) \vee Q$	7 DeM
9.	$(P \vee R) \Rightarrow Q$	8 Impl ■

The last example will demonstrate a usage of the Distributive Law. It will also test our ability to precisely use the rules.

Example. Prove $(P \wedge Q) \vee (R \wedge S) \vdash (P \vee S) \wedge (Q \vee R)$.

Proof.

1.	$(P \wedge Q) \vee (R \wedge S)$	Given
	$\langle$Show $(P \vee S) \wedge (Q \vee R)\rangle$	
2.	$[(P \wedge Q) \vee R] \wedge [(P \wedge Q) \vee S]$	1 Dist
3.	$[R \vee (P \wedge Q)] \wedge [S \vee (P \wedge Q)]$	2 Com
4.	$[(R \vee P) \wedge (R \vee Q)] \wedge [(S \vee P) \wedge (S \vee Q)]$	3 Dist
5.	$[(R \vee Q) \wedge (R \vee P)] \wedge [(S \vee P) \wedge (S \vee Q)]$	4 Com
6.	$(R \vee Q) \wedge ([R \vee P] \wedge [(S \vee P) \wedge (S \vee Q)])$	5 Assoc
7.	$(R \vee Q)$	6 Simp
8.	$[(S \vee P) \wedge (S \vee Q)] \wedge [(R \vee Q) \wedge (R \vee P)]$	5 Com
9.	$(S \vee P) \wedge (S \vee Q)$	8 Simp
10.	$(S \vee P)$	9 Simp
11.	$(S \vee P) \wedge (R \vee Q)$	7, 10 Conj
12.	$(P \vee S) \wedge (Q \vee R)$	11 Com ■

EXERCISES

1. Use truth tables to prove the rules of replacement from Theorem 1.4.2 that we have not proven in the section.
2. Prove the following using truth tables:
   (a) $P \vee {\sim}P \equiv P \Rightarrow P$
   (b) $P \vee {\sim}P \equiv (P \vee Q) \vee {\sim}(P \wedge Q)$
   (c) $P \wedge Q \equiv (P \Leftrightarrow Q) \wedge (P \vee Q)$
   (d) $R \wedge (P \Rightarrow Q) \equiv R \wedge ({\sim}Q \Rightarrow {\sim}P)$
   (e) $(P \wedge Q) \Rightarrow R \equiv (P \wedge {\sim}R) \Rightarrow {\sim}Q$
   (f) $P \Rightarrow (Q \wedge R) \equiv (P \Rightarrow Q) \wedge (P \Rightarrow R)$
   (g) $(P \Rightarrow Q) \Rightarrow (S \Rightarrow R) \equiv [(P \Rightarrow Q) \wedge S] \Rightarrow R$
3. **3.** Identify the rule of replacement used in each of the following:
   (a) $([P \Rightarrow Q] \vee [Q \Rightarrow R]) \vee S \equiv (P \Rightarrow Q) \vee ([Q \Rightarrow R] \vee S)$
   (b) ${\sim}{\sim}P \Leftrightarrow (Q \wedge R) \equiv P \Leftrightarrow (Q \wedge R)$
   (c) $(P \vee Q) \vee R \equiv (Q \vee P) \vee R$
   (d) ${\sim}P \wedge {\sim}(Q \vee R) \equiv {\sim}(P \vee [Q \vee R])$
   (e) $P \Rightarrow (Q \vee R) \equiv {\sim}(Q \vee R) \Rightarrow {\sim}P$
   (f) ${\sim}(P \vee Q) \vee R \equiv (P \vee Q) \Rightarrow R$
   (g) $P \Leftrightarrow (Q \Rightarrow R) \equiv (P \wedge [Q \Rightarrow R]) \vee ({\sim}P \wedge {\sim}[Q \Rightarrow R])$

(h) $(P \vee Q) \Leftrightarrow (Q \wedge Q) \equiv (P \vee Q) \Leftrightarrow Q$
(i) $(P \wedge Q) \wedge R \equiv R \wedge (P \wedge Q)$
(j) $(P \vee Q) \wedge (Q \vee R) \equiv [(P \vee Q) \wedge Q] \vee [(P \vee Q) \wedge R]$
(k) $\sim(\sim Q \Rightarrow P) \equiv \sim(\sim P \Rightarrow \sim\sim Q)$
(l) $(P \wedge [R \Rightarrow Q]) \Rightarrow S \equiv P \Rightarrow ([R \Rightarrow Q] \Rightarrow S)$

4. For each of the following propositional forms find another propositional form to which it is equivalent using the rules of replacement from Theorem 1.4.2 on page 34. (Notice that there are infinitely many possibilities for each!)
(a) $\sim\sim P$
(b) $P \vee Q$
(c) $P \Rightarrow Q$
(d) $\sim(P \wedge Q)$
(e) $(P \Leftrightarrow Q) \wedge (\sim P \Leftrightarrow Q)$
(f) $(P \Rightarrow Q) \vee (Q \Rightarrow S)$
(g) $P \wedge (Q \vee R)$
(h) $(P \Rightarrow Q) \vee P$
(i) $\sim(P \Rightarrow Q)$
(j) $(P \Rightarrow Q) \wedge (Q \Leftrightarrow R)$
(k) $(P \vee \sim Q) \Leftrightarrow T \wedge Q$
(l) $P \vee (\sim Q \Leftrightarrow T) \wedge Q$

5. Identify all mistakes in the following "proofs."
(a) "$P \Rightarrow \sim Q \vdash R \Rightarrow (P \vee Q)$"

Attempted Proof.

1.	$P \Rightarrow \sim Q$	Given
	$\langle$Show $R \Rightarrow (P \vee Q)\rangle$	
2.	$\sim P \Rightarrow Q$	1 Contra
3.	$\sim\sim P \vee Q$	2 Impl
4.	$P \vee Q$	3 DN
5.	$(P \vee Q) \vee \sim\sim R$	4 Add
6.	$R \Rightarrow (P \vee Q)$	5 Impl ■

(b) "$(P \wedge Q) \Rightarrow R \vdash Q \Rightarrow R$"

Attempted Proof.

1.	$(P \wedge Q) \Rightarrow R$	Given
	$\langle$Show $Q \Rightarrow R\rangle$	
2.	$\sim(P \wedge Q) \vee R$	1 Impl
3.	$(\sim P \wedge \sim Q) \vee R$	2 DeM
4.	$\sim P \wedge (\sim Q \vee R)$	3 Com
5.	$\sim Q \vee R$	4 Simp
6.	$Q \Rightarrow R$	5 Imp ■

(c) "$\sim(\sim P \Rightarrow Q),\ \sim Q \vdash P$"

Attempted Proof.

1.	$\sim(\sim P \Rightarrow Q)$	Given
2.	$\sim Q$	Given
	⟨Show P⟩	
3.	$P \Rightarrow Q$	1 DN
4.	$\sim Q \Rightarrow P$	3 Contra
5.	P	2, 4 MP ■

6. Arrange each group of lines into a proof for the given argument and then supply the appropriate reasons.
   (a) $(P \vee Q) \Rightarrow R \vdash (P \Rightarrow R) \wedge (Q \Rightarrow R)$
   - $R \vee (\sim P \wedge \sim Q)$
   - $(R \vee \sim P) \wedge (R \vee \sim Q)$
   - $(\sim P \vee R) \wedge (\sim Q \vee R)$
   - $(\sim P \wedge \sim Q) \vee R$
   - $\sim(P \vee Q) \vee R$
   - $(P \vee Q) \Rightarrow R$
   - $(P \Rightarrow R) \wedge (Q \Rightarrow R)$

   (b) $\sim(P \wedge Q) \Rightarrow (R \vee S),\ \sim P,\ \sim S \vdash R$
   - $\sim(P \wedge Q) \Rightarrow (R \vee S)$
   - $\sim(P \wedge Q)$
   - $\sim S$
   - $\sim P$
   - $S \vee R$
   - $R \vee S$
   - R
   - $\sim P \vee \sim Q$

   (c) $P \Rightarrow (Q \Rightarrow R),\ \sim P \Rightarrow S,\ \sim Q \Rightarrow T,\ R \Rightarrow \sim R \vdash \sim T \Rightarrow S$
   - $S \vee T$
   - $\sim P \vee \sim Q$
   - $\sim R \vee \sim R$
   - $(P \wedge Q) \Rightarrow R$
   - $R \Rightarrow \sim R$
   - $P \Rightarrow (Q \Rightarrow R)$
   - $\sim R$
   - $\sim P \Rightarrow S$
   - $\sim(P \wedge Q)$
   - $\sim Q \Rightarrow T$
   - $T \vee S$
   - $\sim T \Rightarrow S$
   - $\sim\sim T \vee S$
   - $(\sim P \Rightarrow S) \wedge (\sim Q \Rightarrow T)$

7. Give proofs for the following:
   (a) $\sim P \vdash P \Rightarrow Q$
   (b) $P \vdash \sim Q \Rightarrow P$
   (c) $\sim Q \vee (\sim R \vee \sim P) \vdash P \Rightarrow \sim(Q \wedge R)$
   (d) $P \Rightarrow Q \vdash (P \wedge R) \Rightarrow Q$
   (e) $P \Rightarrow (Q \wedge R) \vdash P \Rightarrow Q$
   (f) $(P \vee Q) \Rightarrow R \vdash \sim R \Rightarrow \sim Q$
   (g) $P \Rightarrow (Q \Rightarrow R) \vdash (Q \wedge \sim R) \Rightarrow \sim P$

(h) $P \Rightarrow (Q \Rightarrow R) \vdash Q \Rightarrow (P \Rightarrow R)$
(i) $(P \wedge Q) \vee (R \wedge S) \vdash \sim S \Rightarrow (P \wedge Q)$
(j) $Q \Rightarrow R \vdash P \Rightarrow (Q \Rightarrow R)$
(k) $P \Rightarrow \sim(Q \Rightarrow R) \vdash P \Rightarrow \sim R$
(l) $P \vee (Q \vee R) \vee S \vdash (P \vee Q) \vee (R \vee S)$
(m) $(Q \vee P) \Rightarrow (R \wedge S) \vdash Q \Rightarrow R$
(n) $P \Leftrightarrow (Q \wedge R) \vdash P \Rightarrow Q$
(o) $P \Leftrightarrow (Q \vee R) \vdash Q \Rightarrow P$
(p) $[P \vee (Q \vee R)] \Rightarrow S \vdash Q \Rightarrow S$
(q) $(P \vee Q) \wedge (R \vee S) \vdash [(P \wedge R) \vee (P \wedge S)] \vee [(Q \wedge R) \vee (Q \wedge S)]$
(r) $[P \wedge (Q \vee R)] \Rightarrow (Q \wedge R) \vdash P \Rightarrow (Q \Rightarrow R)$

8. Prove each of the following:
   (a) $P \Leftrightarrow Q, \sim P \vdash \sim Q$
   (b) $P \vee (Q \wedge R), (P \Rightarrow S) \wedge (R \Rightarrow S) \vdash S$
   (c) $(P \Rightarrow Q) \Rightarrow R, \sim R \vdash \sim Q$
   (d) $P \Rightarrow (Q \Rightarrow R), R \Rightarrow (S \wedge T) \vdash P \Rightarrow (Q \Rightarrow T)$
   (e) $P \Rightarrow (Q \Rightarrow R), R \Rightarrow (S \vee T) \vdash (P \Rightarrow S) \vee (Q \Rightarrow T)$
   (f) $P \Rightarrow Q, P \Rightarrow R \vdash P \Rightarrow (Q \wedge R)$
   (g) $(P \vee Q) \Rightarrow (R \wedge S), \sim P \Rightarrow (T \Rightarrow \sim T), \sim R \vdash \sim T$
   (h) $(P \Rightarrow Q) \wedge (R \Rightarrow S), P \vee R, (P \Rightarrow \sim S) \wedge (R \Rightarrow \sim Q) \vdash (Q \Leftrightarrow \sim S)$
   (i) $P \wedge (Q \wedge R), (P \wedge R) \Rightarrow [S \vee (T \vee M)], \sim S \wedge \sim T \vdash M$
   (j) $P \Rightarrow (Q \Rightarrow R), Q \Rightarrow (R \Rightarrow S) \vdash P \Rightarrow (Q \Rightarrow S)$
9. It turns out that the list of logical connectives that we have is redundant. We do not need all of them. To see this, answer the following:
   (a) Find equivalent propositional forms to $P \wedge Q$, $P \Rightarrow Q$, and $P \Leftrightarrow Q$ using only $\vee$ and $\sim$.
   (b) Find equivalent propositional forms to $P \vee Q$, $P \wedge Q$, and $P \Leftrightarrow Q$ using only $\Rightarrow$ and $\sim$.
   (c) Can we obtain similar results if we exclude $\sim$? Explain.
10. Let P be a proposition, T a tautology, and C a contradiction. Prove the following:
   (a) $P \equiv P \wedge T$ (b) $P \equiv P \vee C$

2 Predicates and Proofs

In the previous chapter we discussed what it means for a sentence to be a proposition. We also saw examples that were not propositions. Such an example is a sentence with variables. In this chapter we will work with these sentences and see how they can become propositions, both by substitution and by using words such as *for all* and *there exists*. We will also use them in proofs. The collection of rules that govern these predicates is known as ***first order predicate logic*** and is the subject of this chapter.

2.1. PREDICATES AND SETS

A sentence that has variables into which we can substitute values is called a ***predicate***. We shall represent a predicate by a capital letter (as with propositions) followed by a list of variables within parentheses, such as $P(x)$ or $Q(x, y, z)$. This mimics standard function notation. For example, $Q(x, y, z)$ may represent $x + y + z = 0$. If the predicate has a variable that we can replace with a value, then that variable must be in the list. As with a function, not every variable in the list is necessarily included in the predicate.

Example. Let $P(y)$ denote the predicate $y + 2 = 7$. This is not a proposition, for we do not know the value of y. We may also denote this predicate by $Q(x, y)$, even though x is not a variable in the equation. As with propositions, an assignment of a predicate will be specified with a :=. Hence, in this example we can either assign $P(y)$ or $Q(x, y)$ the predicate $y + 2 = 7$ by writing

$$P(y) := y + 2 = 7$$

or

$$Q(x, y) := y + 2 = 7.$$

If we wanted to view $y + 2 = 7$ as an equation with one variable, we would use $P(y)$. If we wanted to view it as a horizontal line, we would use $Q(x, y)$.

When writing algebraic expressions, we usually want to discover the values that make them true. In other words, we wish to form a proposition by replacing the variables with values. This replacement is called a ***substitution***. When a substitution yields a true proposition, we say that the predicate is ***satisfied*** by the substitution.

Example. Let $P(y)$ and $Q(x, y)$ be defined as above. The substitution $P(6)$ yields $6 + 2 = 7$, a false proposition. However, the substitution $Q(3, 5)$ gives a true proposition, $5 + 2 = 7$. Hence, $x = 5$ and $y = 3$ satisfy $Q(x, y)$. (Notice that the value substituted for x does not appear in the equation, for the predicate never included that variable.)

Example. There is no reason why substitutions must always involve numbers. For instance, make the assignment

$$P(f) := \textit{the linear function } f \textit{ is increasing.}$$

Here the variable is f. If $g(x) = 3x - 5$, then $P(g)$ is

the linear function g is increasing,

a true proposition. However, when $h(x) = -3x + 10$, $P(h)$ is false.

Example. Let $R(x, y, z)$ be $x + y = 2z - 3x$. This predicate has three variables. Since the x occurs twice, we say that is has two ***occurrences***. When making a substitution for a variable, the same replacement must be made for all occurrences of that variable. For instance, $R(1, 0, 2)$ is $1 + 0 = 2(2) - 3(1)$.

Since it is important to know what can be substituted into a variable, we will often determine ahead of time the collection of objects that are allowed for substitution. In mathematics an object can be a number, a function, an ordered pair, or almost anything that we would want to study. A collection of objects is called a ***set***, and the objects are referred to as the set's ***elements*** or ***members***. Picture these sets as boxes with things inside. The possibilities are limitless. For example, suppose we have a box that contains five animals. There is only one box. It contains five animals. Similarly, a set is an abstract structure that may contain many elements.

We will use uppercase letters to label sets, and elements will usually be represented by lowercase letters. The symbol $\in$ (fashioned after the Greek letter *epsilon*—see Appendix C)* is used to mean "element of." So, if A is a set and a an element of A, we write $a \in A$. The notation $a, b \in A$ means $a \in A$ and $b \in A$. If c is not an element of A, then write $c \notin A$. If A contains no elements (think of an empty box), it is the ***empty set***. Its symbol is $\varnothing$.

We can write sets by listing their elements and surrounding them with braces. This is called the ***roster method*** of writing a set, and the list is known as a ***roster***. The braces signify that a set has been defined. For example, the set of all integers between 1 and 10 inclusive is

$$\{1,\ 2,\ 3,\ 4,\ 5,\ 6,\ 7,\ 8,\ 9,\ 10\}.$$

Read this as "the set containing 1, 2, 3, 4, 5, 6, 7, 8, 9, and 10." The set of all integers between 1 and 10 (exclusive) is

$$\{2,\ 3,\ 4,\ 5,\ 6,\ 7,\ 8,\ 9\}.$$

If the roster is too long, use ellipses (...) to simplify matters. When there is a pattern to the elements of the set, write down enough members so that the pattern is clear. Then use the ellipses to represent the continuing pattern. For example, the set of all integers inclusively between 1 and 1,000,000 can be written as

$$\{1,\ 2,\ 3,\ \ldots,\ 999{,}999,\ 1{,}000{,}000\}.$$

Follow this strategy to write infinite sets as rosters. For instance, the set of even integers can be written as

$$\{\ldots,\ -4,\ -2,\ 0,\ 2,\ 4,\ \ldots\}.$$

Example. Let us write some sets as rosters.

1. As a roster, $\{\ \}$ denotes the empty set. Warning: never write $\{\varnothing\}$ for the empty set. This set has one element in it!
2. A set that contains exactly one element is called a ***singleton***. Hence, $\{1\}$, $\{f\}$, and $\{\varnothing\}$ are singletons written in roster form. Furthermore, $1 \in \{1\}$, $f \in \{f\}$, and $\varnothing \in \{\varnothing\}$.
3. The set of linear functions that intersect the origin with an integer slope can be written as:

   $$\{\ldots,\ -2x,\ -x,\ 0,\ x,\ 2x,\ \ldots\}.$$

   (Note: here 0 is representing the function $f(x) = 0$.)

---

*The $\in$ was first used to denote membership in Guiseppe Peano's *Arithmeticae Principia* in 1889. It is an abbreviation of the Greek word meaning "is" ($\varepsilon\sigma\tau\iota\nu$). Peano (Spinetta, Italy, 1858 – Turin, Italy, 1932) is also famous for his axioms of arithmetic.

Example. We should not use the roster method if the set does not have a pattern that is clearly seen. Such an example would be the set of real numbers. Image for a moment trying to write this set as a roster. The problem should be apparent!

Now let A and B be sets. These are ***equal*** if they contain exactly the same elements. The notation for this is $A = B$. What this means is if any element is in A, then it is also in B, and conversely, if an element is in B, then it is in A. To fully understand set equality, consider again the analogy between sets and boxes. Suppose that we have a box containing a carrot and a rabbit. We could describe it with the phrase *the box that contains the carrot and the rabbit*. Alternately, it could be referred to as *the box that contains the orange vegetable and the furry, cotton-tailed animal with long ears*. Although these are different descriptions, they do not refer to different boxes. Similarly, the set $\{1, 3\}$ and *the set containing the solutions of* $(x - 1)(x - 3) = 0$ are the same, for they contain the same elements. Furthermore, the order in which the elements are listed does not matter. The box can just as easily be described as *the box with the rabbit and the carrot*. Likewise, $\{1, 3\} = \{3, 1\}$. Lastly, suppose that the box was described as containing *the carrot, the rabbit, and the carrot*, forgetting that the carrot had already been mentioned. This should not be confusing, for one understands that such mistakes are possible. It is similar with sets. A repeated element does not add to the set. Hence, $\{1, 3\} = \{1, 3, 1\}$.

Although sets may contain many different types of elements, numbers are the most common. Therefore, particular important sets of numbers have been given their own symbols. They are as follows:

2.1.1. Notation

Symbol	Name
$\mathbb{N}$	*the set of natural numbers*
$\mathbb{Z}$	*the set of integers*
$\mathbb{Q}$	*the set of rational numbers*
$\mathbb{R}$	*the set of real numbers*
$\mathbb{C}$	*the set of complex numbers*

As rosters:

$$\mathbb{N} = \{0, 1, 2, \ldots\}$$

and

$$\mathbb{Z} = \{\ldots, -2, -1, 0, 1, 2, \ldots\}.$$

(See Section 3.1, page 99 for definitions of the other sets.) Notice that we define the set of natural numbers to include zero and do not make a distinction between counting numbers and whole numbers.* Instead, use a superscripted $+$ and $-$ to define

$$\mathbb{Z}^+ = \{1, 2, 3, \ldots\}$$

and

$$\mathbb{Z}^- = \{\ldots, -3, -2, -1\}.$$

Example.

- $10 \in \mathbb{Z}^+$, but $0 \notin \mathbb{Z}^+$.
- $4 \in \mathbb{N}$, but $-5 \notin \mathbb{N}$.
- $-5 \in \mathbb{Z}$, but $.65 \notin \mathbb{Z}$.
- $.65 \in \mathbb{Q}$ and $1/2 \in \mathbb{Q}$, but $\pi \notin \mathbb{Q}$.
- $\pi \in \mathbb{R}$, but $3 - 2i \notin \mathbb{R}$.
- $3 - 2i \in \mathbb{C}$.

Of the sets mentioned above, the real numbers are probably the most familiar. It is the set of numbers most frequently used in calculus and is often represented by a number line. The line can be subdivided into ***intervals***. Given two ***endpoints***, an interval includes all real numbers between the endpoints and possibly the endpoints themselves. ***Interval notation*** is used to work with these sets. A parenthesis—"(" or ")"—next to an endpoint means that the endpoint is not included in the set while a bracket—"[" or "]"—means that the endpoint is included. If the endpoints are included, then the interval is ***closed***. If they are excluded, then the interval is ***open***. If one endpoint is included and the other is not, then we say that the interval is ***half-open***. If the interval has only one endpoint, then the set is called a ***ray*** and is defined using the ***infinity symbol*** (∞).

2.1.2. Notation

If $a, b \in \mathbb{R}$ so that $a < b$, then we have the following:

closed interval	$[a, b]$	*closed ray*	$[a, \infty)$
open interval	(a, b)	*closed ray*	$(-\infty, a]$
half-open interval	$[a, b)$	*open ray*	(a, ∞)
half-open interval	$(a, b]$	*open ray*	$(-\infty, a)$

*Sometimes the set of counting numbers is defined as $\{1, 2, 3, \ldots\}$ and the set of whole numbers as $\{0, 1, 2, 3, \ldots\}$.

Example. In English we may describe $(4, 7)$ as

the interval between 4 and 7

and $[4, 7]$ as

all real numbers between 4 and 7 inclusive.

There is not a straightforward way to name the half-open intervals. For $(4, 7]$ we may try

the set of all real numbers x such that $4 < x \leq 7$

or

the set of all real numbers greater than 4 and less than or equal to 7.

The infinity symbol does not represent a real number. Hence, a parenthesis must be used with it. Furthermore, the interval $(-\infty, \infty)$ can be used to denote $\mathbb{R}$.

Example. The interval $(-1, 3]$ contains all real numbers that are greater than -1 but less than or equal to 3. This interval's number line representation would be:

A common mistake is to equate $(-1, 3]$ with $\{0, 1, 2, 3\}$. It is important to remember that $(-1, 3]$ includes all real numbers between -1 and 3. Hence, this set is infinite, as is $(-\infty, 2)$. It contains all real numbers less than 2. Therefore, as a number line this interval is:

When the objects of a set are the elements that are allowed to be substituted into a predicate, the set is called the ***domain*** of the predicate.

Example. Let $P(x) := x + 2 = 7$. Since $P(5)$ is true and no other real number satisfies $P(x)$, exactly one element of $\mathbb{R}$ satisfies $P(x)$. However, no element of $\mathbb{Z}^-$ or $(-\infty, 5)$ satisfies $P(x)$.

Example. If $Q(X) := x \geq 10$, then all elements of $[20, 100]$ satisfy $Q(x)$, some but not all elements of $\mathbb{Q}$ satisfy $Q(x)$, and no elements of $\{1, 2, 3\}$ satisfy $Q(x)$.

EXERCISES

1. Given a predicate, make the indicated substitution and then specify whether the resulting proposition is true or false.

(a) $P(2)$ when $P(x) := x^2 - 1 = 0$
(b) $Q(1)$ when $Q(x) :=$ *x is a natural number*
(c) $R(f)$ when $f(x) = x + 1$ and $R(x) :=$ *x is a linear function*
(d) $S(\mathbb{N})$ when $S(A) := -4 \in A$
(e) $T(6)$ when $T(x) :=$ *there is a natural number n so that* $x = 2^n$
(f) $U(0)$ when $U(x) :=$ *every integer is greater than x*
(g) $P(1, 3)$ when $P(x, y) := x + y = 4$
(h) $Q(1, 3)$ when $Q(x, y) := y = 1$
(i) $R(f, g)$ when $f(x) = x+1$, $g(x) = x$, and $R(x, y) := x - y = 1$
(j) $S(\sin x, e^x)$ when $S(f, g) :=$ *f and g are increasing functions*
(k) $T(\mathbb{Z}, \mathbb{N})$ when $T(A, B) := 0 \in A$ *and* $0 \in B$
(l) $U(\cos x, \sin x, 1)$ when $U(f, g, y) := f^2 + g^2 = y$

2. Determine whether the following are true or false:

(a) $0 \in \mathbb{N}$
(b) $1/2 \in \mathbb{Z}$
(c) $-4 \in \mathbb{Q}$
(d) $4 + \pi \in \mathbb{R}$
(e) $4.34534 \in \mathbb{C}$
(f) $\{1, 2\} = \{2, 1\}$
(g) $\{1, 2\} = \{1, 2, 1\}$
(h) $[1, 2] = \{1, 2\}$
(i) $(1, 3) = \{2\}$
(j) $-1 \in (-\infty, -1)$
(k) $-1 \in [-1, \infty)$
(l) $\varnothing \in (-2, 2)$
(m) $\varnothing \in \varnothing$
(n) $0 \in \varnothing$

3. Indicate whether there is an element in the given domain that satisfies $Q(x) := x + 3.14 = 0$.

(a) $\mathbb{Z}$
(b) $\mathbb{R}$
(c) $\mathbb{R}^+$
(d) $\mathbb{R}^-$
(e) $\mathbb{Q}$
(f) $\mathbb{C}$
(g) $(0, 6)$
(h) $(-\infty, -1)$

4. Write the following sets as rosters:

(a) The set of all integers between 1 and 5 inclusive
(b) The set of all odd numbers
(c) The set of all nonnegative integers
(d) The set of integers in the interval $(-3, 7]$
(e) The set of rational numbers in the interval $(0, 1)$ that can be represented with only two decimal places
(f) The set of all linear equations with a slope of 1 and an integer y-intercept

5. Write the following sets of real numbers using interval notation and then graph on a number line.
   (a) The set of all real numbers greater than 4
   (b) The set of all real numbers between -6 and -5 inclusive
   (c) The set of all real numbers x so that $x < 5$
   (d) The set of all real numbers x such that $10 < x \leq 14$
6. Use a predicate to uniquely describe the elements in the following sets. For example, $a \in \mathbb{N}$ if and only if a satisfies the predicate x *is an integer and* $x \geq 0$.
   (a) $(0, 1)$
   (b) $(-3, 3]$
   (c) $[0, \infty)$
   (d) $\mathbb{Q}$
   (e) $\mathbb{R}$
   (f) $\mathbb{C}$
   (g) $\mathbb{Z}^+$
   (h) $\{\ldots, -2, -1, 0, 1, 2, \ldots\}$
   (i) $\{2a, 4a, 6a, \ldots\}$
   (j) $\{\ldots, 1-3i, 1-2i, 1-i\}$

2.2. QUANTIFICATION

Besides substitution, there is another way to form a proposition from a predicate. Take

$$x + 2 = 7.$$

Assume that its domain is the set of real numbers. We know that it is true when $x = 5$. This means,

there exists $x \in \mathbb{R}$ *such that* $x + 2 = 7$

is a true proposition. The notation for this is

$$(\exists x \in \mathbb{R})(x + 2 = 7).$$

Such a statement is known as an ***existential proposition***. When true, it declares that there is an element of the domain that makes the predicate true. (See Figure 2.2.1.) It is false when no element of the domain satisfies the predicate. In general, if $P(x)$ is a predicate with domain D, the propositional form

$$(\exists x \in D)P(x)$$

means

there exists $x \in D$ *such that* $P(x)$.

Next consider $x + 5 = 5 + x$. This is true for all substitutions of real numbers for x. Figure 2.2.2 illustrates three of those substitutions. Therefore,

for all $x \in \mathbb{R}$, $x + 5 = 5 + x$.

2.2.1. Figure

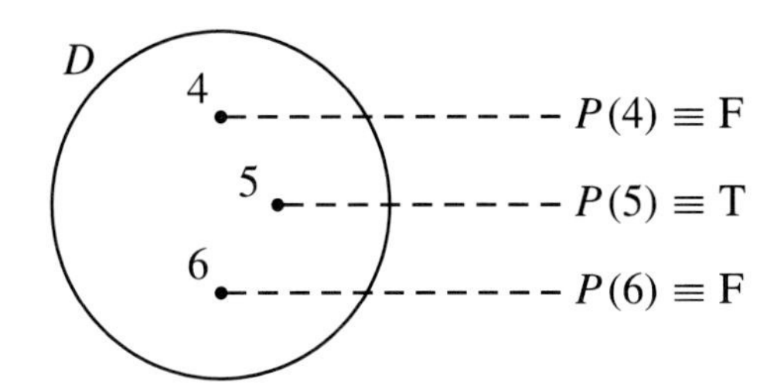

When at least one substitution yields a true proposition, $(\exists x \in D)P(x)$ is true. Otherwise, it is false.

The notation for this is

$$(\forall x \in \mathbb{R})(x + 5 = 5 + x).$$

This is an example of a ***universal proposition***. It would be false if one element of the domain did not satisfy the predicate. Again, generalize by letting $P(x)$ be a predicate and D be its domain. The propositional form

$$(\forall x \in D)P(x)$$

represents the proposition

$$\textit{for all } x \in D,\ P(x).$$

The symbols $\forall$ and $\exists$ are called ***quantifiers***. The $\forall$ is the ***universal quantifier***, and the $\exists$ is the ***existential quantifier***.

The next three examples show how the quantifiers can be written in English.

Example. Both

$$\textit{for all } y \in \mathbb{R},\ y + 0 = y$$

and

$$\textit{there exists } y \in \mathbb{R} \textit{ such that } 8 = 2y$$

are true propositions. The first one is true since a substitution of any real number for y in $y + 0 = y$ yields a true proposition. The second is true because $8 = 2 \cdot 4$ and 4 is a real number. Furthermore, there are many ways that these can be written. The quantifiers can be at the front of the sentence.

2.2.2. Figure

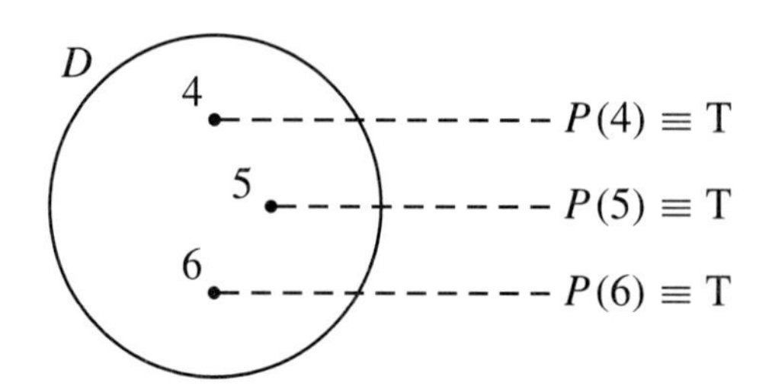

When all substitutions yield true propositions, $(\forall x \in D)P(x)$ *is true. Otherwise, it is false.*

Often this is preferred because it guarantees the sentence beginning with an uppercase letter, as with the following:

For each real number y, $y + 0 = y$.
For every $y \in \mathbb{R}$, $y + 0 = y$.
There is a real number y *so that* $8 = 2y$.
For some $y \in \mathbb{R}$, $8 = 2y$.

However, the quantifiers can be at the back:

$y + 0 = y$ *for all* $y \in \mathbb{R}$.
$y + 0 = y$ *for each real number* y.
$8 = 2y$ *for some real number* y.
8 *equals* $2y$ *for a particular* $y \in \mathbb{R}$.

Example. Typically, when writing propositions with multiple quantifiers that are the same, they are all either put in front or in back. This is not necessarily the case when we have a universal and an existential. Take the following which asserts the existence of additive inverses:

$$(\forall x \in \mathbb{R})(\exists y \in \mathbb{R})(x + y = 0).$$

Both

for all $x \in \mathbb{R}$, *there exists* $y \in \mathbb{R}$ *such that* $x + y = 0$

and

for every real number x, $x + y = 0$ *for some* $y \in \mathbb{R}$

are satisfactory translations.

A list of values can be handled as in the next example.

Example. If $a_1, a_2, \ldots, a_k$ is a list of integers and d divides one of them, we may write

$$d \text{ divides } a_i \text{ for some } i = 1, \ldots, k$$

instead of *there exists* $a \in \{a_1, a_2, \ldots, a_k\}$ *such that* d *divides* a. Similarly, if d divides them all, write

$$d \text{ divides } a_i \text{ for all } i = 1, \ldots, k.$$

There are times when a domain is not given with the quantifier. In this case, assume that the variable can represent any object. However, this can lead to substitutions that do not make sense. For example, $(\forall x)(x + 0 = x)$ is not a proposition, for what is its truth value when x represents a dog? To make this work, we should include within the predicate the fact that the formula holds if x is real and write

$$(\forall x)(x \in \mathbb{R} \Rightarrow x + 0 = x).$$

The existence of a real solution to the equation $x + 1 = 10$ can be written as

$$(\exists x)(x \in \mathbb{R} \wedge x + 1 = 10).$$

This states that there is an object that is both a real number and a solution to the equation. This yields the following relationship between this and our original notation.

2.2.3. Theorem

If D is a domain for a predicate $P(x)$, then:

$$(\forall x \in D)P(x) \equiv (\forall x)[x \in D \Rightarrow P(x)],$$
$$(\exists x \in D)P(x) \equiv (\exists x)[x \in D \wedge P(x)].$$

Example. These equivalences are crucial to understanding the truth values of quantified propositions that have empty domains. Let $P(x)$ be any predicate.

- Since $x \in \varnothing$ is always false, the predicate $x \in \varnothing \Rightarrow P(x)$ will be satisfied by any x. This is because its hypothesis is always false. Therefore, $(\forall x)[x \in \varnothing \Rightarrow P(x)]$ is true, and this implies that the proposition $(\forall x \in \varnothing)P(x)$ is true.
- On the other hand, the conjunction $x \in \varnothing \wedge P(x)$ will be false for every substitution. Hence, $(\exists x \in \varnothing)P(x)$ is false.

Now that we have introduced propositions with one quantifier, let us move to propositions with many quantifiers. For instance, the Associative Law for addition states

$$(x + y) + z = x + (y + z)$$

for all real numbers x, y, and z. Using logic notation, we may rewrite this as

$$(*) \qquad (\forall x \in \mathbb{R})(\forall y \in \mathbb{R})(\forall z \in \mathbb{R})[(x + y) + z = x + (y + z)].$$

Let us examine what this means. Whenever we have a quantifier of the form $(\forall x)$ or $(\exists x)$, there will be a predicate to its right. This is what we have here. The predicate to the right of $(\forall x \in \mathbb{R})$ in $(*)$ is

$$(**) \qquad (\forall y \in \mathbb{R})(\forall z \in \mathbb{R})[(x + y) + z = x + (y + z)].$$

Because of the placement of the brackets, $(\forall y \in \mathbb{R})$ applies to the occurrences of y and $(\forall z \in \mathbb{R})$ applies to the occurrences of z. When a quantifier applies to an occurrence of a variable, we say that the occurrence is ***bound***. The important property of bound occurrences is that we cannot substitute into them. However, the occurrences of x are not bound in $(**)$. No quantifier applies to them. We say that these occurrences are ***free***, and we may substitute into them. Therefore, $(**)$ is a predicate, and we will call it $P(x)$. For example, $P(2)$ is the proposition

$$P(2) := (\forall y \in \mathbb{R})(\forall z \in \mathbb{R})[(2 + y) + z = 2 + (y + z)].$$

Proposition $(*)$ is true because $P(x)$ is true for all substitutions of real numbers for x. Since $P(x)$ contains quantifiers, let us examine it next. If we let

$$Q(x,\ y) := (\forall z \in \mathbb{R})[(x + y) + z = x + (y + z)],$$

we have $P(x) \equiv (\forall y \in \mathbb{R})Q(x,\ y)$, and $(*)$ is true when $Q(x,\ y)$ is satisfied by all real numbers. We may further write $Q(x,\ y)$ as $(\forall z \in \mathbb{R})R(x,\ y,\ z)$ where

$$R(x,\ y,\ z) := [(x + y) + z = x + (y + z)].$$

Hence, $(*)$ is true when $R(x,\ y,\ z)$ is true for every substitution of real numbers into x, y, and z. What we have done is to break apart a proposition that contains multiple quantifiers into a sequence of predicates:

$$(\forall x \in \mathbb{R})\,\overbrace{(\forall y \in \mathbb{R})\,\underbrace{(\forall z \in \mathbb{R})\,\underbrace{[(x + y) + z = x + (y + z)]}_{R(x,\ y,\ z)}}_{Q(x,\ y)}}^{P(x)}.$$

To determine whether it is true, we must show that every real number satisfies $P(x)$. To do this we must demonstrate that $Q(x,\ y)$ is true for all real numbers, and for this it is enough to show that $R(x, y, z)$ is true for every substitution of real numbers.

Example. Consider the equation $y = 2x^2 + 1$. Before we learned the various techniques that make graphing this equation simple, we graphed it by writing a table with one column for the x values and another for the y. We then chose numbers to substitute for x and calculated the corresponding y. Although we did not explicitly write it this way, we learned that

for every $x \in \mathbb{R}$, there exists $y \in \mathbb{R}$ such that $y = 2x^2 + 1$,

which can be written as

$$(\forall x \in \mathbb{R})(\exists y \in \mathbb{R})(y = 2x^2 + 1).$$

Let $P(x)$ denote $(\exists y \in \mathbb{R})(y = 2x^2 + 1)$. All occurrences of y in $P(x)$ are bound, but the occurrence of x is free, and we can make substitutions like $P(3) \equiv (\exists y \in \mathbb{R})(y = 2(3)^2 + 1)$. This is a true proposition because $19 = 2(3)^2 + 1$ and 19 is a real number. (We would then plot $(3, 19)$.) In fact, $P(x)$ is satisfied by every real number.

Example. Define $P(x, y)$ to be the predicate

$$(\forall x)(\exists z)[P(x, y) \wedge Q(z)] \vee (\exists y)[Q(y) \Rightarrow R(x)].$$

The first occurrence of y is free, and since $(\forall x)$ applies only to variables within the brackets surrounding the left disjunct, the last occurrence of x is also free. Therefore, $P(x, y)$ is the disjunction of two predicates that we may call $Q(y)$ and $R(x)$:

$$\overbrace{(\forall x)(\exists z)[P(x, y) \wedge Q(z)]}^{Q(y)} \vee \overbrace{(\exists y)[Q(y) \Rightarrow R(x)]}^{R(x)},$$

and the substitution $P(9, 5)$ is

$$Q(5) \vee R(9) \equiv (\forall x)(\exists z)[P(x, 5) \wedge Q(z)] \vee (\exists y)[Q(y) \Rightarrow R(9)].$$

EXERCISES

1. Rewrite each proposition at least two other ways: once with the quantifier in front and once with at least one quantifier in back.
   (a) *For all $x \in \mathbb{Z}$, $x - x = 0$.*
   (b) *For all $x \in \mathbb{R}$, there exists $y \in \mathbb{R}$ such that $y = 2x$.*
   (c) *There exists $x \in \mathbb{Z}$ so that for all $y \in \mathbb{Z}$, $y + x = x$.*
   (d) *For all integers x and y, there exists $z \in \mathbb{R}$ such that $z = 2x - 3y$.*

2. Define the following predicates:

$$P(x) := x - 3 = 2,$$
$$Q(x) := \sqrt{x} < 100,$$
$$R(x) := x \in \mathbb{Q}.$$

Write the given sentences using $P(x)$, $Q(x)$, or $R(x)$ and quantifiers. Be sure to include the indicated domain in the answer.

(a) *Every integer is rational.*
(b) *There is a real number whose square root is less than 100.*
(c) *There exists a natural number x so that* $x - 3 = 2$.
(d) *For all real numbers x, if* $x - 3 = 2$, *then* $\sqrt{x} < 100$.
(e) *Either* $x \in \mathbb{Q}$ *or there exists an integer x such that* $x \notin \mathbb{Q}$.
(f) *Every natural number x is also rational.*
(g) *For all rational numbers x, there exists* $y < x$ *so that* $\sqrt{y} < 100$.

3. Given the following:

$$P(x) := x > 5,$$
$$Q(x) := x + 5 = 9,$$
$$R(x) := x^2 = 4.$$

Translate the following into English:

(a) $(\forall x \in \mathbb{Z}) P(x)$
(b) $(\exists x \in \mathbb{C})[Q(x) \Rightarrow R(x)]$
(c) $(\forall x \in \mathbb{R})(\exists y \in \mathbb{R})[P(x) \wedge Q(y)]$
(d) $(\exists x \in \mathbb{R}) \sim R(x) \vee (\forall x \in \mathbb{R})[Q(x) \Leftrightarrow \sim P(x)]$

4. Indicate whether the following are true or false.

(a) $(\forall x \in (3, 9])(x > 2)$
(b) $(\exists x \in (3, 9])(x/2 = 4)$
(c) $(\exists x \in (3, 9])(x^2 - 1 = 0)$
(d) $(\forall x \in (3, 9])(x - 5 = 0)$
(e) $(\exists x \in \mathbb{R})(\exists y \in \mathbb{R})(y = 2x^2 + 1)$
(f) $(\forall x \in \mathbb{R})(\forall y \in \mathbb{R})(y = 2x^2 + 1)$
(g) $(\exists x \in \mathbb{R})(\forall y \in \mathbb{R})(y = 2x^2 + 1)$
(h) *For all* $x \in \mathbb{Z}$, $x - x = 0$.
(i) *For all* $x \in \mathbb{R}$, *there exists* $y \in \mathbb{R}$ *such that* $y = 2x$.
(j) *There exists* $x \in \mathbb{Z}$ *so that for all* $y \in \mathbb{Z}$, $y + x = x$.
(k) *For all integers x and y, there exists* $z \in \mathbb{R}$ *such that* $z = 2x - 3y$.

5. Explain why for any nonempty domain D,

$$(\forall x \in D) P(x) \Rightarrow (\exists x \in D) P(x)$$

is true for every predicate $P(x)$. What happens when $D = \varnothing$?

6. Indicate whether $(\forall x \in D)(x^2 > 4)$ is true or false for the following domains.
   (a) $D = \mathbb{N}$
   (b) $D = \mathbb{Q}$
   (c) $D = [2, \infty)$
   (d) $D = (2, 10)$
   (e) $D = (-\infty, -5)$
   (f) $D = \varnothing$
7. Indicate whether $(\exists f \in D)(f'(x) = 3)$ is true or false for the following domains.
   (a) $D =$ the set of linear functions
   (b) $D =$ the set of functions of the form $g(x) = ax + 3$ with $a \in \mathbb{R}$
   (c) $D =$ the set of functions of the form $f(x) = ax^2$ with $a \in \mathbb{R}$
   (d) $D =$ the set of continuous functions
   (e) $D =$ the set of differentiable functions
   (f) $D = \varnothing$
8. If $P(x)$ is true for every real number in $[5, \infty)$, then we may write $(\forall x \in [5, \infty))P(x)$. This can be simplified as $(\forall x \geq 5)P(x)$ with the assumption that $x \in \mathbb{R}$. Use this notation to write the following:
   (a) $(\forall x \in (\infty, 3])P(x)$
   (b) $(\exists x \in (0, \infty))Q(x)$
   (c) $(\forall x \in \mathbb{R}^+)(\exists y \in (\infty, -3])R(x, y)$
   (d) $(\exists x \in (1, \infty))(\forall y \in \mathbb{R})S(x, y)$
9. For each of the following, identify all bound occurrences of x.
   (a) $(\forall x)[P(x) \Rightarrow (\exists y)Q(y)]$
   (b) $(\exists x)P(x, y) \Rightarrow (\forall y)Q(x)$
   (c) $(\exists x)(x > 4) \wedge x < 10$
   (d) $[(\forall x)(x + 3 = 1) \wedge x = 9] \vee (\exists y)(x < 0)$
10. Given the predicate

$$P(x, y, z) := R(x, y) \Rightarrow (\forall x)[S(x) \wedge (\exists y)T(y, z)],$$

    find all free occurrences and make the substitution $P(1, 2, 3)$.
11. For each propositional form, identify the predicate to the right of each of its quantifiers.
   (a) $(\forall x)[P(x) \Rightarrow (\exists y)Q(y)]$
   (b) $(\exists x)P(x, y) \Rightarrow (\forall y)Q(y)$
   (c) $(\forall x)(\exists y)(\exists z)P(x, y, z) \wedge R(w)$
   (d) $P(x) \wedge (\exists x)(\forall y)Q(x, y) \vee (\forall z)R(z)$
12. Which of the following best defines f as being a constant function:
   - $(\forall x)(\exists k)[f(x) = k]$
   - $(\forall k)(\exists x)[f(x) = k]$
   - $(\exists x)(\forall k)[f(x) = k]$
   - $(\exists k)(\forall x)[f(x) = k]$
13. Use Theorem 2.2.3 to show $(\forall x \in D)(\forall y \in D)P(x, y)$ is equivalent to $(\forall x)(x \in D \Rightarrow (\forall y)[y \in D \Rightarrow P(x, y)])$.

14. Let $P(x, y)$ be a predicate. For any domain D, we can combine quantifiers using the following rules:
    - $(\forall x \in D)(\forall y \in D)P(x, y) \equiv (\forall x, y \in D)P(x, y)$
    - $(\exists x \in D)(\exists y \in D)P(x, y) \equiv (\exists x, y \in D)P(x, y)$

    Use these to show that

    $$(\forall x \in \mathbb{R})(\forall y \in \mathbb{R})(\exists z \in \mathbb{Z})(\exists w \in \mathbb{Z})P(x, y, z, w)$$

    is equivalent to

    $$(\forall x, y \in \mathbb{R})(\exists z, w \in \mathbb{Z})P(x, y, z, w).$$

15. One must be careful when combining quantifiers as in the previous problem. They must be the same and must not switch relative to different quantifiers. Explain why $(\forall x, y \in \mathbb{R})(\exists z \in \mathbb{R})(x + y = z)$ has a different meaning than $(\forall x \in \mathbb{R})(\exists z \in \mathbb{R})(\forall y \in \mathbb{R})(x + y = z)$.
16. Combine as many quantifiers as possible.
    (a) $(\forall x)(\forall y)[P(x) \Rightarrow Q(x, y)]$
    (b) $(\exists x)(\exists y)P(x) \wedge (\forall z)Q(z, y)$
    (c) $(\forall x)(\exists y)(\forall z)(\exists w)P(x, z) \vee Q(y, w)$
    (d) $(\exists x \in \mathbb{Z})(\exists y \in \mathbb{R})P(x, y)$
    (e) $(\forall x \in \mathbb{Z})(\exists y \in \mathbb{Q})(\forall z \in \mathbb{R})(\forall w \in \mathbb{R})P(x, y, z, w)$

2.3. NEGATING QUANTIFIERS

Let $P(x)$ be a predicate. We have discussed the meaning of $(\forall x \in D)P(x)$ and $(\exists x \in D)P(x)$. Now let us turn to their negations. Choose a domain D. To see how this works, suppose $D = \{4, 5, 6\}$. If $(\forall x \in D)P(x)$ is true, then

$$P(4) \equiv \mathrm{T},\ P(5) \equiv \mathrm{T},\ \text{and}\ P(6) \equiv \mathrm{T}.$$

On the other hand, if $(\forall x \in D)P(x)$ is to be false, then at least one of the substitutions must be false. Suppose it is $P(5)$. In this case, $(\exists x \in D){\sim}P(x)$ is true. (Figure 2.3.1.) Generalizing to an arbitrary domain D, since the propositional form $(\forall x \in D)P(x)$ means that $P(a)$ is true for all $a \in D$, ${\sim}(\forall x \in D)P(x)$ must mean that there is at least one element of D that makes $P(x)$ false. Hence,

$${\sim}(\forall x \in D)P(x) \equiv (\exists x \in D){\sim}P(x).$$

Now consider $(\exists x \in D)P(x)$. If the domain is again $\{4, 5, 6\}$ and the truth value of $(\exists x \in D)P(x)$ is true, then there is at least one number in D that satisfies $P(x)$. Suppose it is 6. However, if $(\exists x \in D)P(x)$ is to be false, all

2.3.1. Figure

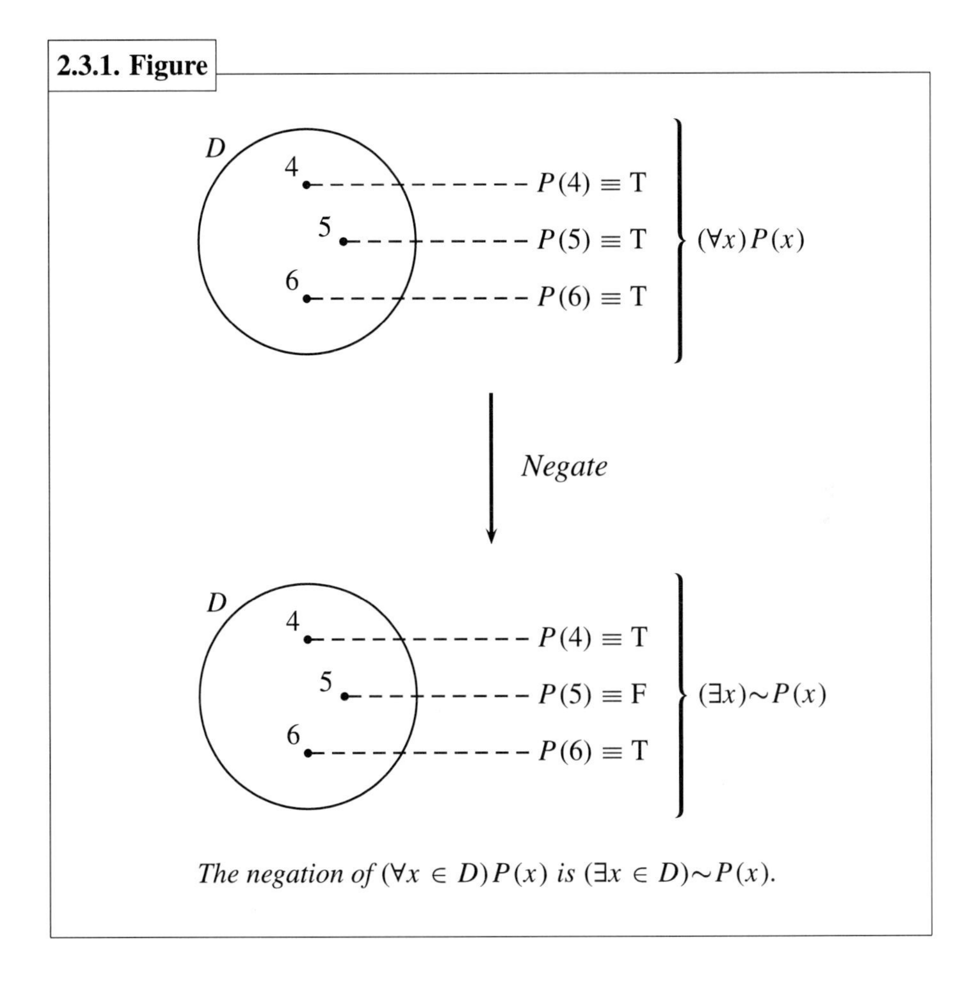

The negation of $(\forall x \in D)P(x)$ *is* $(\exists x \in D)\sim P(x)$.

substitutions for x must yield a false proposition, including $x = 6$. Therefore, $(\forall x \in D)\sim P(x)$ is true, and we have

$$P(4) \equiv \mathrm{F},\ P(5) \equiv \mathrm{F},\ \text{and } P(6) \equiv \mathrm{F}.$$

(See Figure 2.3.2.) In the general case, $\sim(\exists x \in D)P(x)$ must mean that every element of D makes $P(x)$ false because $(\exists x \in D)P(x)$ means there is at least one $a \in D$ such that $P(a)$ is true. Therefore,

$$\sim(\exists x \in D)P(x) \equiv (\forall x \in D)\sim P(x).$$

In summary, to negate a proposition with a quantifier, exchange quantifiers and negate the predicate.

Example. The following are negations of each other:

1. $(\forall x)[P(x) \wedge Q(x)]$ and $(\exists x)\sim[P(x) \wedge Q(x)]$
2. $(\exists x)[R(x) \Rightarrow S(x)]$ and $(\forall x)\sim[R(x) \Rightarrow S(x)]$

2.3.2. Figure

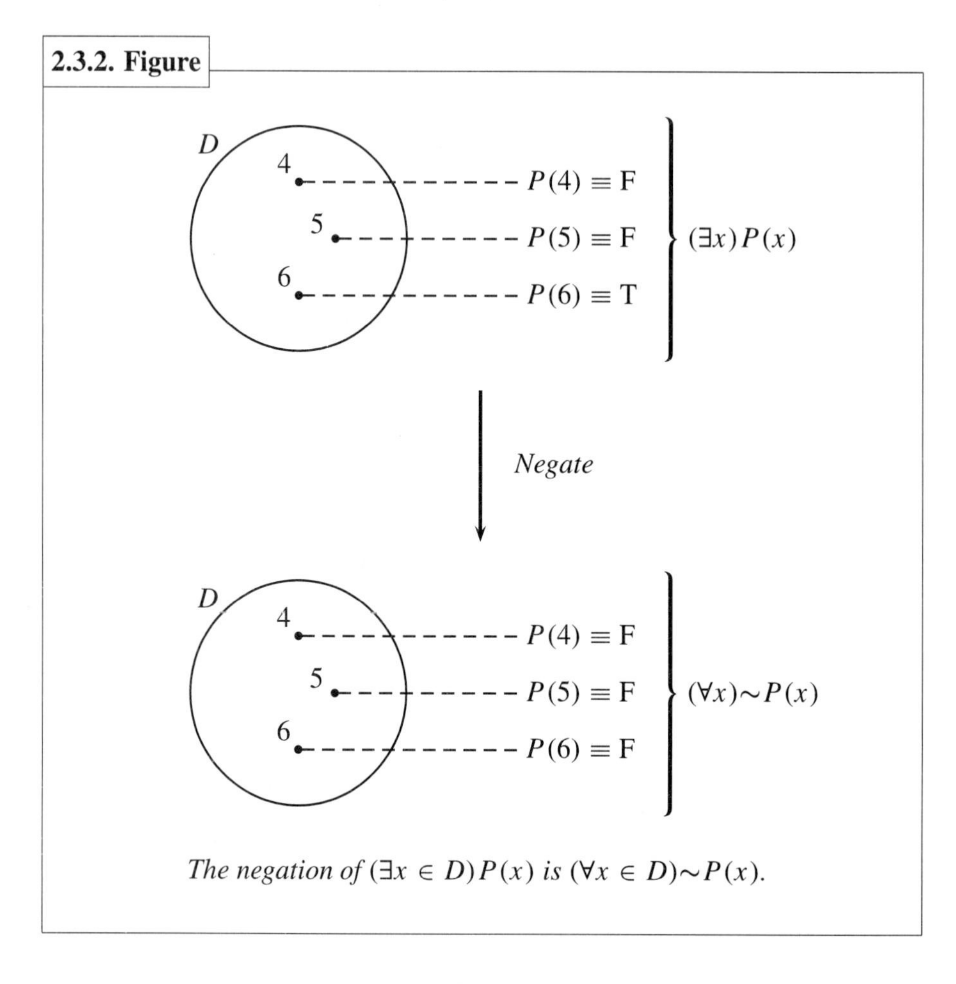

The negation of $(\exists x \in D)P(x)$ *is* $(\forall x \in D){\sim}P(x)$.

Whenever we negate quantifiers, the final form should not have a negation immediately to the left of any quantifier, and using the rules of replacement, the negation should be as far into the predicate as possible. We will say that such negations are in ***positive form***.

Example. Let us find the negation of $(\forall x)[P(x) \wedge Q(x)]$ and put the final answer in positive form.

$$\begin{aligned} {\sim}(\forall x)[P(x) \wedge Q(x)] &\equiv (\exists x){\sim}[P(x) \wedge Q(x)] \\ &\equiv (\exists x)[{\sim}P(x) \vee {\sim}Q(x)]. \end{aligned}$$

The next example will use De Morgan's Law as the last one did. It also needs Material Implication.

Example. The negation of $(\forall x)(\exists y)[P(x) \Rightarrow Q(y)]$ written in positive form is as follows:

$$\begin{aligned} \sim(\forall x)(\exists y)[P(x) \Rightarrow Q(y)] &\equiv (\exists x)\sim(\exists y)[P(x) \Rightarrow Q(y)] \\ &\equiv (\exists x)(\forall y)\sim[P(x) \Rightarrow Q(y)] \\ &\equiv (\exists x)(\forall y)\sim[\sim P(x) \vee Q(y)] \\ &\equiv (\exists x)(\forall y)[P(x) \wedge \sim Q(y)] \end{aligned}$$

Example. Now let us try to negate some propositions that are written in English.

- For the first one, we will translate using logic symbols, do the work, and then translate back. To negate

  there exists a real number x so that $x^2 = 4$,

  first write:

  $$(\exists x \in \mathbb{R})(x^2 = 4).$$

  Its negation is

  $$(\forall x \in \mathbb{R})\sim(x^2 = 4).$$

  Since $\sim(x^2 = 4)$ means $x^2 \neq 4$, the negation can be written as

  for every real number x, $x^2 \neq 4$,

  or in better English,

  the square of every real number is not four.

- The negation of

  for every $x \in \mathbb{R}$, there exists $y \in \mathbb{R}$ such that $x > 0$ and $y = \cos x$

  is

  there exists $x \in \mathbb{R}$ so that for every $y \in \mathbb{R}$, $x \leq 0$ or $y \neq \cos x$.

  This follows by applying the negation rules and De Morgan's Law.

There are many times in mathematics when we must show that a proposition of the form $(\forall x \in D)P(x)$ is false. This can be accomplished by proving $(\exists x \in D)\sim P(x)$ is true, and this is done by finding an element $a \in D$ such that $P(a)$ is false. This element is called a ***counterexample*** to $(\forall x \in D)P(x)$.

Example. To show that $(\forall x \in \mathbb{R})(x + 2 = 7)$ is false, we will show $\sim(\forall x \in \mathbb{R})(x + 2 = 7)$, which is equivalent to

$$(\exists x \in \mathbb{R})(x + 2 \neq 7).$$

This proposition is true because 0 does not satisfy $x + 2 = 7$ and 0 is a real number. Hence, 0 is a counterexample.

Example. We know from algebra that *every quadratic function has a real root* is false. This can be symbolized as

$$(\forall f)(f \text{ is quadratic } \Rightarrow (\exists x \in \mathbb{R})[f(x) = 0]).$$

The counterexample is found by demonstrating

$$(\exists f)(f \text{ is quadratic } \wedge (\forall x \in \mathbb{R})[f(x) \neq 0]).$$

Many functions could be chosen, but we will use a simple one. The function $f(x) = x^2 + 1$ will be our counterexample because its only roots are i and $-i$.

Notice that in both examples a specific object was identified, not some generality. This is important to remember.

EXERCISES

1. Indicate whether or not the following are equivalent.
   (a) $\sim(\forall x)P(x) \equiv (\forall x)\sim P(x)$
   (b) $\sim(\exists x)P(x) \equiv \sim(\forall x)P(x)$
   (c) $(\forall x)\sim P(x) \equiv (\exists x)P(x)$
   (d) $(\exists x)[P(x) \Rightarrow Q(x)] \equiv \sim(\forall x)[\sim P(x) \Rightarrow \sim Q(x)]$
   (e) $\sim(\forall x)(\exists y)P(x, y) \equiv (\exists x)(\forall y)\sim P(x, y)$
   (f) $\sim(\forall x)(\exists y)P(x, y) \equiv (\exists y)(\forall x)P(x, y)$

2. Negate and put into positive form.
   (a) $(\exists x)[Q(x) \Rightarrow R(x)]$
   (b) $(\forall x)(\exists y)[P(x) \wedge Q(y)]$
   (c) $(\exists x)(\exists y)[P(x) \vee Q(x, y)]$
   (d) $(\forall x)(\forall y)[P(x) \vee (\exists z)Q(y, z)]$
   (e) $(\exists x)\sim R(x) \vee (\forall x)[Q(x) \Leftrightarrow \sim P(x)]$
   (f) $(\forall x)(\forall y)(\exists z)(P(x) \Rightarrow [Q(y) \wedge R(z)]) \Rightarrow (\exists x)\sim Q(x)$

3. Determine whether each pair of propositions are negations. If they are not, write the negation of both.
   (a) *Every real number has a square root.*
   Every real number does not have a square root.
   (b) *All derivatives are continuous.*
   Some derivatives are continuous.
   (c) *Every multiple of four is a multiple of two.*
   Some multiples of two are multiples of four.
   (d) *There is a function that has a derivative.*
   All functions do not have a derivative.

(e) *For all $x \in \mathbb{R}$, if x is odd, then x^2 is odd.*
There exists $x \in \mathbb{R}$ such that if x is odd, then x^2 is even.
(f) *There exists an integer x such that $x + 1 = 10$.*
For all integers x, $x + 1 \neq 10$.
(g) *There exists an even function.*
There is no even function.
(h) *For all functions $f(x) = ax + b$, $f'(x) = a$.*
There exists a function $f(x) \neq ax + b$ so that $f'(x) \neq a$.

4. Write the negation of the following sentences in positive form and in English.
   (a) *For all $x \in \mathbb{R}$, there exists $y \in \mathbb{Z}$ such that $y/x = 9$.*
   (b) *There exists $x \in \mathbb{R}$ so that $xy = 1$ for all real numbers y.*
   (c) *There is an function with y-intercept equal to -3.*
   (d) *Every multiple of ten is a multiple of five.*
   (e) *No interval contains a rational number.*
   (f) *All functions have a derivative at $x = 0$.*
5. Provide counterexamples for each of the following false propositions.
   (a) *Every integer is a solution to $x + 1 = 0$.*
   (b) *For every $x \in \mathbb{Z}$, there exists $y \in \mathbb{Z}$ such that $xy = 1$.*
   (c) *The product of any two integers is even.*
   (d) *For every integer n, if n is even, then n^2 is a multiple of eight.*
   (e) *If a function has a minimum, then it also has a maximum.*
   (f) *Every function has a derivative at $x = 0$.*
6. Use the rules in Exercise 14 of Section 2.2 to show the following:
   (a) $\sim(\forall x, y)P(x, y) \equiv (\exists x, y)\sim P(x, y)$
   (b) $\sim(\exists x, y)P(x, y) \equiv (\forall x, y)\sim P(x, y)$
7. A pair of propositions are ***contraries*** if they can be both false but not both true, and two propositions are ***subcontraries*** if they can be both true but not both false. Assuming a nonempty domain, give an example of a pair of contraries and a pair of subcontraries.

2.4. PROOFS WITH QUANTIFIERS

In the previous chapter we studied propositions and valid arguments. These arguments involved general propositional forms without writing any quantifiers. However, in mathematics most propositions that we will be interested in proving involve quantifiers. To prove these propositions, we will need to add to our list of rules of inference. In this section we will see how this is done. We will also begin to write paragraph proofs. A ***paragraph proof*** is a argument written in English or some other language. Its structure is based on a two-column proof, but it usually does not contain all of the detail. Its purpose is to convince the reader that the theorem logically follows from the hypotheses.

We begin by examining the universal quantifier. For our work here, assume that all domains are nonempty, even if the domain is not included with the quantifier. This is not a significant assumption, for we usually deal with nonempty sets.

2.4.1. Universal Instantiation (UI)

If a is a constant from a nonempty domain D, then

$$(\forall x \in D)P(x) \vdash P(a).$$

Universal Instantiation is quickly justified. When $(\forall x \in D)P(x)$ is true, every element of the domain satisfies $P(x)$. Hence, any constant may be substituted into $P(x)$.

Example. If $a \in D$, then the following are legitimate uses of Universal Instantiation:

1. $(\forall x \in D)[P(x) \Rightarrow Q(x)] \vdash P(a) \Rightarrow Q(a)$
2. $(\forall x \in D)[P(x) \vee (\forall y \in D)Q(y)] \vdash P(a) \vee (\forall y \in D)Q(y)$
3. $(\forall x \in D)(\forall y \in D)[Q(x) \vee R(y)] \vdash (\forall y \in D)[Q(a) \vee R(y)]$

Before we can write proofs, we need a rule that will attach a universal quantifier to a predicate. It will be different from Universal Instantiation, for it requires a criterion on the constant. A constant a is ***arbitrary*** if it can represent a randomly selected element of the domain. For example, the expression *let a be a real number* means that a represents any real number without restriction. If a constant is not arbitrary, then it is ***particular***. A particular element has some extra characteristic other than simply being a member of the domain.

Example. Consider the predicate $P(x, y) := y = x + 5$. Pick an element $a \in \mathbb{R}$. This constant is arbitrary. It was chosen randomly. When we substitute a for x, we obtain $P(a, y) := y = a + 5$. If b is another constant that satisfies $y = a + 5$, then we know that $b = a + 5$. It is not arbitrary. It represents a particular real number, the one that makes the equation true.

Example. An arbitrary real number will satisfy $x + 7 = 7 + x$, but only a particular real number will satisfy $2 + x = 10$.

One way to see if a symbol is arbitrary is to check if it has a free occurrence in a hypothesis of a proof. If it does not, then it is arbitrary. If it does, then it is particular because the assumption is declaring that the symbol has a specific property.

Example. In $(\forall x)P(x) \vdash P(a)$, the a is arbitrary, but the a in the following is not, for the occurrence of a in the hypothesis $P(a)$ is free:

1.	$P(a)$	Given
2.	$P(a) \vee Q(a)$	

We may now introduce the rule of inference that allows us to attach universal quantifiers to predicates. The arbitrary elements will usually be represented by a and sometimes b.

2.4.2. Universal Generalization (UG)

Let a be an arbitrary constant symbol from a nonempty domain D. If $P(a)$ contains no particular constants from D, then

$$P(a) \vdash (\forall x \in D)P(x).$$

Since $a \in D$ is arbitrary, $P(a)$ means that $P(x)$ is satisfied by every element of the domain. Therefore, the conclusion $(\forall x \in D)P(x)$ is valid. The restriction against $P(a)$ containing no particular symbols prevents us from making invalid arguments such as concluding $(\forall x \in \mathbb{R})(x + b = 2)$ from $a + b = 2$. In $a + b = 2$, a and b cannot both be arbitrary.

Example. Examine these illegal uses of Universal Generalization.

1. The following is invalid:

1.	$P(c)$	Given
2.	$(\forall x)P(x)$	1 UG

   The constant c in line 1 is a particular constant. It is a element that satisfies $P(x)$. Therefore, Universal Generalization does not apply.

2. The following is an attempt to prove $(\forall x \in \mathbb{R})(\forall y \in \mathbb{R})(x+y = 2x)$ from $(\forall x \in \mathbb{R})(x + x = 2x)$:

1.	$(\forall x \in \mathbb{R})(x + x = 2x)$	Given
2.	$a + a = 2a$	1 UI
3.	$(\forall y \in \mathbb{R})(a + y = 2a)$	2 UG
4.	$(\forall x \in \mathbb{R})(\forall y \in \mathbb{R})(x + y = 2x)$	3 UG

Although the occurrences of a in line 2 are arbitrary, the proof is not valid. The reason is that an illegal substitution was made in line 3. To see this, let

$$P(x) := x + x = 2x.$$

Applying Universal Generalization gives

$$(\forall y \in \mathbb{R})(y + y = 2y)$$

because $P(y) := y + y = 2y$, but in line 3, the variable y was not substituted into every free occurrence of a.

Now for some valid proofs.

Example. We will give two-column proofs for both of the following.

1. $(\forall x)(\forall y)P(x, y) \vdash (\forall y)(\forall x)P(x, y)$

Proof.

1.	$(\forall x)(\forall y)P(x, y)$	Given
	$\langle$Show $(\forall y)(\forall x)P(x, y)\rangle$	
2.	$(\forall y)P(a, y)$	1 UI
3.	$P(a, b)$	2 UI
4.	$(\forall x)P(x, b)$	3 UG
5.	$(\forall y)(\forall x)P(x, y)$	4 UG ■

Since both a and b are arbitrary, Universal Generalization can be applied to both constants.

2. $(\forall x)[P(x) \Rightarrow Q(x)],\ (\forall x)\sim[Q(x) \vee R(x)] \vdash (\forall x)\sim P(x)$

Proof.

1.	$(\forall x)[P(x) \Rightarrow Q(x)]$	Given
2.	$(\forall x)\sim[Q(x) \vee R(x)]$	Given
	$\langle$Show $(\forall x)\sim P(x)\rangle$	
3.	$P(a) \Rightarrow Q(a)$	1 UI
4.	$\sim[Q(a) \vee R(a)]$	2 UI
5.	$\sim Q(a) \wedge \sim R(a)$	4 DeM
6.	$\sim Q(a)$	5 Simp
7.	$\sim P(a)$	3, 6 MT
8.	$(\forall x)\sim P(x)$	7 UG ■

Notice that the a in line 3 did not occur free in a hypothesis. Hence, a is arbitrary throughout the proof.

We will use the ideas behind these proofs to write paragraph proofs for propositions with universal quantifiers. To prove $(\forall x \in D)P(x)$ from a given set of hypotheses, we must show that every element of D satisfies $P(x)$ assuming those hypotheses. This is accomplished by picking an arbitrary element of D by writing what we will call an ***introduction***. This is a sentence that declares the type of object represented by the variable. The following are examples of introductions:

Let $a \in D$.
Take $a \in D$.
Suppose a is in D.

From here we must show that $P(a)$ is true. This process is exemplified by the next diagram:

$$(\forall x \in D)P(x)$$
$$\downarrow$$
$$\text{Let } a \in D$$
$$\langle \text{Show } P(a) \rangle$$

These types of proofs are called ***universal proofs***.

Example. To prove that for all real numbers x, $(x-1)^3 = x^3 - 3x^2 + 3x - 1$, we introduce a real number and then check the equation.

Proof. Let $a \in \mathbb{R}$. Then,

$$\begin{aligned}(a-1)^3 &= (a-1)(a-1)^2 \\ &= (a-1)(a^2 - 2a + 1) \\ &= a^3 - 3a^2 + 3a - 1. \blacksquare\end{aligned}$$

For our next example, we need some definitions. Let a and b be integers such that $a \neq 0$. We say that a ***divides*** b if $b = ak$ for some $k \in \mathbb{Z}$. For example, 4 divides 12 since $12 = 4 \cdot 3$. Therefore, to translate a sentence like *4 divides a*, write

$$a = 4k \textit{ for some } k \in \mathbb{Z}.$$

An integer c is ***even*** if 2 divides c, but c would be ***odd*** if 2 did not divide it. This means:

$$c \text{ is even if and only if } c = 2\ell \text{ for some } \ell \in \mathbb{Z},$$

and

$$c \text{ is odd if and only if } c = 2k + 1 \text{ for some } k \in \mathbb{Z}.$$

We are now ready for the example.

Example. Let us prove the proposition

the square of every even integer is even.

This can be rewritten using a variable:

for all even integers n, n^2 is even.

The proof then goes like this:

Proof. Let n be an even integer. This means there exists $k \in \mathbb{Z}$ such that $n = 2k$. To see that n^2 is even, calculate:

$$n^2 = (2k)^2 = 4k^2 = 2(2k^2).$$

Since $2k^2$ is an integer, n^2 is even. ■

Notice how the definition was used in the proof. After the even number was introduced, a sentence was written that translated the introduction into a form that was easier to use. This was done using the definition of the term.

In the last example, we found the particular element of $\mathbb{Z}$ (it was $2k^2$ in the proof) that when doubled equals the square of n. This means that we proved $(\exists x \in \mathbb{Z})(n^2 = 2x)$. We did this by using the next rule.

2.4.3. Existential Generalization (EG)

Let D be a nonempty domain. If a is an element from D, then

$$P(a) \vdash (\exists x \in D)P(x).$$

This rule states that if $P(a)$ is true for some constant a in the domain, then it is also true that $(\exists x \in D)P(x)$. There is no additional stipulation, for an existential proposition is true exactly when the predicate is satisfied by at least one element of the domain.

Example. Each of the following is a valid use of Existential Generalization.

1. $P(a) \wedge \sim R(a) \vdash (\exists x)[P(x) \wedge \sim R(x)]$
2. $Q(a) \wedge T(b) \vdash (\exists y)[Q(a) \wedge T(y)]$

Before we write some proofs, here is the inference rule that allows us to attach existential quantifiers. It involves a new symbol, $\hat{a}$, that is read "a hat." The a represents an element of the domain. The hat is used in the two-column proof as a reminder that the symbol is particular.

2.4.4. Existential Instantiation (EI)

If a is a constant symbol from a nonempty domain D that has no prior occurrence in the proof, then

$$(\exists x \in D)P(x) \vdash P(\hat{a}).$$

The propositional form $(\exists x \in D)P(x)$ means that $P(x)$ is satisfied by some element of D. We will identify this element by $\hat{a}$. It has a hat since it is particular. If a had already been used in the proof, we must choose another symbol. This is because a used symbol already represents some element of the domain, and we have no reason to assume that this element satisfies $P(x)$. For example, if we know that $a = b - 7$, there is no reason to think that a also equals $c + 2$. Thus, using our new symbol $\hat{a}$, we may then conclude that $P(\hat{a})$ is true.

Example. The following are valid uses of Existential Instantiation:

1. $(\exists x)[P(x) \wedge Q(x)] \vdash P(\hat{a}) \wedge Q(\hat{a})$
2. $(\exists y)[R(a, y, c) \Rightarrow R(a, y, c)] \vdash R(a, \hat{b}, c) \Rightarrow R(a, \hat{b}, c)$
3. $(\exists x)(\forall y)(\exists z)Q(x, y, z) \vdash (\forall y)(\exists z)Q(\hat{a}, y, z)$

Example. Existential Instantiation cannot be used to justify either one of the following:

1. $(\exists z)[P(z) \vee Q(z)] \nvdash P(\hat{b}) \vee Q(z)$—The substitution was not made correctly. The result should have been $P(\hat{b}) \vee Q(\hat{b})$.
2. $(\exists x)(\exists y)T(a, x, y) \nvdash (\exists y)T(a, a, y)$—A new variable was not used with the application of Existential Instantiation, and the hat notation was not used.

The next example illustrates an illegal usage of Universal Generalization after an application of Existential Instantiation.

Example. We will try to prove that $(\forall x)(\exists y)(x + y = 2)$ logically implies $(\exists y)(\forall x)(x + y = 2)$.

Attempted Proof.

1.	$(\forall x)(\exists y)(x + y = 2)$	Given
2.	$(\exists y)(a + y = 2)$	1 UI
3.	$a + \hat{b} = 2$	2 EI
4.	$(\forall x)(x + \hat{b} = 2)$	3 UG
5.	$(\exists y)(\forall x)(x + y = 2)$	4 EG ■

Since $\hat{b}$ is particular, the application of Universal Generalization in line 4 is invalid.

Here are some valid uses of Existential Instantiation and Generalization:

Example. We will prove the following:

1. $(\exists x)[P(x) \wedge Q(x)],\ (\forall x)[P(x) \Rightarrow R(x)] \vdash (\exists x)R(x)$

Proof.

1.	$(\exists x)[P(x) \wedge Q(x)]$	Given
2.	$(\forall x)[P(x) \Rightarrow R(x)]$	Given
	$\langle$Show $(\exists x)R(x)\rangle$	
3.	$P(\hat{c}) \wedge Q(\hat{c})$	1 EI
4.	$P(\hat{c}) \Rightarrow R(\hat{c})$	2 UI
5.	$P(\hat{c})$	3 Simp
6.	$R(\hat{c})$	4, 5 MP
7.	$(\exists x)R(x)$	6 EG ■

This proof would be invalid if lines 3 and 4 were interchanged. Remember that when Existential Instantiation is used, the constant symbol must be new!

2. $(\forall x)(\exists y)[Q(x) \wedge T(y)] \vdash (\forall x)[Q(x) \wedge (\exists y)T(y)]$

Proof.

1.	$(\forall x)(\exists y)[Q(x) \wedge T(y)]$	Given
	$\langle$Show $(\forall x)[Q(x) \wedge (\exists y)T(y)]\rangle$	
2.	$(\exists y)[Q(a) \wedge T(y)]$	1 UI
3.	$Q(a) \wedge T(\hat{c})$	2 EI
4.	$T(\hat{c}) \wedge Q(a)$	3 Com
5.	$T(\hat{c})$	4 Simp
6.	$(\exists y)T(y)$	5 EG
7.	$Q(a)$	3 Simp
8.	$Q(a) \wedge (\exists y)T(y)$	6, 7 Conj
9.	$(\forall x)[Q(x) \wedge (\exists y)T(y)]$	8 UG ■

3. $P(a) \Rightarrow (\exists x)[Q(x) \wedge R(x)],\ P(a) \vdash (\exists x)R(x)$
(In the proof, a is a particular symbol because it appears free in the hypothesis in lines 1 and 2. That is why it has a hat from the beginning of the proof.)

Proof.

1.	$P(\hat{a}) \Rightarrow (\exists x)[Q(x) \wedge R(x)]$	Given
2.	$P(\hat{a})$	Given
	$\langle$Show $(\exists x)R(x)\rangle$	
3.	$(\exists x)[Q(x) \wedge R(x)]$	1, 2 MP
4.	$Q(\hat{b}) \wedge R(\hat{b})$	3 EI
5.	$R(\hat{b}) \wedge Q(\hat{b})$	4 Com
6.	$R(\hat{b})$	5 Simp
7.	$(\exists x)R(x)$	6 EG ■

Suppose that we want to write a paragraph proof for $(\exists x \in D)P(x)$. This means that we must show that there is at least one element of the domain D that satisfies $P(x)$. It will be our job to find the element. This can be done by calculation, educated guess, or even lucky guess! Once we have an element, called a ***candidate***, that we think will satisfy $P(x)$, we must check that it does. This type of a proof is called an ***existential proof***, and its structure is illustrated as follows:

$(\exists x \in D)P(x)$

↓

$\langle$Find $c \in D$ such that $P(c)\rangle$

↓

Choose a candidate

↓

Check the candidate

Example. To prove that there exists $x \in \mathbb{Z}$ such that $x^2 + 2x - 3 = 0$, we must find an integer that satisfies the predicate. A basic factorization yields

$$(x + 3)(x - 1) = 0.$$

Therefore, either $x = -3$ or $x = 1$ will satisfy the predicate.

Example. We will next prove that there is a function f such that the derivative of f is $20x^4$. After a quick mental calculation, we choose $f(x) = 4x^5$ as a candidate. Now to check:

$$\frac{d}{dx}4x^5 = (5)4x^4 = 20x^4.$$

Let us take what we have learned concerning the universal and existential quantifiers and and write paragraph proofs involving both quantifiers. The first example is a simple one from algebra but will nicely illustrate our method.

Example. Prove that for every $x \in \mathbb{R}$, there exists a $y \in \mathbb{R}$ such that $x + y = 2$. This translates to

$$(\forall x \in \mathbb{R})(\exists y \in \mathbb{R})(x + y = 2).$$

Remembering that a universal must apply to all elements of the domain and an existential means that we must find the desired element, we translate the propositional form by writing what we will call a ***quantifier diagram***:

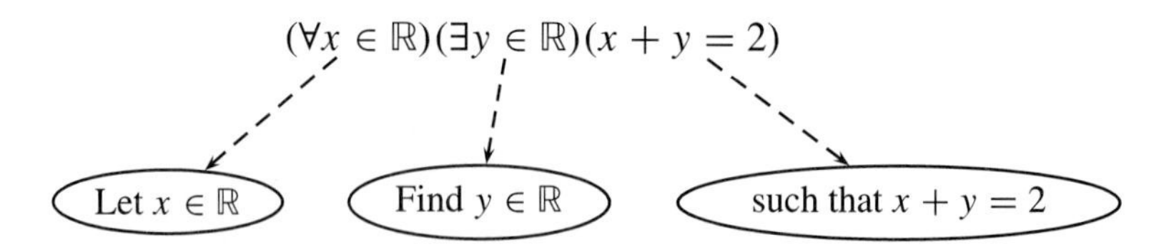

After taking an arbitrary x, our candidate will be $y = 2 - x$.

Proof. Let $x \in \mathbb{R}$. We must find $y \in \mathbb{R}$ so that $x + y = 2$. We choose $y = 2 - x$ and calculate:

$$x + y = x + (2 - x) = 2. \blacksquare$$

Now let us switch the order of the quantifiers.

Example. To see that there exists $x \in \mathbb{R}$ such that for all $y \in \mathbb{R}$, $x + y = y$, we use the following structure:

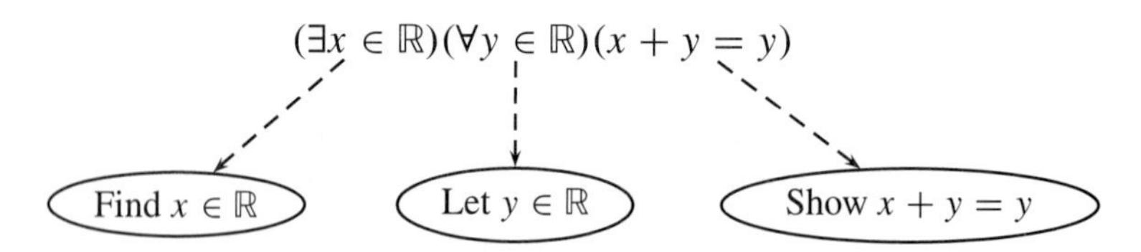

In the proof, the first goal is to identify a candidate. Then we must show that it works with every real number.

Proof. We claim that 0 is the sought after element. To see this, let $y \in \mathbb{R}$. Then $0 + y = y$. $\blacksquare$

(Notice that y is an arbitrary element of $\mathbb{R}$.)

The next example will involve two existential quantifiers. Therefore, we will have to find two candidates.

Example. We will prove that there is a function f and a real number x such that the derivative of f at x is 2. Translating we arrive at:

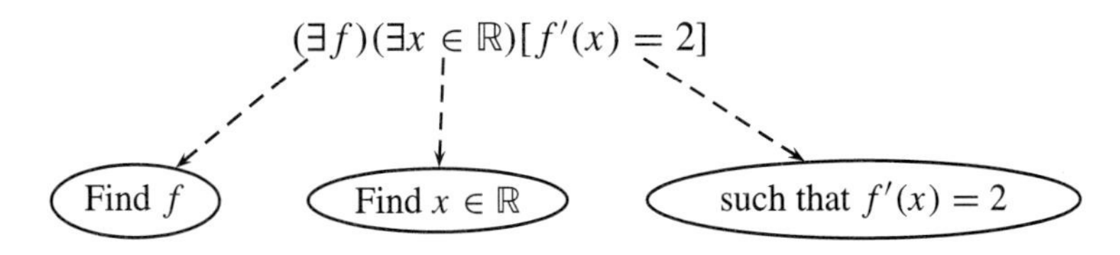

We base our proof on this structure. We will have to choose two candidates, one for f and one for x, and then check.

Proof. Define $f(x) = x^2$. This means $f'(x) = 2x$. Therefore, let $x = 1$ and calculate:

$$f'(x) = f'(1) = 2(1) = 2. \blacksquare$$

EXERCISES

Assume that all domains are nonempty

1. Find all errors in the given proofs.

   (a) "$(\exists x)[P(x) \vee Q(x)], (\exists x)\sim Q(x) \vdash (\exists x)P(x)$"

   Attempted Proof.

1.	$(\exists x)[P(x) \vee Q(x)]$	Given
2.	$(\exists x)\sim Q(x)$	Given
	$\langle$Show $(\exists x)P(x)\rangle$	
3.	$P(c) \vee Q(c)$	1 EI
4.	$\sim Q(c)$	2 EI
5.	$\sim P(c)$	3, 4 DS
6.	$(\exists x)\sim P(x)$	5 EG $\blacksquare$

   (b) "$(\forall x)P(x) \vdash (\exists x)(\forall y)[P(x) \vee Q(y)]$"

   Attempted Proof.

1.	$(\forall x)P(x)$	Given
	$\langle$Show $(\exists x)(\forall y)[P(x) \vee Q(y)]\rangle$	
2.	$P(\hat{c})$	1 UI
3.	$P(\hat{c}) \vee Q(a)$	2 Add
4.	$(\forall y)[P(\hat{c}) \vee Q(y)]$	3 UG
5.	$(\exists y)(\forall x)[P(x) \vee Q(y)]$	4 EG $\blacksquare$

(c) "$(\exists x)P(x),\ (\exists x)Q(x) \vdash (\forall x)[P(x) \wedge Q(x)]$"

Attempted Proof.

1.	$(\exists x)P(x)$	Given
2.	$(\exists x)Q(x)$	Given
	$\langle$Show $(\exists x)[P(x) \wedge Q(x)]\rangle$	
3.	$P(\hat{c})$	1 EI
4.	$Q(\hat{c})$	2 EI
5.	$P(\hat{c}) \wedge Q(c)$	3, 4 Conj
6.	$(\forall x)[P(x) \wedge Q(x)]$	5 UG ■

2. Prove each of the following:
   (a) $(\forall x)P(x) \vdash (\forall x)[P(x) \vee Q(x)]$
   (b) $(\forall x)P(x),\ (\forall x)[Q(x) \Rightarrow \sim P(x)] \vdash (\forall x)\sim Q(x)$
   (c) $(\forall x)[P(x) \Rightarrow Q(x)],\ (\forall x)P(x) \vdash (\forall x)Q(x)$
   (d) $(\forall x)[P(x) \vee Q(x)],\ (\forall x)\sim Q(x) \vdash (\forall x)P(x)$
   (e) $(\exists x)P(x) \vdash (\exists x)[P(x) \vee Q(x)]$
   (f) $(\exists x)(\exists y)P(x, y) \vdash (\exists y)(\exists x)P(x, y)$
   (g) $(\exists x)(\forall y)P(x, y) \vdash (\forall y)(\exists x)P(x, y)$
   (h) $(\forall x)\sim P(x),\ (\exists x)[Q(x) \Rightarrow P(x)] \vdash (\exists x)\sim Q(x)$
   (i) $(\exists x)P(x),\ (\forall x)[P(x) \Rightarrow Q(x)] \vdash (\exists x)Q(x)$
   (j) $(\forall x)[P(x) \Rightarrow Q(x)],\ (\forall x)[R(x) \Rightarrow S(x)],\ (\exists x)[P(x) \vee R(x)] \vdash$ $(\exists x)[Q(x) \vee S(x)]$
   (k) $(\exists x)[P(x) \wedge \sim R(x)],\ (\forall x)[Q(x) \Rightarrow R(x)] \vdash (\exists x)\sim Q(x)$
   (l) $(\forall x)P(x),\ (\forall x)[P(x) \vee Q(x)] \Rightarrow [R(x) \wedge S(x)]) \vdash (\exists x)S(x)$
   (m) $(\exists x)(\exists y)P(x, y) \vdash (\exists y)(\exists x)P(x, y)$
   (n) $(\exists x)P(x),\ (\exists x)P(x) \Rightarrow (\forall x)(\exists x)[P(x) \Rightarrow Q(x)] \vdash$ $(\forall x)Q(x) \vee (\exists x)R(x)$
   (o) $(\exists x)P(x),\ (\exists x)Q(x),\ (\exists x)(\exists y)[P(x) \wedge Q(y)] \Rightarrow (\forall x)R(x) \vdash$ $(\forall x)R(x)$
3. Sometimes these proofs require that a proposition with a quantifier be negated. This is the case with the following. Write two-column proofs for them. When a quantifier must be negated, reference the line number and use the reason QN (***Quantifier Negation***).
   (a) $(\exists x)P(x) \vdash (\forall x)\sim P(x) \Rightarrow (\forall x)Q(x)$
   (b) $(\forall x)[P(x) \Rightarrow Q(x)],\ (\forall x)[Q(x) \Rightarrow R(x)],\ \sim(\forall x)R(x) \vdash$ $(\exists x)\sim P(x)$
   (c) $(\forall x)P(x) \Rightarrow (\forall y)[Q(y) \Rightarrow R(y)],\ (\exists x)[Q(x) \wedge \sim R(x)] \vdash$ $(\exists x)\sim P(x)$
   (d) $(\exists x)P(x) \Rightarrow (\exists y)Q(y),\ (\forall x)\sim Q(x) \vdash (\forall x)\sim P(x)$

4. Write quantifier diagrams for each of the following:
   (a) $(\exists x \in \mathbb{Z})(\forall y \in \mathbb{Z})Q(x, y)$
   (b) $(\forall x \in \mathbb{R})(\forall y \in \mathbb{R})R(x, y)$
   (c) $(\forall x \in \mathbb{Q})(\exists y \in \mathbb{R})(\forall y \in \mathbb{Q})[P(x, y) \wedge Q(z)]$
   (d) $(\forall \varepsilon > 0)(\exists N \in \mathbb{Z}^+)(\forall n \in \mathbb{Z}^+)(n \geq N \Rightarrow |a_n - a| < \varepsilon)$
   (e) $(\forall x \in \mathbb{R})(\forall \varepsilon > 0)(\exists \delta > 0)(|x - x_0| < \delta \Rightarrow |f(x) - f(x_0)| < \varepsilon)$
5. Let $n \in \mathbb{Z}$. Demonstrate that each of the following are divisible by 6.
   (a) 18
   (b) -24
   (c) 0
   (d) $6n + 12$
   (e) $2^3 \cdot 3^4 \cdot 7^5$
   (f) $(2n + 2)(3n + 6)$
6. Let a be a nonzero integer. Write paragraph proofs for the following:
   (a) 1 divides a.
   (b) a divides 0.
   (c) a divides a.
7. Write a universal proof for each of the following:
   (a) For all $x \in \mathbb{R}$, $(x + 2)^2 = x^2 + 4x + 4$.
   (b) For all $x \in \mathbb{Z}$, $x - 1$ divides $x^3 - 1$.
   (c) The square of every even integer is even.
8. Write existential proofs for the following:
   (a) There exists $x \in \mathbb{R}$ such that $x - \pi = 9$.
   (b) There exists $x \in \mathbb{Z}$ such that $x^2 + 2x - 3 = 0$.
   (c) There exists a function whose derivative is $3x^3 + 4$.
   (d) The square of some integer is odd.
9. Prove each of the following by writing a paragraph proof.
   (a) For all $x, y, z \in \mathbb{R}$, $z(x + y)^2 = x^2z + 2xyz + y^2z$.
   (b) There exist $u, v \in \mathbb{R}$ such that $2u + 5v = -29$.
   (c) For all $x \in \mathbb{R}$, there exists $y \in \mathbb{R}$ so that $x - y = 10$.
   (d) There exists $x \in \mathbb{Z}$ such that for all $y \in \mathbb{Z}$, $yx = x$.
   (e) For all $a, b, c \in \mathbb{R}$, there exists $x \in \mathbb{C}$ such that $ax^2 + bx + c = 0$.
   (f) There exists an integer that divides every integer.

2.5. DIRECT AND INDIRECT PROOF

Most propositions that mathematicians prove are implications. For example,

if f is a differentiable function, then f is continuous.

As we know, this means that whenever f is differentiable, it must also be the case that f is continuous. Proofs of conditionals like this are typically very difficult if we are only allowed to use our Rules of Inference and Replacement.

To remedy this, to prove that a conditional is true, we will usually use the following:

2.5.1. Direct Proof (DP)

For propositional forms $h_1, \ldots, h_k, p, q$,

$$h_1, \ldots, h_k \vdash p \Rightarrow q$$
$$\text{if and only if}$$
$$h_1, \ldots, h_k, p \vdash q.$$

The Rule of Direct Proof (sometimes called the ***Deduction Theorem***) is shown to be true by use of Exportation:

$$[h_1 \wedge \cdots \wedge h_k] \Rightarrow (p \Rightarrow q) \equiv [h_1 \wedge \cdots \wedge h_k \wedge p] \Rightarrow q.$$

If these are tautologies, the left-hand side means

$$h_1, \ldots, h_k \vdash p \Rightarrow q,$$

and the right-hand sign means

$$h_1, \ldots, h_k, p \vdash q.$$

Our first use of Direct Proof will be a two-column proof. Carefully note the structure.

Example. Prove $(P \vee Q) \Rightarrow (R \wedge S) \vdash P \Rightarrow R$.

Proof.

1.	$(P \vee Q) \Rightarrow (R \wedge S)$	Given
	$\langle$Show $P \Rightarrow R\rangle$	
2.	→ P	Assumption
	│ $\langle$Show $R\rangle$	
3.	│ $P \vee Q$	2 Add
4.	│ $R \wedge S$	1, 3 MP
5.	│ R	4 Simp
6.	$P \Rightarrow R$	2–5 DP ∎

The proof of $P \Rightarrow R$ is a ***subproof*** of the main proof. Its purpose is to prove the implication. To separate the propositions of the subproof from the rest of the proof, they are indented with a vertical line. The line begins with the assumption of P in line 2. (Hence, its reason is *Assumption*.) This assumption can only be used in the subproof, not elsewhere. Consider it a local hypothesis. It is only used to prove the conditional. If we were allowed to use it in other places of the proof, then we would not be proving our original theorem. Instead, we would be proving a theorem that had the original hypotheses plus the assumption as givens. Similarly, all lines within the subproof cannot be referenced from the outside. We use the indentation to isolate the assumption and all the propositions that follow from it. To indicate what needs to be done, a comment is added. It states the conclusion of the conditional, R. When we arrive at that proposition, we know that we have proven $P \Rightarrow R$. The next line is this implication. It is entered into the proof with the reason DP. The lines that are referenced are the lines of the subproof.

This rule can be used in the proof of any implication, even if that implication is not the conclusion of the theorem. This is seen in the next two examples. The first one is a natural application of Direct Proof, for we are to prove an implication that has an implication as its conclusion.

Example. Prove $P \Rightarrow \sim Q,\ \sim R \vee S \vdash (R \vee Q) \Rightarrow (P \Rightarrow S)$.

Proof.

1.	$P \Rightarrow \sim Q$	Given
2.	$\sim R \vee S$	Given
	$\langle$Show $(R \vee Q) \Rightarrow (P \Rightarrow S)\rangle$	
3.	→ $R \vee Q$	Assumption
	$\langle$Show $P \Rightarrow S\rangle$	
4.	→ P	Assumption
	$\langle$Show $S\rangle$	
5.	$\sim Q$	1, 4 MP
6.	$Q \vee R$	3 Com
7.	R	5, 6 DS
8.	$\sim\sim R$	7 DN
9.	S	2, 8 DS
10.	$P \Rightarrow S$	4–9 DP
11.	$(R \vee Q) \Rightarrow (P \Rightarrow S)$	3–10 DP ■

In the second example we must be more creative. Although the conclusion is not a conditional, one can be found within the givens. It is the hypothesis of an implication that has what we want in its conclusion. Therefore, we will use Direct Proof to prove the hypothesis and then use *Modus Ponens* to complete the proof.

Example. Prove $[(P \wedge Q) \Rightarrow R] \Rightarrow S,\ {\sim}Q \vee R \vdash S$.

Proof.

1.	$[(P \wedge Q) \Rightarrow R] \Rightarrow S$	Given
2.	${\sim}Q \vee R$	Given
	$\langle$Show $S\rangle$	
3.	→ $P \wedge Q$	Assumption
	$\langle$Show $R\rangle$	
4.	$Q \wedge P$	3 Com
5.	Q	4 Simp
6.	${\sim}{\sim}Q$	5 DN
7.	R	2, 6 DS
8.	$(P \wedge Q) \Rightarrow R$	3–7 DP
9.	S	1, 8 MP ■

Direct Proof is the preferred method for proving implications. It says that given certain hypotheses, $h_1, \ldots, h_k$, to prove $p \Rightarrow q$, add p to the list of hypotheses and demonstrate q. In other words:

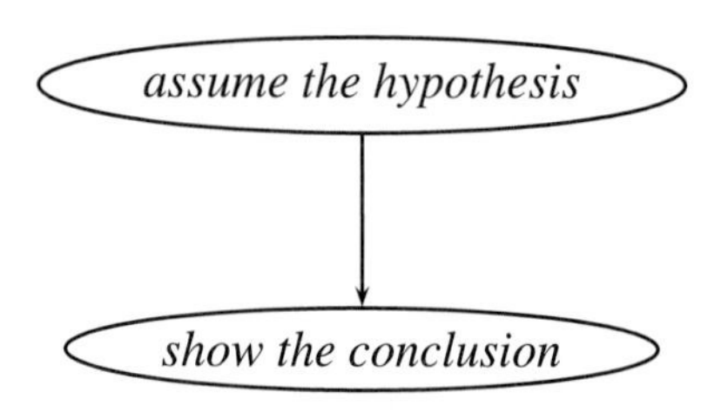

We will now use this to write a paragraph proof. To use Direct Proof, we will follow these steps:

1. identify the hypothesis and conclusion of the implication,
2. assume the hypothesis,
3. translate the hypothesis,
4. write a comment regarding what must be shown,
5. translate the comment, and
6. deduce the conclusion.

Our first example will use some of the terms that we defined in the last section on page 65. Notice the introductions in the proof.

Example. We will use Direct Proof to write a paragraph-style proof of the sentence

for all $x \in \mathbb{Z}$, if 4 divides x, then x is even.

First, choose an arbitrary integer a and then identify the hypothesis and conclusion:

if $\boxed{\textit{4 divides a}}$, *then* $\boxed{\textit{a is even}}$.

Use these to identify the structure of the proof:

4 divides a ⟵	*assume the hypothesis*
This means $a = 4k$ ⟵	*translate hypothesis*
⟨Show a is even⟩ ⟵	*comment conclusion*
⟨Show $a = 2\ell$⟩ ⟵	*translate conclusion*
⋮	
2 divides x ⟵	*deduce the conclusion*

Now write the final version from this structure:

> **Proof.** Let $a \in \mathbb{Z}$ and assume 4 divides a. This means $a = 4k$ for some integer k. We must show that 2 divides a. In other words, we must show $a = 2\ell$ with $\ell \in \mathbb{Z}$. But we know $a = 4k = 2(2k)$. So let $\ell = 2k$. ■

Notice how the translation of the comment was introduced with *in other words*. Other options include *this means*, *note*, *recall*, and *remember* as in *remember that this means* $x = 2\ell$.

The next examples illustrate an important proof-writing task. There will be times when we will want to show that there exists exactly one element that satisfies a given predicate $P(x)$. In other words, there is a ***unique*** element that satisfies $P(x)$. This will be a two step process:

- **Existence:** Show that there is at least one element that satisfies $P(x)$.
- **Uniqueness:** Show that there is at most one element that satisfies $P(x)$. This is usually done by by assuming both a and b satisfy $P(x)$ and then proving $a = b$.

This process is summarized by the ***Uniqueness Rule***. We will use the notation $(\exists! x \in D)P(x)$ to mean that there exists a unique element of D that satisfies $P(x)$. In other words, $(\exists! x \in D)P(x)$ is equivalent to

$$(*) \qquad (\exists x \in D)P(x) \wedge (\forall x \in D)(\forall y \in D)[P(x) \wedge P(y) \Rightarrow x = y].$$

(To help remember the notation, think about being excited that the only one that works was found!)

Example. We will use $(*)$ to prove $2x + 1 = 5$ has a unique real solution.

Proof. We will follow the Uniqueness Rule. We must show

$$(\exists x \in \mathbb{R})(2x + 1 = 5)$$

and

$$(\forall x \in \mathbb{R})(\forall y \in \mathbb{R})[(2x + 1 = 5 \wedge 2y + 1 = 5) \Rightarrow x = y].$$

1. We know that $x = 2$ is a solution since $2(2) + 1 = 5$.
2. Suppose that we have two solutions, a and b. We want to show that they are actually the same. We know that $2a + 1 = 5$ and $2b + 1 = 5$, so we calculate:

$$\begin{aligned} 2a + 1 &= 2b + 1 \\ 2a &= 2b \\ a &= b. \blacksquare \end{aligned}$$

Example. We shall prove that for every real number x, there exists a unique $y \in \mathbb{R}$ such that $x + y = 0$. In symbolic form this is:

$$(\forall x \in \mathbb{R})(\exists! y \in \mathbb{R})(x + y = 0).$$

Proof. We will again follow the two step strategy. Let $x \in \mathbb{R}$.

1. We prove existence by noting that $x + (-x) = 0$.
2. Now let y_1 and y_2 be real numbers that satisfy $x + y = 0$. Then,

$$\begin{aligned} x + y_1 &= x + y_2 \\ y_1 &= y_2. \blacksquare \end{aligned}$$

Sometimes it is difficult to prove a conditional directly. An alternative may be to prove the contrapositive. This is sometimes easier or simply requires fewer lines. The next example shows this method in a paragraph proof.

Example. Let us show that

for all $n \in \mathbb{Z}$, n^2 is odd implies that n is odd.

A direct proof of this is a problem. Instead, we will prove its contrapositive:

if n is not odd, then n^2 is not odd.

In other words, we will prove:

if n is even, then n^2 is even.

This will be done using Direct Proof:

→ Let n be even
So $n = 2k$
⟨Show n^2 is even⟩
⟨Show $n^2 = 2\ell$⟩
⋮
2 divides n^2

This leads to the final proof.

Proof. We will prove the implication by showing its contrapositive. Let $n \in \mathbb{Z}$ and suppose that n is even. This means $n = 2k$ with k being an integer. To see that n^2 is even, calculate:

$$n^2 = (2k)^2 = 4k^2 = 2(2k^2). \blacksquare$$

Notice that this proof is basically the same as the proof for *the square of every even integer is even* on page 66. This illustrates that there is a connection between universal proofs and direct proofs. (See Exercise 12.)

Similar to using the contrapositive is Indirect Proof, although it can also be used to prove propositions that are not implications.

2.5.2. Indirect Proof (IP)

For any propositional forms $h_1, \ldots, h_k, p$,

$$h_1, \ldots, h_k \vdash p$$
if and only if
$$h_1, \ldots, h_k, \sim p \vdash q \wedge \sim q$$

for some propositional form q. This method is also known as ***proof by contradiction*** or **reductio ad absurdum**.

To see why this works, suppose $h_1 \wedge \cdots \wedge h_k$ is true but $h_1 \wedge \cdots \wedge h_k \wedge \sim p$ leads to a contradiction. This means that $h_1 \wedge \cdots \wedge h_k \wedge \sim p$ must be false, but the only way this can happen is for $\sim p$ to be false. Hence, p is true.

To use Indirect Proof, assume each hypothesis and then assume the negation of the conclusion. Then proceed with the proof until a contradiction is reached. (In the theorem, the contradiction is represented by $q \wedge \sim q$.) At this point, deduce the original conclusion.

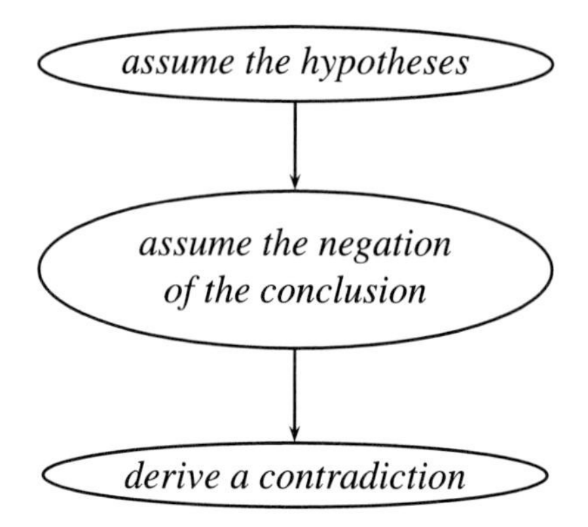

Our first example will be a two-column proof. As with Direct Proof, we will separate the subproof from the rest of the proof with a line. Note that the reason in line 3 is given as *Assumption*. This is because we assume the negation of the conclusion.

Example. Prove $(P \vee Q) \Rightarrow R,\ (R \vee S) \Rightarrow (\sim P \wedge T) \vdash \sim P$.

Proof.

1.	$(P \vee Q) \Rightarrow R$	Given
2.	$(R \vee S) \Rightarrow (\sim P \wedge T)$	Given
	$\langle$Show $\sim P\rangle$	
3.	$\sim\sim P$	Assumption
4.	P	3 DN
5.	$P \vee Q$	4 Add
6.	R	1, 5 MP
7.	$R \vee S$	6 Add
8.	$\sim P \wedge T$	2, 7 MP
9.	$\sim P$	8 Simp
10.	$P \wedge \sim P$	4, 9 Conj
11.	$\sim P$	3–10 IP ■

Indirect Proof can also be nested within another indirect subproof. As with Direct Proof, we cannot appeal to lines within a subproof from outside of it. We see this in the next example.

Example. Show $P \Rightarrow (Q \wedge R),\ Q \Rightarrow S,\ \sim P \Rightarrow S \vdash S$ using nested applications of Indirect Proof.

Proof.

1.	$P \Rightarrow (Q \wedge R)$	Given
2.	$Q \Rightarrow S$	Given
3.	$\sim P \Rightarrow S$	Given
	$\langle$Show $S\rangle$	
4.	$\rightarrow \sim S$	Assumption
5.	$\quad \sim Q$	2, 4 MT
6.	$\quad \rightarrow P$	Assumption
7.	$\quad\quad Q \wedge R$	1, 6 MP
8.	$\quad\quad Q$	7 Simp
9.	$\quad\quad Q \wedge \sim Q$	5, 8 Conj
10.	$\quad \sim P$	6-9 IP
11.	$\quad S$	3, 10 MP
12.	$\quad S \wedge \sim S$	4, 11 Conj
13.	S	4–12 IP ■

Notice that line 11 was not the end of the proof since it was within the first subproof. It followed under the added hypothesis of $\sim S$.

Both proof methods can be nested within each other.

Example. Demonstrate that $P \Rightarrow R \vdash (P \wedge Q) \Rightarrow (R \vee S)$.

Proof.

1.	$P \Rightarrow R$	Given
	$\langle$Show $(P \wedge Q) \Rightarrow (R \vee S)\rangle$	
2.	$\rightarrow P \wedge Q$	Assumption
	$\quad \langle$Show $R \vee S\rangle$	
3.	$\quad \rightarrow \sim R$	Assumption
4.	$\quad\quad \sim P$	1, 3 MT
5.	$\quad\quad P$	2 Simp
6.	$\quad\quad P \wedge \sim P$	4, 5 Conj
7.	$\quad R$	3-6 IP
8.	$\quad R \vee S$	7 Add
9.	$(P \wedge Q) \Rightarrow (R \vee S)$	2–8 DP ■

Our first paragraph example of an indirect proof comes from calculus. We will prove a basic fact about limits.

Example. Define

$$f(x) = \begin{cases} x+1 & \text{if } x \geq 0 \\ x-1 & \text{if } x < 0. \end{cases}$$

We will prove that

f does not have a limit at $x = 0$.

To do this, we will use the following structure:

Assume $f(x) = \ldots$
⟨Show f does not have a limit at $x = 0$⟩
→ Suppose f does have a limit at $x = 0$ ← *negate conclusion*
This means ...
⋮
Contradiction ←——— *we are done!*
f does not have a limit at $x = 0$

The only decision left is to determine which translation to use. We will rely on the calculus fact that if a limit exists, then both its one-sided limits agree. The proof then is as follows:

Proof. Let f be defined as above and suppose that $\lim_{x \to 0} f(x)$ exists. This means

$$\lim_{x \to 0^-} f(x) = \lim_{x \to 0^+} f(x).$$

However, the left-hand limit is -1, and the right-hand limit is 1, a contradiction. Therefore, there is no limit at $x = 0$. ■

Exactly what is the contradiction in the proof? As a result of the assumption, we have

$$\lim_{x \to 0^-} f(x) = \lim_{x \to 0^+} f(x),$$

but we also derived

$$\lim_{x \to 0^-} f(x) \neq \lim_{x \to 0^+} f(x).$$

In the proof, therefore, we may conclude the conjunction of these two, which is impossible.

The structure found in the next example is common in practice. It often appears as an alternative to Contraposition.

Example. Earlier we proved

for every $n \in \mathbb{Z}$, if n^2 is odd, then n is odd by showing its contrapositive. Here we will nest an Indirect Proof within a Direct Proof:

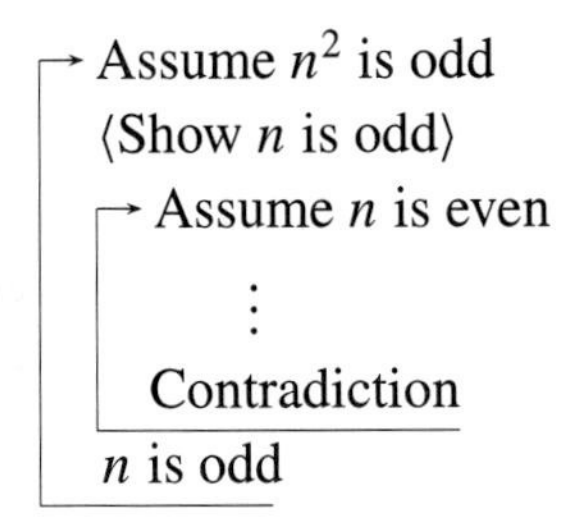

We use this structure to write the paragraph proof.

Proof. Take $n \in \mathbb{Z}$ and let n^2 be odd. In order to obtain a contradiction, assume that n is even. So, $n = 2k$ for some $k \in \mathbb{Z}$. Substituting, we have

$$n^2 = (2k)^2 = 2(2k^2),$$

showing that n^2 is even. This is a contradiction. Therefore, n is an odd integer. ■

EXERCISES

1. Find all mistakes in the given proofs.

   (a) "$(P \vee Q) \Rightarrow \sim R,\ (R \Rightarrow \sim Q) \Rightarrow (S \vee Q) \vdash S$"

Attempted Proof.

1.	$(P \vee Q) \Rightarrow \sim R$	Given
2.	$(R \Rightarrow \sim Q) \Rightarrow (S \vee Q)$	Given
	⟨Show S⟩	
3.	→ R	Assumption
4.	$\sim\sim R$	Assumption
5.	$\sim(P \vee Q)$	1, 4 MT
6.	$\sim P \wedge \sim Q$	5 DeM
7.	$\sim Q$	6 Simp
8.	$R \Rightarrow \sim Q$	3–7 DP
9.	$S \vee Q$	2, 8 MP
10.	$Q \vee S$	9 Com
11.	S	7, 10 DS ■

(b) "$\sim P \vee Q \vdash (P \Rightarrow Q) \Rightarrow R$"

Attempted Proof.

1.	$\sim P \vee Q$	Given
	$\langle$Show $(P \Rightarrow Q) \Rightarrow R\rangle$	
2.	$\to P$	Assumption
3.	$\sim\sim P$	2 DN
4.	Q	1, 3 DS
5.	$P \Rightarrow Q$	2–4 DP
6.	R	MP
7.	$(P \Rightarrow Q) \Rightarrow R$	2–6 DP ■

(c) "$\sim R \wedge S,\ (\sim P \vee Q) \Rightarrow R \vdash (\sim P \vee Q) \Rightarrow Q$"

Attempted Proof.

1.	$\sim R \wedge S$	Given
2.	$(\sim P \vee Q) \Rightarrow R$	Given
	$\langle$Show $(\sim P \vee Q) \Rightarrow Q\rangle$	
3.	$\to \sim P$	Assumption
4.	$\to \sim P \vee Q$	Assumption
5.	R	2, 3 MP
6.	$\sim R$	1 Simp
7.	$R \wedge \sim R$	5, 6 Conj
8.	P	4–7 IP
9.	$\sim\sim P$	8 DN
10.	Q	4, 9 DS
11.	$(\sim P \vee Q) \Rightarrow Q$	3–10 DP ■

2. Prove the following using the method of Direct Proof.
   (a) $P \Rightarrow (Q \wedge R) \vdash P \Rightarrow Q$
   (b) $(P \vee Q) \Rightarrow R \vdash P \Rightarrow R$
   (c) $P \vee (Q \vee R) \Rightarrow S \vdash Q \Rightarrow S$
   (d) $P \Rightarrow Q,\ R \Rightarrow Q \vdash (P \vee R) \Rightarrow Q$
   (e) $P \Rightarrow Q,\ P \Rightarrow R \vdash P \Rightarrow (Q \wedge R)$
   (f) $P \Rightarrow (Q \Rightarrow R) \vdash (Q \wedge \sim R) \Rightarrow \sim P$
   (g) $R \Rightarrow \sim S \vdash (P \wedge Q) \Rightarrow (R \Rightarrow \sim S)$
   (h) $P \Rightarrow (Q \Rightarrow R) \vdash Q \Rightarrow (P \Rightarrow R)$
   (i) $P \Rightarrow (Q \Rightarrow R),\ R \Rightarrow (S \wedge T) \vdash P \Rightarrow (Q \Rightarrow T)$
   (j) $[(P \wedge Q) \vee R] \Rightarrow (Q \wedge R) \vdash P \Rightarrow (Q \Rightarrow R)$

3. Prove the following using Indirect Proof.
   (a) $P \Rightarrow Q,\ Q \Rightarrow R,\ {\sim}R \vdash {\sim}P$
   (b) $P \vee (Q \wedge R),\ P \Rightarrow S,\ Q \Rightarrow S \vdash S$
   (c) $P \vee (Q \wedge {\sim}R),\ P \Rightarrow S,\ Q \Rightarrow R \vdash S$
   (d) $P \Leftrightarrow Q,\ {\sim}P \vdash {\sim}Q$
   (e) $[(P \vee Q) \vee R] \Rightarrow (Q \wedge R) \vdash {\sim}P \vee (Q \wedge R)$
   (f) $P \Rightarrow Q,\ Q \Rightarrow R,\ S \Rightarrow T,\ P \vee R,\ {\sim}R \vdash T$
   (g) $P \Rightarrow {\sim}Q,\ R \Rightarrow {\sim}S,\ T \Rightarrow Q,\ U \Rightarrow S,\ P \vee R \vdash {\sim}T \vee {\sim}U$
   (h) $P \Rightarrow (Q \wedge R),\ (Q \vee S) \Rightarrow (T \wedge U),\ P \vdash T$
4. Prove the following by using both Direct and Indirect Proof.
   (a) $P \Rightarrow Q,\ P \vee (R \Rightarrow S),\ {\sim}Q \vdash R \Rightarrow S$
   (b) $P \Rightarrow {\sim}(Q \Rightarrow {\sim}R) \vdash P \Rightarrow R$
   (c) $P \Rightarrow Q \vdash (P \wedge R) \Rightarrow Q$
   (d) $P \Leftrightarrow (Q \vee R) \vdash Q \Rightarrow P$
5. Demonstrate each of the following by using Direct Proof to prove the contrapositive.
   (a) $P \Rightarrow Q,\ R \Rightarrow S,\ S \Rightarrow T,\ {\sim}Q \vdash {\sim}T \Rightarrow {\sim}(P \vee R)$
   (b) $(P \wedge Q) \vee (R \wedge S) \vdash {\sim}S \Rightarrow (P \wedge Q)$
   (c) $(P \vee Q) \Rightarrow {\sim}R,\ S \Rightarrow R \vdash P \Rightarrow {\sim}S$
   (d) ${\sim}P \Rightarrow {\sim}Q,\ ({\sim}R \vee S) \wedge (R \vee Q) \vdash {\sim}S \Rightarrow P$
6. Assuming $a,\ b,\ c,\ d \in \mathbb{Z}$ with $a \neq 0$ and $c \neq 0$, give paragraph proofs using Direct Proof for the following divisibility results.
   (a) If a divides b, then a divides bd.
   (b) If a divides b and a divides d, then a^2 divides bd.
   (c) If a divides b and c divides d, then ac divides bd.
7. Write paragraph proofs using Direct Proof:
   (a) The sum of two even integers is even.
   (b) The sum of two odd integers is even.
   (c) The sum of an even and an odd is odd.
   (d) The product of two even integers is even.
   (e) The product of two odd integers is odd.
   (f) The product of an even and an odd is even.
8. Prove the theorems in the previous problem indirectly.
9. Let $a,\ b \in \mathbb{Z}$. Write paragraph proofs for the following:
   (a) If a and b are even, then $a^4 + b^4 + 32$ is divisible by 8.
   (b) If a and b are odd, then 4 divides $a^3 + b^3 + 6$.
10. Prove the following by using Direct Proof to prove the contrapositive.
   (a) For every integer n, if n^4 is even, then n is even.
   (b) For all $n \in \mathbb{Z}$, if $n^3 + n^2$ is odd, then n is odd.
   (c) For all $a,\ b \in \mathbb{Z}$, if ab is even, then a is even or b is even.

11. Write paragraph proofs for the following:
    (a) The equation $x - 10 = 23$ has a unique solution.
    (b) The equation $\sqrt{2x - 5} = 2$ has a unique solution.
    (c) For every $y \in \mathbb{R}$, the equation $2x+5y = 10$ has a unique solution.
    (d) The equation $x^2 + 5x + 6 = 0$ has at most two solutions.
12. Show why a proof of $(\forall x)P(x)$ is an application of Direct Proof.

2.6. MORE METHODS

In this last section we will examine proofs that involve Material Equivalence, disjunctions, and cases. Each method will rely on Direct Proof. The first of these takes three forms. It is the usual method of proving biconditionals.

2.6.1. Biconditional Proof (BP)

Let $h_1, \ldots, h_k, p, q$ be propositional forms. Then,

$$h_1, \ldots, h_k \vdash p \Leftrightarrow q$$
if and only if
$$h_1, \ldots, h_k \vdash p \Rightarrow q \text{ and } h_1, \ldots, h_k \vdash q \Rightarrow p.$$

The rule states that a biconditional should be proven by showing both implications. Each is usually proven with Direct Proof. As is seen in the next example, the $p \Rightarrow q$ subproof is introduced by $(\Rightarrow)$ and its converse with $(\Leftarrow)$. Moreover, the conclusions of the two applications of Direct Proof are not stated. Instead, we combine the conclusions in line 7.

Example. Prove $P \Rightarrow Q \vdash (P \wedge Q) \Leftrightarrow P$

Proof.

1.	$P \Rightarrow Q$	Given
	$\langle$Show $(P \wedge Q) \Leftrightarrow P\rangle$	
$(\Rightarrow)$ 2.	$\to P \wedge Q$	Assumption
	$\langle$Show $P\rangle$	
3.	P	2 Simp
$(\Leftarrow)$ 4.	$\to P$	Assumption
	$\langle$Show $P \wedge Q\rangle$	
5.	Q	1, 4 MP
6.	$P \wedge Q$	4, 5 Conj
7.	$(P \wedge Q) \Leftrightarrow P$	2–6 BP ■

Example. Let us use Biconditional Proof to show:

for all $n \in \mathbb{Z}$, n is even if and only if n^3 is even.

Since this is a biconditional, we must show both directions:

$$\textit{if } \boxed{n \text{ is even}}, \textit{ then } \boxed{n^3 \text{ is even}},$$

and

$$\textit{if } \boxed{n^3 \text{ is even}}, \textit{ then } \boxed{n \text{ is even}}.$$

To prove the second conditional, we will need to prove its contrapositive. Therefore, using the pattern of the previous example, the structure is:

($\Rightarrow$) → Assume n is even
Then $n = 2k$, $k \in \mathbb{Z}$
$\langle$Show n^3 is even$\rangle$
$\langle$Show $n^3 = 2\ell$, $\ell \in \mathbb{Z}\rangle$
⋮
n^3 is even

($\Leftarrow$) → Assume n is odd
Then $n = 2k + 1$, $k \in \mathbb{Z}$
$\langle$Show n^3 is odd$\rangle$
$\langle$Show $n^3 = 2\ell + 1$, $\ell \in \mathbb{Z}\rangle$
⋮
n^3 is odd

Proof. Let $n \in \mathbb{Z}$.

($\Rightarrow$) Assume n is even. Then, $n = 2k$ for some $k \in \mathbb{Z}$. We must show n^3 is even. To do this, we calculate:

$$n^3 = (2k)^3 = 2(4k^3),$$

which means that n^3 is even.

($\Leftarrow$) Now suppose that n is odd. This means that $n = 2k + 1$ for some $k \in \mathbb{Z}$. To show n^3 is odd, we again calculate:

$$\begin{aligned} n^3 &= (2k+1)^3 \\ &= 8k^3 + 12k^2 + 6k + 1 \\ &= 2(4k^3 + 6k^2 + 3k) + 1. \end{aligned}$$

Hence, n^3 is odd. ■

Notice two things about the proof:

- Some of the comments were skipped to prevent redundancy.
- Words were chosen carefully to make the proof more readable.

Furthermore, the example could have been written with the words *necessary* and *sufficient* introducing the two subproofs. The ($\Rightarrow$) step could have been introduced with a phrase like

to show sufficiency,

and the ($\Leftarrow$) could have opened with

as for necessity.

Sometimes the steps for one part are simply the steps for the other in reverse order. When this happens, our work is cut in half, and we can use the ***Shorter Rule of Biconditional Proof***. These proofs are simply a sequence of biconditionals without the reasons. This is a good method when only rules of replacement are used as in the next example. (Notice that there are no hypotheses to assume.)

Example. Prove $(P \Rightarrow Q) \Leftrightarrow (P \wedge \sim Q) \Rightarrow \sim P$.

Proof. Proceed as follows:

$$\begin{aligned} P \Rightarrow Q &\Leftrightarrow \sim P \vee Q \\ &\Leftrightarrow (\sim P \vee \sim P) \vee Q \\ &\Leftrightarrow \sim P \vee (\sim P \vee Q) \\ &\Leftrightarrow (\sim P \vee Q) \vee \sim P \\ &\Leftrightarrow (\sim P \vee \sim\sim Q) \vee \sim P \\ &\Leftrightarrow \sim(P \wedge \sim Q) \vee \sim P \\ &\Leftrightarrow (P \wedge \sim Q) \Rightarrow \sim P \ \blacksquare \end{aligned}$$

There will be times when we need to prove a sequence of biconditionals. The propositional forms $p_1, p_2, \ldots, p_k$ are ***pairwise equivalent*** if for all i, j,

$$p_i \Leftrightarrow p_j.$$

In other words,

$$p_1 \Leftrightarrow p_2,\ p_1 \Leftrightarrow p_3,\ \ldots,\ p_2 \Leftrightarrow p_3,\ p_2 \Leftrightarrow p_4,\ \ldots,\ p_{k-1} \Leftrightarrow p_k.$$

To prove all of these, we make use of Transitivity. The result is the Equivalence Rule.

2.6.2. Equivalence Rule

To prove that the propositional forms $p_1, p_2, \ldots, p_k$ are pairwise equivalent, prove:

$$p_1 \Rightarrow p_2, \; p_2 \Rightarrow p_3, \; \ldots, \; p_{k-1} \Rightarrow p_k, \; p_k \Rightarrow p_1.$$

In practice, the Equivalence Rule will typically be used to prove propositions that include the phrase

the following are equivalent.

For instance, recall from algebra that if $f(x)$ is a polynomial with real coefficients, then

$$f(x) = a_n x^n + a_{n-1}x^{n-1} + \cdots + a_1 x + a_0$$

with each $a_i \in \mathbb{R}$ and $n \in \mathbb{N}$. An integer r is a ***root*** of $f(x)$ if $f(r) = 0$, and a polynomial $g(x)$ is a ***factor*** of $f(x)$ if there is another polynomial $h(x)$ such that $f(x) = g(x)h(x)$. Given a polynomial $f(x)$, we can prove that the following are equivalent:

1. r is a root of $f(x)$,
2. r is a solution to $f(x) = 0$, and
3. $x - r$ is a factor of $f(x)$.

To do this, we will prove (1) $\Rightarrow$ (2), (2) $\Rightarrow$ (3), and (3) $\Rightarrow$ (1). This entails proving three conditionals:

$$\textit{if} \; \boxed{r \textit{ is a root of } f(x)}, \textit{ then } \boxed{r \textit{ is a solution to } f(x) = 0},$$

$$\textit{if} \; \boxed{r \textit{ is a solution to } f(x) = 0}, \textit{ then } \boxed{x - r \textit{ is a factor of } f(x)},$$

and

$$\textit{if} \; \boxed{x - r \textit{ is a factor of } f(x)}, \textit{ then } \boxed{r \textit{ is a root of } f(x)}.$$

We will use Direct Proof on each.

Proof. Let $f(x)$ be a polynomial.

(1) $\Rightarrow$ (2) Let r be a root of $f(x)$. By definition this means $f(r) = 0$, so r is a solution to $f(x) = 0$.

(2) $\Rightarrow$ (3) Suppose r is a solution to $f(x) = 0$. To show that $x - r$ is a factor of $f(x)$, divide. The Polynomial Division Algorithm (7.5.3) gives polynomials $q(x)$ and $r(x)$ such that

$$f(x) = q(x)(x - r) + r(x)$$

and the highest exponent of $r(x)$—the ***degree***—is less than the degree of $x - r$. Hence, $r(x)$ is a constant that we will simply write as c. Now,

$$\begin{aligned} 0 = f(r) &= q(r)(r - r) + c \\ &= 0 + c \\ &= c. \end{aligned}$$

Therefore, $c = 0$, and

$$f(x) = q(x)(x - r).$$

Hence, $x - r$ is a factor of $f(x)$.

(3) $\Rightarrow$ (1) Lastly, assume $x - r$ is a factor of $f(x)$. This means there is a polynomial $q(x)$ so that

$$f(x) = (x - r)q(x).$$

Thus,

$$f(r) = (r - r)q(r) = 0,$$

which means r is a root of $f(x)$. ■

The second type of proof in this section is the proof of a disjunction. To prove $p \vee q$, it is standard to assume $\sim p$ and show q. This means that we would be using Direct Proof to show $\sim p \Rightarrow q$. This is what we want because

$$\sim p \Rightarrow q \equiv \sim\sim p \vee q \equiv p \vee q.$$

2.6.3. Proof of Disjunctions

For any propositional forms p and q:

$$\sim p \Rightarrow q \vdash p \vee q.$$

Two points should be emphasized. First, when proving $\sim p \Rightarrow q$, the hypothesis $\sim p$ will be assumed. It will feel like the beginning of an indirect proof, but that is not what is being applied. It is Direct Proof. Second, the intuition behind the strategy goes like this. If we need to prove $p \vee q$ from some hypotheses, it is not reasonable to believe that we can simply prove p and then use Addition to conclude $p \vee q$. Indeed, if we could prove simply p, then we would expect the conclusion of the theorem to be stated as p and not $p \vee q$. Hence, we need to incorporate both disjuncts into the proof. We do this by assuming the negation of a disjunct. If we can prove the other, then the disjunction must be true.

Example. To prove

for all $a,\ b \in \mathbb{Z}$, *if* $ab = 0$, *then* $a = 0$ *or* $b = 0$,

we will assume $ab = 0$ and show that $a \neq 0$ implies $b = 0$. The outline of the proof is given by the following:

→ Assume $ab = 0$
⟨Show $a = 0$ or $b = 0$⟩
→ Suppose $a \neq 0$
⋮
$b = 0$
$a = 0$ or $b = 0$

Now we give the final form.

Proof. Take $a,\ b \in \mathbb{Z}$. Let $ab = 0$ and suppose $a \neq 0$. Then a^{-1} exists. Multiplying both sides of the equation by a^{-1} gives $a^{-1}ab = a^{-1} \cdot 0$, and so $b = 0$. ■

The last type of proof for the section is Proof by Cases. Suppose that we wish to prove $p \Rightarrow q$ and for some reason this is difficult. We notice, however, that p can be broken into cases. Namely, there exist $p_1,\ p_2,\ \ldots,\ p_k$ such that

$$p \equiv p_1 \vee p_2 \vee \cdots \vee p_k.$$

It turns out that if we can prove $p_i \Rightarrow q$ for each i, then we have proven $p \Rightarrow q$. When $k = 2$, $p \equiv p_1 \vee p_2$, and the justification of this is as follows:

$$\begin{aligned}(p_1 \Rightarrow q) \wedge (p_2 \Rightarrow q) &\equiv ({\sim}p_1 \vee q) \wedge ({\sim}p_2 \vee q)\\ &\equiv (q \vee {\sim}p_1) \wedge (q \vee {\sim}p_2)\\ &\equiv q \vee ({\sim}p_1 \wedge {\sim}p_2)\\ &\equiv ({\sim}p_1 \wedge {\sim}p_2) \vee q\\ &\equiv {\sim}(p_1 \vee p_2) \vee q\\ &\equiv (p_1 \vee p_2) \Rightarrow q\\ &\equiv p \Rightarrow q\end{aligned}$$

Example. Let $a \in \mathbb{R}$. Although technically a predicate, we may still write

$$a \in \mathbb{R} \equiv a > 0 \text{ or } a = 0 \text{ or } a < 0.$$

Thus, if we needed to prove a proposition about an arbitrary real number, it would suffice to prove the result for each of the disjuncts.

2.6.4. Proof by Cases (CP)

For any positive integer k, if $p \equiv p_1 \vee \cdots \vee p_k$, then

$$p_1 \Rightarrow q, \ \ldots, \ p_k \Rightarrow q \vdash p \Rightarrow q.$$

Example. Our example of a proof by cases is a well-known one:

for all $a, b \in \mathbb{Z}$, if $a = \pm b$, then a divides b and b divides a.

The hypothesis means

$$a = b \text{ or } a = -b.$$

Therefore, the two disjuncts are the two cases, and we have to show both

if $a = b$, then a divides b and b divides a,

and

if $a = -b$, then a divides b and b divides a.

This leads to the following structure:

→ Suppose $a = \pm b$
Thus $a = b$ or $a = -b$
⟨Show a divides b and b divides a⟩
⟨Show $b = ak$ and $a = b\ell$ with $k, \ell \in \mathbb{Z}$⟩
(Case 1) → Assume $a = b$
⋮
a divides b and b divides a
(Case 2) → Assume $a = -b$
⋮
a divides b and b divides a

The final proof looks something like this:

Proof. Let a and b be integers and suppose $a = \pm b$. To show a divides b and b divides a, we have two cases to prove.
(Case 1) Assume $a = b$. Then, $a = b \cdot 1$ and $b = a \cdot 1$.
(Case 2) Next assume $a = -b$. This means that $a = b \cdot (-1)$ and $b = a \cdot (-1)$.
In both cases we have proven that a divides b and b divides a. ■

EXERCISES

1. Prove the following using the method of Biconditional Proof.
   (a) $(P \vee Q) \Rightarrow {\sim}R,\ S \Rightarrow R,\ ({\sim}P \vee Q) \Rightarrow S \vdash P \Leftrightarrow {\sim}S$
   (b) $P \vee ({\sim}Q \vee P),\ Q \vee ({\sim}P \vee Q) \vdash P \Leftrightarrow Q$
   (c) $(P \Rightarrow Q) \wedge (R \Rightarrow S),\ (P \Rightarrow {\sim}S) \wedge (R \Rightarrow {\sim}Q),\ P \vee R \vdash (Q \Leftrightarrow {\sim}S)$
   (d) $(P \wedge R) \Rightarrow {\sim}(S \vee T),\ ({\sim}S \vee {\sim}T) \Rightarrow (P \wedge R) \vdash S \Leftrightarrow T$
2. Prove the following using the Shorter Rule of Biconditional Proof:
   (a) $P \Rightarrow (Q \wedge R) \Leftrightarrow (P \Rightarrow Q) \wedge (P \Rightarrow R)$
   (b) $P \Rightarrow (Q \vee R) \Leftrightarrow (P \wedge {\sim}Q) \Rightarrow R$
   (c) $(P \vee Q) \Rightarrow R \Leftrightarrow (P \Rightarrow R) \wedge (Q \Rightarrow R)$
   (d) $(P \wedge Q) \Rightarrow R \Leftrightarrow (P \Rightarrow R) \vee (Q \Rightarrow R)$
3. Prove the following:
   (a) $P \wedge (\exists y)Q(y) \Leftrightarrow (\exists y)[P \wedge Q(y)]$
   (b) $P \vee (\forall y)Q(y) \Leftrightarrow (\forall y)[P \vee Q(y)]$
4. Let $a,\ b,\ c,\ d \in \mathbb{Z}$. Write paragraph proofs using Biconditional Proof:
   (a) a is even if and only if a^2 is even.
   (b) a is odd if and only if $a + 1$ is even.
   (c) a is even if and only if $a + 2$ is even.
   (d) $a^3 + a^2 + a$ is even if and only if a is even.
   (e) If $c \neq 0$, then a divides b if and only if ac divides bc.
5. When possible, prove each proposition in the previous problem using the Shorter Rule of Biconditional Proof.
6. Let $a,\ b \in \mathbb{Z}$. Prove that the following are equivalent:
   - a divides b.
   - $-a$ divides b.
   - a divides $-b$.
   - $-a$ divides $-b$.
7. Let $a \in \mathbb{Z}$. Prove that the following are equivalent:
   - a is divisible by 3.
   - $3a$ is divisible by 9.
   - $a + 3$ is divisible by 3.
8. Prove by using Direct Proof but do not use the contrapositive: for all $a,\ b \in \mathbb{Z}$,

   if ab is even, then a is even or b is even.

9. Prove by cases:
   (a) For all $a \in \mathbb{Z}$, if $a = 0$ or $b = 0$, then $ab = 0$.
   (b) The square of every odd integer is of the form $8k + 1$ for some $k \in \mathbb{Z}$. (Hint: Square $2\ell + 1$. Then consider two cases: ℓ is even and ℓ is odd.)
   (c) For every $a \in \mathbb{Z}$, if a is an integer, then $a^2 + a + 1$ is odd.

(d) The fourth power of every odd integer is of the form $16k + 1$ for some $k \in \mathbb{Z}$.
(e) Every nonhorizontal line intersects the x-axis.

10. Let a be an integer. Use cases to prove:
(a) 2 divides $a(a+1)$ (b) 3 divides $a(a+1)(a+2)$

11. Recall for any real number c,

$$|c| = \begin{cases} c & \text{if } c \geq 0 \\ -c & \text{if } c < 0. \end{cases}$$

Let a be a positive real number. Prove the following about absolute value for every $x \in \mathbb{R}$.
(a) $|-x| = |x|$
(b) $|x^2| = |x|^2$
(c) $x \leq |x|$
(d) $|xy| = |x|\,|y|$
(e) $|x| < a$ if and only if $-a < x < a$.
(f) $|x| > a$ if and only if $x > a$ or $x < -a$.

12. For every $a, b \in \mathbb{Z}$, prove:
(a) a divides 1 if and only if $a = \pm 1$.
(b) If $a = \pm b$, then $|a| = |b|$.
(c) If a divides b and b divides a, then $a = \pm b$.

II Main Topics

We will next study topics that will allow us to practice writing basic proofs and put our logic to good use. The topics include set theory, mathematical induction, number theory, relations, and functions.

3 Set Theory

The study of mathematics involves understanding various types of objects such as numbers, functions, relations, points, and lines. Instead of examining particular examples of these objects, we often need to study general collections. These collections are known as sets. We have already seen them in action when we studied the domains of predicates. They also play an important role in all modern mathematics from calculus to algebra. They enable us to discover general properties that can then be applied to specific cases. Although set theory is a subject in its own right as a branch of mathematical logic, in this context it becomes a tool.

3.1. SET BASICS

We used basic set notation when studying logic. We needed the idea of a set and the notion of set membership when defining the domain of a quantifier in Section 2.1. In this section we will use predicates and logical connectives to define sets.

Sometimes it is important to find a precise test to determine whether an element belongs to a set. (Recall Exercise 2.1.6.) Let A be a set. A predicate $P(x)$ with the property

$$a \in A \text{ if and only if } P(a)$$

provides such a test. Notice that $P(x)$ completely describes the members of A. Namely, for a to be an element of A, it must be the case that $P(a)$ is true. Conversely, if $P(a)$ is true, then we know that a is in A. For example, let E be the set of even integers. As a roster,

$$E = \{\ldots, -4, -2, 0, 2, 4, \ldots\}.$$

Let $Q(x)$ denote the predicate $(\exists n \in \mathbb{Z})(x = 2n)$. The even integers are exactly those numbers that satisfy $Q(x)$. In particular, $Q(2)$ and $Q(-4)$ are true, but $Q(5)$ is false. Therefore, 2 and -4 are elements of E, but 5 is not.

We will use predicates to write sets in ***set-builder notation***. Given a predicate $P(x)$,

$$\{x : P(x)\}$$

is the set of all elements that satisfy $P(x)$. Read this by translating the colon as "such that." What comes before are the elements of the set. What comes after is the condition that those elements must satisfy to be in the set (see Figure 3.1.1). Hence, $\{x : P(x)\}$ is read as "the set of all x such that $P(x)$." (Note: Some texts use the vertical bar for *such that*, so $\{x : P(x)\}$ can be written as $\{x \mid P(x)\}$.)

3.1.1. Figure

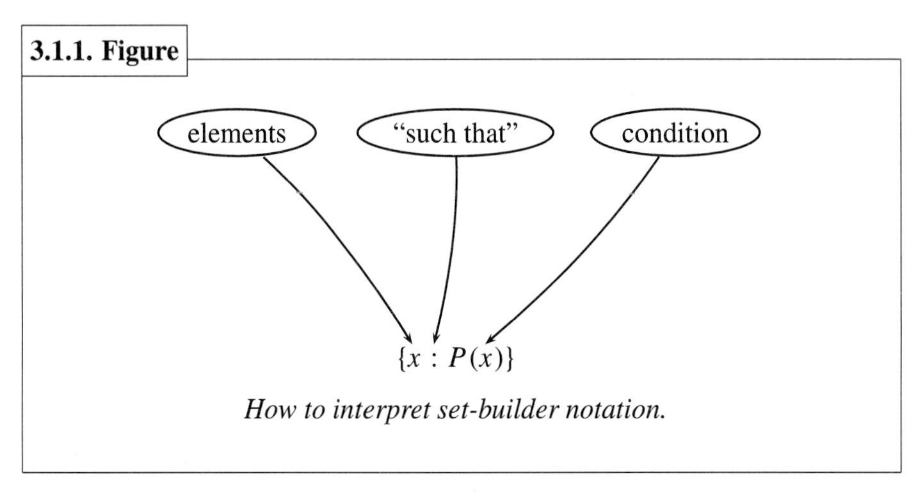

How to interpret set-builder notation.

We can now write E in set-builder notation as

$$\{x : (\exists n \in \mathbb{Z})(x = 2n)\}$$

or as

$$\{x : x = 2n \text{ for some } n \in \mathbb{Z}\}.$$

Read this as "the set of all x such that $x = 2n$ for some integer n." Because the elements of the set are written before the colon, we may simplify this notation by making a substitution for x. Since $x = 2n$, x can be removed from the definition, and then E can be rewritten as

$$\{2n : n \in \mathbb{Z}\}.$$

Read this as "the set of all $2n$ such that n is an integer." This simplified notation is still illustrated by Figure 3.1.1. The elements of E are even. They look like $2n$ for an integer n. These elements are found to the right of the colon. To its left is still the condition on the elements.

Example. We will write the following using set-builder notation.

1. Given the quadratic equation $x^2 - x - 2 = 0$, we know that its solutions are -1 and 2. Thus, its solution set is $\{-1, 2\}$. Using set-builder notation, this may be written as
$$\{x : x^2 - x - 2 = 0 \wedge x \in \mathbb{R}\}.$$
However, similar to our work with quantifiers, we can include the domain of the variable to the left of the colon and write
$$\{x \in \mathbb{R} : x^2 - x - 2 = 0\},$$
a much cleaner notation. Therefore, given an arbitrary equation $f(x) = 0$, its solution set over the real numbers is
$$\{x \in \mathbb{R} : f(x) = 0\}.$$
2. Let C be the set of real-valued, constant functions. It would not be easy to list C as a roster, so let us find a predicate and use set-builder notation. One way to do this is to write
$$C = \{f : f \text{ is a constant}\}.$$
However, it would be more precise to use the predicate
$$(\exists k \in \mathbb{R})(\forall x \in \mathbb{R})(f(x) = k).$$
Hence, C is the set
$$\{f : \text{there exists } k \in \mathbb{R} \text{ so that for all } x \in \mathbb{R},\ f(x) = k\}.$$
3. Since $x \notin \varnothing$ is always true, to write the empty set using set-builder notation, we must use a predicate like $x \neq x$ or a contradiction like $Q \wedge \sim Q$ where Q is a proposition. Then,
$$\varnothing = \{x : x \neq x\} = \{x : Q \wedge \sim Q\}.$$
There are infinitely many ways to do this.

Example. Using the natural numbers as the starting point, $\mathbb{Z}$, $\mathbb{Q}$, $\mathbb{R}$, and $\mathbb{C}$ can be written using set-builder notation:

$$\mathbb{Z} = \{\pm n : n \in \mathbb{N}\}$$
$$\mathbb{Q} = \{a/b : a,\ b \in \mathbb{Z} \wedge b \neq 0\}$$
$$\mathbb{R} = \{x : x \text{ has a decimal representation}\}$$
$$\mathbb{C} = \{a + bi : a,\ b \in \mathbb{R} \wedge i^2 = -1\}$$

Notice the redundancy in the definition of $\mathbb{Q}$. The fraction $1/2$ is named many ways like $2/4$ or $9/18$, but remember that this does not mean the numbers appear infinitely many times in the set. They appear only once.

Example. We can define the open intervals using set-builder notation:

$$(a, b) = \{x \in \mathbb{R} : a < x < b\}$$
$$(a, \infty) = \{x \in \mathbb{R} : a < x < \infty\}$$
$$(-\infty, b) = \{x \in \mathbb{R} : -\infty < x < a\}$$

See Exercise 6 for the closed and half-open intervals.

Finally, when dealing with sets we must often write them in either roster or set-builder notation from an English description.

Example. Each of the following are written in set-builder form and, where appropriate, as a roster.

1. The set of all rational numbers with denominator equal to 3 in roster form is
$$\{\ldots, -3/3, -2/3, -1/3, 0/3, 1/3, 2/3, 3/3, \ldots\}.$$
In set-builder form it is
$$\{a/3 : a \in \mathbb{Z}\}.$$
2. The set of all linear polynomials with leading coefficient equal to 1 and integer y-intercept in roster form is
$$\{\ldots, x-2, x-1, x, x+1, x+2, \ldots\}.$$
In set-builder form it is
$$\{x - a : a \in \mathbb{Z}\}.$$
3. The set of all polynomials of degree at most 5 can be written as
$$\{a_5x^5 + \cdots + a_1x + a_0 : a_i \in \mathbb{R} \text{ and } i = 0, \ldots, 5\}.$$
(Note: for our purposes the exponents of polynomials will always be natural numbers.)
4. The set of all two by two ***diagonal matrices*** is given by
$$\left\{\begin{bmatrix} a & 0 \\ 0 & b \end{bmatrix} : a, b \in \mathbb{R}\right\}.$$

We will now use connectives to define the ***set operations***. These will allow us to build new sets from given ones. To do this, first assume that all elements are members of a fixed domain U. In set theory this set is usually called a ***universe***. It is supposed to be the set of all possible elements. Let A and B be sets. Our first set operation is called ***union***. The symbol for this operation is $\cup$. It is defined using the $\vee$ connective:

$$A \cup B = \{x : x \in A \vee x \in B\}.$$

Read this as "A union B." The union of two sets can be viewed as the combination of all elements from both sets. On the other hand, the ***intersection*** of two sets is defined with $\wedge$ and uses the symbol $\cap$:

$$A \cap B = \{x : x \in A \wedge x \in B\}.$$

It is read as "A intersect B" and can be considered as the overlap between the two.

> **Example.** If $A = \{1, 2, 3, 4\}$ and $B = \{3, 4, 5, 6\}$, then $A \cup B = \{1, 2, 3, 4, 5, 6\}$ and $A \cap B = \{3, 4\}$.

The operations of union and intersection can be illustrated with pictures called ***Venn Diagrams***.* These use circles to represent sets and a rectangle to symbolize the universe. The elements exist in the space inside these shapes. Shading is used to represent the set that results from applying some set operations. If U is a universe, then the Venn Diagrams for union and intersection are illustrated in Figure 3.1.2.

> **Example.** Let A and B be any two sets. If $A \cap B = \varnothing$, call the sets ***disjoint*** or ***mutually exclusive***. The sets $\{1, 2, 3\}$ and $\{6, 7\}$ are disjoint. The Venn Diagram for two disjoint sets is illustrated by the following:

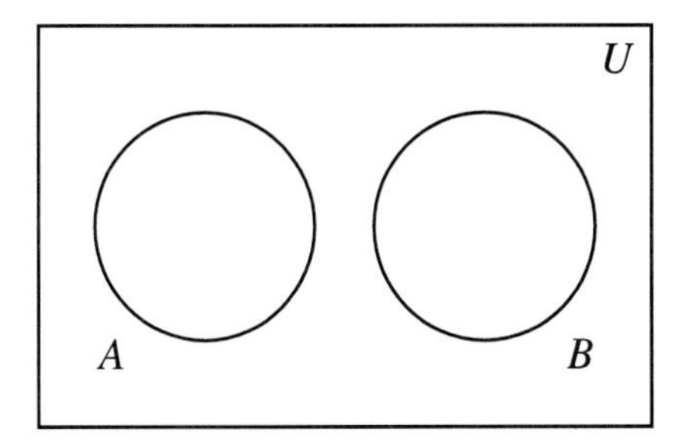

These operations are quite ordinary in mathematics. The next one may not be as familiar. ***Set difference*** or ***set minus*** forms a new set by taking all of the elements of one that are not in the other. It is defined using the backslash symbol and uses *not*:

$$A \setminus B = \{x \in A : x \notin B\}.$$

---

*John Venn (Hull, England, 1834 – Cambridge, England, 1923): Venn studied logic and probability, and he developed a method of analyzing propositions with diagrams. These diagrams now bear his name.

3.1.2. Figure

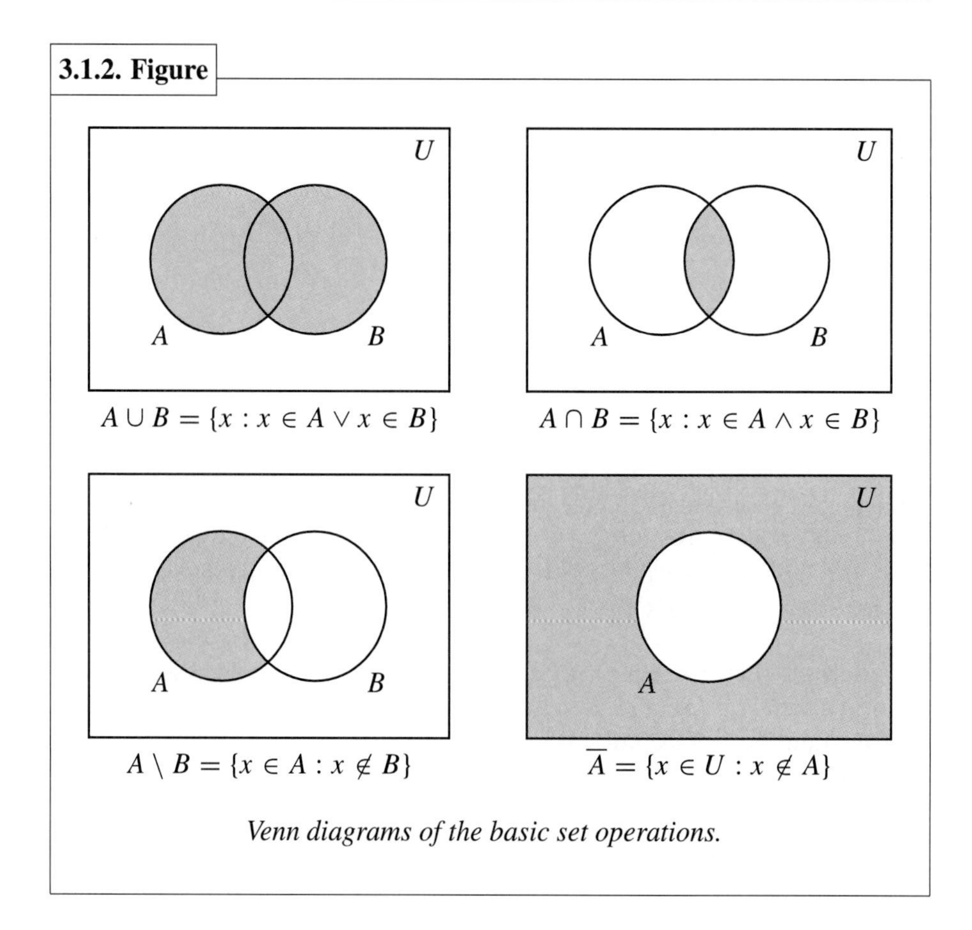

$A \cup B = \{x : x \in A \vee x \in B\}$ $A \cap B = \{x : x \in A \wedge x \in B\}$

$A \setminus B = \{x \in A : x \notin B\}$ $\overline{A} = \{x \in U : x \notin A\}$

Venn diagrams of the basic set operations.

Read this as "A minus B" or "A without B." If a universe U is chosen, then the more common ***complement*** is obtained:

$$\overline{A} = U \setminus A = \{x \in U : x \notin A\}.$$

See Figure 3.1.2 for the Venn Diagrams of set difference and the complement.

Example. Let us find the union, intersection, set difference, and complement of each of the following:

1. Define the universe to be

$$U = \{1, 2, \ldots, 10\}.$$

We will use a Venn Diagram to find the results of the set operations on

$$A = \{1, 2, 3, 4, 5\}$$

and

$$B = \{3, 4, 5, 6, 7, 8\}.$$

Each element will be represented as a point within a circle and labeled with a number.

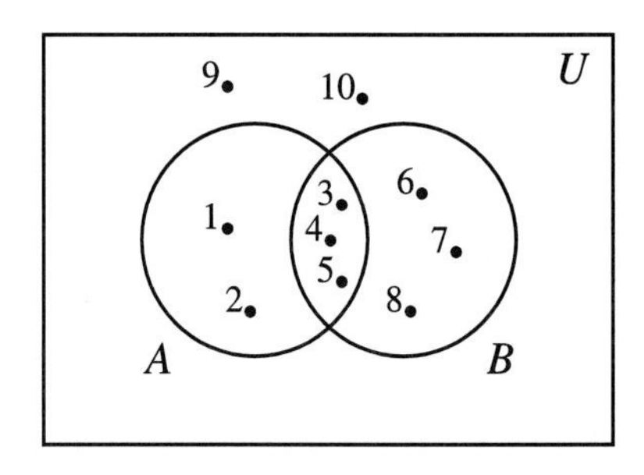

(a) $A \cup B = \{1, 2, 3, 4, 5, 6, 7, 8\}$
(b) $A \cap B = \{3, 4, 5\}$
(c) $A \setminus B = \{1, 2\}$
(d) $\overline{A} = \{6, 7, 8, 9, 10\}$

2. Let $C = (-4, 2)$, $D = [-1, 3]$, and $U = \mathbb{R}$. Use the following diagram to perform the set operations:

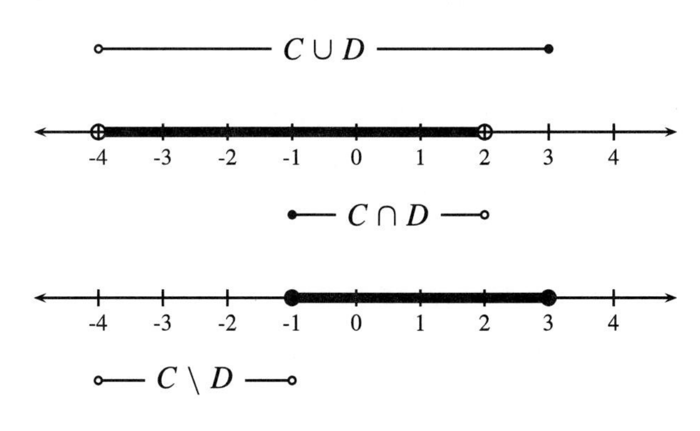

(a) $C \cup D = (-4, 3]$
(b) $C \cap D = [-1, 2)$
(c) $C \setminus D = (-4, -1)$
(d) $\overline{C} = (-\infty, -4] \cup [2, \infty)$

Example. The following equalities use set difference:

1. $\mathbb{N} = \mathbb{Z} \setminus \mathbb{Z}^-$
2. $\mathbb{R} \setminus \mathbb{Q} =$ the set of ***irrational numbers***.
3. $\mathbb{R} = \mathbb{C} \setminus \{a + bi : a,\ b \in \mathbb{R} \text{ and } b \neq 0\}$

3.1.3. Figure

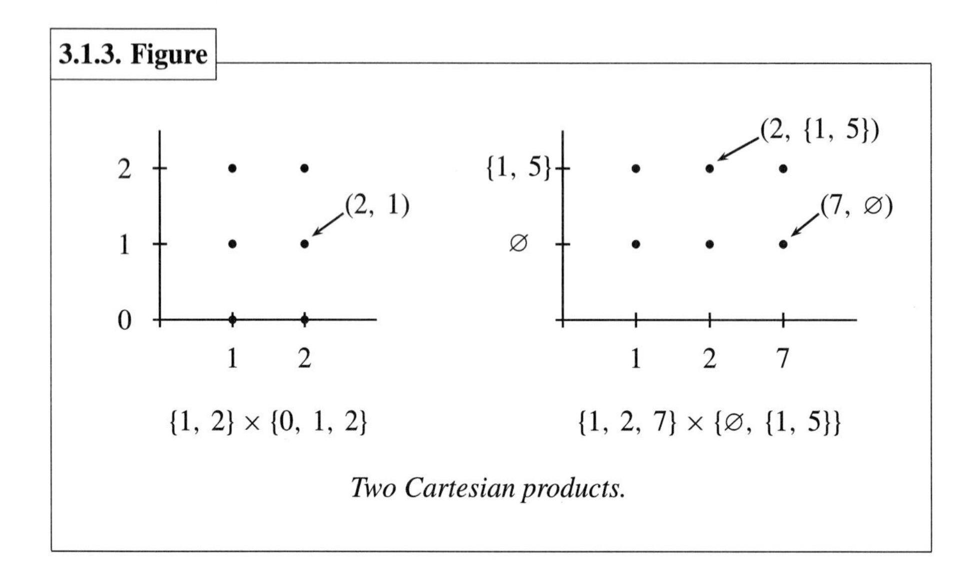

Two Cartesian products.

The last set operation is not related to the logic connectives as the others, but it is nonetheless very important to mathematics. Let A and B be sets. Given two elements $a \in A$ and $b \in B$, we call (a, b) an ***ordered pair***. In this context, a and b are called ***coordinates***. It is similar to the set $\{a, b\}$ except that the order matters, so we must define what it means for two ordered pairs to be ***equal***:

$$(a, b) = (a', b') \text{ if and only if } a = a' \text{ and } b = b'.$$

The set of all ordered pairs with the first coordinate from A and the second from B is called the ***Cartesian product*** of A and B, named after René Descartes.* In set-builder notation it is defined as

$$A \times B = \{(a, b) : a \in A \wedge b \in B\}.$$

The product $\mathbb{R} \times \mathbb{R}$ is the set of ordered pairs of the Cartesian plane. We will often write $\mathbb{R}^2$ for $\mathbb{R} \times \mathbb{R}$ and $\mathbb{R}^3$ for $\mathbb{R} \times \mathbb{R} \times \mathbb{R}$.

Example. We will write the following Cartesian products as rosters.

1. Since $(1, 2) \neq (2, 1)$, the product $\{1, 2\} \times \{0, 1, 2\}$ equals

$$\{(1, 0), (1, 1), (1, 2), (2, 0), (2, 1), (2, 2)\}.$$

Visually we can represent this set on a grid as in Figure 3.1.3.

---

*René Descartes (La Haye, France 1596 – Stockholm, Sweden, 1650): Descartes was a mathematician and philosopher who studied geometry algebraically using a coordinate system. Although buried in Stockholm, his remains (except his right hand) were returned to France in 1667.

2. If $A = \{1, 2, 7\}$ and $B = \{\varnothing, \{1, 5\}\}$, then $A \times B$ is
$$\{(1, \varnothing), (1, \{1, 5\}), (2, \varnothing), (2, \{1, 5\}), (7, \varnothing), (7, \{1, 5\})\}.$$
Again, see Figure 3.1.3.
3. For any set A,
$$\varnothing \times A = \varnothing$$
since $x \in \varnothing \wedge y \in A$ is always false. Similarly, $A \times \varnothing = \varnothing$.

As with the logical connectives, we need an order of operations to make sense of expressions that involve many operations. To do this, we note the association between the set operations and certain logical connectives:

$$\begin{array}{ll} \left.\begin{array}{c} \overline{A} \\ A \setminus B \end{array}\right\} & \sim \\ A \cap B & \wedge \\ A \cup B & \vee \end{array}$$

From this we derive the order for the set operations:

3.1.4. Order of Operations

To find a set determined by set operations, read from left to right and use the following precedence:

1. sets within parentheses (innermost first),
2. complements,
3. set differences,
4. intersections,
5. unions.

However, as with propositional forms, we will often rely on parentheses for clarity.

Example. If the universe is taken to be $\{1, 2, 3, 4, 5\}$, then
$$\begin{aligned} \{5\} \cup \overline{\{1, 2\}} \cap \{2, 3\} &= \{5\} \cup \{3, 4, 5\} \cap \{2, 3\} \\ &= \{5\} \cup \{3\} \\ &= \{3, 5\} \end{aligned}$$

This set can also be written using parentheses: $\{5\} \cup (\overline{\{1, 2\}} \cap \{2, 3\})$.

Example. The order of operations allows us to make sense of sets written like $A \cup B \cup C \cup D$. For example, if $A = \{1\}$, $B = \{2\}$, $C = \{3\}$, and $D = \{4\}$, then $A \cup B \cup C \cup D = \{1, 2, 3, 4\}$. Written with parentheses,

$$A \cup B \cup C \cup D = ([(A \cup B) \cup C] \cup D).$$

With the given assignments, $A \cap B \cap C \cap D$ is empty.

Notice that the Cartesian product was not included in the order of operations. If we use only the Cartesian product, we do not need parentheses. For instance, using the assignments from the last example,

$$A \times B \times C \times D = \{(1, 2, 3, 4)\}.$$

This set does not contain an ordered pair, but it contains an ordered 4-tuple. When it has has n coordinates, it is called an ***ordered n-tuple***. We will need to use parentheses if it is used with the other set operations. For example,

$$(A \times B) \cup (C \times D) = \{(1, 2), (3, 4)\},$$

but

$$A \times (B \cup C) \times D = \{(1, 2, 4), (1, 3, 4)\}.$$

EXERCISES

1. Determine whether the following are true or false:
 (a) $1 \in \{x : P(x)\}$ when $P(1)$ is false
 (b) $7 \in \{x \in \mathbb{R} : x^2 - 5x - 14 = 0\}$
 (c) $7x^2 - 0.5x \in \{a_2x^2 + a_1x + a_0 : a_i \in \mathbb{Q}\}$
 (d) $xy \in \{2k : k \in \mathbb{Z}\}$ if x is even and y is odd.
 (e) $\cos\theta \in \{a\cos\theta + b\sin\theta : a, b \in \mathbb{R}\}$
 (f) $f \in \{g : g \text{ is a continuous function}\}$ if $f(x) = |x|$
 (g) $f \in \{g : g \text{ is a differentiable function}\}$ if $f(x) = |x|$
 (h) $\{1, 3\} = \{x : (x-1)(x-3) = 0\}$
 (i) $\{1, 3\} = \{x : (x-1)(x-3)^2 = 0\}$
 (j) $\left\{\left[\begin{smallmatrix} a & 0 \\ 0 & 0 \end{smallmatrix}\right] : a \in \mathbb{R}\right\} = \left\{\left[\begin{smallmatrix} x & 0 \\ 0 & y \end{smallmatrix}\right] : x \in \mathbb{R} \text{ and } y = 0\right\}$

2. Each of the given sentences are false. Replace the underlined word with another word to make each sentence true.
 (a) Intersection is defined using a disjunction.
 (b) Set diagrams are used to illustrate set operations.
 (c) $\mathbb{R} \setminus (\mathbb{R} \setminus \mathbb{Q})$ is the set of irrational numbers.
 (d) Set difference has a higher order of precedence than complements.
 (e) The complement of A is equal to $\mathbb{R}$ set minus A.
 (f) The intersection of two intervals is always an interval.

(g) The union of two intervals is never an interval.
(h) $A \times B$ is equal to $\varnothing$ if B does not contain ordered pairs.

3. Write the following in roster form:
   (a) $\{-3n : n \in \mathbb{Z}\}$
   (b) $\{0 \cdot n : n \in \mathbb{R}\}$
   (c) $\{n \cos x : n \in \mathbb{Z}\}$
   (d) $\{ax^2 + ax + a : a \in \mathbb{N}\}$
   (e) $\left\{\left[\begin{smallmatrix} n & 0 \\ 0 & 0 \end{smallmatrix}\right] : n \in \mathbb{Z}\right\}$
4. Write each set in set-builder notation:
   (a) All odd integers
   (b) All positive rational numbers
   (c) All multiples of 7
   (d) All integers that have a remainder of 1 when divided by 3
   (e) All ordered pairs in which the x-coordinate is positive and the y-coordinate is negative
   (f) All differentiable functions
   (g) All complex numbers whose real part is 0
   (h) All closed intervals that contain π
   (i) All open intervals that do not contain a rational number
   (j) All closed rays that contain no numbers in $(-\infty, 3]$
   (k) All 2×2 matrices with real entries that have a diagonal sum of 0
   (l) All polynomials with real coefficients of degree at most 3
   (m) All polynomials with real coefficients with degree at most 7 and at least 5
5. Let $A = \{0, 2, 4, 6\}$, $B = \{3, 4, 5, 6\}$, $C = \{0, 1, 2\}$, and $U = \{0, 1, \ldots, 9, 10\}$. Write the following sets in roster notation:
   (a) $A \cup B$
   (b) $A \cap B$
   (c) $A \setminus B$
   (d) $B \setminus A$
   (e) $\overline{A}$
   (f) $A \times B$
   (g) $A \cup B \cap C$
   (h) $A \cap B \cup C$
   (i) $\overline{A} \cap B$
   (j) $A \cup \overline{A \cap B}$
   (k) $A \setminus B \setminus C$
   (l) $A \cup B \setminus A \cap C$
6. Let $a, b \in \mathbb{R}$. Write the following intervals using set-builder notation:
   (a) $[a, b]$
   (b) $[a, \infty)$
   (c) $(-\infty, b]$
   (d) $(a, b]$
7. Write each of these sets using interval notation:
   (a) $[2, 3] \cap (5/2, 7]$
   (b) $(-\infty, 4) \cup (-6, \infty)$
   (c) $(-2, 4) \cap [-6, \infty)$
   (d) $\varnothing \cup (4, 12]$

8. What set is $\varnothing \times \varnothing$?

9. Identify each of the the following sets.
   (a) $[6,\ 17] \cap [17,\ 32)$
   (b) $[6,\ 17) \cap [17,\ 32)$
   (c) $[6,\ 17] \cup (17,\ 32)$
   (d) $[6,\ 17) \cup (17,\ 32)$
10. Draw Venn Diagrams for the following sets.
    (a) $A \cap \overline{B}$
    (b) $\overline{A \cup B}$
    (c) $A \cap \overline{B} \cup \overline{C}$
    (d) $(A \cup B) \setminus \overline{C}$
    (e) $A \cap C \cap \overline{B}$
    (f) $\overline{A \setminus B} \cap C$
11. Match each Venn Diagram to as many sets as possible.

A)

B) 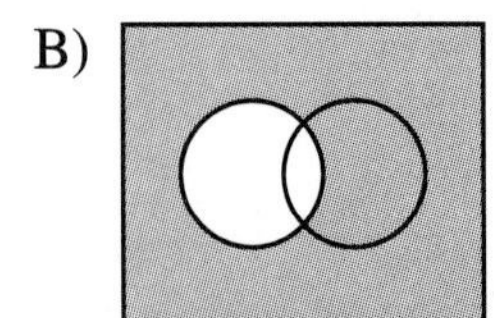

C) 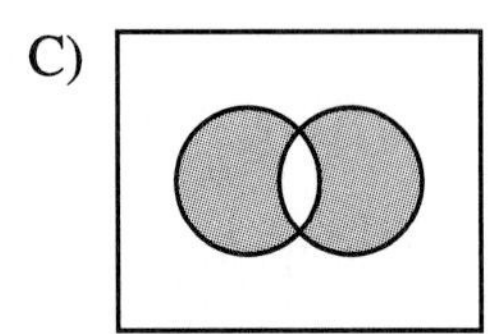

D) 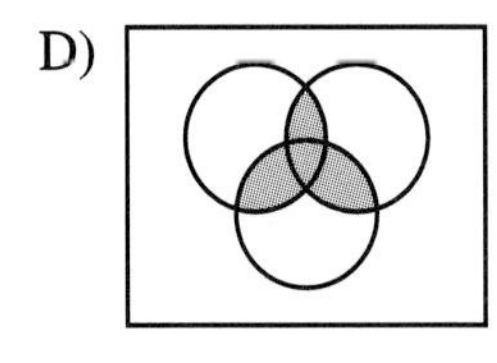

E) 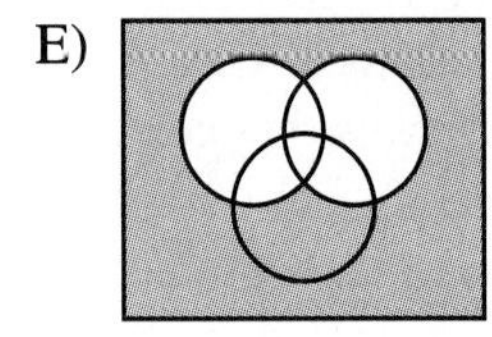

F) 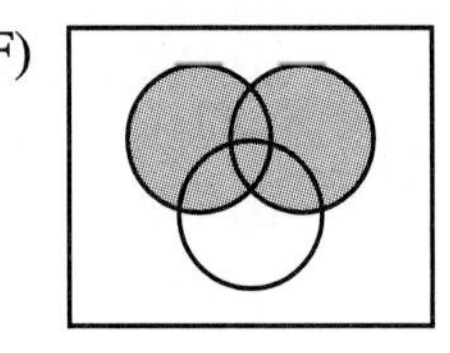

(a) $A \cup B$
(b) $A \setminus B$
(c) $A \cap \overline{B}$
(d) $(A \cup B) \setminus (A \cap B)$
(e) $(A \cap B) \cup (A \cap C) \cup (B \cap C)$
(f) $\overline{(A \cap B) \cup (A \setminus B)}$
(g) $(A \cup B) \cap (\overline{A} \cup \overline{B})$
(h) $[(A \cup B) \cap \overline{C}] \cup [(A \cup B) \cap C]$
(i) $[\overline{A} \cap \overline{B} \cap C] \cup [\overline{A} \cap \overline{B} \cap \overline{C}]$
(j) $\overline{A \setminus (B \cup C)} \cap \overline{B \setminus (A \cup C)} \cap \overline{C \setminus (A \cup B)}$

3.2. SUBSETS

An important relation between any two sets is when one is contained within another. Let A and B be sets. A is a ***subset*** of B exactly when every element of A is also in B, in symbols $A \subseteq B$. The relationship $A \subseteq B$ is called an ***inclusion***. This is represented in a Venn Diagram by the circle for A being within the circle for B. (See Figure 3.2.2.)

3.2.1. Definition

For any two sets A and B,

$$A \subseteq B \text{ if and only if } (\forall x)(x \in A \Rightarrow x \in B).$$

The notation $A \subset B$ will mean $A \subseteq B$ but $A \neq B$. In this case, A is a ***proper subset*** of B, and B is the ***improper subset*** of B. This notion can also be turned around. Given the same condition above, B is a ***superset*** of A. This is written as $B \supseteq A$, and $B \supset A$ means $B \supseteq A$ but $B \neq A$.

Example.

1. $\{1, 2, 3\} \subset \{1, 2, 3, 4, 5\}$ and $\{1, 2, 3\} \subseteq \{1, 2, 3\}$
2. $\mathbb{N} \subset \mathbb{Z} \subset \mathbb{Q} \subset \mathbb{R} \subset \mathbb{C}$
3. $(4, 5) \subset (4, 5] \subset [4, 5]$
4. $\{1, 2, 3\}$ is not a subset of $\{1, 2, 4\}$

Each of these can be rewritten as supersets.

Example.

1. $\{1, 2, 3, 4, 5\} \supset \{1, 2, 3\}$ and $\{1, 2, 3\} \supseteq \{1, 2, 3\}$
2. $\mathbb{C} \supset \mathbb{R} \supset \mathbb{Q} \supset \mathbb{Z} \supset \mathbb{N}$
3. $[4, 5] \supset (4, 5] \supset (4, 5)$
4. $\{1, 2, 4\}$ is not a superset of $\{1, 2, 3\}$

(Note: We will only use $\subset$ if we know that the set is a proper subset.)

3.2.2. Figure

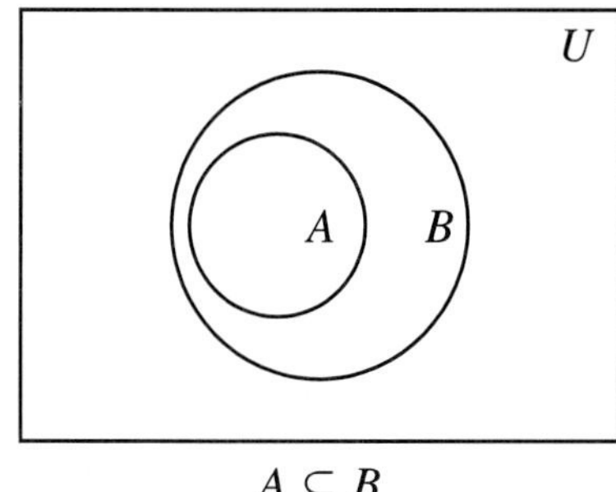

$A \subseteq B$

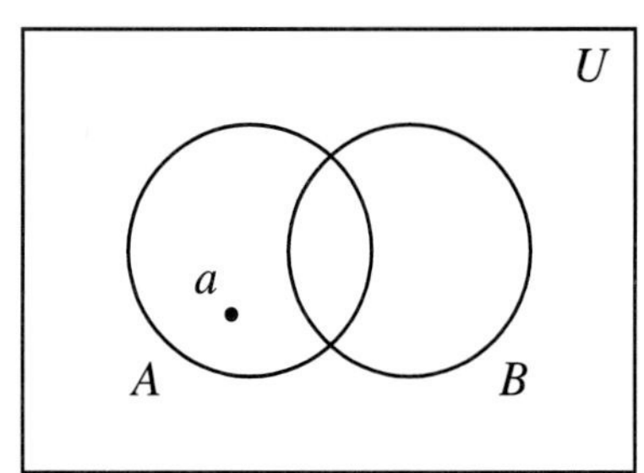

$A \nsubseteq B$ with $a \notin B$

The Venn Diagrams for subsets and non-subsets.

If A is not a subset of B, write $A \not\subseteq B$ (or $B \not\supseteq A$). This is represented in a Venn Diagram by A overlapping B with a point in A but not within B. This is illustrated in Figure 3.2.2. Logically this means

$$\sim(\forall x)(x \in A \Rightarrow x \in B) \equiv (\exists x)(x \in A \wedge x \notin B).$$

Thus, to show $A \not\subseteq B$, we must find an element in A that is not in B. For example, if $A = \{1,\ 2,\ 3\}$ and $B = \{1,\ 2,\ 5\}$, then $A \not\subseteq B$ because $3 \in A$ but $3 \notin B$.

Example. The proposition *for all sets A and B, $A \cup B \subseteq A$* is false. To see this, we must prove that there exists A and B such that $A \cup B \not\subseteq A$. We can take $A = \{1\}$ and $B = \{2\}$ as our candidates. There is no reason to make the sets complicated when simple ones will do. Since $A \cup B = \{1,\ 2\}$ and $2 \notin A$, we have $A \cup B \not\subseteq A$.

Our study of subsets begins by looking at the empty set. The first result is somewhat surprising. It states that the empty set is a subset of every set. This proposition translates to $(\forall A)(\varnothing \subseteq A)$, which in turn means

$$(\forall A)(\forall x)(x \in \varnothing \Rightarrow x \in A).$$

As in the previous chapter, the proposition tells us how to proceed:

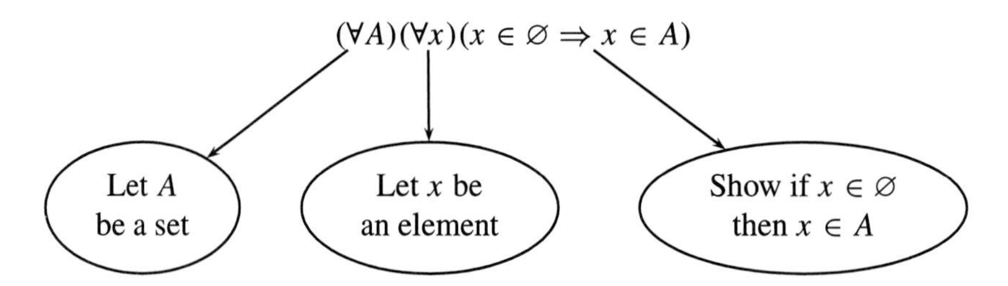

But to prove the implication, we must first assume $x \in \varnothing$. This is always false. Hence, the implication is true for every x independent of the choice of A. This means we may write the proof:

3.2.3. Theorem

If A is any set, then $\varnothing \subseteq A$.

Proof. Let A be a set. Since $x \in \varnothing$ is false for every x, the implication

$$x \in \varnothing \Rightarrow x \in A$$

is always true. Therefore, $\varnothing \subseteq A$. ■

On the other extreme, every set is a subset of the universe. To see this, prove

$$(\forall A)(\forall x)(x \in A \Rightarrow x \in U).$$

Following Direct Proof, suppose $x \in A$. This means x is an object. Since U contains all possible objects, $x \in U$.

Although it was short, the previous paragraph has a typical proof structure. Since the definition of a subset is based on an implication, Direct Proof is the primary tool in these proofs. Specifically, to prove that a set A is a subset of a set B, take an element $x \in A$ and show $x \in B$:

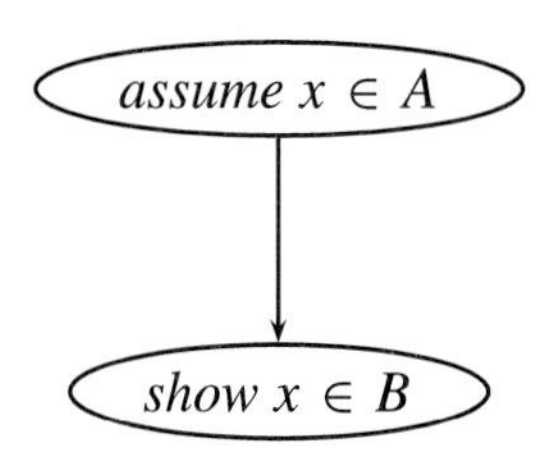

Example. We will write paragraph proofs for the following:

1. $\mathbb{Z} \setminus \mathbb{N} \subseteq \mathbb{Z}$.

   Proof. Let $x \in \mathbb{Z} \setminus \mathbb{N}$ and show $x \in \mathbb{Z}$. By definition, $x \in \mathbb{Z}$ but $x \notin \mathbb{N}$. Simplifying we see that $x \in \mathbb{Z}$. ■

2. $\{x : (x - 2n)(x - 2n - 2) = 0 \text{ for some } n \in \mathbb{Z}\} \subseteq \{2n : n \in \mathbb{Z}\}$.

   Proof. Let $A = \{x : (x - 2n)(x - 2n - 2) = 0 \text{ for some } n \in \mathbb{Z}\}$ and $B = \{2n : n \in \mathbb{Z}\}$. To show $A \subseteq B$, take $x \in A$. This means

   $$(x - 2n)(x - 2n - 2) = 0$$

   for some $n \in \mathbb{Z}$. Hence,

   $$x - 2n = 0 \text{ or } x - 2n - 2 = 0.$$

   We have two cases to check.
   (Case 1) If $x - 2n = 0$, then $x = 2n$. This means $x \in B$.
   (Case 2) If $x - 2n - 2 = 0$, then $x = 2n + 2 = 2(n + 1)$, which also means $x \in B$ since $n + 1$ is an integer.
   In either case, we have $A \subseteq B$. ■

3. If $a, b \in \mathbb{Z}$, then $\{na : n \in \mathbb{Z}\} \subseteq \{na + mb : n, m \in \mathbb{Z}\}$.

   Proof. Fix $a, b \in \mathbb{Z}$. Let $x = na$, some $n \in \mathbb{Z}$. This means $x = na + 0b$, and we are done. ■

This next result is based only on the definitions and the rules of inference. This further links set theory with logic.

3.2.4. Theorem

Let A, B, C, and D be sets.

1. If $A \subseteq B$ and $x \in A$, then $x \in B$.
2. If $A \subseteq B$ and $x \notin B$, then $x \notin A$.
3. If $A \subseteq B$ and $B \subseteq C$, then $A \subseteq C$.
4. Let $A \subseteq B$ and $C \subseteq D$. If $x \in A$ or $x \in C$, then $x \in B$ or $x \in D$.
5. Let $A \subseteq B$ and $C \subseteq D$. If $x \notin B$ or $x \notin D$, then $x \notin A$ or $x \notin C$.

Proof. (1) and (4) are left as exercises.

2. Assume $A \subseteq B$ and let $x \notin B$. Then using Modus Tolens, $x \notin A$.
3. Assume $A \subseteq B$ and $B \subseteq C$. By definition, $x \in A$ implies $x \in B$ and $x \in B$ implies $x \in C$. Therefore by Transitivity, if $x \in A$, then $x \in C$. In other words, $A \subseteq C$.
5. Let $A \subseteq B$ and $C \subseteq D$. Assume $x \notin B$ or $x \notin D$. Then by the Destructive Dilemma, $x \notin A$ or $x \notin C$. ■

Although the theorem has a transitivity result, it is not the case that for all sets A and B, $A \subseteq B \Rightarrow B \subseteq A$. To prove this, we must find two sets for which the condition fails. If we choose $A = \varnothing$ and $B = \{0\}$, then $A \subseteq B$ yet $B \nsubseteq A$ because $0 \in B$ but $0 \notin \varnothing$.

EXERCISES

1. Answer true or false for the following:

(a) $\varnothing \in \varnothing$
(b) $\varnothing \subseteq \{1\}$
(c) $1 \in \mathbb{Z}$
(d) $1 \subseteq \mathbb{Z}$
(e) $1 \in \varnothing$
(f) $\{1\} \subseteq \varnothing$
(g) $0 \in \varnothing$
(h) $\{1\} \in \mathbb{Z}$
(i) $\{1\} \subseteq \mathbb{Z}$
(j) $\varnothing \subseteq \overline{\varnothing}$

2. Answer true or false for the following. For each false proposition, find one element that is in the first set but is not in the second.

(a) $\mathbb{Z}^+ \subseteq \mathbb{C}$
(b) $\mathbb{Q}^+ \subseteq \mathbb{Z}^+$
(c) $\mathbb{Q} \setminus \mathbb{R} \subseteq \mathbb{Z}$
(d) $\mathbb{R} \setminus \mathbb{Q} \subseteq \mathbb{Z}$
(e) $\mathbb{Z} \cap (-1, 1) \subseteq \mathbb{Q}$
(f) $(0, 1) \subseteq \mathbb{Q}^+$
(g) $(0, 1) \subseteq \{0, 1, 2\}$
(h) $(0, 1) \subseteq (0, 1]$

3. Supply the missing phrases to each proof.
   (a) If $A \subseteq B$ and $x \in A$, then $x \in B$.

   Proof. _______ and _______.
   This means if $x \in A$, then $x \in B$.
   Therefore, _______ because of _______. ■

   (b) Let $A \subseteq B$ and $C \subseteq D$. If $x \in A$ or $x \in C$, then $x \in B$ or $x \in D$.

   Proof. Let $A \subseteq B$ and $C \subseteq D$.
   This means _______ and _______.
   Assume _______.
   Therefore, _______ by _______. ■

4. Show that the following are true.
   (a) $\{x \in \mathbb{R} : x^2 - 3x + 2 = 0\} \subseteq \mathbb{N}$
   (b) $(0, 1) \subseteq [0, 1]$
   (c) $[0, 1] \not\subseteq (0, 1)$
   (d) $\mathbb{Z} \times \mathbb{Z} \not\subseteq \mathbb{Z} \times \mathbb{N}$
   (e) $(0, 1) \cap \mathbb{Q} \not\subseteq [0, 1] \cap \mathbb{Z}$
5. Let A, B, C, and D be sets. Prove each of the following.
   (a) $A \subseteq A$
   (b) $A \subseteq U$
   (c) $A \subseteq A \cup B$
   (d) $A \cap B \subseteq A$
   (e) $A \setminus B \subseteq A$
   (f) If $A \subseteq B$, then $A \cup C \subseteq B \cup C$.
   (g) If $A \subseteq B$, then $A \cap C \subseteq B \cap C$.
   (h) If $A \subseteq B$, then $C \setminus B \subseteq C \setminus A$.
   (i) If $A \neq \varnothing$, then $A \not\subseteq \overline{A}$.
   (j) If $A \subseteq \overline{B}$, then $B \subseteq \overline{A}$.
   (k) If $A \subseteq B$, then $\{1\} \times A \subseteq \{1\} \times B$.
   (l) If $A \subseteq C$ and $B \subseteq D$, then $A \times B \subseteq C \times D$.
   (m) If $A \subseteq C$ and $B \subseteq D$, then $\overline{C \times D} \subseteq \overline{A \times B}$.
6. Prove that $A \subseteq B$ if and only if $\overline{B} \subseteq \overline{A}$.
7. **Show that the proposition**

   for all sets A and B, $A \cap B \neq \varnothing \Rightarrow A \not\subseteq A \cap B$,

   is false.
8. Let A and B be sets. Under what conditions are the following true?
   (a) $A \cup B \subseteq A$
   (b) $A \cup B \subseteq A \cap B$
   (c) $A \cap B \subseteq A \setminus B$
   (d) $\overline{A \cap B} \subseteq \overline{A} \cap \overline{B}$

9. Prove each of the following about sets of complex numbers.
   (a) $\mathbb{R} \subseteq \mathbb{C}$
   (b) $\{bi : b \in \mathbb{R}\} \subseteq \mathbb{C}$
   (c) $\mathbb{C} \nsubseteq \mathbb{R}$
10. Prove: $(A \times B) \cup (C \times D) \subseteq (A \cup C) \times (B \cup D)$.
11. Let $P(x)$ and $Q(x)$ be predicates. Assuming each of the following, prove $\{x : P(x)\} \subseteq \{x : Q(x)\}$.
    (a) $(\forall x)[P(x) \Rightarrow Q(x)]$
    (b) $(\forall x)[P(x) \Rightarrow Q(x) \wedge R(x)]$
    (c) $(\forall x)[P(x) \vee R(x) \Rightarrow Q(x)]$
    (d) $(\forall x)Q(x)$
    (e) $(\forall x){\sim}P(x)$

3.3. EQUALITY OF SETS

For two sets to be equal, they must contain exactly the same elements. We met this notion in Section 2.1. This can be stated more precisely using the idea of a subset.

3.3.1. Definition

Take two sets, A and B:

$$A = B \text{ if and only if } A \subseteq B \text{ and } B \subseteq A.$$

To prove that two sets are equal, we must show both inclusions. Let us do this to prove $\overline{A \cup B} = \overline{A} \cap \overline{B}$. This is one of ***De Morgan's Laws***. To prove it we must demonstrate both

$$\overline{A \cup B} \subseteq \overline{A} \cap \overline{B}$$

and

$$\overline{A} \cap \overline{B} \subseteq \overline{A \cup B}.$$

This amounts to proving a biconditional, which means we will use the Rule of Biconditional Proof. Look at the first direction:

$x \in \overline{A \cup B}$	Given
${\sim}(x \in A \cup B)$	Definition of complement
${\sim}(x \in A \vee x \in B)$	Definition of union
$x \notin A \wedge x \notin B$	De Morgan's Law
$x \in \overline{A} \wedge x \in \overline{B}$	Definition of complement
$x \in \overline{A} \cap \overline{B}$	Definition of intersection

Now read backward through those steps. Each follows logically when read in this order. This means the steps are reversible. Hence, we have a series of biconditionals, and we may use the shorter rule:

Proof. Let A and B be sets. Then,

$$\begin{aligned} x \in \overline{A \cup B} &\Leftrightarrow \sim(x \in A \cup B) \\ &\Leftrightarrow \sim(x \in A \vee x \in B) \\ &\Leftrightarrow x \notin A \wedge x \notin B \\ &\Leftrightarrow x \in \overline{A} \wedge x \in \overline{B} \\ &\Leftrightarrow x \in \overline{A} \cap \overline{B}. \end{aligned}$$

Hence, $\overline{A \cup B} = \overline{A} \cap \overline{B}$. ■

We must be careful, though, when writing these types of proofs since it is easy to confuse the notation. Take two examples.

- The correct translation for $x \in A \cap B$ is $x \in A \wedge x \in B$. Common mistakes for this translation include using predicates with set operations:

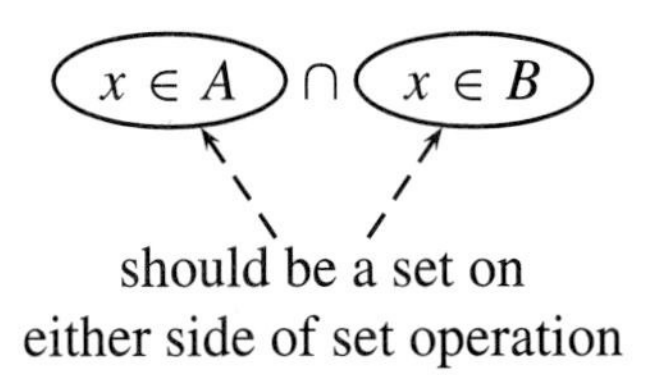

  ... and sets with connectives:

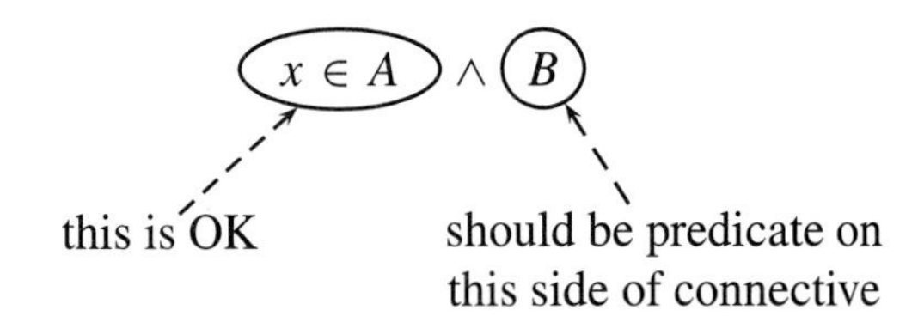

  Remember that connectives connect predicates or propositions and set operations connect sets.

- Negations also pose problems. The predicate $x \notin A \cup B$ can neither be written as $x \notin A \cup x \notin B$ nor $x \notin A \vee x \notin B$. Instead, translate using De Morgan's Law as we did above. Similarly, if a complement is used, first translate using

$$x \in \overline{A} \Leftrightarrow \sim(x \in A) \Leftrightarrow x \notin A$$

  and then proceed with the proof.

We can now prove many basic properties about set operations. Notice how the following are closely related to their corresponding rules of replacement.

3.3.2. Theorem

Let A, B, and C be sets.

Commutative Laws
$A \cap B = B \cap A$
$A \cup B = B \cup A$

Associative Laws
$(A \cap B) \cap C = A \cap (B \cap C)$
$(A \cup B) \cup C = A \cup (B \cup C)$

De Morgan's Laws
$\overline{A \cup B} = \overline{A} \cap \overline{B}$
$\overline{A \cap B} = \overline{A} \cup \overline{B}$

Distributive Laws
$A \cap (B \cup C) = (A \cap B) \cup (A \cap C)$
$A \cup (B \cap C) = (A \cup B) \cap (A \cup C)$

Tautology Laws
$A \cap A = A$
$A \cup A = A$

Example. We will use the Shorter Rule of Biconditional Proof to prove the equality

$$(A \cap B) \cap C = A \cap (B \cap C).$$

Proof. If A, B, and C are sets, then:

$$\begin{aligned} x \in (A \cap B) \cap C &\Leftrightarrow (x \in A \cap B) \wedge x \in C \\ &\Leftrightarrow (x \in A \wedge x \in B) \wedge x \in C \\ &\Leftrightarrow x \in A \wedge (x \in B \wedge x \in C) \\ &\Leftrightarrow x \in A \wedge (x \in B \cap C) \\ &\Leftrightarrow x \in A \cap (B \cap C). \end{aligned}$$

Thus, $(A \cap B) \cap C = A \cap (B \cap C)$. ■

Another way to prove it is to use a chain of equal signs. The logic of the proof will be the same, but it will look different.

Proof. If A, B, and C are sets, then:

$$\begin{aligned} (A \cap B) \cap C &= \{x : (x \in A \cap x \in B) \wedge x \in C\} \\ &= \{x : (x \in A \wedge x \in B) \wedge x \in C\} \\ &= \{x : x \in A \wedge (x \in B \wedge x \in C)\} \\ &= \{x : x \in A \wedge (x \in B \cap C)\} \\ &= A \cap (B \cap C). \end{aligned}$$ ■

We have to be careful when proving equality. If two sets are equal, there are always proofs for both inclusions. However, the steps needed for the one direction may not simply be the steps for the other in reverse. The next proposition is an example. It is always true that $A \cap B \subseteq A$. However, the hypothesis is needed to show the other inclusion. We must use Biconditional Proof here. However, instead of introducing both directions with ($\Rightarrow$) and ($\Leftarrow$), we use ($\subseteq$) and ($\supseteq$).

3.3.3. Theorem

If A and B are sets so that $A \subseteq B$, then $A \cap B = A$.

Proof. Assume $A \subseteq B$.

($\subseteq$) Let $x \in A \cap B$. This means that $x \in A$ and $x \in B$. Then by Simplification, $x \in A$.

($\supseteq$) Assume that x is an element of A. Since $A \subseteq B$, x is also an element of B. Hence, $x \in A \cap B$. ■

As with subsets, let us now prove some results concerning the empty set and the universe. We will use two strategies:

- Let A be a set. We know that

  $$A = \varnothing \text{ if and only if } (\forall x)(x \notin A).$$

  Therefore, to prove that A is empty, take an arbitrary a and show $a \notin A$. This can sometimes be shown directly, but more often an indirect proof is better. To do this, assume $a \in A$ and derive a contradiction. Since the contradiction arose simply by assuming $a \in A$, this statement must be the problem. Hence, A can have no elements.
- To prove $A = U$, we must show that $A \subseteq U$ and $U \subseteq A$. The first conjunct is always true. To prove the second, take an arbitrary element and show that it belongs to A. This works because U contains all possible elements.

We will use both of these strategies in the next example.

Example. Let A be a set. We will demonstrate each of the following. (Watch for the use of the Disjunctive Syllogism!)

1. $A \cap \varnothing = \varnothing$

   Proof. Suppose $x \in A \cap \varnothing$. Then $x \in \varnothing$, which is impossible. Therefore, $A \cap \varnothing = \varnothing$. ■

2. $A \cup \varnothing = A$

   Proof. Certainly $A \subseteq A \cup \varnothing$, so we must show the opposite inclusion. Take $x \in A \cup \varnothing$. Since $x \notin \varnothing$, it must be the case that $x \in A$. Thus, $A \cup \varnothing \subseteq A$, and we have proven the equality. ■

3. $A \cap U = A$

   Proof. From Section 3.2, Exercise 5d, we know $A \cap U \subseteq A$. We are left with showing the other inclusion. Let $x \in A$. This means x must also belong to the universe. Hence, $x \in A \cap U$. ■

4. $A \cup U = U$

   Proof. Certainly $A \cup U \subseteq U$. Moreover, by Exercise 5c of Section 3.2 we have the other inclusion. Thus, $A \cup U = U$. ■

Example. If we want to prove that a set A is not equal to the empty set, we must show $\sim(\forall x)(x \notin A)$, but this is equivalent to

$$(\exists x)(x \in A).$$

For instance, let

$$A = \{x \in \mathbb{R} : x^2 + 6x + 5\}.$$

We know that A is nonempty since $-1 \in A$. (Also, $-5 \in A$, but we only need one element.)

Example. Let

$$A = \{(a, b) \in \mathbb{R} \times \mathbb{R} : a + b = 0\}$$

and

$$B = \{(0, b) : b \in \mathbb{R}\}.$$

These sets are not disjoint since $(0, 0)$ is an element of both A and B. However, $A \neq B$ because $(1, -1) \in A$ but $(1, -1) \notin B$.

Let us combine the two strategies to show a relationship between $\varnothing$ and U. (A Venn Diagram may help in understanding the proof.)

3.3.4. Theorem

For all sets A and B, the following are equivalent:

1. $A \subseteq B$
2. $\overline{A} \cup B = U$
3. $A \cap \overline{B} = \varnothing$

Proof. Let A and B be two sets.

(1 ⇒ 2) Assume $A \subseteq B$ and show $\overline{A} \cup B = U$. To prove that an arbitrary x is an element of $\overline{A} \cup B$, suppose $x \notin \overline{A}$. Then $x \in A$. Hence, $x \in B$, which gives us $x \in \overline{A} \cup B$.

(2 ⇒ 3) Suppose $\overline{A} \cup B = U$. In order to obtain a contradiction, take $x \in A \cap \overline{B}$. Then $x \in \overline{B}$. But also since $x \in A$, the supposition gives $x \in B$, a contradiction. Therefore, $A \cap \overline{B} = \varnothing$.

(3 ⇒ 1) Let $A \cap \overline{B} = \varnothing$. Assume $x \in A$. We want to show $x \in B$. By hypothesis, x cannot be a member of $\overline{B}$, otherwise the intersection would be nonempty. Hence, $x \in B$. ■

The following theorem is a generalization of the corresponding result concerning subsets. Although it would be legitimate to write the proof by using the Shorter Rule of Biconditional Proof or by appealing to Lemma 3.2.4 (see Exercise 12), it was written in this manner to make clear the role of the hypotheses. Notice the use of the word *similar* in the second part of the proof. This means that the logic in the second part is essentially the same to that of the first.

3.3.5. Theorem

Let A, B, and C be sets. If $A = B$ and $B = C$, then $A = C$.

Proof. Assume $A = B$ and $B = C$. This means $A \subseteq B$, $B \subseteq A$, $B \subseteq C$, and $C \subseteq B$. We must show that $A = C$:

($\subseteq$) Let $x \in A$. By hypothesis, x is then an element of B, which then means $x \in C$.

($\supseteq$) Similar. ■

The last result of the section involves the Cartesian product. The first part is illustrated in Figure 3.3.7. The sets B and C are illustrated along the vertical axis, and A is along the horizontal. The Cartesian products are represented as boxes. The other parts of the theorem can be similarly visualized.

3.3.6. Theorem

If A, B, C, and D are sets, then

1. $A \times (B \cap C) = (A \times B) \cap (A \times C)$,
2. $A \times (B \cup C) = (A \times B) \cup (A \times C)$,
3. $A \times (B \setminus C) = (A \times B) \setminus (A \times C)$, and
4. $(A \times B) \cap (C \times D) = (A \cap C) \times (B \cap D)$.

3.3.7. Figure

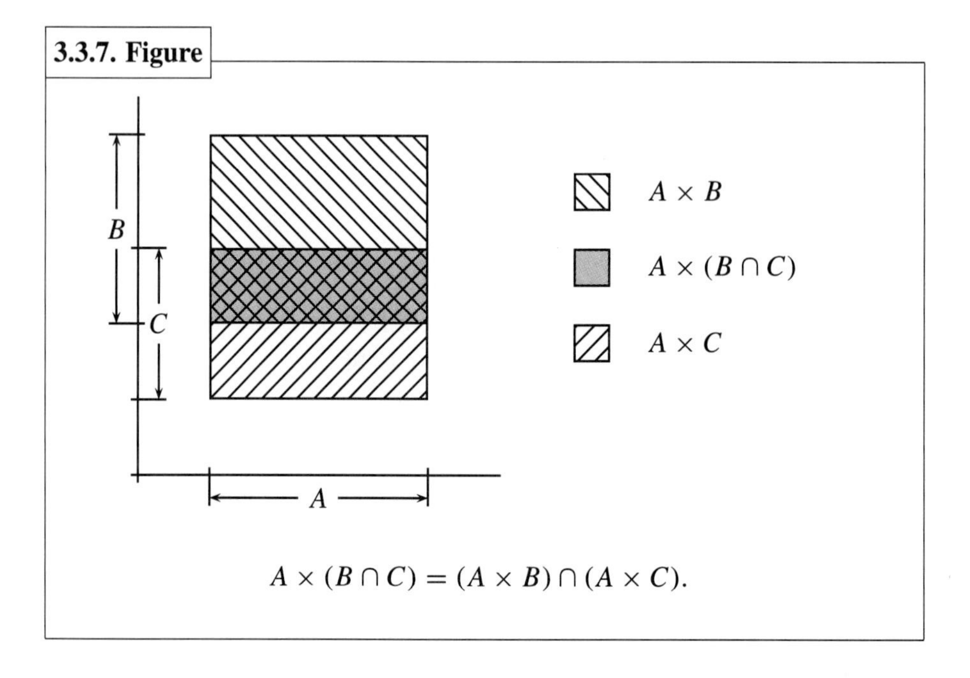

$$A \times (B \cap C) = (A \times B) \cap (A \times C).$$

Proof. We will prove the first one. The last three are left as exercises. Take three sets A, B, and C. Then

$$\begin{aligned}
&(a, b) \in A \times (B \cap C) \\
&\quad \Leftrightarrow a \in A \wedge b \in B \cap C \\
&\quad \Leftrightarrow a \in A \wedge b \in B \wedge a \in A \wedge b \in C \\
&\quad \Leftrightarrow (a, b) \in A \times B \wedge (a, b) \in A \times C \\
&\quad \Leftrightarrow (a, b) \in (A \times B) \cap (A \times C). \blacksquare
\end{aligned}$$

Inspired by the last result, we may try proving $(A \times B) \cup (C \times D)$ equals $(A \cup C) \times (B \cup D)$. No such proof exists. To show this, we must find an example with the property that one of the inclusions is false. Take $A = B = \{1\}$ and $C = D = \{2\}$. Then

$$(A \times B) \cup (C \times D) = \{(1, 1)\} \cup \{(2, 2)\} = \{(1, 1), (2, 2)\},$$

but

$$(A \cup C) \times (B \cup D) = \{1, 2\} \times \{1, 2\} = \{(1, 1), (1, 2), (2, 1), (2, 2)\}.$$

Hence, $(A \times B) \cup (C \times D) \not\supseteq (A \cup C) \times (B \cup D)$. Notice, however, that the opposite inclusion is always true. (See Exercise 3.2.10.)

EXERCISES

1. Indicate which of the following are legitimate expressions and which are not. For those that are not, explain the problem.

(a) $x \in \{1, 2\} \cap x \in \{2, 3\}$

(b) $x \in \{1, 2\} \cup \{2, 3\}$

(c) $x \in \{1, 2\} \wedge \{2, 3\}$

(d) $x \in \{1, 2\} \vee x \in \{2, 3\}$

(e) $\overline{x \in \{1, 2\}} \cap \overline{x \in \{2, 3\}}$

(f) $x \in \overline{\{1, 2\}} \wedge x \in \overline{\{2, 3\}}$

(g) $x \in \overline{\{1, 2\} \vee \{2, 3\}}$

(h) $x \in \overline{\{1, 2\} \cup \{2, 3\}}$

2. Supply the missing lines to each proof.

(a) $\overline{A} \cap B = B \setminus A$

Proof. For every x,

$$\begin{aligned} x \in \overline{A} \cap B &\Leftrightarrow \underline{\qquad\qquad} \\ &\Leftrightarrow x \notin A \text{ and } x \in B \\ &\Leftrightarrow \underline{\qquad\qquad}. \ \blacksquare \end{aligned}$$

(b) $A \setminus (B \setminus C) = A \cap (\overline{B} \cup C)$

Proof. Let x be an arbitrary element.

$$\begin{aligned} \underline{\qquad\qquad} &\Leftrightarrow x \in A \text{ and } x \notin B \setminus C \\ &\Leftrightarrow \underline{\qquad\qquad} \\ &\Leftrightarrow \underline{\qquad\qquad} \\ &\Leftrightarrow x \in A \text{ and } (x \notin B \text{ or } x \in C) \\ &\Leftrightarrow \underline{\qquad\qquad} \\ &\Leftrightarrow \underline{\qquad\qquad} \\ &\Leftrightarrow x \in A \cap (\overline{B} \cup C). \ \blacksquare \end{aligned}$$

(c) $A \subseteq B$ if and only if $A \cup B = B$.

Proof. Let A and B be two sets.

($\Rightarrow$) Assume $A \subseteq B$ and show ______.
It is clear that ______, so let $x \in A \cup B$.
This means ______, and we have two cases to examine.
(Case 1) If ______, then we are immediately done.
(Case 2) Now let ______. By hypothesis we have ______.
In either case, $x \in B$.

($\Leftarrow$) Suppose ______ and assume $x \in A$
From this we conclude ______, which by the supposition means ______. ■

3. Prove the unproven parts of Theorem 3.3.2.
4. Prove each equality.
   (a) $\overline{\varnothing} = U$
   (b) $\overline{U} = \varnothing$
   (c) $A \cap \overline{A} = \varnothing$
   (d) $A \cup \overline{A} = U$
   (e) $\overline{\overline{A}} = A$
   (f) $A \setminus A = \varnothing$
   (g) $A \setminus \varnothing = A$
   (h) $\overline{A \cup B} \cap B = \varnothing$
   (i) $A \setminus B = A \cap \overline{B}$
   (j) $A \cap (B \setminus A) = \varnothing$
5. Sketch a Venn Diagram for each problem and then write a proof. (Hint: for each proof work on the more complicated side first.)
   (a) $A = (A \cap B) \cup (A \cap \overline{B})$
   (b) $A \cup B = A \cup (\overline{A} \cap B)$
   (c) $(A \cap B) \cup (A \cap \overline{B}) \cup (\overline{A} \cap B) = A \cup B$
   (d) $A \setminus (A \cap B) = A \setminus B$
   (e) $(A \cup B) \setminus C = (A \setminus C) \cup (B \setminus C)$
   (f) $A \setminus (B \setminus C) = (A \setminus B) \cup (A \cap C)$
   (g) $(A \setminus B) \setminus C = A \setminus (B \cup C)$
6. Prove each of the following:
   (a) If $A \subseteq B$, then $A \setminus B = \varnothing$.
   (b) If $A \subseteq \varnothing$, then $A = \varnothing$.
   (c) If $U \subseteq A$, then $A = U$.
   (d) If $A \subseteq B$, then $B \setminus (B \setminus A) = A$.
   (e) $A \times B = \varnothing$ if and only if $A = \varnothing$ or $B = \varnothing$.
7. Let $a,\ c,\ m \in \mathbb{Z}$ and define two sets:

$$A = \{a + mk : k \in \mathbb{Z}\}$$

and

$$B = \{a + m(c + k) : k \in \mathbb{Z}\}.$$

Show $A = B$.
8. Define

$$A = \left\{\left[\begin{smallmatrix} a & 0 \\ 0 & b \end{smallmatrix}\right] : a + b = 0 \text{ and } a,\ b \in \mathbb{R}\right\}$$

and

$$B = \left\{\left[\begin{smallmatrix} a & b \\ c & d \end{smallmatrix}\right] : a = -d,\ b^2 + c^2 = 0, \text{ and } a,\ b,\ c,\ d \in \mathbb{R}\right\}.$$

Prove $A = B$.
9. Prove the following:
   (a) $\mathbb{Q} \neq \mathbb{Z}$
   (b) $\mathbb{C} \neq \mathbb{R}$
   (c) $\{0\} \times \mathbb{Z} \neq \mathbb{Z}$
   (d) $\mathbb{R} \times \mathbb{Z} \neq \mathbb{Z} \times \mathbb{R}$

(e) If A is the set of constant functions and B the set of linear functions, then $A \neq B$.

(f) If $A = \{ax^3 + b : a, b \in \mathbb{R}\}$ and $B = \{x^3 + b : b \in \mathbb{R}\}$, then $A \neq B$.

(g) If $A = \{ax^3 + b : a, b \in \mathbb{Z}\}$ and $B = \{ax^3 + b : a, b \in \mathbb{C}\}$, then $A \neq B$.

(h) If $A = \left\{\left[\begin{smallmatrix} a & 0 \\ 0 & b \end{smallmatrix}\right] : a, b \in \mathbb{R}\right\}$ and $B = \left\{\left[\begin{smallmatrix} a & b \\ c & d \end{smallmatrix}\right] : a, b, c, d \in \mathbb{R}\right\}$, then $A \neq B$.

10. For Parts 2 through 4 of Proposition 3.3.6.
    (a) Draw diagrams as in Figure 3.3.7.
    (b) Prove the results.
11. Is it possible for $A = \overline{A}$? Explain.
12. Prove Proposition 3.3.5:
    (a) using the Shorter Rule of Biconditional Proof.
    (b) by directly appealing to Theorem 3.2.4, part 3.
13. Prove that the following are equivalent.
    - $A \subseteq B$
    - $A \setminus B = \varnothing$
    - $A \cup B = B$
    - $A \cap B = A$
14. Prove that the following are equivalent.
    - $A \cap B = \varnothing$
    - $A \subseteq \overline{B}$
    - $A \setminus \overline{B} = \varnothing$
15. Find sets to illustrate the following:
    (a) $A \setminus B \neq B \setminus A$
    (b) $(A \times B) \times C \neq A \times (B \times C)$
    (c) $A \times B \neq B \times A$
16. Find an example of sets A, B, and C that shows that $A \cap B = A \cap C$ does not imply that $B = C$.
17. Is the implication

$$(A \cup B = A \cup C) \Rightarrow (B = C)$$

true or false for all sets A and B? Explain.

3.4. FAMILIES OF SETS

The elements of a set may be sets themselves. We call such a collection a ***family of sets*** and usually use capital script letters to name them. For example, let

$$\mathscr{E} = \{\{1, 2, 3\}, \{2, 3, 4\}, \{3, 4, 5\}\}.$$

The set $\mathscr{E}$ has three elements: $\{1, 2, 3\}$, $\{2, 3, 4\}$, and $\{3, 4, 5\}$.

Example. The following are true:

1. $\{1, 2, 3\} \in \{\{1, 2, 3\}, \{1, 4, 9\}\}$
2. $1 \notin \{\{1, 2, 3\}, \{1, 4, 9\}\}$
3. $\{1, 2, 3\} \nsubseteq \{\{1, 2, 3\}, \{1, 4, 9\}\}$
4. $\{\{1, 2, 3\}\} \subseteq \{\{1, 2, 3\}, \{1, 4, 9\}\}$

Example. The following are also true:

1. Since $\varnothing \in \{\varnothing, \{\varnothing\}\}$, $\{\varnothing\} \subseteq \{\varnothing, \{\varnothing\}\}$.
2. Since $\{\varnothing\} \in \{\varnothing, \{\varnothing\}\}$, $\{\{\varnothing\}\} \subseteq \{\varnothing, \{\varnothing\}\}$.

Families may have infinitely many elements. For example, let

$$\mathscr{F} = \{[n, n+1] : n \in \mathbb{Z}\}.$$

In roster notation,

$$\mathscr{F} = \{\ldots, [-2, -1], [-1, 0], [0, 1], [1, 2], \ldots\}.$$

Notice that in this case the set-builder notation is more convenient. For each integer n, the closed interval $[n, n+1]$ is in $\mathscr{F}$. The set $\mathbb{Z}$ plays the role of an ***index set***, a set whose only purpose is to enumerate the elements of the family. Each element of an index set is called an ***index***. If we let $I = \mathbb{Z}$ and $A_i = [i, i+1]$, then the family can be written as

$$\mathscr{F} = \{A_i : i \in I\}.$$

Example. To write the family $\mathscr{E}$ from page 123 using an index set, let $I = \{1, 2, 3\}$ and define

$$A_1 = \{1, 2, 3\},$$
$$A_2 = \{2, 3, 4\},$$
$$A_3 = \{3, 4, 5\}.$$

Then,

$$\mathscr{E} = \{A_i : i \in I\} = \{A_1, A_2, A_3\}.$$

This family is illustrated in Figure 3.4.1.

There is no reason why I must be $\{1, 2, 3\}$. Any three-element set will do. The order in which the sets are defined is also irrelevant. For instance, we could have defined $I = \{w, \pi, 99\}$ and

$$A_w = \{3, 4, 5\},$$
$$A_\pi = \{2, 3, 4\},$$
$$A_{99} = \{1, 2, 3\}.$$

3.4.1. Figure

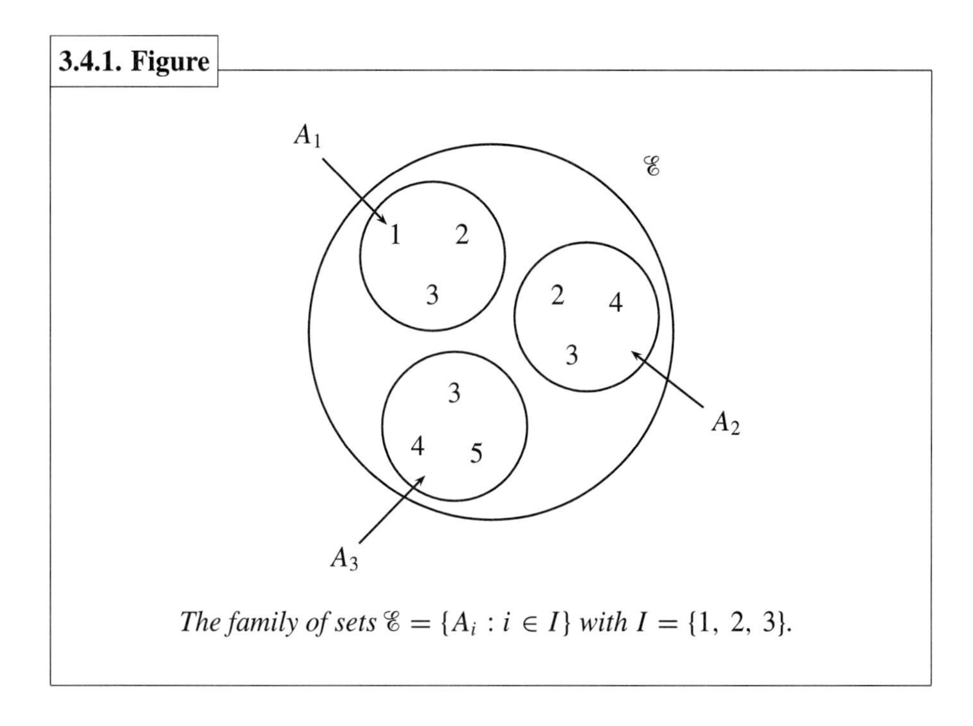

The family of sets $\mathscr{E} = \{A_i : i \in I\}$ *with* $I = \{1, 2, 3\}$.

The goal is to have each set in the family referenced or ***indexed*** by exactly one element of the index set. In this case, we still have $\mathscr{E} = \{A_i : i \in I\}$ with a similar diagram. See Figure 3.4.2.

Example. Let us rewrite $\mathscr{F}$ using $\mathbb{N}$ as the index set instead of $\mathbb{Z}$. We must be clever when using this index set. The strategy that we will follow is to use the even natural numbers to index those intervals that begin with a non-negative integer. The odd natural numbers will then index those intervals that begin with a negative. This is as follows:

$$\begin{gathered}
\vdots \\
B_5 = [-3, -2] \\
B_3 = [-2, -1] \\
B_1 = [-1, 0] \\
B_0 = [0, 1] \\
B_2 = [1, 2] \\
B_4 = [2, 3] \\
\vdots
\end{gathered}$$

3.4.2. Figure

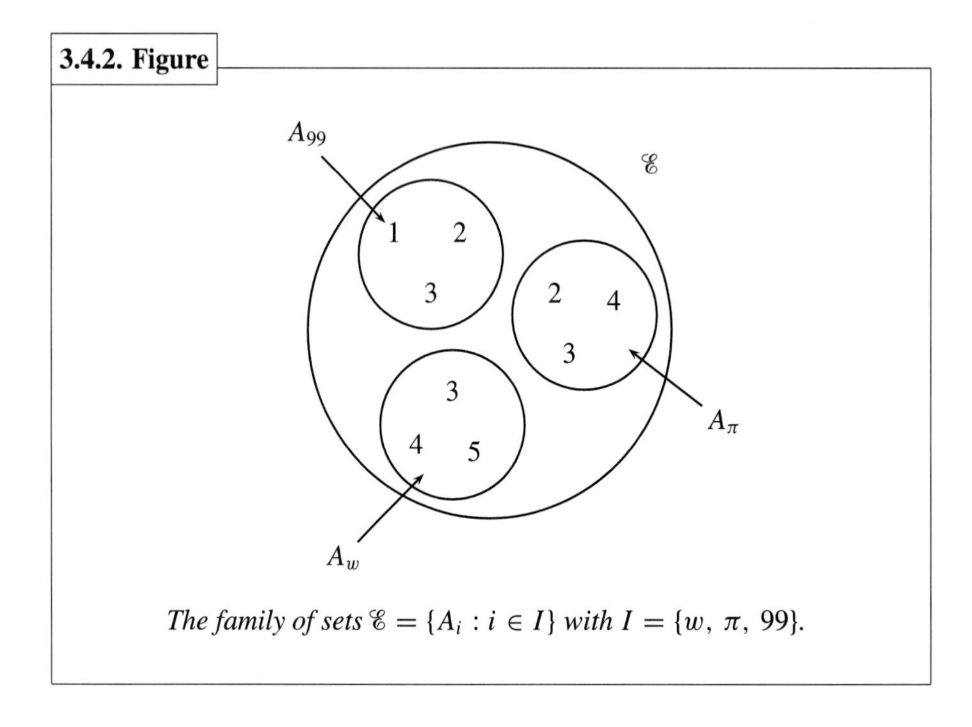

The family of sets $\mathscr{E} = \{A_i : i \in I\}$ *with* $I = \{w, \pi, 99\}$.

Use $2n + 1$ to represent the odd naturals and $2n$ to represent the evens ($n \in \mathbb{N}$). Then,

$$\begin{aligned} &\vdots \\ B_{2(2)+1} &= [-2 - 1, -2] \\ B_{2(1)+1} &= [-1 - 1, -1] \\ B_{2(0)+1} &= [-0 - 1, -0] \end{aligned}$$

and

$$\begin{aligned} B_{2(0)} &= [0, 0 + 1] \\ B_{2(1)} &= [1, 1 + 1] \\ B_{2(2)} &= [2, 2 + 1] \\ &\vdots \end{aligned}$$

Therefore, define for all natural numbers n,

$$B_{2n+1} = [-n - 1, -n]$$

and

$$B_{2n} = [n, n + 1].$$

Upon first glance it appears that each natural number is used twice, once for a "positive" interval and once for a "negative" one. This is not the case. Instead, we have indexed the elements of $\mathscr{F}$ as follows:

$$\begin{aligned} B_0 &= [0, 1] \\ B_1 &= [-1, 0] \\ B_2 &= [1, 2] \\ B_3 &= [-2, -1] \\ B_4 &= [2, 3] \\ B_5 &= [-3, -2] \\ &\vdots \end{aligned}$$

So under this definition, $\mathscr{F} = \{B_j : j \in \mathbb{N}\}$.

There is a natural way to form a family of sets. Take a set A. The collection of all subsets of A is called its ***power set***. It is represented by $\mathbf{P}(A)$. This set is defined using set-builder notation as follows.

3.4.3. Definition

For any set A:

$$\mathbf{P}(A) = \{B : B \subseteq A\}.$$

Since $\varnothing \subseteq A$ for any set A, $\varnothing \in \mathbf{P}(A)$. We see this in the next example.

Example. We will find the power set for each of the following:

1. $A = \{1, 2, 3\}$

$$\mathbf{P}(A) = \{\varnothing, \{1\}, \{2\}, \{3\}, \{1, 2\}, \{1, 3\}, \{2, 3\}, \{1, 2, 3\}\}$$

2. $B = \{\varnothing, \{\varnothing\}\}$

$$\mathbf{P}(B) = \{\varnothing, \{\varnothing\}, \{\{\varnothing\}\}, \{\varnothing, \{\varnothing\}\}\}$$

3. $C = \mathbb{N}$

$$\mathbf{P}(C) = \{\varnothing, \{0\}, \{1\}, \ldots, \{0, 1\}, \{0, 2\}, \ldots\}$$

The last theorem of the section needs the next lemma.

3.4.4. Lemma

If A and B are sets, then $A \subseteq B$ if and only if $\mathbf{P}(A) \subseteq \mathbf{P}(B)$.

Proof.

($\Rightarrow$) Let $A \subseteq B$. Assume $X \in \mathbf{P}(A)$. Then $X \subseteq A$, which gives $X \subseteq B$ by Theorem 3.2.4. Hence, $X \in \mathbf{P}(B)$. We have proven $\mathbf{P}(A) \subseteq \mathbf{P}(B)$.

($\Leftarrow$) Assume $\mathbf{P}(A) \subseteq \mathbf{P}(B)$. This time we will show that $A \subseteq B$. So let $x \in A$. In other words, $\{x\} \subseteq A$, but this means that $\{x\} \in \mathbf{P}(A)$. Hence, $\{x\} \in \mathbf{P}(B)$ by hypothesis, which yields $\{x\} \subseteq B$. Thus, $x \in B$. ■

Example. To see the lemma in action, let $A = \{2, 6\}$ and $B = \{2, 6, 10\}$. Then $A \subseteq B$. Examining the power sets of each, we find that

$$\mathbf{P}(A) = \{\varnothing, \{2\}, \{6\}, \{2, 6\}\}$$

and

$$\mathbf{P}(B) = \{\varnothing, \{2\}, \{6\}, \{10\}, \{2, 6\}, \{2, 10\}, \{6, 10\}, \{2, 6, 10\}\}.$$

Hence, $\mathbf{P}(A) \subseteq \mathbf{P}(B)$.

The definition of set equality and the lemma are used to prove the next theorem. Its proof is left as an exercise.

3.4.5. Theorem

For any sets A and B, $A = B$ if and only if $\mathbf{P}(A) = \mathbf{P}(B)$.

EXERCISES

1. Let $I = \{1, 2, 3, 4, 5\}$ and define $A_1 = \{1, 2\}$, $A_2 = \{3, 4\}$, $A_3 = \{1, 4\}$, $A_4 = \{3, 4\}$, and $A_5 = \{1, 3\}$. Write the following as a rosters.

(a) $\{A_i : i \in I\}$
(b) $\{A_i : i \in \{2, 4\}\}$
(c) $\{A_i : i = 1\}$
(d) $\{A_i : i = 1, 2\}$
(e) $\{A_i : i \in \varnothing\}$
(f) $\{A_i : i \in A_5\}$

2. Answer true or false for the following:

(a) $1 \in \{\{1\}, \{2\}, \{1, 2\}\}$
(b) $\{1\} \in \{\{1\}, \{2\}, \{1, 2\}\}$

(c) $\{1\} \subseteq \{\{1\}, \{2\}, \{1, 2\}\}$
(d) $\{1, 2\} \in \{\{\{1, 2\}, \{3, 4\}\}, \{1, 2\}\}$
(e) $\{1, 2\} \subseteq \{\{\{1, 2\}, \{3, 4\}\}, \{1, 2\}\}$
(f) $\{3, 4\} \in \{\{\{1, 2\}, \{3, 4\}\}, \{1, 2\}\}$
(g) $\{3, 4\} \subseteq \{\{\{1, 2\}, \{3, 4\}\}, \{1, 2\}\}$
(h) $\varnothing \in \{\{\{1, 2\}, \{3, 4\}\}, \{1, 2\}\}$
(i) $\varnothing \subseteq \{\{\{1, 2\}, \{3, 4\}\}, \{1, 2\}\}$
(j) $\{\varnothing\} \in \{\varnothing, \{\varnothing, \{\varnothing\}\}\}$
(k) $\{\varnothing\} \subseteq \{\varnothing, \{\varnothing, \{\varnothing\}\}\}$
(l) $\varnothing \in \{\varnothing, \{\varnothing, \{\varnothing\}\}\}$
(m) $\varnothing \subseteq \{\varnothing, \{\varnothing, \{\varnothing\}\}\}$
(n) $\{\varnothing\} \subseteq \varnothing$
(o) $\{\varnothing\} \subseteq \{\varnothing, \{\varnothing\}\}$
(p) $\{\varnothing\} \subseteq \{\{\varnothing, \{\varnothing\}\}\}$
(q) $\{\{\varnothing\}\} \subseteq \{\varnothing, \{\varnothing\}\}$
(r) $\{1\} \in \mathbf{P}(\mathbb{Z})$
(s) $\{1\} \subseteq \mathbf{P}(\mathbb{Z})$
(t) $\varnothing \in \mathbf{P}(\varnothing)$
(u) $\{\varnothing\} \in \mathbf{P}(\varnothing)$
(v) $\{\varnothing\} \subseteq \mathbf{P}(\varnothing)$

3. Given $\{A_n : n \in \mathbb{N}\}$ where $A_n = \{0, 1, 2, \dots, n\}$, find:
(a) A_3
(b) $A_5 \cup A_8$
(c) $A_2 \cap A_4$
(d) $\overline{A_3}$
(e) $A_0 \cup A_1 \cup A_2$
(f) $A_0 \cap A_1 \cap A_2$

4. Given $\{B_n : n \in \mathbb{Z}^+\}$ where $B_n = \{nk : k \in \mathbb{Z}\}$, find:
(a) B_3
(b) $B_5 \cup B_8$
(c) $B_2 \cap B_4$
(d) $\overline{B_3}$
(e) $B_0 \cup B_1 \cup B_2$
(f) $B_0 \cap B_1 \cap B_2$

5. Given $\{C_n : n \in \mathbb{N}\}$ where $C_n = \{1 + nk : k \in \mathbb{Z}^+\}$, find:
(a) C_3
(b) $C_5 \cup C_8$
(c) $C_2 \cap C_4$
(d) $\overline{C_3}$
(e) $C_0 \cup C_1 \cup C_2$
(f) $C_0 \cap C_1 \cap C_2$

6. Let $\mathscr{C} = \{A_n : n \in \mathbb{Z}^+\}$ be a family of sets. Write $\mathscr{C}$ as a family of sets $\{B_i : i \in I\}$ where I is:
(a) $\mathbb{N}$
(b) $\{n\pi : n \in \mathbb{Z}^+\}$
(c) $\{6, 7, 8, \dots\}$
(d) $\{\dots, -4, -3, -2\}$
(e) $\{a, aa, aaa, aaaa, \dots\}$
(f) $\mathbb{Z}$

7. Find sets that make the following implication false: *if* $I \cap J = \varnothing$, *then* $\{A_i : i \in I\}$ *and* $\{A_j : j \in J\}$ *are disjoint.*

8. Let $\{A_i : i \in K\}$ be a family of sets and let I and J be subsets of K. Define $\mathscr{E} = \{A_i : i \in I\}$ and $\mathscr{F} = \{A_j : j \in J\}$ and prove the following:
   (a) If $I \subseteq J$, then $\mathscr{E} \subseteq \mathscr{F}$.
   (b) $\mathscr{E} \cup \mathscr{F} = \{A_i : i \in I \cup J\}$
   (c) $\mathscr{E} \cap \mathscr{F} \supseteq \{A_i : i \in I \cap J\}$
   (d) $\mathscr{E} \setminus \mathscr{F} \subseteq \{A_i : i \in I \setminus J\}$
9. Using the same notation as in the previous problem, find a family of sets $\{A_i : i \in K\}$ and subsets I and J of K such that:
   (a) $\mathscr{E} \cap \mathscr{F} \not\subseteq \{A_i : i \in I \cap J\}$
   (b) $\mathscr{E} \setminus \mathscr{F} \not\supseteq \{A_i : i \in I \setminus J\}$
10. Show $\{A_i : i \in I\} = \varnothing$ if and only if $I = \varnothing$.
11. Let A be a finite set. How many elements are in $\mathbf{P}(A)$ if A has
   (a) 0 elements?
   (b) 1 element?
   (c) 2 elements?
   (d) 3 elements?
   (e) n elements? Explain.
12. Find the indicated power sets.
   (a) $\mathbf{P}(\{1, 2\})$
   (b) $\mathbf{P}(\mathbf{P}(\{1, 2\}))$
   (c) $\mathbf{P}(\varnothing)$
   (d) $\mathbf{P}(\mathbf{P}(\varnothing))$
   (e) $\mathbf{P}(\{\{\varnothing\}\})$
   (f) $\mathbf{P}(\{\varnothing, \{\varnothing\}, \{\varnothing, \{\varnothing\}\}\})$
13. Prove Theorem 3.4.5.
14. Prove for all sets A, $A \in \mathbf{P}(A)$.
15. For each of the following equalities, prove or show false. If one is false, prove any true inclusion.
   (a) $\mathbf{P}(A \cup B) = \mathbf{P}(A) \cup \mathbf{P}(B)$
   (b) $\mathbf{P}(A \cap B) = \mathbf{P}(A) \cap \mathbf{P}(B)$
   (c) $\mathbf{P}(A \setminus B) = \mathbf{P}(A) \setminus \mathbf{P}(B)$
   (d) $\mathbf{P}(A \times B) = \mathbf{P}(A) \times \mathbf{P}(B)$
16. Prove $\mathbf{P}(A) \subseteq \mathbf{P}(B)$ implies $A \subseteq B$ by using the fact that $A \in \mathbf{P}(A)$.

3.5. GENERALIZED UNION AND INTERSECTION

We will now generalize our set operations to be used with families. Let $\mathscr{F} = \{A_i : i \in I\}$ be a family of sets. Define the ***union*** of $\mathscr{F}$ to be the set of all elements that belong to at least one member of the family. Think of this as a generalized disjunction. This union is denoted by

$$\bigcup \mathscr{F} \text{ or } \bigcup_{i \in I} A_i$$

and can be defined using set-builder notation.

3.5.1. Definition

Let $\mathscr{F}$ be a family of sets.

$$\bigcup \mathscr{F} = \{x : x \in A \text{ for some } A \in \mathscr{F}\}.$$

If the family is indexed so that $\mathscr{F} = \{A_i : i \in I\}$, then we may use the alternate notation:

$$\bigcup \mathscr{F} = \bigcup_{i \in I} A_i = \{x : x \in A_i \text{ for some } i \in I\}.$$

The notation chosen depends on the specific circumstances. Typically if the focus is on the family, then then first form is used. If the members of the family are central to the problem, probably the second form will be used.

We generalize the notion of intersection similarly. The ***intersection*** of $\mathscr{F}$ is the set of all elements that belong to each member of the family. It is represented by

$$\bigcap \mathscr{F} \text{ or } \bigcap_{i \in I} A_i$$

and defined by:

3.5.2. Definition

Let $\mathscr{F}$ be a family of sets.

$$\bigcap \mathscr{F} = \{x : x \in A \text{ for every } A \in \mathscr{F}\}.$$

If the family is indexed so that $\mathscr{F} = \{A_i : i \in I\}$, then we may use the alternate notation:

$$\bigcap \mathscr{F} = \bigcap_{i \in I} A_i = \{x : x \in A_i \text{ for every } i \in I\}.$$

We can consider this as a generalized conjunction. Furthermore, notice that both definitions are indeed generalizations of the operations learned earlier, for

$$\bigcup \{A, B\} = A \cup B,$$

and

$$\bigcap \{A, B\} = A \cap B.$$

See Exercise 8.

We find the union and intersection of the families as before. The union of a family is the set obtained by combining all of the sets together. The intersection is determined by finding the elements that are common to all of the sets.

Example. Define $\mathscr{E} = \{[n, n+1] : n \in \mathbb{Z}\}$:

1. $\bigcup \mathscr{E} = \mathbb{R}$—When all of these intervals are combined, the result is the real line.
2. $\bigcap \mathscr{E} = \varnothing$—There is not one element that is common to all of the intervals. Hence, the intersection is empty.

The next example illustrates how to write the union or intersection of a family of sets as a roster.

Example. Let $\mathscr{F} = \{\{1, 2, 3\}, \{2, 3, 4\}, \{3, 4, 5\}\}$:

1. $\bigcup \mathscr{F} = \{1, 2, 3, 4, 5\}$—Since 1 is in the first set of $\mathscr{F}$, $1 \in \bigcup \mathscr{F}$. The others can be explained similarly. Notice that mechanically this amounts to removing the braces around the sets of the family and setting the union to the resulting set:

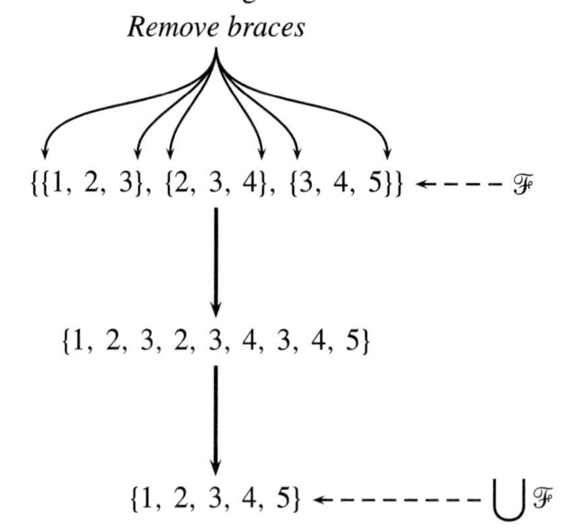

2. $\bigcap \mathscr{F} = \{3\}$—The generalized intersection is simply the overlap of all of the sets. Hence, 3 is the only element of $\bigcap \mathscr{F}$:

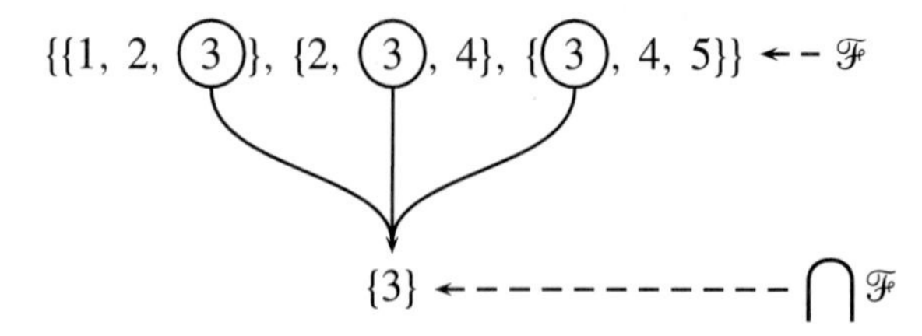

Since a family of sets may be empty, we must be able to take the union and intersection of the empty set.

Example. Let us show that $\bigcup \varnothing = \varnothing$. In order to obtain a contradiction, let $x \in \bigcup \varnothing$. This means there exists $A \in \varnothing$ such that $x \in A$, but this is impossible.

We leave the fact that $\bigcap \varnothing = U$ as an exercise.

The notation for indexed families of sets can be written using a style similar to summations in calculus. (See Appendix B.) For the following we will consider only intersection. A similar notation is also used for union. For example, if the index set is $\mathbb{N}$, then write:

$$\bigcap_{i \in \mathbb{N}} A_i = A_0 \cap A_1 \cap A_2 \cap \cdots = \bigcap_{i=0}^{\infty} A_i.$$

If $I = \{0, 1, 2, \ldots, k\}$, for some natural number k (in other words, the family is finite), then we may write

$$\bigcap_{i \in I} A_i = A_0 \cap A_1 \cap \cdots \cap A_k = \bigcap_{i=0}^{k} A_i.$$

Example. Let $\mathscr{F} = \{A_i : i \in \mathbb{Z}\}$.

1. $A_2 \cup A_3 \cup A_4 \cup A_5 = \bigcup_{i=2}^{5} A_i$
2. $A_7 \cap A_8 \cap A_9 \cap \cdots = \bigcap_{i=7}^{\infty} A_i$
3. $\ldots A_2 \cup A_3 \cup A_4 = \bigcup_{-\infty}^{4} A_i$
4. $A_6 \cap A_8 \cap A_{10} \cap \cdots = \bigcup_{i=0}^{\infty} A_{2i+6}$
5. $\bigcap \mathscr{F} = \bigcap_{-\infty}^{\infty} A_i$

The next theorem generalizes the Distributive Laws. Exercise 2.6.3 plays an important role in its proof. It allows us to move the quantifier.

3.5.3. Theorem

Let $\{A_i : i \in I\}$ be a family of sets. For any set B:

1. $B \cup \bigcap_{i \in I} A_i = \bigcap_{i \in I} (B \cup A_i)$
2. $B \cap \bigcup_{i \in I} A_i = \bigcup_{i \in I} (B \cap A_i)$

Proof. We leave the second part as an exercise. Let B be any set. Then

$$\begin{aligned}
x \in B \cup \bigcap_{i \in I} A_i &\Leftrightarrow x \in B \vee x \in \bigcap_{i \in I} A_i \\
&\Leftrightarrow x \in B \vee (\forall i \in I)(x \in A_i) \\
&\Leftrightarrow (\forall i \in I)(x \in B \vee x \in A_i) \\
&\Leftrightarrow (\forall i \in I)(x \in B \cup A_i) \\
&\Leftrightarrow x \in \bigcap_{i \in I} B \cup A_i. \blacksquare
\end{aligned}$$

To understand the next result, consider the following example. Choose the universe to be $\{1, 2, 3, 4, 5\}$ and perform some set operations on the family $\{\{1, 2, 3\}, \{2, 3, 4\}, \{2, 3, 5\}\}$:

$$\overline{\bigcap\{\{1, 2, 3\}, \{2, 3, 4\}, \{2, 3, 5\}\}} = \overline{\{2, 3\}} = \{1, 4, 5\},$$

and

$$\begin{aligned}
\bigcup\{\overline{\{1, 2, 3\}}, \overline{\{2, 3, 4\}}, \overline{\{2, 3, 5\}}\} &= \bigcup\{\{4, 5\}, \{1, 5\}, \{1, 4\}\} \\
&= \{1, 4, 5\}.
\end{aligned}$$

This leads us to the next generalization of De Morgan's Laws:

3.5.4. Theorem

Let $\{A_i : i \in I\}$ be a family of sets.

1. $\overline{\bigcap_{i \in I} A_i} = \bigcup_{i \in I} \overline{A_i}$
2. $\overline{\bigcup_{i \in I} A_i} = \bigcap_{i \in I} \overline{A_i}$

The proof will be left as an exercise. For a hint on how to proceed, look back to the proof at the top of page 115.

The next theorem is proven in Exercise 17 for arbitrary families.

3.5.5. Theorem

Let $\{A_i : i \in I\}$ be a family of sets and take $I \subseteq J$.

1. $\bigcup_{i \in I} A_i \subseteq \bigcup_{i \in J} A_i$
2. $\bigcap_{i \in J} A_i \subseteq \bigcap_{i \in I} A_i$

Proof. We will leave the second part as an exercise. To prove the first, let $x \in \bigcup_{i \in I} A_i$. This means $x \in A_j$ for some $j \in I$. But $I \subseteq J$, which means $j \in J$. Therefore, $x \in \bigcup_{i \in J} A_i$. ■

If $\mathcal{F}$ is a family of sets, then $\mathcal{F}$ is ***disjoint*** when $\bigcap \mathcal{F} = \varnothing$. A stronger condition is $\mathcal{F}$ being ***pairwise disjoint***. This means for all $A, B \in \mathcal{F}$,

$$\text{if } A \neq B, \text{ then } A \cap B = \varnothing.$$

Example. Let $\mathcal{F} = \{\{1, 2\}, \{3, 4\}, \{5, 6\}\}$. This set is both disjoint and pairwise disjoint. Its elements have no common members.

Example. Let A be any set. We see that

$$\begin{aligned} \bigcup \mathbf{P}(A) &= \{x : x \in B \text{ for some } B \in \mathbf{P}(A)\} \\ &= \{x : x \in B \text{ for some } B \subseteq A\} \\ &= A. \end{aligned}$$

However, $\mathbf{P}(A)$ is a disjoint family of sets because

$$\begin{aligned} \bigcap \mathbf{P}(A) &= \{x : x \in B \text{ for all } B \in \mathbf{P}(A)\} \\ &= \{x : x \in B \text{ for all } B \subseteq A\}. \end{aligned}$$

But $\varnothing \subseteq A$. Therefore, the predicate, if satisfied, would yield $x \in \varnothing$, a contradiction. Thus, $\bigcap \mathbf{P}(A) = \varnothing$.

If the family is indexed, we can use another test to determine if it is pairwise disjoint. Let $\mathcal{F} = \{A_i : i \in I\}$ be a family of sets. If for all $i, j \in I$,

$$(*) \qquad i \neq j \text{ implies } A_i \cap A_j = \varnothing,$$

then $\mathcal{F}$ is pairwise disjoint. To prove this, let $\mathcal{F}$ be a family that satisfies condition $(*)$. To prove that $\mathcal{F}$ is pairwise disjoint, take $A_i, A_j \in \mathcal{F}$ for some $i, j \in I$ and assume $A_i \neq A_j$. Therefore, $i \neq j$, for otherwise they would be the same. Hence, $A_i \cap A_j = \varnothing$ by $(*)$.

Example. The family $\mathcal{E} = \{B_1, B_2, B_3\}$ where

$$\begin{aligned} B_1 &= \{1, 2, 3\} \\ B_2 &= \{4, 5, 6\} \\ B_3 &= \{7, 8, 9\} \end{aligned}$$

is pairwise disjoint.

The next result illustrates the relationship between these two terms. One must be careful, though. The converse is false. (See Exercise 7.)

3.5.6. Theorem

Let $\mathcal{F}$ be a family of sets with at least two members. If $\mathcal{F}$ is pairwise disjoint, then it is disjoint.

Proof. Assume $\mathcal{F}$ is pairwise disjoint. To show that $\bigcap \mathcal{F}$ is empty, take $A,\ B \in \mathcal{F}$ such that $A \neq B$. Then by assumption, $A \cap B = \varnothing$. Using Exercise 16, $\bigcap \mathcal{F} \subseteq A \cap B$. Hence, $\mathcal{F}$ is disjoint since the only subset of the empty set is the empty set. ■

Our last example is an important one. Let $\mathcal{C}$ be a family of sets. We call $\mathcal{C}$ a ***chain*** if and only if for all $A,\ B \in \mathcal{C}$,

$$A \subseteq B \text{ or } B \subseteq A.$$

If the family can be written as $\mathcal{C} = \{C_i : i \in I\}$, then this condition can be written as

$$C_i \subseteq C_j \text{ or } C_j \subseteq C_i$$

for all $i,\ j \in I$.

Example. The following are examples of chains:

1. $\{\{1,\ 2\},\ \{1,\ 2,\ 3,\ 4\},\ \{1,\ 2,\ 3,\ 4,\ 5,\ 6\}\}$
2. $\{\varnothing,\ \{2,\ 4\},\ \{2,\ 4,\ 6\},\ \{2,\ 4,\ 6,\ 8\},\ \dots\}$
3. $\{A_k : k \in \mathbb{Z}^+\}$ where $A_k = \{x \in \mathbb{Z} : (x-1)(x-2)\cdots(x-k) = 0\}$

To see that the last example is a chain, take $m,\ n \in \mathbb{Z}^+$. By definition,

$$A_m = \{1,\ 2,\ \dots,\ m\}$$

and

$$A_n = \{1,\ 2,\ \dots,\ n\}.$$

If $m \leq n$, then $A_m \subseteq A_n$, otherwise $A_n \subseteq A_m$.

Example. Let $\mathcal{C}$ be the chain of intervals $\{(0,\ n) : n \in \mathbb{Z}^+\}$. Then

$$\bigcup \mathcal{C} = (0,\ \infty),$$

and

$$\bigcap \mathcal{C} = (0,\ 1).$$

EXERCISES

1. Write the following sets in roster form:

(a) $\bigcup\{\{1, 2\}, \{1, 2\}, \{1, 3\}, \{1, 4\}\}$

(b) $\bigcap\{\{1, 2\}, \{1, 2\}, \{1, 3\}, \{1, 4\}\}$

(c) $\bigcap \mathbf{P}(\varnothing)$

(d) $\bigcup \mathbf{P}(\varnothing)$

(e) $\bigcup\bigcup\{\{\{1\}\}, \{\{1, 2\}\}, \{\{1, 3\}\}, \{\{1, 4\}\}\}$

(f) $\bigcup\bigcap\{\{\{1\}\}, \{\{1, 2\}\}, \{\{1, 3\}\}, \{\{1, 4\}\}\}$

(g) $\bigcap\bigcup\{\{\{1\}\}, \{\{1, 2\}\}, \{\{1, 3\}\}, \{\{1, 4\}\}\}$

(h) $\bigcap\bigcap\{\{\{1\}\}, \{\{1, 2\}\}, \{\{1, 3\}\}, \{\{1, 4\}\}\}$

(i) $\bigcup\bigcup\varnothing$

(j) $\bigcap\bigcup\varnothing$

2. Given $\{A_i : i \in \mathbb{N}\}$ where $A_i = \{0, 1, 2, \ldots, i\}$, find:

(a) $\bigcup_{i=0}^{3} A_i$ (c) $\bigcup_{i=0}^{\infty} A_i$ (e) $\bigcup_{i=0}^{3} \mathbb{N} \setminus A_i$

(b) $\bigcap_{i=0}^{3} A_i$ (d) $\bigcap_{i=0}^{\infty} A_i$ (f) $\bigcap_{i=0}^{3} \mathbb{N} \setminus A_i$

3. Given $\{B_i : i \in \mathbb{Z}^+\}$ where $B_i = \{ik : k \in \mathbb{Z}\}$, find:

(a) $\bigcup_{i=1}^{3} B_i$ (c) $\bigcup_{i=1}^{\infty} B_i$ (e) $\bigcup_{i=1}^{3} \mathbb{Z} \setminus B_i$

(b) $\bigcap_{i=1}^{3} B_i$ (d) $\bigcap_{i=1}^{\infty} B_i$ (f) $\bigcap_{i=1}^{3} \mathbb{Z} \setminus B_i$

4. Given $\{C_i : i \in \mathbb{N}\}$ where $C_i = \{1 + ik : k \in \mathbb{Z}^+\}$, find:

(a) $\bigcup_{i=0}^{3} C_i$ (c) $\bigcup_{i=0}^{\infty} C_i$ (e) $\bigcup_{i=0}^{3} \mathbb{Z}^+ \setminus C_i$

(b) $\bigcap_{i=0}^{3} C_i$ (d) $\bigcap_{i=0}^{\infty} C_i$ (f) $\bigcap_{i=0}^{3} \mathbb{Z}^+ \setminus C_i$

5. Let $\{D_i : i \in \mathbb{N}\}$ be a family of sets and $k, \ell \in \mathbb{N}$. Assuming $k \leq \ell$, prove the following:

(a) $\bigcup_{i=0}^{k+1} D_i = \bigcup_{i=0}^{k} D_i \cup D_{k+1}$ **(d)** $\bigcap_{i=0}^{\ell} D_i \subseteq \bigcap_{i=0}^{k} D_i$

(b) $\bigcap_{i=0}^{k+1} D_i = \bigcap_{i=0}^{k} D_i \cap D_{k+1}$ (e) $\bigcup_{i=0}^{k} D_i \subseteq \bigcup_{i=0}^{\infty} D_i$

(c) $\bigcup_{i=0}^{k} D_i \subseteq \bigcup_{i=0}^{\ell} D_i$ (f) $\bigcap_{i=0}^{\infty} D_i \subseteq \bigcap_{i=0}^{k} D_i$

6. Draw a Venn Diagram for:
   (a) a disjoint family of sets that is not pairwise disjoint.
   (b) a pairwise disjoint family of sets.
7. Show by example that a disjoint family of sets may not be pairwise disjoint.
8. Prove for any sets A and B:
   (a) $\bigcup\{A, B\} = A \cup B$
   (b) $\bigcap\{A, B\} = A \cap B$
9. Prove Theorem 3.5.6 indirectly.
10. Let $\{A_i : i \in I\}$ be a family of sets. Demonstrate:
   (a) If $B \subseteq A_i$ for some $i \in I$, then $B \subseteq \bigcup_{i \in I} A_i$.
   (b) If $B \subseteq \bigcap_{i \in I} A_i$, then $B \subseteq A_i$ for all $i \in I$.
11. Prove the second parts of Theorem 3.5.3 and Theorem 3.5.5.
12. Prove Theorem 3.5.4.
13. Show $\bigcap \varnothing = U$ where U is the universe.
14. Let $\mathscr{F}$ be a family of sets such that $\varnothing \in \mathscr{F}$. Prove $\bigcap \mathscr{F} = \varnothing$.
15. Find families of sets $\mathscr{E}$ and $\mathscr{F}$ so that $\bigcup \mathscr{E} = \bigcup \mathscr{F}$ but $\mathscr{E} \neq \mathscr{F}$. Can this be repeated by replacing union with intersection?
16. Let $\mathscr{F}$ be a family of sets and let $A \in \mathscr{F}$. Prove $\bigcap \mathscr{F} \subseteq A \subseteq \bigcup \mathscr{F}$.
17. Let $\mathscr{E}$ and $\mathscr{F}$ be families of sets. Show the following:
   (a) $\bigcup\{\mathscr{F}\} = \mathscr{F}$
   (b) $\bigcap\{\mathscr{F}\} = \mathscr{F}$
   (c) If $\mathscr{E} \subseteq \mathscr{F}$, then $\bigcup \mathscr{E} \subseteq \bigcup \mathscr{F}$.
   (d) If $\mathscr{E} \subseteq \mathscr{F}$, then $\bigcap \mathscr{F} \subseteq \bigcap \mathscr{E}$.
18. Suppose that $\mathscr{E}$ and $\mathscr{F}$ are families of sets. Prove:
   (a) $\bigcup(\mathscr{E} \cup \mathscr{F}) = \bigcup \mathscr{E} \cup \bigcup \mathscr{F}$ (b) $\bigcap(\mathscr{E} \cup \mathscr{F}) = \bigcap \mathscr{E} \cup \bigcap \mathscr{F}$
19. Find families of sets $\mathscr{E}$ and $\mathscr{F}$ that make the following false:
   (a) $\bigcap(\mathscr{E} \cap \mathscr{F}) = \bigcap \mathscr{E} \cap \bigcap \mathscr{F}$ (b) $\bigcup(\mathscr{E} \cap \mathscr{F}) = \bigcup \mathscr{E} \cap \bigcup \mathscr{F}$
20. Let $\mathscr{F}$ be a family of sets. For each of the following equalities, prove that it is true in general or show that it is false by finding a counter-example. If one is false, identify when it is true.
   (a) $\bigcup \mathbf{P}(\mathscr{F}) = \mathscr{F}$ (c) $\mathbf{P}(\bigcup \mathscr{F}) = \mathscr{F}$
   (b) $\bigcap \mathbf{P}(\mathscr{F}) = \mathscr{F}$ (d) $\mathbf{P}(\bigcap \mathscr{F}) = \mathscr{F}$
21. Prove that the following are chains:
   (a) $\{[0, n] : n \in \mathbb{Z}^+\}$
   (b) $\{(2^n)\mathbb{Z} : n \in \mathbb{N}\}$ where $(2^n)\mathbb{Z} = \{2^n \cdot k : k \in \mathbb{Z}\}$
   (c) $\{B_n : n \in \mathbb{N}\}$ where $B_n = \bigcup_{i=0}^{n} A_i$ and A_i is a set for all $i \in \mathbb{N}$
22. Can a chain be disjoint or pairwise disjoint? Explain.

CHAPTER EXERCISES

1. By definition, $1/2$, $13/5$, $2 = 2/1$, and $4\frac{3}{7} = 31/7$ are all rational. Since $.23457 = 23457/100000$, $.23457 \in \mathbb{Q}$. Indeed, any terminating decimal is rational. Any repeating decimal is also rational. For example, to see $.\overline{47} \in \mathbb{Q}$, let $n = .\overline{47}$. Then $100n = 47.\overline{47}$. Subtracting these equations yields $99n = 47$. Hence, $n = 47/99 \in \mathbb{Q}$. Use this to show:
   (a) $.76352 \in \mathbb{Q}$
   (b) $.\overline{3} \in \mathbb{Q}$
   (c) $13.3\overline{97} \in \mathbb{Q}$
   (d) $.453\overline{23} \in \mathbb{Q}$
2. Explain why $.345934599345999\ldots \notin \mathbb{Q}$.
3. Let A and B be sets. The ***symmetric difference*** of A and B is denoted by $A \,\Delta\, B$ and defined as

$$A \Delta B = (A \setminus B) \cup (B \setminus A).$$

   Find the following sets:
   (a) $\{1, 2, 3, \ldots, 10\} \,\Delta\, \{2, 4, 6, 8, 10\}$
   (b) $\{1, 2, 3\} \,\Delta\, \{1, 2, 3\}$
   (c) $[0, 1] \,\Delta\, (0, 1)$
   (d) $[0, 1] \,\Delta\, [2, 3]$
   (e) $[0, 2] \,\Delta\, [1, 3]$
4. Let U be the universe. Prove for all sets A, B, and C:
   (a) $A \,\Delta\, \varnothing = A$
   (b) $A \,\Delta\, U = \overline{A}$
   (c) $A \,\Delta\, B = B \,\Delta\, A$
   (d) $A \,\Delta\, (B \,\Delta\, C) = (A \,\Delta\, B) \,\Delta\, C$
   (e) $A \Delta B = (A \cup B) \setminus (A \cap B)$
   (f) $(A \,\Delta\, B) \cap C = (A \cap C) \,\Delta\, (B \cap C)$
   (g) $A \,\Delta\, B = \varnothing$ if and only if $A = B$
   (h) If A and B are disjoint, then $A \Delta B = A \cup B$
5. For the following, prove true or show false.
   (a) $(A \,\Delta\, B) \cup C = (A \cup C) \,\Delta\, (B \cup C)$.
   (b) $\overline{A \,\Delta\, B} = \overline{A} \,\Delta\, \overline{B}$
6. Prove these alternate forms of De Morgan's Laws.
   (a) $A \setminus \bigcup_{i \in I} B_i = \bigcap_{i \in I} A \setminus B_i$
   (b) $A \setminus \bigcap_{i \in I} B_i = \bigcup_{i \in I} A \setminus B_i$
7. Prove for all sets A, B, and C:
   (a) If $A \cup B \subseteq A \cap B$, then $A = B$.
   (b) If $A \cap B = A \cap C$ and $A \cup B = A \cup C$, then $B = C$.
   (c) $A \subseteq C$ and $B \subseteq C$ if and only if $A \cup B \subseteq C$.
   (d) $A \subseteq B$ and $A \subseteq C$ if and only if $A \subseteq B \cap C$.

8. Prove the following:
   (a) If $A \cup B \neq \varnothing$, then $A \neq \varnothing$ or $B \neq \varnothing$.
   (b) If $A \cap B \neq \varnothing$, then $A \neq \varnothing$ and $B \neq \varnothing$.
9. When does $A \times B = C \times D$ imply that $A = C$ and $B = D$?
10. Prove that there is a cancellation law with the Cartesian product. Namely, if $A \neq \varnothing$, then $A \times B = A \times C$ implies $B = C$.
11. Given a family of sets $\{A_i : i \in I\}$, there exists another family $\mathcal{B} = \{B_i : i \in I\}$ such that $\{A_i \times B_i : i \in I\}$ is pairwise disjoint. Find $\mathcal{B}$. Is $\mathcal{B}$ unique?
12. Sometimes we need two index sets to describe a family. Let $I_n = [n, n+1]$ and $J_n = [n, n+1]$, $n \in \mathbb{Z}$. Define

$$\mathcal{F} = \{I_n \times J_m : n, m \in \mathbb{Z}\}.$$

   Now show $\bigcup \mathcal{F} = \mathbb{R}^2$ and $\bigcap \mathcal{F} = \varnothing$. Can $\mathcal{F}$ be proven to be pairwise disjoint?
13. Find a family of subsets of $\mathbb{R}^2$ whose union is the following grid:

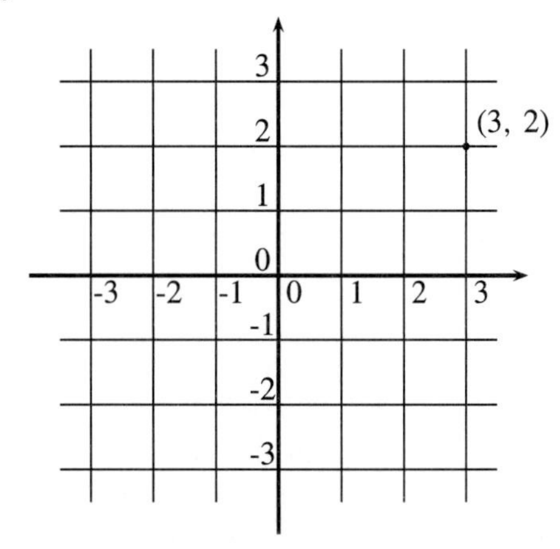

14. Assume that the universe U contains only sets such that for all $A \in U$, $A \subseteq U$ and $\mathbf{P}(A) \in U$. Prove the following:
   (a) $\bigcup U = U$
   (b) $\bigcap U = \varnothing$
15. Let $S \subseteq \mathbb{R}$. S is said to be ***closed*** under addition if and only if $x, y \in S$ implies $x + y \in S$. (Warning: this definition does not say that x must be different from y!)
   (a) Prove that if the sets A and B are closed under addition, then $A \cap B$ is closed under addition.
   (b) Show that if the sets A and B are closed under addition, then $A \cup B$ may not be closed under addition.

(c) Let $\mathscr{C} = \{A_n : n \in \mathbb{N}\}$ be a chain of sets. Prove that if A_n is closed under addition for each $n \in \mathbb{N}$, then $\bigcup \mathscr{C}$ is closed under addition.

16. The ordered pair (x, y) can be defined to be the set $\{\{x\}, \{x, y\}\}$. Answer the following:
    (a) Find $(1, 2) \cup (2, 1)$.
    (b) Find $(1, 2) \cap (2, 1)$.
    (c) Find $(1, 2) \setminus (2, 1)$.
    (d) With this definition, prove that $(x, y) = (x', y')$ if and only if $x = x'$ and $y = y'$.
17. Suppose that $\{A_i : i \in \mathbb{N}\}$ is a chain of sets such that for all $i \leq j$, $A_i \subseteq A_j$. Prove for all natural numbers k:
    (a) $\bigcup_{i=0}^{k} A_i = A_k$ (b) $\bigcap_{i=0}^{k} A_i = A_0$
18. Let $\{A_n : n \in \mathbb{N}\}$ be a family of sets. For every natural number m, define $B_m = \bigcup_{i=0}^{m} A_i$. Show that $\{B_m : m \in \mathbb{N}\}$ is a chain.
19. Let $\mathscr{C}_i$ be a chain for all $i \in I$. Prove that $\bigcap_{i \in I} \mathscr{C}_i$ is a chain. Is $\bigcup_{i \in I} \mathscr{C}_i$ necessarily a chain? Explain.
20. Let $P(x)$ be a predicate. Prove the following:
    (a) $(\forall x \in A \cup B)P(x) \Rightarrow (\forall x \in A \cap B)P(x)$
    (b) $(\forall x \in A \cup B)P(x) \Leftrightarrow (\forall x \in A)P(x) \wedge (\forall x \in B)P(x)$
    (c) $(\exists x \in A \cap B)P(x) \Rightarrow (\exists x \in A)P(x) \wedge (\exists x \in B)P(x)$
    (d) $(\exists x \in A \cup B)P(x) \Leftrightarrow (\exists x \in A)P(x) \vee (\exists x \in B)P(x)$
21. Find a predicate $P(x)$ to show that the following are false:
    (a) $[(\forall x \in A)P(x) \vee (\forall x \in B)P(x)] \Rightarrow (\forall x \in A \cup B)P(x)$
    (b) $[(\exists x \in A)P(x) \wedge (\exists x \in B)P(x)] \Rightarrow (\exists x \in A \cap B)P(x)$
22. Can any predicate be used to describe a set with set-builder notation? (Hint: Consider $A = \{x : x \notin x\}$. Is $A \in A$? This "set" was introduced by Bertrand Russell.*)

*Bertrand Russell (Trelleck, Wales, 1872 – Penrhyndeudraeth, Wales, 1970): Russell contributed much to mathematical logic. He and Alfred North Whitehead (Ramsgate, England, 1861 – Cambridge, Massachusetts, 1947) authored the *Principia Mathematica* in 1913. The goal of this work was to show that mathematics and logic are identical.

4 Mathematical Induction

Suppose that we have a pile of dominoes on a table. We carefully stand them on their ends. Each one is placed within a half inch of the next one. Perhaps we have even lined them up to make a design. When all of the dominoes are in place, we tip over the first one. What happens? They all fall. With only two steps—lining up the dominoes and tipping over the first one—we have caused all to come crashing down. Keep this picture in mind. It will serve as the intuition for this chapter because we will be studying the topic of mathematical induction.

4.1. THE FIRST PRINCIPLE

Let $S = \{n \in \mathbb{Z}^+ : P(n)\}$ for some predicate $P(n)$. Suppose we want to prove $S = \mathbb{Z}^+$. Since S is already a subset of $\mathbb{Z}^+$, we need to show $\mathbb{Z}^+ \subseteq S$. Our standard method is to take an arbitrary positive integer n and show $n \in S$. If this is not possible, we may be tempted to try proving $P(n)$ for each $n \in \mathbb{Z}^+$ individually. This is impossible because it would take infinitely many steps. Therefore, we should follow the strategy of the dominoes. First prove $1 \in S$. Then show that $n \in S$ implies $n + 1 \in S$ for every positive integer n. This means that by *Modus Ponens*, $P(2)$ is true, then $P(3)$ is true, and so forth. Each falling like dominoes:

$$\begin{array}{llll} P(1) & & & \\ P(1) \Rightarrow P(2) \quad \therefore P(2) & & & \\ & P(2) \Rightarrow P(3) \quad \therefore P(3) & & \\ & & P(3) \Rightarrow P(4) \quad \therefore P(4) \ldots & \end{array}$$

This method is called the ***First Principle of Mathematical Induction***. It is denoted by (MI).

4.1.1. Axiom (MI)

Let $S \subseteq \mathbb{Z}^+$. If

1. $1 \in S$ and
2. for all $n \in \mathbb{Z}^+$, $n \in S$ implies $n + 1 \in S$,

then $S = \mathbb{Z}^+$.

It is important to note that (MI) is an axiom. Think a moment about how we would try to prove this. We would have to assume the two hypotheses and show $S = \mathbb{Z}^+$, but this brings us back to our original problem. Furthermore, we cannot use (MI) to prove (MI). That would be arguing in circles! Now, there are ways to prove it by choosing a different axiom, but many of these proofs use propositions that are logically equivalent to (MI), and others are too complicated for our purposes here. Therefore, we will choose to take (MI) as an axiom, for we are confident that it is true.

Example. Let us show for all positive integers n,

$$\sum_{i=1}^{n} i^2 = 1^2 + 2^2 + \cdots + n^2 = \frac{n(n+1)(2n+1)}{6}.$$

Proceed by induction on n. Let

$$S = \{n \in \mathbb{Z}^+ : 1^2 + 2^2 + \cdots + n^2 = n(n+1)(2n+1)/6\}$$

and show $S = \mathbb{Z}^+$. To do this, satisfy the hypotheses of (MI).

1. First show $1 \in S$. This is clear because

$$1(1+1)(2 \cdot 1 + 1)/6 = 1(2)(3)/6 = 1.$$

2. Now for the implication. Assume that $n \in S$. This means

$$(*) \qquad 1^2 + 2^2 + \cdots + n^2 = \frac{n(n+1)(2n+1)}{6}.$$

We must show $n + 1 \in S$. In other words, we must show

$$1^2 + 2^2 + \cdots + n^2 + (n+1)^2 = \frac{(n+1)([n+1]+1)(2[n+1]+1)}{6}.$$

Adding $(n+1)^2$ to both sides of equation $(*)$ gives

$$1^2 + 2^2 + \cdots + n^2 + (n+1)^2 = \frac{n(n+1)(2n+1)}{6} + (n+1)^2$$

$$= \frac{(n+1)[n(2n+1)+6(n+1)]}{6}$$
$$= \frac{(n+1)(2n^2+7n+6)}{6}$$
$$= \frac{(n+1)(n+2)(2n+3)}{6}$$
$$= \frac{(n+1)([n+1]+1)(2[n+1]+1)}{6}.$$

Therefore, by (MI) we have proved our result.

This structure is typical for proofs using mathematical induction. Indicate that mathematical induction will be used and identify the variable. (Use a phrase like *by induction on n*.) Once the set S is defined, induction becomes a two step process:

1. First show that $1 \in S$. This is called the ***basis case***. It is typically the easiest part of the proof but should, nonetheless, be explicitly shown.
2. Next prove the implication $n \in S \Rightarrow n+1 \in S$. For this use Direct Proof. Assume $n \in S$ and show $n+1 \in S$. The assumption is called the ***induction hypothesis***. In the last example, equation $(*)$ is the induction hypothesis.

(The diagram in Figure 4.1.3 represents this process.) Once both steps are completed, a concluding remark can be made such as *we are done by induction*, although these are often omitted.

Since the main work of the proof focused on the predicate, we will often skip the explicit definition of the set S and focus on the predicate. This is the purpose of the next theorem. It can also be referred to as "induction."

4.1.2. Theorem

Let $P(n)$ be a predicate. If

1. $P(1)$ and
2. $P(n) \Rightarrow P(n+1)$ for all $n \in \mathbb{Z}^+$,

then $P(n)$ is true for all $n \in \mathbb{Z}^+$.

Proof. Suppose $P(1)$ and $P(n) \Rightarrow P(n+1)$ for all $n \in \mathbb{Z}^+$. Define S to be the set $\{n \in \mathbb{Z}^+ : P(n)\}$. We will use (MI) to show $S = \mathbb{Z}^+$.

1. 1 is an element of S since $P(1)$ is true.
2. Assume $n \in S$. Then we have $P(n)$. Therefore, by the assumption, $P(n+1)$ is true, which means that $n+1 \in S$.

By induction $S = \mathbb{Z}^+$. Hence, $P(n)$ is true for all $n \in \mathbb{Z}^+$. ■

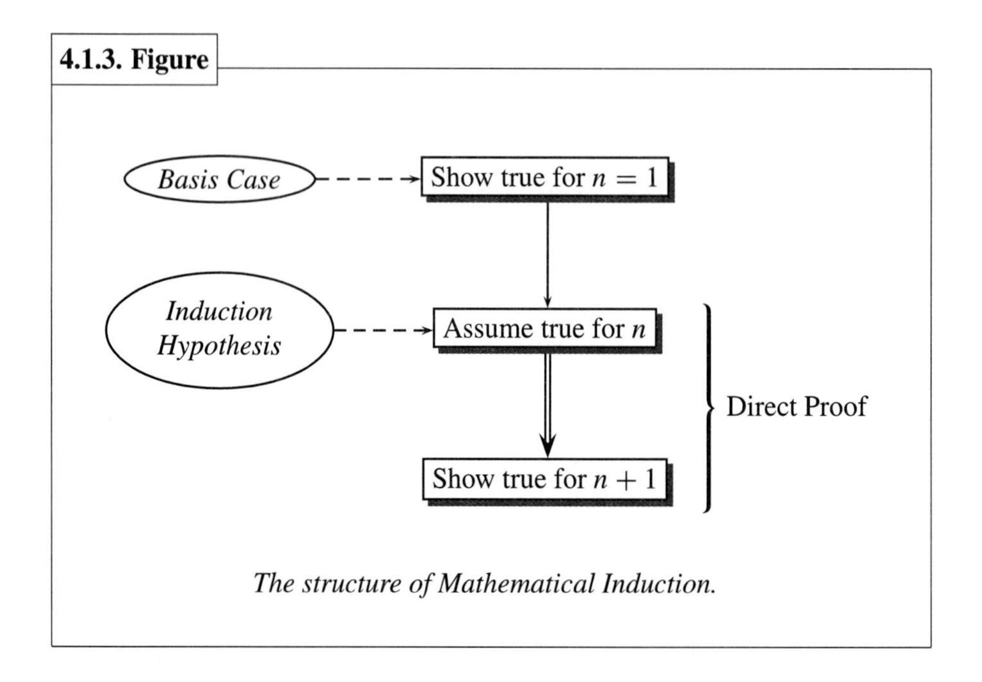

The structure of Mathematical Induction.

Example. Let $x \geq 0$. Prove that for any positive integer n,

$$(x+1)^n \geq x^n + 1.$$

Proof. We will proceed by induction on n.

1. For the basis case we have

$$(x+1)^1 = x + 1 = x^1 + 1.$$

2. Now let $n \in \mathbb{Z}^+$ and assume that $(x+1)^n \geq x^n + 1$. We need to show $(x+1)^{n+1} \geq x^{n+1} + 1$. By multiplying both sides of our given inequality by $x + 1$ we have

$$\begin{aligned}(x+1)^n(x+1) &\geq (x^n+1)(x+1)\\ &= x^{n+1} + x^n + x + 1\\ &\geq x^{n+1} + 1.\end{aligned}$$

The first inequality is true by the induction hypothesis and since $x + 1$ is positive, and the last one holds because $x \geq 0$. ■

Thus far our examples of mathematical induction have been problems that have been given to us. In practice we may need to find a formula and then use mathematical induction to prove that it works. For example, suppose we want to find an equation for the sum

$$1 \cdot 2 + 2 \cdot 3 + \cdots + n(n+1)$$

where $n \in \mathbb{Z}^+$. Can we find its formula using only n and constants? The best way to start is to try some examples:

$$\begin{gathered}1 \cdot 2 = 2 \\ 1 \cdot 2 + 2 \cdot 3 = 8 \\ 1 \cdot 2 + 2 \cdot 3 + 3 \cdot 4 = 20 \\ 1 \cdot 2 + 2 \cdot 3 + 3 \cdot 4 + 4 \cdot 5 = 40 \\ 1 \cdot 2 + 2 \cdot 3 + 3 \cdot 4 + 4 \cdot 5 + 5 \cdot 6 = 70\end{gathered}$$

Upon examining the pattern and working some equations, we note that if the last term of the nth series, $n(n+1)$, is multiplied by the last number of the next series, $n+2$, and then divided by three, the sum is obtained. We conjecture that the sum is given by the formula

$$\sum_{i=1}^{n} i(i+1) = \frac{n(n+1)(n+2)}{3}.$$

Since we can only check finitely many of the equations, let us prove that the formula works by induction:

1. The basis case is clear since $1 \cdot (1+1) \cdot (1+2)/3 = 6/3 = 2 = 1(1+1)$.
2. Let $n \in \mathbb{Z}^+$ and assume

   $$1 \cdot 2 + 2 \cdot 3 + \cdots + n(n+1) = \frac{n(n+1)(n+2)}{3}.$$

   We must show

   $$1 \cdot 2 + 2 \cdot 3 + \cdots + (n+1)(n+2) = \frac{(n+1)(n+2)(n+3)}{3}.$$

   By adding $(n+1)(n+2)$ to both sides of the induction hypothesis we have:

   $$\begin{aligned}1 \cdot 2 + \cdots + n(n+1) &+ (n+1)(n+2) \\ &= \frac{n(n+1)(n+2)}{3} + (n+1)(n+2) \\ &= \frac{n(n+1)(n+2)}{3} + \frac{3(n+1)(n+2)}{3} \\ &= \frac{n(n+1)(n+2) + 3(n+1)(n+2)}{3} \\ &= \frac{(n+1)(n+2)(n+3)}{3}.\end{aligned}$$

Therefore, we have proven the formula by induction.

Think back to the picture of the dominoes at the beginning of the chapter. What would happen if they were all lined up, but instead of tipping over the first, another was toppled? Every domino past that one would fall. This is the idea behind the following theorem. It is a revised version of (MI). Instead of beginning at $n = 1$, an arbitrary integer N is fixed as the starting point for the induction. Then, if the induction step is as before, the result will be proven for all integers greater than or equal to N:

$$\begin{array}{lll} P(N) & & \\ P(N) \Rightarrow P(N+1) & \therefore P(N+1) & \\ & P(N+1) \Rightarrow P(N+2) & \therefore P(N+2)\dots \end{array}$$

The result is as follows:

4.1.4. Theorem

Let $P(n)$ be a predicate. For any integer N, $P(n)$ is true for all integers $n \geq N$ if

1. $P(N)$ is true, and
2. $P(n)$ implies $P(n+1)$ for all integers $n \geq N$.

Proof. Assume that $P(N)$ is true and $P(n)$ implies $P(n+1)$ for all integers $n \geq N$. Define a new predicate

$$Q(n) := P(N+n-1).$$

Notice that $Q(1) \equiv P(N)$, $Q(2) \equiv P(N+1)$, and so on. We now use induction on n to prove $Q(n)$ for all positive integers n.

1. By hypothesis, $P(N)$ is true. Therefore, $Q(1)$ is also true.
2. Assume $Q(n)$. We must show $Q(n+1)$. But $Q(n) \equiv P(N+n-1)$. So by hypothesis, $P(N+n) \equiv Q(n+1)$ is true.

Therefore, by (MI) we are done. Hence, by definition of $Q(n)$ we have $P(n)$ for all integers $n \geq N$. ■

The strategy behind the theorem is to shift the values from N, $N+1$, $N+2$, ... to 1, 2, 3, ... so that our original axiom can be used. For instance, if $N = 3$, we have:

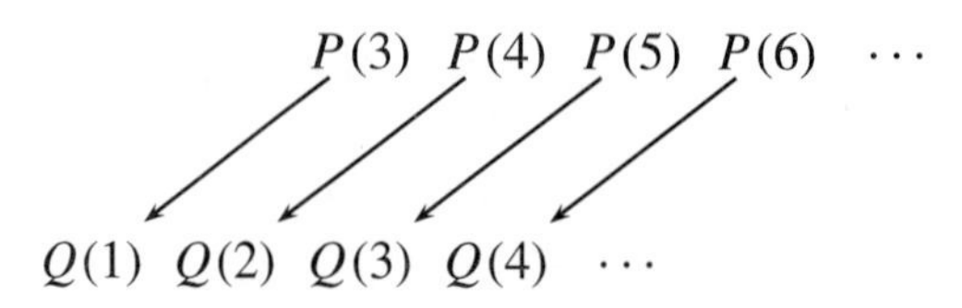

Theorem 4.1.4 is applied in the same way as (MI). The only difference is the starting point. For this reason we will also refer to this theorem as (MI).

Example. Prove $n^3 < n!$ for all integers $n \geq 6$. (Recall that the ***factorial*** is defined as

$$n! = n(n-1)(n-2)\cdots 2 \cdot 1,$$

and $0! = 1$. We will use this definition extensively in the next section.)

Proof. By induction on $n \geq 6$.

1. We must first show that the inequality holds for $n = 6$:

$$6^3 = 216 < 720 = 6!.$$

2. Assume $n^3 < n!$ and show $(n+1)^3 < (n+1)!$. We note that since $n \geq 6$, the induction hypothesis yields the following three inequalities:
   - $3n^2 < n \cdot n^2 = n^3 \leq n!$,
   - $3n < n \cdot n = n^2 < n^3 \leq n!$, and
   - $1 < n!$.

   Therefore we have:

$$\begin{aligned} (n+1)^3 &= n^3 + 3n^2 + 3n + 1 \\ &< n! + n! + n! + n! \\ &= 4n! \\ &< (n+1)n! \\ &= (n+1)!. \end{aligned}$$

By (MI) we are done. ■

Finally, we state a result of Theorem 4.1.4 that looks at the dominoes in a slightly different fashion. The dominoes are set up so that two successive dominoes are close enough. We could view them as one knocking down the next. Another possibility is that we see each as knocked down by its predecessor. This is the case in the corollary. We leave the proof as an exercise.

4.1.5. Corollary

Fix an integer N and let $P(n)$ be a predicate. $P(n)$ is true for all integers $n > N$ if

1. $P(N)$ is true, and
2. $P(n-1)$ implies $P(n)$ for all integers $n > N$.

EXERCISES

1. For each of the following, identify $P(1)$, and $P(n+1)$.
   (a) $P(n) := 1 + 2 \cdot 2 + 3 \cdot 2^2 + \cdots + n2^{n-1}$
   (b) $P(n) := 1 \cdot 1! + 2 \cdot 2! + \cdots + n \cdot n! = (n+1)! - 1$
   (c) $P(n) := \sum_{i=1}^{n} \frac{1}{i^2} \leq 2 - \frac{1}{n}$
   (d) $P(n) := n! < n^n$.
   (e) $P(n) :=$ *If A has n elements, then* $\mathbf{P}(A)$ *has* 2^n *elements.*
   (f) $P(n) :=$ *Every set of size n contains a.*
2. Use mathematical induction to prove for all $n \in \mathbb{Z}^+$:
   (a) $1 + 2 + 3 + \cdots + n = \frac{n(n+1)}{2}$
   (b) $1 + 3 + 5 + \cdots + (2n-1) = n^2$
   (c) $1^2 + 3^2 + 5^2 + \cdots + (2n-1)^2 = \frac{n(2n-1)(2n+1)}{3}$
   (d) $1^3 + 2^3 + 3^3 + \cdots + n^3 = \left[\frac{n(n+1)}{2}\right]^2$
   (e) $1 + r + r^2 + \cdots + r^n = \frac{1 - r^{n+1}}{1 - r}$ $(r \neq 1)$
   (f) $1 \cdot 1! + 2 \cdot 2! + \cdots + n \cdot n! = (n+1)! - 1$
   (g) $\frac{1}{2!} + \frac{2}{3!} + \cdots + \frac{n}{(n+1)!} = 1 - \frac{1}{(n+1)!}$
   (h) $2 \cdot 6 \cdot 10 \cdot 14 \cdot \cdots \cdot (4n-2) = \frac{(2n)!}{n!}$
3. Let $n \in \mathbb{Z}$. Find formulas for the following expressions and prove the conjecture using mathematical induction.
   (a) $(1 - \frac{1}{2}) \cdot (1 - \frac{1}{3}) \cdots (1 - \frac{1}{n})$ for $n > 1$
   (b) $1 + 2 + 2^2 + \cdots + 2^n$ for $n \geq 1$
   (c) $1 + 2 \cdot 2 + 3 \cdot 2^2 + \cdots + n2^{n-1}$ for $n \geq 1$
4. Prove that the following inequalities are true for all $n \in \mathbb{Z}^+$.
   (a) $n < 2^n$
   (b) $n! < n^n$.
   (c) $\sum_{i=1}^{n} \frac{1}{i^2} \leq 2 - \frac{1}{n}$
   (d) $\frac{1}{2} + \frac{2}{2^2} + \frac{3}{2^3} + \cdots + \frac{n}{2^n} \leq 2 - \frac{n}{2^n}$

5. Prove using mathematical induction:

$$\sum_{i=1}^{n} \frac{1}{(2i-1)(2i+1)} = \frac{n}{2n+1}$$

6. Let n be an integer. Prove:
   (a) $n^2 < 2^n$ for all $n \geq 5$.
   (b) $n^3 < 2^n$ for all $n \geq 10$.
   (c) $2^n < n!$ for all $n \geq 4$.
   (d) $n^2 < n!$ for all $n \geq 4$.
7. Prove that for every positive integer n, $(2n)! < 2^{2n}(n!)^2$.
8. Prove Corollary 4.1.5.
9. Sometimes the basis case is so trivial it seems unimportant to state. However, as we know, it is crucial to the argument as the following demonstrates: Let $P(n) := n = n + 1$.
   (a) Prove $P(n) \Rightarrow P(n+1)$ for all $n \in \mathbb{Z}^+$.
   (b) Prove that $(\forall n \in \mathbb{Z}^+)P(n)$ is false.
   (c) If $P(1)$ is true, then by mathematical induction $(\forall n \in \mathbb{Z}^+)P(n)$ must also be true. Therefore, $P(1)$ must be false. Show this.

   (Moral of the story: never forget to show the basis step!)

4.2. COMBINATORICS

A plentiful source of problems that involve mathematical induction is the field of combinatorics. ***Combinatorics*** is the study of the properties that a set has based only on its number of elements. This obviously includes the size of a set, known as its ***cardinality***. We will examine three formulas that are basic to finding cardinalities of finite sets.

Given a set of objects, a ***permutation*** of that set is simply a rearrangement of the objects. The number of permutations of the collection $\{a, b, c, d, e, f\}$ is 720. If we were to write all of the permutations in a list, it would look like the following:

$$\begin{array}{cccccc} a & b & c & d & e & f \\ a & b & c & d & f & e \\ a & b & c & e & d & f \\ & & & \vdots & & \\ f & e & d & c & a & b \\ f & e & d & c & b & a \end{array}$$

The number of permutations is calculated using the factorial. The number of permutations of $\{a, b, c, d, e, f\}$ is $6! = 720$. In general, the number of permutations of an n-element set is $n!$.

There are a number of ways to prove this fact about the factorial. We will use induction on n:

1. There is only one way to write the elements of one-element set. Since $1! = 1$, we have proven the basis case.
2. Assume that the number of permutations of an n-element set is $n!$. Using this we will show that the number of permutations of an $(n+1)$-element set is $(n+1)!$. Let $A = \{a_1, a_2, \ldots, a_{n+1}\}$ be an $(n+1)$-element set. By induction, there are $n!$ permutations of the set $\{a_1, a_2, \ldots, a_n\}$. After writing the permutations in a list, notice that there are $n + 1$ columns before, between, and after each element of the permutations:

$$\begin{array}{|c|c||c||c||c||c|c|} & a_1 & & a_2 & & \cdots & & a_n & \\ & a_1 & & a_2 & & \cdots & & a_{n-1} & \\ & \vdots & & \vdots & & & & \vdots & \\ & a_n & & a_{n-1} & & \cdots & & a_1 & \end{array}$$

   To form the permutations of A, place a_{n+1} into all of the positions of each column. For example, if a_{n+1} is put into the first column, we form the following permutations:

$$\begin{array}{ccccc} a_{n+1} & a_1 & a_2 & \cdots & a_n \\ a_{n+1} & a_1 & a_2 & \cdots & a_{n-1} \\ \vdots & \vdots & \vdots & & \vdots \\ a_{n+1} & a_n & a_{n-1} & \cdots & a_1 \end{array}$$

   Since there are $n!$ rows with $n + 1$ ways to add a_{n+1} to each row, we have $(n + 1)n! = (n + 1)!$ permuations of A.

For our second equation, suppose that we do not want to rearrange the entire set but only subsets of it. For example, let $S = \{a, b, c, d, e\}$. To see all of the 3-element permutations of S, look at the list:

$$\begin{array}{cccccc} abc & acb & bac & bca & cab & cba \\ abd & adb & bad & bda & dab & dba \\ abe & aeb & bae & bea & eab & eba \\ acd & adc & cad & cda & dac & dca \\ ace & aec & cae & cea & eac & eca \\ ade & aed & dae & dea & ead & eda \\ bcd & bdc & cbd & cdb & dbc & dcb \\ bce & bec & ceb & cbe & ebc & ecb \\ bde & bed & dbe & deb & ebd & edb \\ cde & ced & dce & dec & ecd & edc \end{array}$$

There are 60 arrangements because there are 5 choices for the first entry. Once that is chosen there are only 4 left for the second, and then 3 for the last. So we calculate:

$$60 = 5 \cdot 4 \cdot 3 = \frac{5 \cdot 4 \cdot 3 \cdot 2 \cdot 1}{2 \cdot 1} = \frac{5!}{(5-3)!}.$$

Generalizing, the number of r-element permutations of an n-element set is

$${}_nP_r = \frac{n!}{(n-r)!}.$$

The third formula counts subsets instead of rearrangements. Let n and r be natural numbers with $n \geq r$. Define

$$\binom{n}{r} = \frac{n!}{r!(n-r)!}.$$

This number is called a ***binomial coefficient*** and is read as "n choose r." (Its name will be justified in a moment.) It counts the number of r-element subsets of an n-element set. For example, the set $\{a, b, c, d, e\}$ has ten 3-element subsets. They are

$$\begin{array}{ccccc} \{a, b, c\} & \{a, b, d\} & \{a, b, e\} & \{a, c, d\} & \{a, c, e\} \\ \{a, d, e\} & \{b, c, d\} & \{b, c, e\} & \{b, d, e\} & \{c, d, e\} \end{array}$$

This number could have been calculated using the binomial coefficient:

$$\binom{5}{3} = \frac{5!}{3!2!} = \frac{5 \cdot 4 \cdot 3 \cdot 2 \cdot 1}{3 \cdot 2 \cdot 1 \cdot 2 \cdot 1} = 10.$$

To see why this formula holds, consider the next grid.

$$\underbrace{\left.\begin{array}{cccccc} abc & acb & bac & bca & cab & cba \\ abd & adb & bad & bda & dab & dba \\ abe & aeb & bae & bea & eab & eba \\ acd & adc & cad & cda & dac & dca \\ ace & aec & cae & cea & eac & eca \\ ade & aed & dae & dea & ead & eda \\ bcd & bdc & cbd & cdb & dbc & dcb \\ bce & bec & ceb & cbe & ebc & ecb \\ bde & bed & dbe & deb & ebd & edb \\ cde & ced & dce & dec & ecd & edc \end{array}\right\}}_{3! \text{ columns}} \text{10 rows}$$

We know that there are ${}_5P_3$ 3-element permutations of this set. They are found as the entries in the grid. However, since we are looking at subsets, we do not want to count abc as different from acb because $\{a, b, c\} = \{a, c, b\}$.

For this reason, all elements in any given row of the grid are considered as one combination. Each row has $6 = 3!$ entries because that is the number of permutations of a 3-element set. Hence, multiplying the number of rows by the number of columns gives

$${}_5P_3 = 10(3!).$$

But by definition, $\binom{5}{3}$ is the number of rows. Therefore,

$$\binom{5}{3} = \frac{{}_5P_3}{3!} = \frac{5!}{3!(5-3)!} = \frac{5!}{3!2!} = 10.$$

A generalization of this argument will justify the formula for an arbitrary binomial coefficient (see Exercise 7).

There are many interesting and involved problems that arise when studying combinatorics. It is not our intention here to study them in great detail. Instead, we will use this definition as a source of problems to which we can apply mathematical induction.

Before we can begin, we will need the following equation originally due to Pascal*: Let $n, r \in \mathbb{Z}^+$ so that $n \geq r$. Then

$$\binom{n}{r-1} + \binom{n}{r} = \binom{n+1}{r}.$$

This equation is called ***Pascal's Identity***. Simply add the fractions to prove it:

$$\begin{aligned}
\binom{n}{r-1} + \binom{n}{r} &= \frac{n!}{(r-1)!(n-r+1)!} + \frac{n!}{r!(n-r)!} \\
&= \frac{rn!}{r(r-1)!(n-r+1)!} + \frac{(n-r+1)n!}{(n-r+1)r!(n-r)!} \\
&= \frac{(n+1)n!}{r!(n-r+1)!} \\
&= \frac{(n+1)!}{r!(n+1-r)!} \\
&= \binom{n+1}{r}.
\end{aligned}$$

There is an easy way to calculate binomial coefficients by using the triangle found in Figure 4.2.1. It is known as ***Pascal's Triangle*** and is formed by first

*Blaise Pascal (Clermont-Ferrand, France 1623 – Paris, France 1662): Pascal worked in physics and mathematics and contributed to the early work of probability. This study included the triangle which now bears his name (although it was earlier known to the Chinese). Later in his life, a religious experience prompted Pascal to completely devote his studies to Christianity. After his death, fragments of these writings were collected into a compilation known as the *Pensées* (the thoughts).

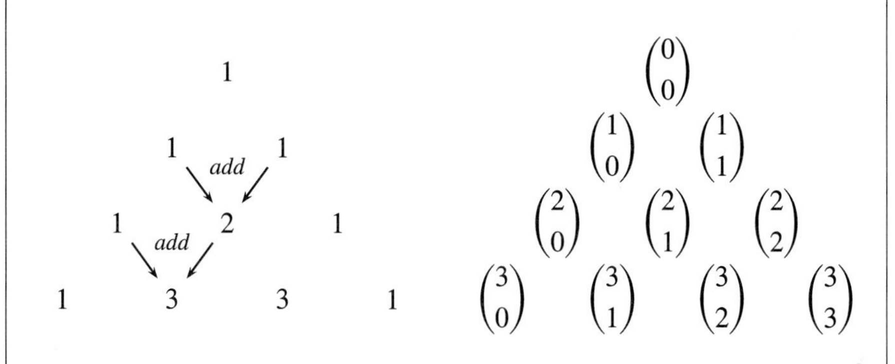

The first four rows of Pascal's Triangle.

putting ones down the two sides. Each remaining number is the sum of the two integers above it in the triangle. Finally, let the vertex of the triangle be row 0, entry 0. Because of Pascal's Identity and the way the triangle is formed, $\binom{n}{r}$ is the number in row n, entry r. Hence, we can use the triangle to calculate binomial coefficients. For example, examine the triangle to see that $\binom{3}{1} = 3$.

This next theorem will make clear why the binomial coefficient received its name.

4.2.2. Binomial Theorem

Let n be a positive integer. Then

$$(x+y)^n = \sum_{r=0}^{n} \binom{n}{r} x^{n-r} y^r.$$

Proof. By induction on n:

1. Since $\binom{1}{0} = \binom{1}{1} = 1$,

$$(x+y)^1 = \binom{1}{0}x + \binom{1}{1}y = \sum_{r=0}^{1} \binom{1}{r} x^{1-r} y^r.$$

Therefore, the basis case is true.

2. Assume

$$(x+y)^n = \sum_{r=0}^{n} \binom{n}{r} x^{n-r} y^r$$

for a positive integer n and show the $n+1$ case. To do this, first factor out $(x+y)^n$ so that we can use the induction hypothesis:

$$(x+y)^{n+1} = (x+y)(x+y)^n = (x+y)\sum_{r=0}^{n}\binom{n}{r}x^{n-r}y^r.$$

Multiplying the $(x+y)$ term through yields

$$\sum_{r=0}^{n}\binom{n}{r}x^{n-r+1}y^r + \sum_{r=0}^{n}\binom{n}{r}x^{n-r}y^{r+1}.$$

Taking out the $n+1$ terms and shifting the index on the second summation gives

$$x^{n+1} + \sum_{r=1}^{n}\binom{n}{r}x^{n-r+1}y^r + \sum_{r=1}^{n}\binom{n}{r-1}x^{n-r+1}y^r + y^{n+1},$$

which using Pascal's Identity equals

$$x^{n+1} + \sum_{r=1}^{n}\binom{n+1}{r}x^{n-r+1}y^r + y^{n+1},$$

and this is

$$\sum_{i=0}^{n+1}\binom{n+1}{r}x^{n+1-r}y^r. \blacksquare$$

Example. We can use the binomial theorem to expand $(x+1)^3$:

$$\begin{aligned}(x+1)^3 &= \sum_{r=0}^{3}\binom{3}{r}x^{3-r}1^r \\ &= \binom{3}{0}x^3 + \binom{3}{1}x^2 + \binom{3}{2}x + \binom{3}{3} \\ &= x^3 + 3x^2 + 3x^2 + 1.\end{aligned}$$

EXERCISES

1. Find the eighth row of Pascal's Triangle.

2. Show for all integers $n \geq 0$:

(a) $\binom{n}{0} = \binom{n}{n} = 1.$

(b) $\binom{n}{r} = \binom{n}{n-r}.$

3. Use the Binomial Theorem to expand each of the following:
   (a) $(x+y)^7$
   (c) $(x-1)^5$
   (b) $(x-y)^7$
   (d) $(1-x)^5$
4. Let $n \in \mathbb{N}$. Prove the following equations using the Binomial Theorem:
   (a) $\binom{n}{0} + \binom{n}{1} + \cdots + \binom{n}{n} = 2^n.$
   (b) $\binom{n}{0} - \binom{n}{1} + \cdots \pm \binom{n}{n} = 0$
5. Let n and r be positive integers and $n \geq r$. Use induction to show the following:
   (a) $\binom{r}{r} + \binom{r+1}{r} + \cdots + \binom{n}{r} = \binom{n+1}{r+1}$
   (b) $1^2 + 3^2 + 5^2 + \cdots + (2n-1)^2 = \binom{2n+1}{3}$
6. Prove for all integers $n \geq 2$:
   (a) $\sum_{r=1}^{n} r\binom{n}{r} = n2^{n-1}$
   (b) $\sum_{r=1}^{n} (-1)^{r-1} r\binom{n}{r} = 0$
7. Prove
   $$\binom{n}{r} = \frac{n!}{r!(n-r)!}$$
   by generalizing the justification that $\binom{5}{3} = 10$ as given in the section.
8. Let $n, r \in \mathbb{Z}^+$ such that $n > r$. Prove by induction:
   $$\frac{\binom{n}{r}}{\binom{n}{r+1}} = \frac{r+1}{n-r}.$$

4.3. THE SECOND PRINCIPLE

If there is a first principle of induction, we would expect a second. It goes like this. Remember the domino picture that we used to explain how mathematical induction works. The first domino is tipped causing the second to fall, which in turn causes the third to fall. By the time the sequence of falls reaches domino $n+1$, n dominoes have fallen. This means that propositions $P(1)$ through $P(n)$ have been proven true. It is at this point that $P(n+1)$ is proven. This is the intuition behind for the ***Second Principle of Mathematical Induction***, which we will denote by (MI2). This theorem is sometimes called ***strong induction***.

4.3.1. Theorem (MI2)

For any predicate $P(n)$, $(\forall n \in \mathbb{Z}^+)P(n)$ is true if

1. $P(1)$ is true, and
2. for all positive integers n, if $P(1) \wedge P(2) \wedge \cdots \wedge P(n)$, then $P(n+1)$.

Proof. Assume $P(1)$ and

$$[P(1) \wedge P(2) \wedge \cdots \wedge P(n)] \Rightarrow P(n+1)$$

for all $n \in \mathbb{Z}^+$. We will use the set-theoretic version of (MI). Define

$$S = \{n \in \mathbb{Z}^+ : P(m) \text{ for all positive integers } m \leq n\}.$$

Then we proceed with the induction.

1. Since $P(1)$ holds, $1 \in S$.
2. Assume $n \in S$. By definition of S, $P(1)$ through $P(n)$ are true. Thus, $P(n+1)$ is true by hypothesis. This means that $n+1 \in S$.

Therefore by (MI), $S = \mathbb{Z}^+$, which means $P(n)$ holds for all positive integers n. ■

The following illustrates strong induction:

$$\begin{array}{lll} P(1) & & \\ P(1) \Rightarrow P(2) & \therefore P(1) \wedge P(2) & \\ & P(1) \wedge P(2) \Rightarrow P(3) & \therefore P(1) \wedge P(2) \wedge P(3) \ldots \end{array}$$

Example. Define a sequence of numbers:

$$a_1 = 3$$
$$a_n = 2a_{n-1} \text{ for all } n \in \mathbb{Z}^+.$$

Any sequence in which each term is defined based on previous terms is said to be defined ***recursively*** or ***inductively***. Now, find a formula for the nth term of this sequence. As before, look at some examples:

$$a_1 = 3,\ a_2 = 6,\ a_3 = 12,\ a_4 = 24,\ a_5 = 48,\ a_6 = 96,\ \ldots$$

Based on this, conjecture that $a_n = 3 \cdot 2^{n-1}$ for all positive integers n. We will prove this by using Corollary 4.1.5.

Proof. By induction on n:

1. For the basis case, $a_1 = 3 \cdot 2^0 = 3$.

2. Let $n > 1$ and assume $a_{n-1} = 3 \cdot 2^{n-2}$. To show the nth term, calculate

$$a_n = 2a_{n-1} = 2 \cdot 3 \cdot 2^{n-2} = 3 \cdot 2^{n-1}.$$

We are done by induction. ■

Strong induction is used when at least two previous cases are needed to prove the $n + 1$ case. Many good applications of this come from a study of the ***Fibonacci sequence***. In one of his works, Fibonacci* asked a question about how a certain population of rabbits can increase. The rules that govern the population are:

1. no rabbits die;
2. the population starts with a pair of rabbits that are at least two months old; and
3. each pair that is at least two months old will bear a new pair.

We will call a pair that has just been born a baby pair while the others are adult pairs. The population then grows according to the following table:

Month	Adult Pairs	Baby Pairs
1	1	1
2	2	1
3	3	2
4	5	3
5	8	5
6	13	8

It appears that the number of baby pairs at month n is given by the sequence:

$$1, 1, 2, 3, 5, 8, 13, 21, 34, \ldots$$

This sequence now bears his name, and each number is called a ***Fibonacci number***. Let F_n denote the nth Fibonacci number. So,

$$F_1 = 1, F_2 = 1, F_3 = 2, F_4 = 3, \ldots$$

The sequence can be defined recursively as

$$F_1 = 1, F_2 = 2, \text{ and } F_n = F_{n-1} + F_{n-2}.$$

Since we have only checked a few terms, we have not proven that F_n is equal to the number of baby pairs in the nth month. To show this, we need to use strong induction. Since the recursive definition starts with two basis cases, the basis

*Leonardo Fibonacci (Pisa, Italy, c. 1170 – Pisa, Italy, c. 1240) Fibonacci worked in algebra and number theory. He was responsible for introducing Arabic numerals to the West replacing the Roman system.

case for the induction will prove that the formula holds for $n = 1$ and $n = 2$. The induction step will proceed as usual, but it will actually be used to prove the formula for $n > 2$ as opposed to $n > 1$. (See Exercise 5.)

1. From the table, in each of the first two months there is exactly one baby pair of rabbits. This coincides with the first two Fibonacci numbers.
2. Let $n > 1$ and assume that F_k equals the number of baby pairs in the kth month for all $k \leq n$. Because of the third rule, the number of pairs of babies in any month is the same as the number of adult pairs in the previous month. Therefore:

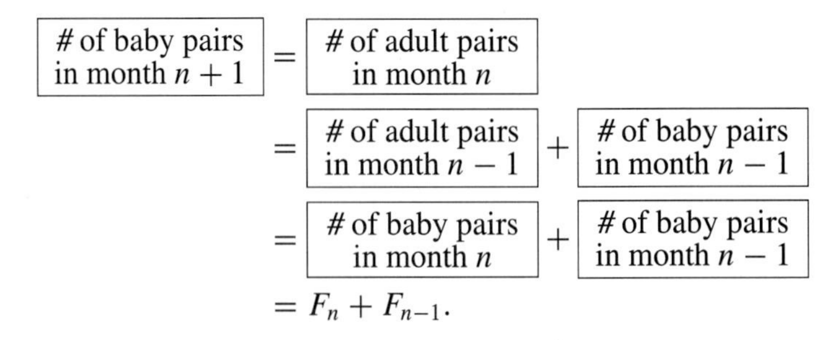

The last equality follows from the induction hypothesis. We then conclude F_{n+1} is the number of baby pairs in month $n+1$ since by definition of the Fibonacci sequence,

$$F_{n+1} = F_n + F_{n-1}.$$

It turns out that this sequence is closely related to another famous object of study in mathematics. Define a sequence

$$a_n = \frac{F_{n+1}}{F_n}.$$

The first seven terms of this sequence are:

$$\begin{aligned}
a_1 &= 1/1 = 1 \\
a_2 &= 2/1 = 2 \\
a_3 &= 3/2 = 1.5 \\
a_4 &= 5/3 \approx 1.667 \\
a_5 &= 8/5 = 1.6 \\
a_6 &= 13/8 = 1.625 \\
a_7 &= 21/13 \approx 1.615.
\end{aligned}$$

4.3.2. Figure

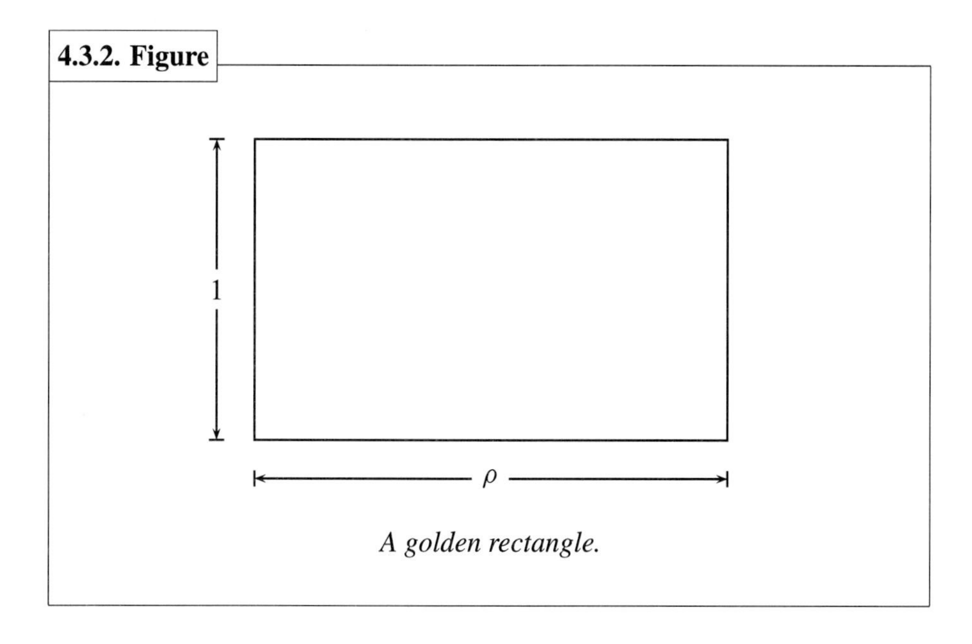

A golden rectangle.

It turns out that this sequence has a limit that we will call ρ. To find this limit, note:

$$\frac{F_{n+1}}{F_n} = \frac{F_n + F_{n-1}}{F_n} = 1 + \frac{F_{n-1}}{F_n}.$$

Because $a_{n-1} = F_n/F_{n-1}$,

$$a_n = 1 + \frac{1}{a_{n-1}},$$

and therefore,

$$a_n - 1 - \frac{1}{a_{n-1}} = 0.$$

We can substitute ρ into this equation since $\lim_{n\to\infty} a_n = \lim_{n\to\infty} a_{n-1} = \rho$. Once this is done, multiply through by ρ to obtain the equation

$$\rho^2 - \rho - 1 = 0.$$

An application of the quadratic formula shows that ρ is $(1 \pm \sqrt{5})/2$, but since $F_{n+1}/F_n > 0$, we must take the positive root. Thus,

$$\rho = \frac{1 + \sqrt{5}}{2}.$$

This number is called the ***golden ratio***. It was considered by the ancient Greeks to represent the ratio of the sides of the most beautiful rectangle (see Figure 4.3.2).

Example. Prove $F_n \le \rho^{n-1}$ when $n \ge 2$ using the Second Principle of Mathematical Induction. As with regular induction, we can start at basis cases other than $n = 1$. (Again, see Exercise 5.)

Proof. By strong induction on n.

1. Since $F_2 = 1 < \rho^1 \approx 1.618$, the inequality holds for $n = 2$.
2. Let $n \ge 3$ and assume $F_i \le \rho^{i-1}$ for all i such that $2 \le i \le n$. The induction hypothesis gives
$$F_n + F_{n-1} \le \rho^{n-1} + \rho^{n-2} = \rho^n(\rho^{-1} + \rho^{-2}).$$
Because $\rho^{-1} = (\sqrt{5} - 1)/2$ and $\rho^{-2} = (3 - \sqrt{5})/2$,
$$\rho^{-1} + \rho^{-2} = 1.$$
We then conclude $F_{n+1} \le \rho^{n+1-1}$.

Thus, the proof is complete by strong induction. ■

EXERCISES

1. Identify the basis case and induction hypothesis for the proof of the Unique Factorization Theorem (5.3.4).
2. Let $n \in \mathbb{Z}^+$. Write $n!$ as a recursive definition and use induction to prove the formula.
3. Given the recursively defined sequence, show that its nth term is as indicated.
   (a) If $a_1 = -1$ and $a_n = -a_{n-1}$, then $a_n = (-1)^n$.
   (b) If $a_1 = 1$ and $a_n = 1/3a_{n-1}$, then $a_n = (1/3)^{n-1}$.
   (c) If $a_1 = 0$, $a_2 = -6$, and $a_n = 5a_{n-1} - 6a_{n-2}$, then
   $$a_n = 3 \cdot 2^n - 2 \cdot 3^n.$$
   (d) If $a_1 = 4$, $a_2 = 12$, and $a_n = 4a_{n-1} - 2a_{n-2}$, then
   $$a_n = (2 + \sqrt{2})^{n-1} + (2 - \sqrt{2})^{n-1}.$$
   (e) If $a_1 = 3$, $a_2 = -3$, $a_3 = 9$, and $a_n = a_{n-1} + 4a_{n-2} - 4a_{n-3}$, then $a_n = 1 - (-2)^n$.
   (f) If $a_1 = 3$, $a_2 = 10$, $a_3 = 21$, and $a_n = 3a_{n-1} - 3a_{n-2} + a_{n-3}$, then $a_n = n + 2n^2$.
4. Show the following concerning Fibonacci numbers are true for all positive integers n.
   (a) $F_{n+2} > \rho^n$
   (b) $\sum_{i=1}^{n} F_i = F_{n+2} - 1$

5. Take a predicate $P(n)$ and let $S = \{k, k+1, k+2, \ldots\}$ where $k \in \mathbb{Z}$. Prove $(\forall n \in S)P(n)$ is true if and only if $P(k)$ and $P(k) \wedge \cdots \wedge P(n) \Rightarrow P(n+1)$ for all integers $n \geq k$.
6. State and prove the result that is strong induction's counterpart to to Corollary 4.1.5.
7. Prove (MI2) $\Rightarrow$ (MI).

4.4. THE WELL-ORDERING PRINCIPLE

Remember that we assumed the First Principle of Mathematical Induction as an axiom. We also noted that there are other propositions that are logically equivalent to it. Each of these could have been taken as an axiom instead of (MI). In this section we will introduce one of those alternatives, but we first need some preliminaries.

Let $A \subseteq \mathbb{R}$. An element $m \in A$ is a ***minimum*** (or sometimes called a ***least element***) if and only if

$$(\forall x \in A)(m \leq x).$$

The idea is that every element of A is at least as large as m.

Example. With this definition any finite set of reals has a least element. The sets $\mathbb{Z}^+$ and $\{5n : n \in \mathbb{Z}^+\}$ also have least elements, but the sets $\{5n : n \in \mathbb{Z}^-\}$ and $\mathbb{Z}^-$ do not.

To show that $S = \{5n : n \in \mathbb{Z}^+\}$ has a minimum, proceed as follows: We know that the minimum of $\mathbb{Z}^+$ is 1. Therefore, the minimum of S is 5 because $5 \in S$ and $5(1) \leq 5n$ for all $n \in \mathbb{Z}^+$.

The property that we will need to define in order to replace (MI) with another axiom is that every nonempty subset of $\mathbb{Z}^+$ must have a least element. We use this in the next definition.

4.4.1. Definition

Let $A \subseteq \mathbb{R}$ and $A \neq \varnothing$. We say that A is ***well-ordered*** if

$$(\forall B)[(B \subseteq A \wedge B \neq \varnothing) \Rightarrow B \text{ has a least element}].$$

Warning: if a set is well-ordered then it has a least element, but the converse is not true. To see this, consider $S = \{0, 1/2, 1/3, 1/4, \ldots\}$. It has a least element, namely 0, but S also contains a ***strictly decreasing sequence***, namely

$$1/2 > 1/3 > 1/4 > \cdots$$

Therefore, S has a subset without a minimum, so it is not well-ordered. We summarize this observation with the following theorem, and leave its proof as an exercise.

4.4.2. Theorem

A nonempty set of real numbers is not well-ordered if and only if it contains a strictly decreasing sequence.

Example. The following sets are not well-ordered.

1. $(0, 1)$
2. $[0, 1]$
3. $\mathbb{Q}$
4. $\{1/n : n \in \mathbb{Z}^+\}$

Example. Any finite set of real numbers is well-ordered. To prove this, suppose A is finite but not well-ordered. This means that there exists a strictly decreasing sequence in A:

$$a_1 > a_2 > a_3 > \dots$$

Therefore, A contains infinitely many elements. This is a contradiction.

Two notes should be made at this point. First, it is necessary to know that a set is not empty in order to show that it is well-ordered. It is a common mistake to forget this. Second, although there are infinitely many well-ordered sets, the examples will flow from the next theorem. It is called the ***Well-Ordering Principle*** and will be designated by (WO). It will be proven using strong induction.

4.4.3. Theorem (WO)

$\mathbb{Z}^+$ is well-ordered.

Proof. Define the following predicate $P(n)$:

for any nonempty set $B \subseteq \mathbb{Z}^+$,
if $n \in B$, then B has a least element.

We will use strong induction to prove $P(n)$ for all $n \in \mathbb{Z}^+$. Let $B \subseteq \mathbb{Z}^+$ and $B \neq \varnothing$.

1. If $1 \in B$, then B could have no smaller element. Hence $P(1)$ is true.
2. Assume $P(1)$ through $P(n)$ are true for some $n \in \mathbb{Z}^+$. To show $P(n+1)$, assume $n+1 \in B$. We must show that B has a smallest element.

(Case 1) $\{1, \ldots, n\} \cap B \neq \varnothing$. By the induction hypothesis, B would have a least element.

(Case 2) $\{1, \ldots, n\} \cap B = \varnothing$. Since $n + 1 \in B$, this must be its least element.

This concludes the induction.

To finish, take $B \subseteq \mathbb{Z}^+$ so that $B \neq \varnothing$. Then $n \in B$ for some positive integer n. Since $P(n)$ is true, B has a least element. ■

The resulting corollary will be left as an exercise. Its proof will be reminiscent to the one for Theorem 4.1.4.

4.4.4. Corollary

For any $n \in \mathbb{Z}$, the set $\{n, n + 1, n + 2, \ldots\}$ is well-ordered. In particular, $\mathbb{N}$ is well-ordered.

We have the following examples.

Example. Let $S = \{n\pi : n \in \mathbb{Z}^+\}$. To prove that S is well-ordered, take $T \subseteq S$ such that $T \neq \varnothing$. This means there exists a nonempty subset A of $\mathbb{Z}^+$ such that $T = \{n\pi : n \in A\}$. Since $\mathbb{Z}^+$ is well-ordered, A has a least element. Call it m. We claim that $m\pi$ is the minimum of T. To see this, let $a \in T$. Then, $a = n\pi$ for some $n \in A$. Since m is the minimum of A, $m \leq n$. Therefore, $m\pi \leq n\pi = a$.

Example. Any subset of a well-ordered set is well-ordered (see Exercise 4). Thus, $\mathbb{Z} \cap (5, \infty)$ is well-ordered because it is a subset of $\mathbb{Z}^+$.

Example. By Corollary 4.4.4, $\{-4, -3, -2, -1, 0, 1, \ldots\}$ is well-ordered. Therefore, as a subset of this set, $\{-4, -2, 0, 2, \ldots\}$ is well-ordered.

Example. The corollary allows us to define the ***ceiling function***. Let $x \in \mathbb{R}$ and define $\lceil x \rceil$ to be the least integer $\geq x$. For example, $\lceil 5 \rceil = 5$, $\lceil 1.4 \rceil = 2$, and $\lceil -3.4 \rceil = -3$. The definition of $\lceil x \rceil$ would have been nonsense if it was not known that the set $\{n \in \mathbb{Z} : x \leq n\}$ has a least element.

Let $S \subseteq \mathbb{R}$ such that $S \neq \varnothing$. We say that S has an ***upper bound*** if there exists $u \in \mathbb{R}$ so that $x \leq u$ for all $x \in S$. Furthermore, S has a ***maximum*** (or ***greatest element***) if there exists $m \in S$ so that m is an upper bound for S.

Example. The maximum of $[0, 1]$ is 1, but it has infinitely many upper bounds, such as 5 and 100.

The Well-ordering Principle can be used to prove the existence of a maximum under certain conditions.

4.4.5. Theorem

Let S be a nonempty set of integers. If S has an upper bound, then S has a maximum.

Proof. In order to arrive at a contradiction, suppose S does not have a maximum. Then there exists a ***strictly increasing sequence***

$$x_0 < x_1 < x_2 < \cdots$$

so that $x_i \in S$ for all $i \in \mathbb{N}$. Now let u denote an upper bound for S and define a sequence by

$$y_n = u - x_n.$$

Since $u \geq x_n$, $y_n \geq 0$ for all n. Hence, we have a strictly decreasing sequence of natural numbers:

$$y_0 > y_1 > y_2 > \cdots$$

This contradicts Corollary 4.4.4. ■

The Well-ordering Property can replace Mathematical Induction as an axiom. We have already seen that (MI) yields (WO). Here we prove the converse and show that they are logically equivalent.

4.4.6. Theorem

(MI) $\Leftrightarrow$ (WO)

Proof. Since (MI) implies (MI2), we only need to prove necessity by Theorem 4.4.3. Assume (WO). To prove (MI), let S be a set of positive integers such that $1 \in S$ and $n \in S \Rightarrow n + 1 \in S$ for all $n \in \mathbb{Z}^+$. We must show that $S = \mathbb{Z}^+$. Define

$$T = \{n \in \mathbb{Z}^+ : n \notin S\}.$$

In order to obtain a contradiction, assume that $T \neq \varnothing$. By (WO), T has a least element m. Now, 1 is not an element of T since $1 \in S$. Hence, $m > 1$ which means $m - 1 \in \mathbb{Z}^+$. However, $m - 1 \notin T$ because m is the minimum of T. Hence, $m - 1 \in S$, and since $m \in S \Rightarrow m + 1 \in S$, m is in S and not in T, a contradiction. Thus, $T = \varnothing$ from which follows $S = \mathbb{Z}^+$. ■

EXERCISES

1. Indicate whether each statement is true or false.
   (a) Every well-ordered set contains a minimum.
   (b) Every well-ordered set contains a maximum.
   (c) Every subset of a well-ordered set contains a minimum.
   (d) Every subset of a well-ordered set contains a maximum.
   (e) Every well-ordered set contains a strictly decreasing sequence.
   (f) Every well-ordered set contains a strictly increasing sequence.

2. Find a strictly decreasing sequence in each of the given sets.
   (a) $(0, 1)$
   (b) $[0, 1]$
   (c) $\mathbb{Q}$
   (d) $\{1/n : n \in \mathbb{Z}^+\}$

3. For each of the following sets, indicate whether or not it is well-ordered. If it is, prove it. If it is not, then find a strictly decreasing sequence of elements of the set.
   (a) $\{\sqrt{2}, 5, 6, 10.56, 17, -100\}$
   (b) $\{2n : n \in \mathbb{N}\}$
   (c) $\{\pi/n : n \in \mathbb{Z}^+\}$
   (d) $\{\pi/n : n \in \mathbb{Z}^-\}$
   (e) $\{-4, -3, -2, -1, \dots\}$
   (f) $\mathbb{Z} \cap (\pi, \infty)$
   (g) $\mathbb{Z} \cap (-7, \infty)$
   (h) $\mathbb{R}$

4. Let A and B be nonempty sets such that $B \subseteq A \subseteq \mathbb{R}$. Prove if A is well-ordered then B is well-ordered.

5. Show that a well-ordered set has a unique minimum.

6. We know that (WO) $\Rightarrow$ (MI) $\Rightarrow$ (MI2). Prove (WO) implies (MI2) without appealing to the middle step.

7. Prove that every nonempty subset of $\mathbb{Z}^-$ has a maximum.

8. Prove that y_n in the proof of Theorem 4.4.5 is strictly decreasing.

9. Give a full proof for Theorem 4.4.2.

10. Prove Corollary 4.4.4.

CHAPTER EXERCISES

1. Prove that the cube of any positive integer is the sum of two squares. [Hint: use an exercise from Section 1 and the fact that n^3 equals

$$(1^3 + 2^3 + 3^3 + \cdots + n^3) - (1^3 + 2^3 + 3^3 + \cdots + (n-1)^3).]$$

2. Use induction to prove for all integers $n > 1$,

$$\frac{d}{dx}x^n = nx^{n-1}.$$

3. Let $n \geq 0$ and $r \neq 1$. Show that the sum of the first n terms of the geometric series $\sum_{i=0}^{n} ar^i$ is

$$\frac{ar^{n+1} - a}{r - 1}.$$

4. Let $n \geq 1$.
   (a) Show:
   $$a^{n+1} - 1 = (a+1)(a^n - 1) - a(a^{n-1} - 1).$$
   (b) Use (a) and strong induction to prove
   $$a^n - 1 = (a-1)(a^{n-1} + a^{n-2} + \cdots + a + 1).$$
5. For $n \in \mathbb{Z}^+$, prove if A has n elements, then $\mathbf{P}(A)$ has 2^n elements.
6. Use mathematical induction to prove that the number of lines in a truth table with n variables is 2^n.
7. Mathematical induction can actually be stated as a biconditional. In this problem prove the converse of mathematical induction, namely if $(\forall n \in \mathbb{Z}^+)P(n)$, then $P(1)$ and $P(n) \Rightarrow P(n+1)$ for all $n \in \mathbb{Z}^+$.
8. Prove: for all $n \in \mathbb{Z}^+$,
   $$(\cos\theta + i\sin\theta)^n = \cos n\theta + i\sin n\theta.$$
   This result is known as ***De Moivre's Theorem***.*
9. Let $n, r \in \mathbb{N}$ so that $n \geq r$. Use mathematical induction to prove the following:
   (a) The number of r-element permutations of an n-element set is ${}_nP_r$.
   (b) The number of r-element subsets of an n-element set is $\binom{n}{r}$.
10. Prove for all $n \in \mathbb{Z}^+$, $\dfrac{(2n)!}{2^n n!} \in \mathbb{Z}$.
11. Prove for all $n \geq r$, $\dbinom{n}{r} \in \mathbb{Z}$.
12. Use strong induction to prove:
   $$\sum_{i=1}^{n} i^3 = (1 + 2 + \cdots + n)^2.$$
13. Let $\rho = (1+\sqrt{5})/2$ and $\sigma = (1-\sqrt{5})/2$. Prove:
   $$F_n = \frac{\rho^n - \sigma^n}{\sqrt{5}}.$$

---

*Abraham De Moivre (Vitry, France, 1667 – London, England, 1754) De Moivre made early contributions to analytic geometry and probability. In 1685 he immigrated to England after the expulsion of the Huguenots. Later in 1710 he helped in the investigation on whether Newton or Leibnitz discovered the calculus.

14. Let $n,\ r \in \mathbb{Z}^+$. Define

$$\binom{n}{r} = 0 \text{ if } r > n.$$

Using this definition, prove that Pascal's Identity still holds if the $n \geq r$ condition is dropped.

15. Use the definition of the previous problem to show the following relationship between the binomial coefficient and Fibonacci numbers:

$$\binom{n}{0} + \binom{n-1}{1} + \binom{n-2}{2} + \cdots + \binom{1}{n-1} = F_{n+1}.$$

16. Let $A \subseteq \mathbb{R}$. We have already defined what it means for A to have an upper bound. The set A has a ***lower bound*** ℓ if $\ell \in \mathbb{R}$ and

$$\text{for all } n \in A, \ell \leq n.$$

Demonstrate that every set of integers with a lower bound has a least element.

17. Let P and $P_1, \ldots, P_n$ be propositions. Define:

$$\bigwedge_{i=1}^{n} P_i \equiv P_1 \wedge \cdots \wedge P_n$$

and

$$\bigvee_{i=1}^{n} P_i \equiv P_1 \vee \cdots \vee P_n.$$

Show the following by induction on $n \geq 1$:

(a) $P \vee \bigvee_{i=1}^{n} P_i \equiv \left(\bigvee_{i=1}^{n} P_i\right) \vee P$

(b) $\bigvee_{i=1}^{n+1} P_i \equiv \left(\bigvee_{i=1}^{n} P_i\right) \vee P_{n+1}$

(c) $P \wedge \bigvee_{i=1}^{n} P_i \equiv \bigvee_{i=1}^{n} (P \wedge P_i)$

(d) $\sim \bigvee_{i=1}^{n} P_i \equiv \bigwedge_{i=1}^{n} \sim P_i$

18. A ***triangular number*** is the number of dots the can be evenly arranged in a equilateral triangle. If we let T_n ($n \in \mathbb{Z}^+$) denote the nth triangular number, then:

$T_1 = 1$ $\quad$ $T_2 = 3$ $\quad$ $T_3 = 6$

Answer the following:
   (a) Find T_4, T_5, and T_6.
   (b) Write a recursive definition for T_n.
   (c) Prove $T_n = \binom{n+1}{2}$.
   (d) Prove $T_n = \sum_{i=1}^{n} i$.
   (e) Show $T_{n-1} + T_n = n^2$ for $n \geq 2$.
   (f) Demonstrate that the sum of the first n triangular numbers is
$$\frac{n(n+1)(n+2)}{6}.$$

19. The following is a famous nonproof. It comes in various forms.

    We will use induction to prove that all horses are the same color. More precisely, we will prove for every $n \in \mathbb{Z}^+$, n horses are the same color. The basis case is clear. All horses in a collection of size one will be of the same color. For the induction step, assume that every n horses are of the same color. Take a collection of $n+1$ horses. Label the set as $H = \{h_1, h_2, h_3, \ldots, h_{n+1}\}$. Now, take h_1 out of the collection. Then by the induction hypothesis, every horse in $H \setminus \{h_1\}$ must be of the same color. Next, put h_1 back and take out h_2. Again, all of the horses in $H \setminus \{h_2\}$ are the same color. Hence, h_1 and h_2 must be the same color. Thus, all horses in H are the same color.

    We know that there are horses of different colors! So, find the error in the proof.

5 Number Theory

The study of the integers is called ***number theory***. This chapter will serve to introduce some basic topics in the subject. These include divisibility, prime numbers, and congruences. These topics will also serve to put our proof-writing techniques to the test. At first, it will appear that there is nothing to number theory. The topics are familiar, many learned in grade school. However, the ease by which results come early in our study will soon disappear. Our solid foundation in logic and set theory will become crucial.

5.1. AXIOMS

To begin our study of number theory, let us set down the basics. We will start by identifying two sets of axioms: those for equality and those for $\mathbb{Z}$. We will then prove some fundamental properties. These properties are usually taken for granted, but here we will be careful to follow the axioms closely to prove the results. First let us list the axioms for the equals relation:

5.1.1. Axioms

Let a, b, and c represent integers.

Reflexive
$a = a$

Symmetric
If $a = b$, then $b = a$.

Transitive
If $a = b$ and $b = c$, then $a = c$.

Equality
If $a = b$, then $a + c = b + c$.
If $a = b$, then $ac = bc$.

The symbols $+$ and $\cdot$ represent the two standard operations of ***addition*** and ***multiplication***. We follow the custom of writing ab for $a \cdot b$, and when a is raised by a positive integer exponent k,

$$a^k = \underbrace{a \cdot a \cdots a}_{k \text{ times}}.$$

Furthermore, $a^0 = 1$. For simplicity, we also assume that multiplication has precedence over addition. The set of integers with these two operations is assumed to satisfy the next collection of axioms. All of them should be familiar, and some we have already seen.

5.1.2. Axioms

Let $x,\ y,\ z \in \mathbb{Z}$.

Closure of $\mathbb{Z}$ [Clo]
$x + y \in \mathbb{Z}$
$xy \in \mathbb{Z}$

Commutative [Com]
$x + y = y + x$
$xy = yx$

Distributive [Dist]
$x(y + z) = xy + xz$

Additive Identity [Add Id]
$0 + x = x$

Additive Inverse [Add Inv]
$(\exists a \in \mathbb{Z})(x + a = 0)$

Multiplicative Identity [Mult Id]
$1 \cdot x = x$

Closure of $\mathbb{Z}^+$ [Clo$^+$]
$x,\ y \in \mathbb{Z}^+ \Rightarrow x + y \in \mathbb{Z}^+$
$x,\ y \in \mathbb{Z}^+ \Rightarrow xy \in \mathbb{Z}^+$

Associative [Assoc]
$x + (y + z) = (x + y) + z$
$x(yz) = (xy)z$

Cancellation [Can]
$(xz = yz \wedge z \neq 0) \Rightarrow x = y$

Well-Ordering [WO]
$\mathbb{Z}^+$ is well-ordered

Trichotomy [Tri]
Exactly one of the following hold:
$x < 0$, $x = 0$, or $x > 0$

Example. In the equation,

$$(3 + 4) + 8 = (8 + 3) + 4,$$

two axioms are being applied. First, $(3 + 4) + 8 = 8 + (3 + 4)$ by the Commutative Axiom (Com). The Associative Axiom (Assoc) is then used to conclude that $8 + (3 + 4) = (8 + 3) + 4$.

Usually when working with equations, several axioms will be applied for each line in the solution. The next example shows what happens when each axiom is explicitly cited.

Example. We shall demonstrate that $(x+2)(x+3) = x^2 + 5x + 6$. In so doing we will see that the standard methods for multiplying polynomials (for example, "FOIL") and factoring polynomials are simply applications of the Distributive Axiom:

$$\begin{aligned}
(x+2)(x+3) &= (x+2)\cdot x + (x+2)\cdot 3 && \text{[Dist]}\\
&= x(x+2) + 3(x+2) && \text{[Com]}\\
&= (x^2+2x) + (3x+6) && \text{[Dist]}\\
&= x^2 + [2x + (3x+6)] && \text{[Assoc]}\\
&= x^2 + [(2x+3x)+6] && \text{[Assoc]}\\
&= x^2 + 5x + 6. && \text{[Assoc]}
\end{aligned}$$

Since the Associative Axiom allows us to regroup, we can drop the parentheses in the last line.

One proposition that many expect to find in the list of axioms is our first theorem. Since the proof can only appeal to the axioms, we will follow them carefully and list our reasons.

5.1.3. Zero Property (ZP)

For all $x \in \mathbb{Z}$, $0x = 0$.

Proof. Take an integer x. Then:

$$\begin{aligned}
0x + 0x &= (0+0)x && \text{[Dist]}\\
&= 0x && \text{[Add Id]}
\end{aligned}$$

Therefore by adding the additive inverse of $0x$ to both sides and using the Equality Axiom, $0x = 0$. ■

This proposition shows that 0 plays a specific role in $\mathbb{Z}$. There are other important elements to this set. To talk about them, we need Definition 5.1.4. This definition is very abstract. It works with any arbitrary set that has an operation defined for it. A ***binary operation*** $*$ on S (sometimes simply called an ***operation***) is a rule that given two elements of S, say a and b, returns a unique element of the set. This element is identified by $a * b$. To be precise, for all $s,\ s',\ t,\ t' \in S$,

$$\text{if } s = s' \text{ and } t = t', \text{ then } s * t = s' * t'.$$

Our main examples of operations are addition and multiplication of integers. For example, given 3 and 6, addition returns 9, and only 9. This result is labeled as $3 + 6$. There are many other examples of operations.

Example. To show that $x * y = 2x - y$ is an operation on $\mathbb{Z}$, take $a,\ a',\ b,\ b' \in \mathbb{Z}$ and assume $a = a'$ and $b = b'$. Then,

$$a * b = 2a - b = 2a' - b' = a' * b'.$$

Therefore, $a * b = a' * b'$ by the Transitive Axiom.

Example. Almost any set can have an operation. Take $S = \{e,\ a,\ b,\ c\}$ and define $*$ by the table:

$*$	e	a	b	c
e	e	a	b	c
a	a	b	c	e
b	b	c	e	a
c	c	e	a	b

The table is read from left to right, so $b * c = a$. The table makes $*$ into a binary operation since every pair of elements of S is assigned a unique element of S.

We will return to the notion of a binary operation in Section 6.4, but for now we are ready for the definition. It is the reason behind some of the terminology in 5.1.2.

5.1.4. Definition

Given an operation $*$ on a set S,

1. an ***identity*** (with respect to $*$) is an element e of S with the property

$$\text{for all } x \in S,\ e * x = x * e = x;$$

2. an ***inverse*** of an element x (with respect to $*$) is some $y \in S$ such that

$$x * y = y * x = e.$$

Notice that the identity, if it exists, must be unique. To prove this, suppose both e and e' are identities. These must be equal because $e = e * e' = e'$. So, if a set has an identity with respect to an operation, we may refer to it as *the* identity of the set.

Similarly, inverses are unique for operations that are associative. Suppose that both y and y' are inverses of x. This means

$$x * y = y * x = e$$

and

$$x * y' = y' * x = e.$$

Therefore,

$$y = y * e = y * (x * y') = (y * x) * y' = e * y' = y'.$$

Note the following:

1. Every nonzero integer has an inverse with respect to multiplication. These are called ***multiplicative inverses***. However, these inverses are not necessarily elements of $\mathbb{Z}$. For instance, $5 \in \mathbb{Z}$ and its multiplicative inverse is $1/5$, but $1/5 \notin \mathbb{Z}$.
2. Let $S = \{e, a, b, c\}$ and define $*$ by the table in the previous example. We can then check that e is the identity and every element has an inverse. For example, the inverse of c is a, for $a * c = c * a = e$.

Definition 5.1.4 explains the naming of many of the parts of Axiom 5.1.2. In $\mathbb{Z}$, 0 is the ***additive identity***. This is because when 0 is added to any integer, the result is that integer. Likewise, 1 is the ***multiplicative identity***. Also, every integer n has an ***additive inverse*** that is designated by $-n$, and we follow the longstanding tradition of writing $a + (-b)$ as $a - b$.

Example. Take $a \in \mathbb{Z}$. To prove $-a = (-1)a$, by definition we must show $a + (-1)a = 0$:

$$\begin{aligned} a + (-1)a &= 1 \cdot a + (-1)a && \text{[Mult Id]} \\ &= a \cdot 1 + a \cdot (-1) && \text{[Com]} \\ &= a(1 + -1) && \text{[Dist]} \\ &= a \cdot 0 && \text{[Add Inv]} \\ &= 0 && \text{[ZP]} \end{aligned}$$

This is important, for it will allow us to find solutions to equations. For instance, to find the solution to $x + 8 = 10$, we were taught to subtract 2 from both sides of the equation. (In other words, add the additive inverse of 2 to both sides!) This yields $x = 2$. To do this we applied the Equality and Additive Inverse Axioms.

To find the solutions to more complicated equations, we need additional results. For example, we need Theorem 5.1.5 to work with quadratics. Take the equation $(x-3)(x+8)=0$. To find its solutions, we know that we first conclude $x-3=0$ or $x+8=0$, and then the Equality Axiom gives $x=3$ or $x=-8$. The first step follows by the next theorem.

5.1.5. Theorem

Let $a,\ b \in \mathbb{Z}$. If $ab=0$, then $a=0$ or $b=0$.

Proof. Assume $ab=0$ for two integers a and b. Further suppose $a \neq 0$. Then,

$$\begin{aligned} ab &= a \cdot 0 && [\text{ZP}] \\ b &= 0 && [\text{Can}]. \end{aligned}$$

Therefore, $a=0$ or $b=0$. ■

Now let us look at inequalities. Let $a,\ b \in \mathbb{Z}$ and define the ***less-than*** relation:

$$a < b \text{ if and only if } b - a \in \mathbb{Z}^+.$$

Similarly, the ***greater-than*** relation is defined by

$$a > b \text{ if and only if } a - b \in \mathbb{Z}^+.$$

For example, because $29 - 12 \in \mathbb{Z}^+$, both $12 < 29$ and $29 > 12$ are true. We now prove the following famous results.

5.1.6. Theorem

Let $a,\ b,\ c \in \mathbb{Z}^+$.

1. $a < b$ implies $a + c < b + c$.
2. $a < b$ implies $a - c < b - c$.

Proof. Let a, b, c be integers and suppose $a < b$. Then by definition, $b - a \in \mathbb{Z}^+$. We need to show that $(b+c)-(a+c) \in \mathbb{Z}^+$. If we calculate, we see that

$$\begin{aligned} (b+c)-(a+c) &= b + c - a - c \\ &= b - a, \end{aligned}$$

and we are done. The second part is left as an exercise. ■

Example. Let $S = \{x \in \mathbb{Z} : x + 5 < 3 \text{ or } 9 < x + 3\}$. Let us write S using interval notation. Using the last theorem, we have:

$$\begin{aligned} S &= \{x \in \mathbb{Z} : x < -2 \text{ or } 6 < x\} \\ &= \{x \in \mathbb{Z} : x < -2\} \cup \{x \in \mathbb{Z} : 6 < x\} \\ &= [\mathbb{Z} \cap (-\infty, -2)] \cup [\mathbb{Z} \cap (6, \infty)] \\ &= \mathbb{Z} \cap [(-\infty, -2) \cup (6, \infty)] \end{aligned}$$

The next theorem justifies the old rule of switching the inequality when multiplying (or dividing) by a negative, at least for the integers!

5.1.7. Theorem

Let a, b, c, and d be integers.

1. If $a < b$ and $c > 0$, then $ac < bc$.
2. If $a < b$ and $c < 0$, then $ac > bc$.

Proof. Assume $a, b, c, d \in \mathbb{Z}$.

1. Let $a < b$ and $c > 0$. We must show $bc - ac \in \mathbb{Z}^+$. We know by hypothesis $b - a \in \mathbb{Z}^+$ and $c \in \mathbb{Z}^+$. Applying the Closure Axiom of $\mathbb{Z}^+$ (5.1.2),

$$(b - a)c \in \mathbb{Z}^+.$$

   Then by the Distributive and Commutative Axioms, $bc - ac \in \mathbb{Z}^+$.
2. Now suppose $a < b$ but $c < 0$. By Theorem 5.1.6,

$$\begin{aligned} c &< 0 \\ c - c &< 0 - c \\ 0 &< -c. \end{aligned}$$

   Hence, $-c \in \mathbb{Z}^+$. So again by closure, $-c(b - a) \in \mathbb{Z}^+$, which means $ac - bc \in \mathbb{Z}^+$ by Exercise 7. Thus, $ac > bc$. ■

We define $a \leq b$ to mean

$$a < b \text{ or } a = b.$$

We can then define $a \geq b$ as $b \leq a$. Both of the previous lemmas can be proven with $\leq$ and $\geq$. The proofs would be nearly identical. The only difference would be that additional work must be done for equality. (See Exercise 12.)

Example. The inequality $7 \leq 2x - 5 \leq 13$ means $7 \leq 2x - 5$ and $2x - 5 \leq 13$. Therefore, to write the set $T = \{x \in \mathbb{Z} : 7 \leq 2x - 5 \leq 13\}$ using interval notation, we need to use the $\leq$ form of Theorems 5.1.6 and 5.1.7 and proceed as follows:

$$\begin{aligned} T &= \{x \in \mathbb{Z} : 7 \leq 2x - 5 \text{ and } 2x - 5 \leq 13\} \\ &= \{x \in \mathbb{Z} : 12 \leq 2x \text{ and } 2x \leq 18\} \\ &= \{x \in \mathbb{Z} : 6 \leq x \text{ and } x \leq 9\} \\ &= \{x \in \mathbb{Z} : x \in [6, 9]\} \\ &= \mathbb{Z} \cap [6, 9] \end{aligned}$$

We close the section with the following helpful theorem.

5.1.8. Theorem

For all $a, b \in \mathbb{Z}$, if $a \leq b$ and $b \leq a$, then $a = b$.

Proof. Take two integers a and b and assume $a \leq b$ and $b \leq a$. This means

$$(a < b \text{ or } a = b) \text{ and } (b < a \text{ or } a = b),$$

but this is equivalent to

$$(a < b \text{ and } b < a) \text{ or } (a < b \text{ and } a = b) \text{ or}$$
$$(a = b \text{ and } b < a) \text{ or } (a = b \text{ and } a = b).$$

By Trichotomy, the first three disjuncts are all false. Therefore, $a = b$ is true by use of the Disjunction Syllogism followed by Simplification. ■

EXERCISES

Assume that all variables represent integers.

1. Each of these sentences is false. Replace each underlined section with the appropriate word or phrase to make the sentence true.

(a) The integers contain exactly one multiplicative <u>inverse</u>.

(b) For all <u>integers</u> a, b, c, if $ab = ac$, then $b = c$.

(c) The product of an element with its inverse is an <u>inverse</u>.

(d) The <u>Commutative Axiom</u> states that $(a + b) + c = a + (b + c)$.

(e) $(a + b) \cdot c = c \cdot (a + b)$ follows from the <u>Distributive Axiom</u>.

(f) The sum of two integers is an integer since $\mathbb{Z}$ is <u>well-ordered</u>.

(g) The Trichotomy Axiom guarantees that for any two integers m and n, <u>$m < n$ or $n < m$</u>.

2. Identify the axiom that is applied in each sentence.

(a) $(4+3)+9=9+(4+3)$
(b) $4+(-4)=0$
(c) $8+3 \in \mathbb{Z}^+$
(d) $(6+3)+4=(3+6)+4$
(e) $3<4$ or $3 \geq 4$
(f) $1 \cdot 10 = 10$
(g) $1 \cdot 10 = 10 \cdot 1$
(h) Since $4=8$, $1=2$.
(i) $4+(-9) \in \mathbb{Z}$
(j) $0+9=9$

3. Derive the solution(s) to each equation by explicitly stating each step with the appropriate reason.

(a) $x+4=10$
(b) $x-1=0$
(c) $2x=4$
(d) $x^2=0$
(e) $x^2-x+6=12$
(f) $x^3-x^2-2x=0$

4. Prove:

(a) $a+(b+c)=(c+a)+b$
(b) $(a-b)+(b-c)+(c-a)=0$
(c) $(a+b)+(c+d)=a+(b+c)+d$

5. Demonstrate these derivations of the Distributive Axiom.

(a) $(a+b)^2=a^2+2ab+b^2$
(b) $(a-b)^2=a^2-2ab+b^2$
(c) $(a+b)(a-b)=a^2-b^2$

6. Why is 4 not the additive inverse of 5, and why is 7 not the multiplicative inverse of 8?

7. Prove these results using the definition of an additive inverse. (Recall: the notation $-(ab)$ refers to the additive inverse of ab.)

(a) $a(-b)=-(ab)$
(b) $(-a)(-b)=ab$
(c) $-(a+b)=(-a)+(-b)$
(d) $-0=0$

8. Prove that $(ab)^{-1}=a^{-1}b^{-1}$. (Note: a^{-1} denotes the multiplicative inverse of a.)

9. Define $*$ by $x*y=x+y+2$.

(a) Show that $*$ is an operation on $\mathbb{Z}$.
(b) Prove that -2 is the identity of $\mathbb{Z}$ with respect to $*$.
(c) For every $n \in \mathbb{Z}$, show that $-n-4$ is the inverse of n with respect to $*$.

10. Define the operation $*$ by $x*y=2x-y$.

(a) Is there an integer that serves as an identity with respect to $*$?
(b) Does every integer have an inverse with respect to $*$?

11. Show the following:

(a) $2<5$
(b) $3 \leq 3$
(c) $8 \geq 5$
(d) $6 \not< 6$
(e) $5 \not> 0$
(f) $8 \not\leq 3$

12. Prove the implications:
    (a) If $a < b$, then $a - c < b - c$.
    (b) If $a \leq b$, then $a + c \leq b + c$.
    (c) If $a < b$, then $a - c < b - c$.
    (d) If $a \leq b$, then $a - c \leq b - c$.
    (e) If $a < b$ and $b < c$, then $a < c$.
    (f) If $a \leq b$ and $b \leq c$, then $a \leq c$.
    (g) If $a \leq b$ and $c \geq 0$, then $ac \leq bc$.
    (h) If $a \leq b$ and $c \leq 0$, then $ac \geq bc$.
13. Prove:
    (a) $a < b$ if and only if $b > a$.
    (b) $a \leq b$ if and only if $b \geq a$. (Use part (a) to prove this.)
14. Prove these inequalities.
    (a) $a^2 \geq 0$
    (b) $a^3 \leq 0$ if $a \leq 0$
    (c) $a^{2k} \geq 0$
    (d) $a^{2k+1} \leq 0$ if $a \leq 0$
15. For all $a, b, c \in \mathbb{Z}$, define $a < b < c$ as $a < b$ and $b < c$. Similar definitions can be made for $>$, $\leq$, and $\geq$. Prove that if $0 < a < c$ and $0 < b < c$, then $-c < a - b < c$.

5.2. DIVISIBILITY

If one takes any two integers and adds, subtracts, or multiplies them together, the result is an integer. We say that the set of integers is ***closed*** under these operations. This is not true for division. When 18 is divided by 6 the result is 3, but when that same number is divided by 8, the answer is 2.25, which is not an integer. For this reason our first definition is important and familiar (see Section 2.4).

5.2.1. Definition

If a and b are integers with $a \neq 0$, then a ***divides*** b if and only if there exists $k \in \mathbb{Z}$ such that $b = ak$. The notation $a \mid b$ means "a divides b."

Therefore, 6 divides 18, but 8 does not divide 18. In this case write $6 \mid 18$ and $8 \nmid 18$. If $a \mid b$, then we also say b is ***divisible*** by a, a is a ***divisor*** or a ***factor*** of b, or b is a ***multiple*** of a.

Example. Compare this example with Theorem 3.2.4 and Exercise 8. Let a, b, and c be integers with $a \neq 0$ and $b \neq 0$.

- Since $a = a \cdot 1$, $a \mid a$.
- The statement $a \mid b \Rightarrow b \mid a$ is false because $2 \mid 4$ but $4 \nmid 2$.
- To prove

  if $a \mid b$ and $b \mid c$, then $a \mid c$,

  proceed as follows:

  Proof. Assume $a \mid b$ and $b \mid c$. This means $b = a\ell$ and $c = bk$ for some $\ell,\ k \in \mathbb{Z}$. By substitution,

  $$c = bk = (a\ell)k = a(\ell k).$$

  Since $\ell k \in \mathbb{Z}$, $a \mid c$. ■

Most of this section will be spent studying two common topics of number theory. The proof of the first one uses the Well-Ordering Principle. The strategy will be to define a nonempty subset of a well-ordered set. Its minimum r will be a number that we want. This minimum will also need to have a particular property, say $P(r)$. To show that it has that property, assume $\sim P(r)$ and use this to find another element of the set that is less than r. This contradicts the minimality of r allowing us to conclude $P(r)$.

5.2.2. Division Algorithm

Let $m,\ n \in \mathbb{Z}$ and $m > 0$. If m is divided into n, then there are unique integers q and r so that $0 \leq r < m$ and $n = mq + r$. The value q is called the ***quotient*** and r is the ***remainder***.

Proof. Define the set S to be

$$\{k \in \mathbb{N} : k = n - m\ell,\ \text{some } \ell \in \mathbb{Z}\}.$$

Notice that if ℓ is any integer at most n/m, then $m\ell \leq n$. Hence, $k = n - m\ell$ is nonnegative and $S \neq \varnothing$. Therefore, S has a least element by Corollary 4.4.4. Call it r and write $r = n - mq$ for some $q \in \mathbb{Z}$. The integers r and q are the ones we seek. Indeed, $r \geq 0$ by definition of S. We will use the minimality of r to show that $r < m$. Assume $r \geq m$, which means $r - m \geq 0$. Also,

$$r - m = n - mq - m = n - m(q + 1).$$

Thus, $r - m \in S$. But, $r > r - m$ since m is positive. This means that r is not the minimum of S, a contradiction. So we have found r and q so that $n = mq + r$ with $0 \leq r < m$.

As for uniqueness, suppose there are integers r, q, r', and q' so that

$$n = mq + r \text{ and } n = mq' + r'$$

with

$$0 \leq r < m \text{ and } 0 \leq r' < m.$$

By Exercise 15 of Section 5.1, $-m < r - r' < m$. Hence,

$$|r - r'| < m.$$

Furthermore, $mq + r = mq' + r'$, and this gives

$$r - r' = m(q' - q).$$

If $q' - q \neq 0$, we can divide and find

$$m = \frac{r - r'}{q' - q}.$$

This means $|r - r'| \geq m$ by Exercise 4, which is impossible. Hence, $q' - q = 0$, which implies $q = q'$ and $r = r'$. ■

Example. If we divide 5 into 17, the Division Algorithm returns a quotient of 3 and a remainder of 2, so we may write:

$$17 = 5(3) + 2.$$

Notice that $0 \leq 2 < 3$.

Example. Let $a,\ b \in \mathbb{Z}$ such that $a > 0$.

- Say $a \mid b$. Then $b = ak$ for some $k \in \mathbb{Z}$, and we may write

  $$b = ak + 0.$$

  The Division Algorithm also applies here. It says that there exist unique integers q and r so that $0 \leq r < a$ and

  $$b = aq + r.$$

  Since there is only one way to write this equation with these assumptions, $k = q$ and $r = 0$.
- Now suppose that $b = aq + 0$, some $q \in \mathbb{Z}$. This means $a \mid b$.

We have just demonstrated that a divides b if and only if the remainder is zero when b is divided by a using the Division Algorithm.

The second of the two subjects requires some prerequisites that include the next definition.

5.2.3. Definition

Let $a,\ b \in \mathbb{Z}$. An integer n is a ***linear combination*** of a and b if

$$n = ua + vb$$

for some $u,\ v \in \mathbb{Z}$.

Example. Since $5 = 7(2) - 1(9)$, 5 is a linear combination of 2 and 9.

Example. Let $u,\ v \in \mathbb{Z}$. If d is an integer and $d \mid a$ and $d \mid b$, then $d \mid ua + vb$. To see this, write $a = d\ell$ and $b = dk$ for some $\ell,\ k \in \mathbb{Z}$. Then,

$$ua + vb = ud\ell + vdk = d(u\ell + vk),$$

and this means $d \mid ua + vb$.

Take two integers a and b. A ***common divisor*** of a and b is an integer c so that $c \mid a$ and $c \mid b$. For example, 4 is a common divisor of 48 and 36, but it is not the largest. The following is our second topic.

5.2.4. Definition

Suppose $a,\ b \in \mathbb{Z}$ with $a \neq 0$ or $b \neq 0$. An integer g is the ***greatest common divisor*** of a and b if

1. it is a common divisor of a and b, and
2. for every common divisor e, $e \leq g$.

In this case write $g = \gcd(a,\ b)$.

We may write $12 = \gcd(48,\ 36)$ and $7 = \gcd(0,\ 7)$. (We will see a method that will allow the computation of greatest common divisors in the next section.) Notice that it is important that at least one of the integers is not zero. The $\gcd(0,\ 0)$ is undefined since all nonzero integers divide 0. Further notice that the greatest common divisor is positive. This is because $a \mid b$ implies $-a \mid b$.

Making a definition does not guarantee that it will be useful. The conditions may be impossible to satisfy. Therefore, we will demonstrate that the greatest common divisor of two integers always exists and is unique. To see uniqueness, suppose that g and g' are greatest common divisors of some two integers not both zero. Then, $g \leq g'$ and $g' \leq g$ by definition. Therefore, $g = g'$ by Theorem 5.1.8.

Existence is more complicated. Take $a, b \in \mathbb{Z}$ not both zero and define S to be the set of all positive common divisors of a and b. In other words,

$$S = \{x \in \mathbb{Z}^+ : x \mid a \text{ and } x \mid b\}.$$

Then, S is nonempty because 1 divides both a and b. Moreover, $\max(|a|, |b|)$ is an upper bound for S, where for all $m, n \in \mathbb{Z}$,

$$\max(m, n) = \begin{cases} m & \text{if } m \geq n \\ n & \text{otherwise.} \end{cases}$$

To see this, let $x \in S$. Then, $x \mid a$ and $x \mid b$, which yields $x \leq |a|, |b|$ (see Exercise 7a). Since $\max(|a|, |b|)$ is either $|a|$ or $|b|$, we must have $x \leq \max(|a|, |b|)$. We may now use Theorem 4.4.5 to conclude that S has a maximum which must be $\gcd(a, b)$.

Now that we have proven that the greatest common divisor of two positive integers exists and is unique, let us prove some results. The next lemma is used to prove the Euclidean Algorithm* (Chapter Exercise 7). The algorithm provides a method for finding greatest common divisors. In the proof of the lemma, we will use a common strategy.

5.2.5. Lemma

If $a, b \in \mathbb{Z}$ are not zero, then $\gcd(a + nb, b) = \gcd(a, b)$ for any $n \in \mathbb{Z}$.

Proof. If both pairs of integers have the same common divisors, then their greatest common divisors must be equal. So, define

$$S = \{d \in \mathbb{Z} : d \mid a + nb \text{ and } d \mid b\}$$

and

$$T = \{d \in \mathbb{Z} : d \mid a \text{ and } d \mid b\}.$$

To show that the greatest common divisors are equal, prove $S = T$:

($\subseteq$) Let $d \in S$. Then $d \mid a + nb$ and $d \mid b$. This means $a + nb = d\ell$ and $b = dk$ for some $\ell, k \in \mathbb{Z}$. We are left to show $d \mid a$. By substitution, $a + ndk = d\ell$. Hence

$$a = d\ell - ndk = d(\ell - nk).$$

Since $\ell - nk \in \mathbb{Z}$, $d \mid a$.

*Euclid (Greece[?], 335 – Alexandria, Egypt, 270 BC): Euclid was the author of the *Elements*. This work was an attempt to derive all known mathematics from a small set of postulates (axioms). These subjects included geometry and number theory.

($\supseteq$) Now take $d \in T$. This means $d \mid a$ and $d \mid b$. Hence, there exists $\ell, k \in \mathbb{Z}$ such that $a = d\ell$ and $b = dk$. Then

$$a + nb = d\ell + ndk = d(\ell + nk).$$

Therefore, $d \mid a + nb$ and $d \in S$. ■

Example. We will use the lemma to find $\gcd(90, 12)$. Since $90 = 6+12\cdot 7$, we have:

$$\gcd(90, 12) = \gcd(6 + 12 \cdot 7, 12) = \gcd(6, 12) = 6.$$

The greatest common divisor is an important notion in number theory. The next result often is involved when the greatest common division is part of a theorem's hypotheses. The strategy in the proof is to define a set whose elements look like the expression in the conclusion to the theorem. An application of the Well-Ordering Principle and the Division Algorithm will then finish the proof.

5.2.6. Theorem

Suppose $a, b \in \mathbb{Z}$ with $a \neq 0$ or $b \neq 0$. If $c = \gcd(a, b)$, then there are integers m and n so that $c = ma + nb$.

Proof. Define

$$T = \{xa + yb : x, y \in \mathbb{Z} \text{ and } xa + yb > 0\}.$$

Notice $T \subseteq \mathbb{Z}^+$ and is not empty because $a^2 + b^2 \in T$. By the Well-Ordering Principle, T has a least element. Call it d and write $d = ma + nb$ for some $m, n \in \mathbb{Z}$. We will use the definition to demonstrate that the greatest common divisor of a and b is d.

1. Show that d is a common divisor by dividing d into both a and b. For a, the Division Algorithm yields $a = dq + r$ for some $q, r \in \mathbb{Z}$ and $0 \leq r < d$. To reach a contradiction, assume $r > 0$. Then

$$\begin{aligned} r &= a - dq \\ &= a - (ma + nb)q \\ &= (1 - mq)a - (nq)b. \end{aligned}$$

This means $r \in T$. This is a contradiction, for $r < d$ and d is the least element of T. Therefore, $r = 0$ and $d \mid a$. Similarly, $d \mid b$.

2. To show that d is the greatest of the common divisors, suppose $s \mid a$ and $s \mid b$ with $s \in \mathbb{Z}^+$. By definition, $a = sk$ and $b = s\ell$ for some $k, \ell \in \mathbb{Z}$. Hence,

$$\begin{aligned} d &= ma + nb \\ &= msk + ns\ell \\ &= s(mk + n\ell). \end{aligned}$$

Since $mk + n\ell \in \mathbb{Z}$, s divides d, and thus $s \leq d$ by Exercise 7a. ■

The previous proof is typical. When showing a common divisor g is the greatest common divisor, an arbitrary common divisor—s in the proof—is usually proven to divide g. Then the conclusion of $s \leq g$ is reached.

Example. Earlier we saw that $\gcd(90, 12) = 6$ and $6 = 1 \cdot 90 - 7 \cdot 12$, confirming the theorem.

Now let a and b be integers with $a \neq 0$ or $b \neq 0$. We say that a and b are ***relatively prime*** when $\gcd(a, b) = 1$. For example, 7 and 11 are relatively prime and so are 4 and 15. Applying Theorem 5.2.6, we find that we can write any integer as a linear combination of any two relatively prime integers. This will be used in the next theorem.

5.2.7. Theorem

Let a, b, and c be integers and $a \neq 0$. If $a \mid bc$ with a and b relatively prime, then $a \mid c$.

Proof. Assume $a \mid bc$ and $\gcd(a, b) = 1$. Then, $bc = ak$ for some $k \in \mathbb{Z}$. By Theorem 5.2.6, there exist $m, n \in \mathbb{Z}$ so that

$$1 = ma + nb.$$

Therefore, multiplying both sides of this equation by c gives:

$$c = cma + cnb = cma + nak = a(cm + nk).$$

Since $cm + nk \in \mathbb{Z}$, $a \mid c$. ■

The strategy is used again in the next proof. We will need to show that two integers are relatively prime. To do this we will take a positive common divisor and show that it must equal one.

5.2.8. Theorem

Let $a, b, m \in \mathbb{Z}$. If $\gcd(a, m) = 1$ and $\gcd(b, m) = 1$, then ab and m are relatively prime.

Proof. Take $d \in \mathbb{Z}^+$ and assume $d \mid ab$ and $d \mid m$. So, $ab = d\ell$ and $m = dk$, some $\ell, k \in \mathbb{Z}$. We will show $d = 1$. Since a and m are relatively prime, there are integers u and v so that $1 = ua + vm$. Multiply both sides of the equality by b. This yields

$$b = uab + vmb.$$

Substituting gives

$$b = ud\ell + vdkb = d(u\ell + vkb).$$

So, $d \mid b$. Because b and m are relatively prime, d must equal 1. ■

EXERCISES

Assume that all lowercase variables represent integers and $\gcd(a, b)$ *presupposes that a and b are not both zero.*

1. Given two integers m and n with $m > 0$, divide n by m and write the answer in the form $n = mq + r$ with $0 \le r < m$.

(a) $n = 0, m = 4$
(b) $n = 24, m = 5$
(c) $n = -5, m = 7$
(d) $n = -100, m = 14$

2. Find the greatest common divisors of each pair of integers.

(a) 12 and 18
(b) 3 and 9
(c) 14 and 0
(d) 7 and 15
(e) -21 and 14
(f) -50 and -75

3. Fill in the blanks with the appropriate conclusion for each conditional.

(a) If $\{d \in \mathbb{Z} : d \mid a\} = \{d \in \mathbb{Z} : d \mid b\}$, then ______.
(b) If $a \mid b$ and $a \nmid c$, then ______.
(c) If $a = qb + r$ and $r = 0$, then ______.
(d) If $\gcd(a, b) = n$, then there exist $u, v \in \mathbb{Z}$ such that ______.
(e) If ab and m are not relatively prime, then $\gcd(a, m) \ne 1$ or ______.
(f) If $a \mid bc$ and $\gcd(a, b) = 1$, then ______.
(g) If $a, b \in \mathbb{Z}$, then there exist $q, r \in \mathbb{Z}$ such that $a = qb + r$ and ______.

4. Show that if $m = (r - r')/(q' - q)$, then $|r - r'| \ge m$.

5. Where does the proof of the Division Algorithm (5.2.2) go wrong if $m = 0$?
6. Prove this extended version of the Division Algorithm: Let $m, n \in \mathbb{Z}$ and $m \neq 0$. If m is divided into n, then there are unique integers q and r so that $0 \leq r < |m|$ and $n = mq + r$.
7. Let $a, b \neq 0$. Prove the following:
   (a) $a \mid b$ implies $a \leq |b|$.
   (b) If $a \mid b$ and $b \mid a$, then $a = \pm b$.
8. Let $A = \{n : n \mid a\}$ and $C = \{n : n \mid c\}$. Suppose $a \mid b$ and $b \mid c$. Prove $A \subseteq C$.
9. Let $U = \{2k + 1 : k \in \mathbb{Z}\}$, the set of odd integers. Define

$$S = \{d \in U : d \mid a \text{ and } d \mid b\}$$

and

$$T = \{d \in U : d \mid a + b \text{ and } d \mid a - b\}.$$

Prove $S = T$.
10. Let a and b not both be zero. Prove that if $S = \{\ell \gcd(a, b) : \ell \in \mathbb{Z}\}$ and $T = \{ua + vb : u, v \in \mathbb{Z}\}$, then $S = T$.
11. Prove: if $d \mid a$ and $d \mid b$, then $d \mid \gcd(a, b)$.
12. Show that $\gcd(a, b) = \gcd(-a, b) = \gcd(a, -b) = \gcd(-a, -b)$.
13. Let a be a positive integer. Find
   (a) $\gcd(a, a + 1)$
   (b) $\gcd(a, 2a)$
   (c) $\gcd(a, a^2)$
   (d) $\gcd(a, 0)$
14. The converse of Theorem 5.2.6 is false.
   (a) Prove this by finding $a, b \in \mathbb{Z}$, not both zero, and integers m and n such that $ma + nb \neq \gcd(a, b)$.
   (b) The converse is true under an additional hypothesis. What is it?
15. Let $k > 1$. Show $\gcd(a, m) = k$ and $\gcd(b, m) = k$ does not imply that $\gcd(ab, m) = k$.
16. Prove $d = \gcd(c, m)$ implies $\gcd(c/d, m/d) = 1$.
17. Demonstrate that if $\gcd(a, c) = \gcd(b, c) = 1$, then ab and c are relatively prime.
18. Assume that a and b are relatively prime. Show:
   (a) If $a \mid n$ and $b \mid n$, then $ab \mid n$.
   (b) $\gcd(a + b, b) = \gcd(a + b, a) = 1$.
   (c) $\gcd(a + b, a - b) = 1$ or $\gcd(a + b, a - b) = 2$.
   (d) If $c \mid a$, then $\gcd(b, c) = 1$.
   (e) If $c \mid a + b$, then $\gcd(a, c) = \gcd(b, c) = 1$.
   (f) If $d \mid ac$ and $d \mid bc$, then $d \mid c$.

19. The fact that $\sqrt{2} \notin \mathbb{Q}$ is something that everyone should know. Fill in the missing lines to its proof:

> **Proof.** In order to obtain a contradiction, suppose _______. In other words, there exist _______ such that $\sqrt{2} = a/b$ with $b \neq 0$. We may further assume that the fraction is reduced. This means, $\gcd(a, b) = 1$. Now, $2 = a^2/b^2$. Hence, _______, which means that a^2 is even. Therefore, _______, and we may write $a = 2k$ for some integer k. Now substitute into a to arrive at the equation _______. Cancelling, we have equation _______. Thus, b^2 is even, and this implies that _______. Hence, a/b is not reduced, so it must be the case that $\sqrt{2} \notin \mathbb{Q}$. ■

5.3. PRIMES

Any discussion of divisibility will eventually lead to the notion of a prime number. Notice that 1 has been excluded from the definition. This is simply to make it easier to state theorems.

5.3.1. Definition

- An integer n is ***prime*** if and only if $n > 1$ and its only positive divisors are 1 and n.
- An integer n is ***composite*** if and only if $n > 1$ and n is not prime.

From the definition, if n is composite, there exist $a, b \in \mathbb{Z}$ such that $n = ab$ and $1 < a, b < n$. For example, 2, 11, and 97 are prime, and 4 and 87 are composite. We will follow the convention of using p to denote a prime. If other primes are needed, we will use letters like q or r to represent them.

Suppose a is an integer and p is a prime that does not divide a. We wish to show that a and p are relatively prime. Take $d \in \mathbb{Z}^+$ and assume $d \mid a$ and $d \mid p$. Since p is prime, $d = 1$ or $d = p$. Since $p \nmid a$, d must equal 1, which means $\gcd(a, p) = 1$. We will use this to prove the next result attributed to Euclid.

5.3.2. Euclid's Lemma

An integer $p > 1$ is prime if and only if for all $a, b \in \mathbb{Z}$, $p \mid ab$ implies $p \mid a$ or $p \mid b$.

Proof. Suppose a and b are two integers.

($\Rightarrow$) Let p be prime. Suppose $p \mid ab$ but $p \nmid a$. Then, $\gcd(a, p) = 1$. Therefore, $p \mid b$ by Theorem 5.2.7.

($\Leftarrow$) Take $p \in \mathbb{Z}$ with $p > 1$. Suppose p satisfies the condition,

$$(\forall a, b \in \mathbb{Z})(p \mid ab \Rightarrow p \mid a \text{ or } p \mid b).$$

Assume p is not prime. This means there are integers c and d so that $p = cd$ with $1 < c, d < p$. Hence, $p \mid cd$. By hypothesis, $p \mid c$ or $p \mid d$. However, since $c, d < p$, p can divide neither c nor d. This is a contradiction. Hence, p must be prime. ■

Example. Since 6 divides $3 \cdot 4$ but $6 \nmid 3$ and $6 \nmid 4$, the lemma tells us that 6 is not prime. On the other hand, if p is a prime that divides 12, then p divides 4 or 3. This means that $p = 2$ or $p = 3$.

The next theorem is a generalization of Euclid's Lemma.

5.3.3. Theorem

Let p be prime and $a_i \in \mathbb{Z}$ for $i = 1, \ldots, n$ ($n \in \mathbb{Z}^+$). If $p \mid a_1 a_2 \cdots a_n$, then $p \mid a_j$ for some $j = 1, \ldots, n$.

Proof. By induction on n.

1. The case when $n = 1$ is trivial because then p divides a_1 by definition of the product.
2. Assume if $p \mid a_1 a_2 \cdots a_n$ then $p \mid a_j$ for some $j = 1, \ldots, n$. We must show

$$\text{if } p \mid a_1 a_2 \cdots a_{n+1} \text{ then } p \mid a_j$$

for some $j = 1, \ldots, n + 1$. Suppose $p \mid a_1 a_2 \cdots a_{n+1}$. Then,

$$p \mid a_1 a_2 \cdots a_n \text{ or } p \mid a_{n+1}$$

by Lemma 5.3.2. If $p \mid a_{n+1}$ then we are done. Otherwise, p divides $a_1 a_2 \cdots a_n$. Hence by the induction hypothesis, p divides one of the a_i. ■

The previous theorem leads to a famous result. Take an integer, say 126. Find a factorization. The theorem states that if a prime divides the integer, then it divides one of the factors. It appears reasonable that any integer can then be written as a product that includes all of its prime divisors. We know this already. We can write 126 as $2 \cdot 3 \cdot 3 \cdot 7$, and this is essentially the only way that we can write 126 as a product of primes. All of this is summarized in the next theorem. It is also known as the ***Fundamental Theorem of Arithmetic***. It is the reason the primes are important. They are the building blocks of the integers. (Note: the proof uses Exercise 4.3.6.)

5.3.4. Unique Factorization Theorem

If $n > 1$ is an integer, then there exist unique primes $p_1 \leq p_2 \leq \cdots \leq p_k$ ($k \in \mathbb{Z}^+$) so that $n = p_1 p_2 \cdots p_k$.

Proof. We will prove existence with strong induction on n.

1. When $n = 2$ we are done since it is already prime.
2. Assume for all k such that $2 \leq k < n$, k can be written as the product of primes as described above. If n is prime, then we are done as in the basis case. So suppose n is composite. Then there exist $a, b \in \mathbb{Z}$ such that $n = ab$ and $1 < a, b < n$. By the induction hypothesis, we can write

$$a = q_1 q_2 \cdots q_u$$

and

$$b = r_1 r_2 \cdots r_v$$

where the q_i and r_j are primes. Now place these primes together in increasing order and relabel them as

$$p_1 \leq p_2 \leq \cdots \leq p_k$$

with $k = u + v$. Then $n = p_1 p_2 \cdots p_k$ as desired.

For uniqueness, suppose that there are two sets of primes

$$p_1 \leq p_2 \leq \cdots \leq p_k \text{ and } q_1 \leq q_2 \leq \cdots \leq q_\ell$$

so that

$$n = p_1 p_2 \cdots p_k = q_1 q_2 \cdots q_\ell.$$

By canceling if necessary, we may assume the sides have no common primes. If the cancellation yields $1 = 1$, then the sets of primes are the same. In order to obtain a contradiction, assume that there is at least one prime remaining on the left-hand side. Suppose it is p_1. If the product on the right equals 1, then $p_1 \mid 1$, which is impossible. If there are primes remaining on the right, then p_1 divides one of them by Lemma 5.3.2. This is also a contradiction, since the sides have no common prime factors because of the cancellation. Hence, the two sequences must be the same. ■

Unique Factorization allows us to make the following definition. Let $n \in \mathbb{Z}$ and $n > 1$. Then there exist distinct primes $p_1, p_2, \ldots, p_k$ and natural numbers $r_1, r_2, \ldots, r_k$ such that

$$n = p_1^{r_1} p_2^{r_2} \cdots p_k^{r_k}.$$

This expression is called a ***prime power decomposition*** of n.

Example. Consider the integer 360. It has $2^3 \cdot 3^2 \cdot 5^1$ as a prime power decomposition. If the exponents are limited to positive integers, then the expression is unique. In this sense we can say that $2^3 \cdot 3^2 \cdot 5^1$ is *the* prime power decomposition of 360. However, there are times when primes need to be included in the product that are not factors of the integer. By setting the exponent to zero, these primes can be included. For example, we can also write 360 as $2^3 \cdot 3^2 \cdot 5^1 \cdot 7^0$.

Example. Suppose $n > 1$. We will use the Unique Factorization Theorem to prove that n is a perfect square if and only if all powers in a prime power decomposition of n are even.

($\Rightarrow$) Let n be a perfect square. This means $n = k^2$ for some integer $k > 1$. Write the prime power decompositon for k:

$$k = p_1^{r_1} \cdots p_k^{r_k}.$$

Therefore,

$$n = k^2 = p_1^{2r_1} \cdots p_k^{2r_k}.$$

($\Leftarrow$) Assume all the powers are even in a prime power decomposition of n. Namely,

$$n = p_1^{r_1} \cdots p_k^{r_k}$$

where there exists $\ell_i \in \mathbb{Z}$ so that $r_i = 2\ell_i$ for $i = 1, \ldots, k$. Thus,

$$n = p_1^{2\ell_1} \cdots p_k^{2\ell_k} = (p_1^{\ell_1} \cdots p_k^{\ell_k})^2,$$

a perfect square.

We can use prime power decompositions to calculate greatest common divisors. Take two positive integers a and b. Let $p_1, p_2, \ldots p_k$ be a sequence of primes ($k \in \mathbb{Z}^+$) so that

$$a = p_1^{r_1} p_2^{r_2} \cdots p_k^{r_k}$$

and

$$b = p_1^{s_1} p_2^{s_2} \cdots p_k^{s_k}$$

where $r_i, s_i \in \mathbb{N}$ with $i = 1, \ldots, k$. Define

$$d = p_1^{\min(r_1, s_1)} p_2^{\min(r_2, s_2)} \cdots p_k^{\min(r_k, s_k)}$$

where $\min(r_i, s_i)$ is the ***minimum*** of r_i and s_i and is defined similar to max:

$$\min(r_i, s_i) = \begin{cases} r_i & \text{if } r_i \leq s_i \\ s_i & \text{otherwise.} \end{cases}$$

We will show that $d = \gcd(a, b)$:

1. To see that d is a common divisor, we note that $\min(r_i, s_i) \leq r_i, s_i$ for all i. Then let
$$\ell = p^{r_1 - \min(r_1, s_1)} p^{r_2 - \min(r_2, s_2)} \cdots p^{r_k - \min(r_k, s_k)}.$$
Notice that $\ell \in \mathbb{Z}$ because $r_i \geq \min(r_i, s_i)$ for all i, and this means $r_i - \min(r_i, s_i) \in \mathbb{N}$. Then, $d \mid a$ since
$$a = p_1^{r_1} p_2^{r_2} \cdots p_k^{r_k} = d\ell.$$
(Check this by adding exponents.) We can show $d \mid b$ similarly.
2. To prove that d is the greatest of the common divisors, assume e is a positive integer so that $e \mid a$ and $e \mid b$. This means that there exist natural numbers $t_1, t_2, \ldots t_k$ such that
$$e = p_1^{t_1} p_2^{t_2} \cdots p_k^{t_k}$$
and $t_i \leq s_i, r_i$ for all i. This follows by Lemma 5.3.2 and Unique Factorization. Hence, $t_i \leq \min(s_i, r_i)$ for each i. Thus, $e \leq d$.

We will use a prime power factorization in the next example to find a greatest common divisor.

Example. To find the $\gcd(360, 3150)$, we may write $360 = 2^3 \cdot 3^2 \cdot 5^1 \cdot 7^0$ and $1575 = 2^0 \cdot 3^2 \cdot 5^2 \cdot 7^1$. Hence,

$$\begin{aligned} \gcd(360, 3150) &= 2^{\min(3, 0)} \cdot 3^{\min(2, 2)} \cdot 5^{\min(1, 2)} \cdot 7^{\min(0, 1)} \\ &= 2^0 \cdot 3^2 \cdot 5^1 \cdot 7^0 \\ &= 45. \end{aligned}$$

EXERCISES

Assume that all lowercase variables represent integers and $\gcd(a, b)$ *presupposes that a and b are not both zero.*

1. For each of the following integers write prime power decompositions that include the primes 2, 3, 5, 7, 11, and 13.

(a) 7
(b) 30
(c) 12
(d) 100
(e) 4,070
(f) 1,000,000

2. Take $n \in \mathbb{Z}$ and $n > 1$. Prove that $n^3 + 1$ is composite.
3. Prove for all positive integers n, there exist n consecutive composite integers. (Hint: show $(n+1)!+2,\ (n+1)!+3,\ \ldots,\ (n+1)!+n+1$ are all composite)
4. Assume p is prime, $n \in \mathbb{Z}^+$, and $p \mid a^n$. Prove $p^n \mid a^n$.
5. If p is prime and a and b are integers such that $a + b = p$, prove that a and b are relatively prime.
6. Follow this outline to prove indirectly that there are infinitely many primes. When asked "why," justify that step.
    - Suppose there are only finitely many primes and list them in a set: $S = \{p_1,\ p_2,\ \ldots,\ p_n\}$.
    - Let $m = p_1 p_2 \cdots p_n + 1$.
    - Then there exists $p \in S$ such that $p \mid m$. (Why?)
    - Therefore, $p \mid 1$. (Why?) This is the contradiction.
7. Assume $e \mid a$ and $e \mid b$. Write prime power decompositions for a and b:

$$a = p_1^{r_1} p_2^{r_2} \cdots p_k^{r_k}$$

and

$$b = p_1^{s_1} p_2^{s_2} \cdots p_k^{s_k}.$$

Prove that there exist natural numbers $t_1,\ t_2,\ \ldots t_k$ such that

$$e = p_1^{t_1} p_2^{t_2} \cdots p_k^{t_k}$$

and $t_i \le r_i,\ s_i$ for all $i = 1,\ \ldots,\ k$.
8. Prove $a^3 \mid b^2$ implies $a \mid b$.
9. A positive integer greater than one is ***square free*** if it is not divisible by any perfect square greater than one. For example, 5 and 10 are square free, but 12 is not. Prove the following:
    (a) If $n \in \mathbb{Z}$ and $n > 1$, then n is square free if and only if it can be factored into distinct primes.
    (b) Every integer greater than 1 is the product of a square free integer and a perfect square.
10. Let $a \in \mathbb{Z}^+$. If a has the property that for all primes p,

$$\text{if } p \mid a, \text{ then } p^2 \mid a,$$

then a is called ***powerful***. Show that every powerful integer is the product of a perfect square and a perfect cube.
11. The ***least common multiple*** of nonzero integers a and b is an integer ℓ with the properties:
    - $a \mid \ell$ and $b \mid \ell$, and
    - if $a \mid k$ and $b \mid k$, then $\ell \le k$.

We will denote the least common multiple of a and b as $\operatorname{lcm}(a, b)$. Find the following:

(a) $\operatorname{lcm}(4, 8)$
(b) $\operatorname{lcm}(6, 9)$
(c) $\operatorname{lcm}(3, 4)$
(d) $\operatorname{lcm}(24, 60)$

12. Let $a, b \neq 0$. Prove that the least common multiple of a and ab is ab.
13. If $a = p_1^{r_1} p_2^{r_2} \cdots p_k^{r_k}$ and $b = p_1^{s_1} p_2^{s_2} \cdots p_k^{s_k}$ are prime power decompositions, prove

$$\operatorname{lcm}(a, b) = p_1^{\max(r_1, s_1)} p_2^{\max(r_2, s_2)} \cdots p_k^{\max(r_k, s_k)}.$$

14. Prove:
    (a) $\gcd(a, b) \cdot \operatorname{lcm}(a, b) = ab$ for any $a, b \in \mathbb{Z}^+$.
    (b) If $a \mid k$ and $b \mid k$, then $\operatorname{lcm}(a, b) \mid k$.
15. Show $\operatorname{lcm}(a, b) = ab$ if and only if a and b are relatively prime.

5.4. CONGRUENCES

The first two sections of this chapter focused on what happens when an integer divides another integer. We are not thwarted, however, if they do not divide evenly. We may appeal to the Division Algorithm and obtain a remainder. For example, divide 7 by 3. The result is a quotient of 2 and a remainder of 1. We write

$$7 = 3(2) + 1.$$

Now, 7 is not the only integer that has a remainder of one when divided by 3. Thirteen is another such number:

$$13 = 3(4) + 1.$$

To determine a test for when two integers have the same remainder when divided by 3, we proceed with some algebra. The previous two equations give $7 - 3(2) = 13 - 3(4)$. Hence, $3(4) - 3(2) = 13 - 7$, and then $3(2) = 13 - 7$. This means that $3 \mid 13 - 7$. Therefore, it looks like two integers have the same remainder when divided by 3 if 3 divides their difference. (See Exercise 8.) We can easily generalize this routine to arbitrary integers and arrive at the next definition.

5.4.1. Definition

Fix $m \in \mathbb{Z}^+$ and let a and b be integers. If $m \mid a - b$, we say that a is ***congruent*** to b ***modulo*** m. In this case we write $a \equiv b \pmod{m}$.

Example. The above shows $13 \equiv 7 \pmod 3$. Also, $2 \equiv 11 \pmod 3$ and $27 \equiv 0 \pmod 3$. But, $2 \not\equiv 9 \pmod 3$ and $25 \not\equiv 0 \pmod 3$.

Another method used to show that two integers are congruent is given by the next lemma.

5.4.2. Lemma

Let m be a positive integer and $a, b \in \mathbb{Z}$. There exists $k \in \mathbb{Z}$ such that $a = b + mk$ if and only if $a \equiv b \pmod m$.

Proof. Let $a, b \in \mathbb{Z}$.

$$\begin{aligned}(\exists k \in \mathbb{Z})(a = b + mk) &\Leftrightarrow (\exists k \in \mathbb{Z})(a - b = mk)\\ &\Leftrightarrow m \mid a - b\\ &\Leftrightarrow a \equiv b \pmod m. \blacksquare\end{aligned}$$

Example. Using the definition we see that $14 \equiv 26 \pmod 6$ because $6 \mid 14 - 26$. The lemma provides another test. We search for an integer k so that $14 = 26 + 6k$. Certainly, $k = -2$.

It turns out that congruence shares many properties with equality. Because of this, many may think that the next theorem is unnecessary or obvious. Be careful! Congruence is similar but not identical to equality. We may not simply rely on this similarity as a justification. They may differ exactly when we think that they are the same. For this reason we write the proof to be sure.

5.4.3. Theorem

Take $m \in \mathbb{Z}^+$. For all integers a, b, and c:

1. $a \equiv a \pmod m$ [***Reflexive***];
2. if $a \equiv b \pmod m$, then $b \equiv a \pmod m$ [***Symmetric***];
3. if $a \equiv b \pmod m$ and $b \equiv c \pmod m$, then $a \equiv c \pmod m$ [***Transitive***].

Proof. Fix $m \in \mathbb{Z}^+$ and let $a, b, c \in \mathbb{Z}$. We will use the two tests of Lemma 5.4.2.

1. Since m divides $a - a = 0$, $a \equiv a \pmod m$.
2. Assume $a \equiv b \pmod m$. This means that $m \mid a - b$. By Exercise 2.6.6, $m \mid b - a$. Hence $b \equiv a \pmod m$.

3. Let $a \equiv b \pmod{m}$ and $b \equiv c \pmod{m}$. We may then write $a = b + mk$ and $b = c + m\ell$ for some $k, \ell \in \mathbb{Z}$. Substitution yields

$$a = c + m\ell + mk = c + m(\ell + k).$$

Since $\ell + k \in \mathbb{Z}$, $a \equiv c \pmod{m}$. ■

Example. Fix $m \in \mathbb{Z}^+$. Let $a, b, c \in \mathbb{Z}$ and assume $a \equiv b \pmod{m}$ and $c \equiv b \pmod{m}$. By symmetry, $b \equiv c \pmod{m}$, and hence by transitivity, $a \equiv c \pmod{m}$. For instance, since $25 \equiv 11$ and $32 \equiv 11 \pmod{7}$, $25 \equiv 32 \pmod{7}$.

Congruence is also similar to equality when it comes to other properties. The next theorem will allow us to "solve" congruences.

5.4.4. Theorem

Fix a positive integer m and take $a, b, c, d \in \mathbb{Z}$. If $a \equiv b \pmod{m}$ and $c \equiv d \pmod{m}$, then the following are true:

1. $a + c \equiv b + d \pmod{m}$
2. $a - c \equiv b - d \pmod{m}$
3. $ac \equiv bd \pmod{m}$

Proof. We will prove the first and leave the other two as exercises. Assume $a \equiv b \pmod{m}$ and $c \equiv d \pmod{m}$. This means $a = b + m\ell$ and $c = d + mk$ for some $\ell, k \in \mathbb{Z}$. Therefore,

$$a + c = b + d + m(\ell + k),$$

and this means $a + c \equiv b + d \pmod{m}$. ■

Example. We know that $23 \equiv 7 \pmod{8}$ and $89 \equiv 1 \pmod{8}$. Thus,

$$23 \cdot 89 \equiv 7 \cdot 1 \pmod{8},$$

and we have $2047 \equiv 7 \pmod{8}$.

Let $a, b, c \in \mathbb{Z}$ and $m \in \mathbb{Z}^+$. A congruence of the form $ax + b \equiv c \pmod{m}$ is called a ***linear congruence***. We now have enough to solve these when $a = 1$.

Example. Let us find all solutions to $x + 5 \equiv 11 \pmod{14}$. The last proposition allows us to subtract 5 from both sides giving $x \equiv 6 \pmod{14}$. Therefore by Lemma 5.4.2, $x = 6 + 14k$ for $k \in \mathbb{Z}$. This means that $\{6 + 14k : k \in \mathbb{Z}\}$ is the solution set.

To generalize the solution set in the example, take $m \in \mathbb{Z}^+$. For any integer a, define

$$\begin{aligned}[a]_m &= \{x \in \mathbb{Z} : x \equiv a \pmod{m}\} \\ &= \{a + km : k \in \mathbb{Z}\}.\end{aligned}$$

This set is called the ***congruence class*** of a modulo m. It consists of all integers that are congruent to a. The set of all congruence classes modulo m is denoted by $\mathbb{Z}_m$.

Example. $\mathbb{Z}_6 = \{[0]_6, [1]_6, [2]_6, [3]_6, [4]_6, [5]_6\}$, where:

$$\begin{aligned}[0]_6 &= \{\dots, -6, 0, 6, 12, \dots\} \\ [1]_6 &= \{\dots, -5, 1, 7, 13, \dots\} \\ [2]_6 &= \{\dots, -4, 2, 8, 14, \dots\} \\ [3]_6 &= \{\dots, -3, 3, 9, 15, \dots\} \\ [4]_6 &= \{\dots, -2, 4, 10, 16, \dots\} \\ [5]_6 &= \{\dots, -1, 5, 11, 17, \dots\}\end{aligned}$$

If a congruence class can be written in the form $[a]_m$, then a is said to be a ***representative*** of the class. A congruence class usually has many representatives as $[0]_6 = [6]_6 = [42]_6$ shows.

Linear equations of the form $ax + b = c$ have exactly one solution when both a and b are not zero. Although linear congruences may have infinitely many solutions, they all come from the same congruence class.

The previous example shows that many integers may represent the same congruence class. However, two congruence classes are either disjoint or equal. To see this, let m be a positive integer and $a, b \in \mathbb{Z}$. Suppose $[a]_m \cap [b]_m \neq \varnothing$. Then there exists $c \in \mathbb{Z}$ so that $c \equiv a$ and $c \equiv b \pmod{m}$. Therefore, we have by Theorem 5.4.3,

$$\begin{aligned}x \in [a]_m &\Leftrightarrow x \equiv a \pmod{m} \\ &\Leftrightarrow x \equiv c \\ &\Leftrightarrow x \equiv b \\ &\Leftrightarrow x \in [b]_m.\end{aligned}$$

Therefore, $[a]_m = [b]_m$.

Example. We have already noted that $25 \equiv 11$ and $32 \equiv 11 \pmod 7$. This means $11 \in [25]_7$ and $11 \in [32]_7$. Therefore, $[25]_7 = [32]_7$.

The next result is used to show that we can raise both sides of a congruence by a natural number and preserve the congruence.

5.4.5. Theorem

Let $m \in \mathbb{Z}^+$ and $a, b \in \mathbb{Z}$ so that $a \equiv b \pmod m$. If k is a natural number, then $a^k \equiv b^k \pmod m$.

Proof. By induction on k.

1. When $k = 0$, both a^k and b^k equal 1. Hence, they are congruent modulo m.
2. Assume $a^k \equiv b^k \pmod m$. By Theorem 5.4.4, $aa^k \equiv bb^k \pmod m$. Hence, $a^{k+1} \equiv b^{k+1} \pmod m$. ■

For every $a \in \mathbb{Z}$, there exists a unique integer r such that $0 \leq r < m$ and $a \equiv r \pmod m$ by the Division Algorithm. This r is called the ***least residue*** of a (modulo m). For example, the least residue of 47 modulo 5 is 2.

Example. We will find the remainder of 4^{500} when divided by 7. To do this, we must compute the least residue of $4^{500} \pmod 7$. Calculate using Theorems 5.4.4 and 5.4.5 modulo 7:

$$\begin{aligned} 4^2 &\equiv 2 \\ 4^3 &\equiv 1 \\ (4^3)^{166} &\equiv 1^{166} \\ 4^{498} &\equiv 1 \\ 4^{500} &\equiv 2. \end{aligned}$$

Therefore, the remainder is 2.

The next theorem shows how division is handled modulo m. Again, division is a problem.

5.4.6. Theorem

Let m be a positive integer and $a, b, c \in \mathbb{Z}$. Define $d = \gcd(c, m)$. If $ac \equiv bc \pmod m$, then $a \equiv b \pmod{m/d}$.

Proof. Assume $ac \equiv bc \pmod{m}$, which means $ac = bc + m\ell$, some $\ell \in \mathbb{Z}$. Since d divides both c and m, c/d and m/d are integers. Thus,

$$a\left(\frac{c}{d}\right) = b\left(\frac{c}{d}\right) + \ell\left(\frac{m}{d}\right),$$

which gives

$$(a-b)\left(\frac{c}{d}\right) = \ell\left(\frac{m}{d}\right).$$

This means m/d divides $(a-b)(c/d)$. But $\gcd(c/d, m/d) = 1$ by Exercise 5.2.16. Hence, m/d divides $a-b$, and $a \equiv b \pmod{m/d}$. ■

Example. To illustrate the theorem, consider the following:

- We know that $28 \equiv 42 \pmod{14}$. It is then tempting to divide both sides by 2 and conclude $14 \equiv 21 \pmod{14}$. This is false. Apply the proposition instead. Since $\gcd(2, 14) = 2$, we must conclude that $14 \equiv 21 \pmod{7}$.
- Since 2 and 15 are relatively prime, we may divide both sides of $28 \equiv 58 \pmod{15}$ by two and conclude $14 \equiv 29 \pmod{15}$.

EXERCISES

Let $m \in \mathbb{Z}^+$ and suppose all lower-case variable are integers.

1. Answer true or false to the following.

(a) $4 \equiv 4 \pmod{7}$
(b) $10 \equiv 15 \pmod{20}$
(c) $-4 \equiv 12 \pmod{8}$
(d) $-18 \equiv -3 \pmod{5}$
(e) $9 \equiv 0 \pmod{18}$
(f) $0 \equiv 27 \pmod{9}$

2. Give the least residue modulo 8 of each integer.

(a) 5
(b) −5
(c) 12
(d) 42
(e) 88
(f) 100

3. Answer true or false for the following implications.

(a) If $a \equiv b \pmod{m}$, then $[a^4]_m = [b^4]_m$.
(b) If $a \equiv b \pmod{m}$, then $[a]_m = [b+m]_m$.
(c) If $a \equiv b \pmod{m}$ and $c \neq 0$, then $[a/c]_m = [b/c]_m$.
(d) If $[a]_m = [b]_m$, then $a = b$.
(e) If $[a]_m = [b]_m$, then $a \mid b$ or $b \mid a$.
(f) If $[a]_m = [b]_m$, then $m \mid a$ or $m \mid b$.
(g) If $[a]_m = [b]_m$ and $[b]_m = [c]_m$, then $a = c$.
(h) If the remainder of 5^{1000} divided by 4 is k, then the remainder of 5^{1001} divided by 4 is congruent to $5k$ modulo 4.

4. Find the remainder of 7^{3001} when divided by each of the following:
   (a) 5 (b) 6 (c) 7
5. Under what circumstances will the method outlined in the section for finding remainders of large numbers fail?
6. Let $b, c \in [a]_m$. Prove $b \equiv c \pmod{m}$.
7. Prove: $m \mid a$ if and only if $a \equiv 0 \pmod{m}$.
8. Show $a \equiv b \pmod{m}$ if and only if the remainder is m when b is divided by a.
9. Prove the following:
   (a) If a is even, then $a \equiv 0 \pmod{2}$.
   (b) If a is odd, then $a \equiv 1 \pmod{2}$.
   (c) If a is even, then $a^2 \equiv 0 \pmod{4}$.
   (d) If a is odd, then $a^2 \equiv 1 \pmod{4}$.
10. Let n be a positive integer. Prove if $a_i \equiv b_i \pmod{m}$ for all $i = 1, \ldots, n$, then:
    (a) $a_1 + a_2 + \cdots + a_n \equiv b_1 + b_2 + \cdots + b_n \pmod{m}$.
    (b) $a_1 a_2 \cdots a_n \equiv b_1 b_2 \cdots b_n \pmod{m}$.
11. Use induction to prove $6^n \equiv 1 + 5n \pmod{25}$ for all $n \in \mathbb{Z}^+$.
12. Find all solutions to the given congruences.
    (a) $x \equiv 5 \pmod{7}$
    (b) $x \equiv 0 \pmod{9}$
    (c) $x + 4 \equiv 6 \pmod{8}$
    (d) $x - 8 \equiv 2 \pmod{5}$
    (e) $3x + 5 \equiv 9 \pmod{11}$
    (f) $4x - 3 \equiv 1 \pmod{9}$
13. Find:
    (a) $[3]_5$
    (b) $[12]_6$
    (c) $[2]_5 \cup [27]_5$
    (d) $[4]_7 \cap [5]_7$
    (e) $\bigcup_{n \in \mathbb{Z}} [n]_{13}$
    (f) $\bigcap_{n \in \mathbb{Z}} [n]_{13}$
14. Take $n \in \mathbb{Z}$ and let r be the remainder obtained when n is divided by m. Prove $[n]_m = [r]_m$.
15. Suppose c and m are relatively prime. Prove:
    (a) There exists $b \in \mathbb{Z}$ such that $bc \equiv 1 \pmod{m}$. (b is called the ***multiplicative inverse*** of c modulo m.)
    (b) If $ca \equiv cb \pmod{m}$, then $a \equiv b \pmod{m}$.
    (c) Prove that the previous implication is false if $\gcd(c, m) \neq 1$.
16. Let $m, n \in \mathbb{Z}^+$. Prove the following:
    (a) If $a \equiv b \pmod{m}$ and $n \mid m$, then $a \equiv b \pmod{n}$.
    (b) If $a \equiv b \pmod{m}$, then $\gcd(a, m) = \gcd(b, m)$.
17. Let p be prime and a be an integer such that $1 \leq a < p$. Show that if $a^2 \equiv 1 \pmod{p}$, then $a = 1$ or $a = p - 1$.
18. Suppose p is prime. Prove that if $a^2 \equiv a \pmod{p}$, then $a \equiv 0$ or $a \equiv 1 \pmod{p}$.

CHAPTER EXERCISES

1. Let $m, n \in \mathbb{Z}$. Show $n\mathbb{Z} \subseteq m\mathbb{Z}$ if and only if $m \mid n$.
2. Prove the following for all integers a:
   (a) $3 \mid a$ if and only if the sum of the digits of a is divisible by 3.
   (b) $9 \mid a$ if and only if the sum of the digits of a is divisible by 9.
3. Prove that if $a \in \mathbb{Z}^+$, then $a \cdot \gcd(b, c) = \gcd(ab, ac)$.
4. Let $a, b, c \in \mathbb{Z}$. Show that if $\gcd(a, b) = \gcd(a, c) = 1$, then $\gcd(ab, c) = 1$.
5. Let $a_1, \ldots, a_n, b \in \mathbb{Z}$. Prove if $\gcd(a_i, b) = 1$ for $i = 1, \ldots, n$, then $\gcd(a_1 \cdot a_2 \cdots a_n, b) = 1$). (Hint: use induction and Exercise 5.2.17.)
6. The following is used to find greatest common divisors. It is known as the ***Euclidean Algorithm***:

   Given $a, b \in \mathbb{Z}$ such that $a \geq b > 0$, define $r_0 = a$ and $r_1 = b$. If the Division Algorithm is successively applied to find a sequence of integers, $r_0, r_1, \ldots, r_n, r_{n+1}$, such that for all $i = 1, \ldots, n-1$,

   $$r_i = r_{i+1}q_{i+1} + r_{i+2}$$

   with $q_{i+1} \in \mathbb{Z}$, $0 < r_{i+2} < r_{i+1}$, and $r_{n+1} = 0$, then $\gcd(a, b) = r_n$. To illustrate, let us find the greatest common divisor of 1620 and 168 using the Algorithm (here $n = 6$):

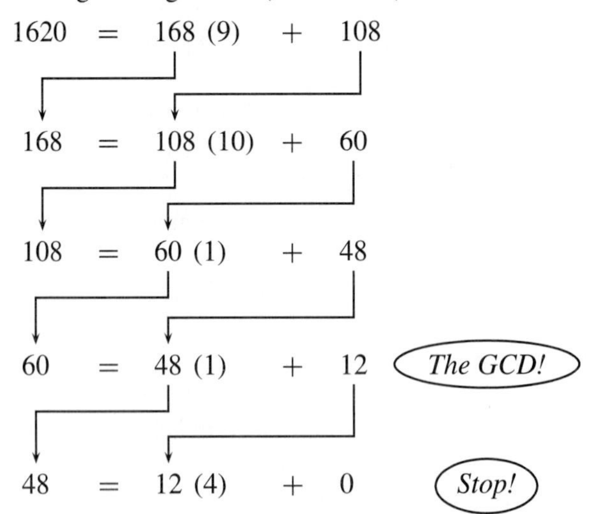

   Therefore, $\gcd(1620, 168) = 12$. Use the Euclidean Algorithm to find the greatest common divisor of each pair of integers.
   (a) 27 and 63
   (b) 3,600 and 5,400
   (c) 13,013 and 77,077
   (d) 110,250 and 18,900

7. Use Lemma 5.2.5 to prove the Euclidean Algorithm.
8. Use the Euclidean Algorithm to show that F_n and F_{n+1} are relatively prime for $n \in \mathbb{Z}^+$. (Recall that F_n is the nth Fibonacci number.)
9. Prove that F_n and F_{n+2} are relatively prime for all $n \in \mathbb{Z}^+$.
10. Let $a,\ b \in \mathbb{Z}$ be relatively prime. Prove if ab is a perfect square, then a and b are perfect squares.
11. Prove that every integer can be written as a linear combination of any two relatively prime integers.
12. Let $a_1,\ \ldots,\ a_k$ ($k \in \mathbb{Z}^+$) be positive integers. We will define $g = \gcd(a_1,\ \ldots,\ a_k)$ to mean that g is the largest integer such that $g \mid a_i$ for all $i = 1,\ \ldots,\ k$. Prove the following:
    (a) $\gcd(a_1,\ \ldots,\ a_k) = \gcd(a_1,\ \ldots,\ a_{k-2},\ \gcd(a_{k-1},\ a_k))$
    (b) $\gcd(ca_1,\ \ldots,\ ca_k) = |c| \gcd(a_1,\ \ldots,\ a_k)$ for all $c \neq 0$
13. Use mathematical induction to prove the following are true for all positive integers n:
    (a) $5 \mid n^5 - n$
    (b) $9 \mid n^3 + (n+1)^3 + (n+2)^3$
    (c) $8 \mid 5^{2n} + 7$
    (d) $5 \mid 3^{3n+1} + 2^{n+1}$
14. Let A and B be sets of integers. Prove:
    (a) $\max A \cup B = \max(\max A,\ \max B)$
    (b) $\min A \cup B = \min(\min A,\ \min B)$
15. Prove: if n is composite, then it has a prime factor p such that $p \leq \sqrt{n}$.
16. The previous problem is used to justify the ***sieve of Eratosthenes***.* The sieve is a method of determining which numbers are prime. It works like this. Take a positive integer n and list all positive integers greater than 1 but less than or equal to n. Next, find all primes $p_1,\ p_2,\ \ldots,\ p_k$ less than or equal to $\sqrt{n}$. If an integer $m \leq n$ is composite, then the previous problem states that it must have a prime factor in our list of primes. Thus, for each prime p_i, cross out all multiples ℓp_i ($\ell > 1$) from the list. After this is done for every $\ell = 1,\ \ldots,\ k$, each number not crossed out must be prime. For $n = 10$, $\sqrt{n} \approx 3.16$. Therefore, the prime list is only 2 and 3. Hence, the sieve gives us the list:

    2, 3, ~~4~~, 5, ~~6~~, 7, ~~8~~, ~~9~~, ~~10~~.

    Use the sieve of Eratosthenes to find all primes not exceeding 100.
17. Prove that if $a^n - 1$ is prime, then $a = 2$ and n is prime.

---

*Eratosthenes (Cyrene, North Africa, c. 285 BC – Alexandria, Egypt, c. 205 BC): Besides his sieve, Eratosthenes is known for his remarkably accurate computation of the earth's circumference.

18. Find the least residue of the following. Explain your answers.
    (a) $1! + 2! + \cdots + 100! \pmod 2$
    (b) $1! + 2! + \cdots + 100! \pmod 7$
    (c) $1! + 2! + \cdots + 100! \pmod{12}$
    (d) $1! + 2! + \cdots + 100! \pmod{25}$
    (e) $1! + 2! + \cdots + 100! \pmod{50}$
    (f) $1! + 2! + \cdots + 100! \pmod{100}$
19. For any nonempty set S of integers, S is ***pairwise incongruent*** modulo m if for all $a,\ b \in S$,

$$a \neq b \text{ implies } a \not\equiv b \pmod m.$$

The sets $\{0,\ 1,\ 2,\ 3,\ 4,\ 5\}$ and $\{12,\ 13,\ 20,\ -3,\ 10,\ 29\}$ are both pairwise incongruent modulo 6.
    (a) Let $\{n_0,\ n_1,\ \ldots n_{m-1}\}$ be pairwise incongruent modulo m. Show that there is a sequence $i_0,\ i_1,\ \ldots i_{m-1}$ such that

$$n_{i_j} \equiv j \pmod m,$$

for all $j = 0,\ \ldots,\ m-1$.
    (b) Prove if $\{n_1,\ n_2 \cdots,\ n_{m-1}\}$ is pairwise incongruent and no n_i is divisible by m, then

$$n_1 n_2 \cdots n_{m-1} \equiv 1 \cdot 2 \cdots (m-1) \pmod m.$$

20. Use mathematical induction to prove the following congruences are true for all positive integers n.
    (a) $4^n \equiv 1 + 3n \pmod 9$
    (b) $5^n \equiv 1 + 4n \pmod{16}$
21. Show $\binom{2n}{n} \equiv 0 \pmod p$ for any prime p such that $n < p < 2n$.
22. Let $a,\ b \in \mathbb{Z}$ and $k,\ \ell \in \mathbb{N}$. Prove the following using induction on k:
    (a) $a^k a^\ell = a^{k+\ell}$
    (b) $a^k b^k = (ab)^k$
    (c) $(a^k)^\ell = a^{k\ell}$
23. Prove Wilson's Theorem*: If $p \in \mathbb{Z}^+$ is a prime, then

$$(p-1)! \equiv -1 \pmod p.$$

24. Use Wilson's Theorem to confirm that 8 is not prime.
25. Prove the converse of Wilson's Theorem.

---

*John Wilson (Applethwaite, England, 1741 – Kendal, England, 1793): Wilson conjectured that the theorem named after him was true, but it was Joseph-Louis Lagrange (Turin, Sardinia, 1736 - Paris, France, 1813) who proved it in 1770.

26. Let $m \in \mathbb{Z}^+$. The ***Euler phi-function***,[†] $\phi(m)$, is the number of positive integers less than m that are relatively prime to m. For example, $\phi(9) = 6$, because 1, 2, 4, 5, 7, and 8 are all relatively prime to 9.
    (a) Find $\phi(1)$, $\phi(2)$, $\phi(6)$, $\phi(13)$, $\phi(15)$, and $\phi(44)$.
    (b) Prove that $\phi(p) = p - 1$ if and only if p is prime.
    (c) Show $\phi(p^r) = p^r - p^{r-1}$
    (d) Prove Euler's Theorem: If $a \in \mathbb{Z}$ such that m and a are relatively prime, then
    $$a^{\phi(m)} \equiv 1 \pmod{m}.$$
    (e) Use Euler's Theorem to find the solutions to $5x \equiv 4 \pmod{9}$.
27. Prove Fermat's Little Theorem[‡]: If p is prime and $a \in \mathbb{Z}^+$ so that $p \nmid a$, then
    $$a^{p-1} \equiv 1 \pmod{p}.$$
    (Hint: Use Euler's Theorem.)
28. Let p be a prime. Show $(p-2)! \equiv 1 \pmod{p}$.
29. Show for any prime p and integer a, $a^{p^k} \equiv a \pmod{p}$ for all $k \in \mathbb{N}$.

[†]Leonhard Euler (Basel, Switzerland, 1707 – St. Petersburg, Russia, 1783): a prolific mathematical writer who studied nearly every branch of mathematics, including the calculus, number theory, and topology.

[‡]Pierre de Fermat (Beaumont-de-Lomagne, France, 1601 – Castres, France, 1665): This lawyer and jurist was certainly the greatest amateur mathematician of all time. He is famous for his work in number theory but more famous for his "Last Theorem." It states that the equation $x^n + y^n = z^n$ has no integer solutions when n is an integer greater than 2. This longstanding conjecture was finally proven by Andrew Wiles (Cambridge, England, 1953 –) in 1994. The "Little Theorem" of Fermat is named so as to distinguish it from the "Last Theorem."

6 Relations and Functions

We now arrive at the topic that is most familiar to all graduates of calculus. That subject is the study of functions. Functions are found throughout mathematics. There is not a mathematical subject that does not at least touch on this idea. In this chapter we will deal with the basics. We will start with a quick look at relations and then see that a function is a particular type of relation. From there we will look at some important properties. All the while we will see functions used in proofs.

6.1. RELATIONS

A relation is an association between objects. A book on a table is an example of the relation of one object being *on* another. It is especially common to speak of relations among people. For example, one person may be the daughter of another. In mathematics there are many relations such as equals and less-than. To formalize this idea, we make the next definition:

6.1.1. Definition

A set R is a ***relation*** if there exist sets A and B such that $R \subseteq A \times B$. Moreover, if $R \subseteq A \times A$, then R is a ***relation on*** A.

The relation *on* can be represented as a subset of the cartesian product of the set of all books and the set of all tables. We could then write (*dictionary*, *desk*) to mean that the dictionary is on the desk. According to the definition, the set $\{(2, 4), (7, 3), (0, 0)\}$ is a relation because it is a subset of $\mathbb{Z} \times \mathbb{Z}$. The ordered pair $(2, 4)$ means that 2 is associated with 4. Likewise, $\mathbb{R} \times \mathbb{Q}$ is a relation. It may seem odd, but according to the definition, the empty set is also a relation. This is because $\varnothing = \varnothing \times \varnothing$. It is important to count the empty set as a relation because there are times when a set of ordered pairs can be written in which no elements are in the set, as in $\{(x, y) \in \mathbb{R}^2 : \sqrt{x} = -|y|\}$.

Example. The definition of the equality relation involves a subtlety. Formally, there is a difference between equality of reals and equality of integers. This is due to the definition of a relation. Equality on $\mathbb{R}$ is

$$I_{\mathbb{R}} = \{(x, x) : x \in \mathbb{R}\},$$

but equality on $\mathbb{Z}$ is

$$I_{\mathbb{Z}} = \{(x, x) : x \in \mathbb{Z}\}.$$

In general, $I_A = \{(a, a) : a \in A\}$ for any set A. These sets are designated by I because they are usually referred to as ***identity relations***.

Example. The less-than relation on $\mathbb{Z}$ appears to be easy to define. Let

$$L = \{(a, b) : a, b \in \mathbb{Z} \text{ and } a < b\}.$$

Another approach is to use membership in the set of positive integers as our test. Define

$$L = \{(a, b) : a, b \in \mathbb{Z} \text{ and } b - a \in \mathbb{Z}^+\}.$$

Hence, $(4, 7) \in L$ because $7 - 4 \in \mathbb{Z}^+$.

When a relation $R \subseteq A \times B$ is defined, all elements in A or B may not be used. For this reason, it is important to identify the sets that comprise all possible values for the two coordinates of the relation.

6.1.2. Definition

Let $R \subseteq A \times B$ be a relation for the sets A and B.

1. The ***domain*** of R is the set

$$\text{dom}(R) = \{x \in A : (x, y) \in R, \text{ some } y \in B\}.$$

2. The ***range*** of R is the set

$$\text{ran}(R) = \{y \in B : (x, y) \in R, \text{ some } x \in A\}.$$

Intuitively, the domain is the set of all possible x-values of the relation, and the range is the set of all possible y-values.

6.1.3. Figure

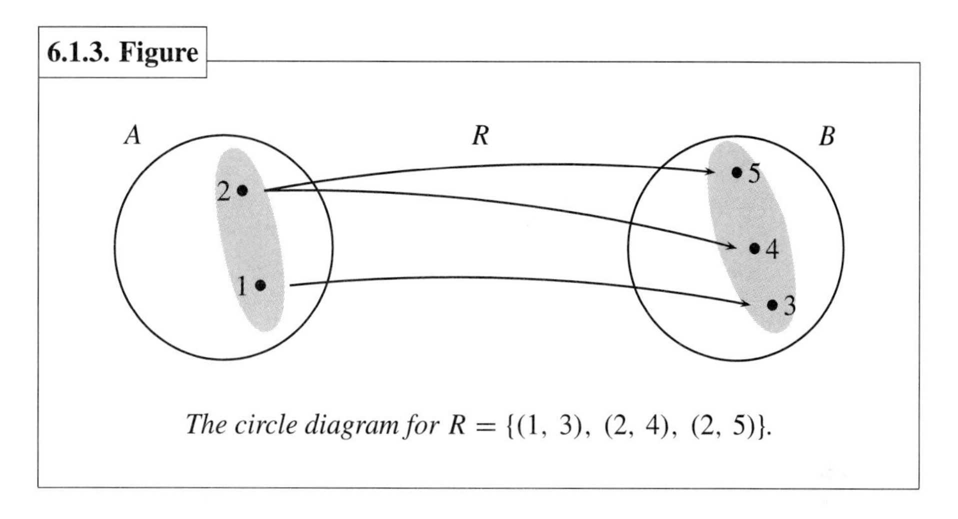

The circle diagram for $R = \{(1, 3), (2, 4), (2, 5)\}$.

Each relation from the next pair of examples has a domain that is different from its range.

Example. If $R = \{(1, 3), (2, 4), (2, 5)\}$, then $\operatorname{dom}(R) = \{1, 2\}$ and $\operatorname{ran}(R) = \{3, 4, 5\}$. We can represent this situation with a picture as in Figure 6.1.3. We will call these ***circle diagrams***. The diagram illustrates the relation as a subset of $A \times B$, where in this case A and B are sets so that $\{1, 2\} \subseteq A$ and $\{3, 4, 5\} \subseteq B$. The ordered pair $(1, 3)$ is denoted by an arrow pointing from 1 to 3. Furthermore, both $\operatorname{dom}(R)$ and $\operatorname{ran}(R)$ are shaded, but this is not always done.

Example. Let $S = \{(x, y) : |x| = y\}$. Notice that both $(2, 2)$ and $(-2, 2)$ are elements of S. As for its domain and range, $\operatorname{dom}(S) = \mathbb{R}$ and $\operatorname{ran}(S) = [0, \infty)$.

The domain and range of a relation can be the same set as in the next two examples.

Example. If $R = \{(0, 1), (0, 2), (1, 0), (2, 0)\}$, then R is a relation on $\{0, 1, 2\}$ with $\operatorname{dom}(R) = \operatorname{ran}(R) = \{0, 1, 2\}$.

Example. For the relation

$$S = \{(x, y) \in \mathbb{R}^2 : \sqrt{x^2 + y^2} \geq 1\},$$

both the domain and range is $\mathbb{R}$. This is seen by the graph in Figure 6.1.4. Since a picture is not a proof, to show $\operatorname{dom}(S) = \mathbb{R}$, first note that it is clear by the definition of S that $\operatorname{dom}(S) \subseteq \mathbb{R}$. For the other inclusion, take $x \in \mathbb{R}$. Since $(x, 1) \in S$, $x \in \operatorname{dom}(S)$.

6.1.4. Figure

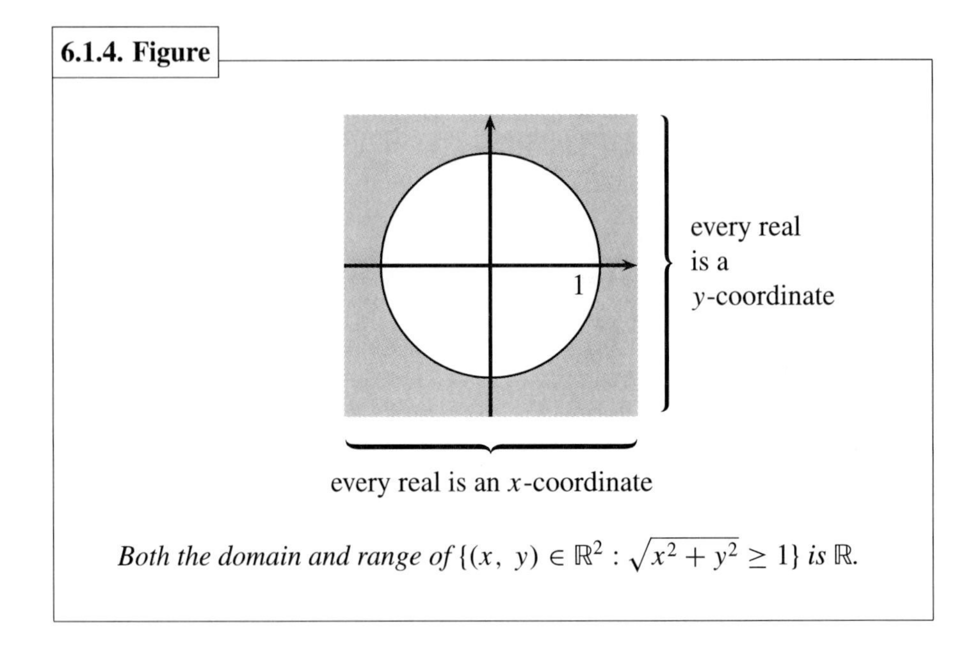

Both the domain and range of $\{(x, y) \in \mathbb{R}^2 : \sqrt{x^2 + y^2} \geq 1\}$ *is* $\mathbb{R}$.

Let R and S be relations with $R \subseteq A \times B$ and $S \subseteq B \times C$. Take $(a, b) \in R$ and $(b, c) \in S$. We will use this to define a new relation that associates a with c. Notice that the notation in the definition may seem backward. The reason for this will be made clear later in the chapter when we study functions.

6.1.5. Definition

Let $R \subseteq A \times B$ and $S \subseteq B \times C$ be relations. The ***composition*** of S and R is denoted by $S \circ R$ and defined as

$$S \circ R = \{(x, z) : (x, y) \in R \text{ and } (y, z) \in S \text{ for some } y \in B\}.$$

Notice $S \circ R \subseteq A \times C$. (See Figure 6.1.6.)

Example. To clarify the definition, let $R = \{(2, 4), (1, 3), (2, 5)\}$ and $S = \{(0, 1), (1, 0), (0, 2), (2, 0)\}$. This means $R \subseteq \mathbb{Z} \times \mathbb{Z}$ and $S \subseteq \mathbb{Z} \times \mathbb{Z}$. Therefore, we may form the compositions:

- $R \circ S = \{(0, 3), (0, 4), (0, 5)\}$. To illustrate, the pair $(0, 3)$ is in $R \circ S$ since $(0, 1) \in S$ and $(1, 3) \in R$.
- $S \circ R = \varnothing$ since $\text{ran}(R)$ and $\text{dom}(S)$ are disjoint.

6.1.6. Figure

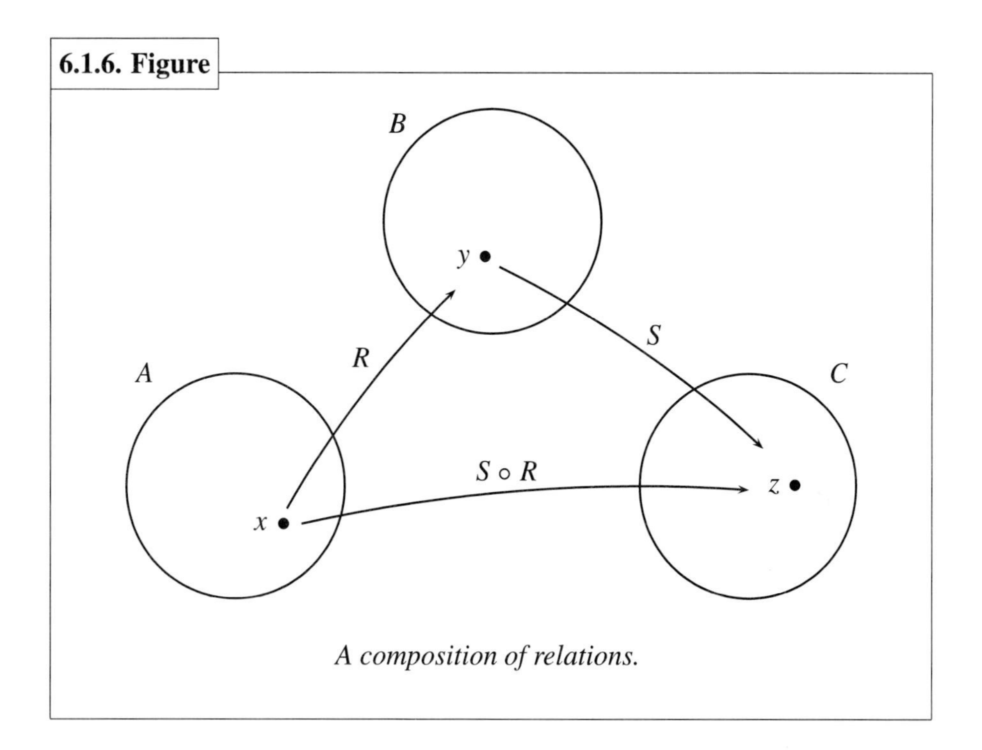

A composition of relations.

Example. Define

$$R = \{(x,\ y) \in \mathbb{R}^2 : x^2 + y^2 = 1\}$$

and

$$S = \{(x,\ y) \in \mathbb{R}^2 : y = x + 1\}.$$

Notice that R is the unit circle centered at the origin and S is the line with slope of 1 and y-intercept of $(0,\ 1)$. Let us find $R \circ S$:

$$\begin{aligned} R \circ S &= \{(x,\ z) : (x,\ y) \in S \text{ and } (y,\ z) \in R, \text{ some } y \in \mathbb{R}\} \\ &= \{(x,\ z) : y = x + 1 \text{ and } y^2 + z^2 = 1, \text{ some } y \in \mathbb{R}\} \\ &= \{(x,\ z) : (x + 1)^2 + z^2 = 1\}. \end{aligned}$$

Therefore, $R \circ S$ is a circle with center $(-1,\ 0)$ and radius 1.

Composition of relations can be considered an operation between two relations. This is because given any two relations, there exists a unique relation that is the composition of the two (see Exercise 13). We may, therefore, consider whether composition is commutative or associative and whether there are identities and inverses with respect to it. From our work with number theory, we know that if there is an identity, it must be unique. If there are inverses, they, too, must be unique. We know this, because we proved the uniqueness of identities and inverses for arbitrary operations (see page 175).

Let us first consider commutativity and associativity. The example on page 210 shows that composition is not commutative, for there $S \circ R \neq R \circ S$. However, we are more fortunate with the associative law.

6.1.7. Theorem

If $R \subseteq A \times B$, $S \subseteq B \times C$, and $T \subseteq C \times D$ are relations, then

$$T \circ (S \circ R) = (T \circ S) \circ R.$$

Proof. If $(a, d) \in A \times D$, then

$$\begin{aligned}
&(a, d) \in T \circ (S \circ R) \\
&\quad \Leftrightarrow (a, c) \in S \circ R \wedge (c, d) \in T, \text{ some } c \in C \\
&\quad \Leftrightarrow [(a, b) \in R \wedge (b, c) \in S] \wedge (c, d) \in T, \text{ some } b \in B,\ c \in C \\
&\quad \Leftrightarrow (a, b) \in R \wedge [(b, c) \in S \wedge (c, d) \in T], \text{ some } b \in B,\ c \in C \\
&\quad \Leftrightarrow (a, b) \in R \wedge (b, d) \in T \circ S, \text{ some } b \in B \\
&\quad \Leftrightarrow (a, d) \in (T \circ S) \circ R.
\end{aligned}$$

Therefore, $T \circ (S \circ R) = (T \circ S) \circ R$. ■

We have already seen the candidate for the identity. It is the identity relation. However, if $R \subseteq A \times B$ is a relation, we need two different identity relations depending on which side it is composed. Specifically, $R \circ I_A = R$ and $I_B \circ R = R$. If we want exactly one relation to act as an identity, we would like $A = B$ so that R is a relation on A. Then,

$$R \circ I_A = I_A \circ R = R.$$

For example, recall the relation $R = \{(2, 4), (1, 3), (2, 5)\}$. We can view R as a relation on $\mathbb{Z}$. We may conclude that $(1, 3) \in R \circ I_{\mathbb{Z}}$. This is because of the following:

$$\begin{aligned}
(1, 3) \in R &\Leftrightarrow (1, 1) \in I_{\mathbb{Z}} \text{ and } (1, 3) \in R \\
&\Leftrightarrow (1, 3) \in R \circ I_{\mathbb{Z}}.
\end{aligned}$$

A generalization of this argument can be used to show that $I_{\mathbb{Z}}$ is an identity for composition of relations on $\mathbb{Z}$ (see Exercise 6).

Example. Not any identity relation will work on both sides. Using the same definition of R as above, $R \circ I_{\{1, 2\}} = R$, but $I_{\{1, 2\}} \circ R = \varnothing$.

For the moment, we will only be able to give a partial answer to the inverse question. We begin with a definition.

6.1.8. Definition

Let R be a relation so that $R \subseteq A \times B$. The ***inverse*** of R is the set

$$R^{-1} = \{(y, x) : (x, y) \in R\}.$$

For a relation S, we say that R and S are ***inverse relations*** if $R^{-1} = S$.

In the first example, we write the inverse of a relation as a roster.

Example. Define $R = \{(2, 1), (2, 3), (3, 1)\}$. Then

$$R^{-1} = \{(1, 2), (3, 2), (1, 3)\}$$

and

$$R \circ R^{-1} = \{(1, 1), (1, 3), (3, 1), (3, 3)\}.$$

The next example shows that two common relations are in fact inverses of each other.

Example. Let $L = \{(x, y) \in \mathbb{R} : x < y\}$ be the less-than relation on $\mathbb{R}$. Then

$$\begin{aligned} L^{-1} &= \{(y, x) \in \mathbb{R} : (x, y) \in L\} \\ &= \{(y, x) \in \mathbb{R} : x < y\} \\ &= \{(y, x) \in \mathbb{R} : y > x\}. \end{aligned}$$

This shows that less-than and greater-than are inverse relations.

Notice that the identity on $\operatorname{ran}(R)$ is a subset of $R \circ R^{-1}$ in the above example. In fact, it is always true that

$$I_{\operatorname{ran}(R)} \subseteq R \circ R^{-1}$$

and

$$I_{\operatorname{dom}(R)} \subseteq R^{-1} \circ R.$$

To see the second inclusion, take (x, x) where $x \in \operatorname{dom}(R)$. By definition, there exists $y \in \operatorname{ran}(R)$ so that $(x, y) \in R$. Hence, $(y, x) \in R^{-1}$, which gives $(x, x) \in R^{-1} \circ R$.

We close this section with the following result that combines inverses and compositions.

6.1.9. Theorem

If $R \subseteq A \times B$ and $S \subseteq B \times C$, then

$$(S \circ R)^{-1} = R^{-1} \circ S^{-1}.$$

Proof. If $(c, a) \in C \times A$, then

$$\begin{aligned}(c, a) \in R^{-1} \circ S^{-1} &\Leftrightarrow (c, b) \in S^{-1} \text{ and } (b, a) \in R^{-1}, \text{ some } b \in B \\ &\Leftrightarrow (b, c) \in S \text{ and } (a, b) \in R, \text{ some } b \in B \\ &\Leftrightarrow (a, b) \in R \text{ and } (b, c) \in S, \text{ some } b \in B \\ &\Leftrightarrow (a, c) \in S \circ R \\ &\Leftrightarrow (c, a) \in (S \circ R)^{-1}. \blacksquare\end{aligned}$$

EXERCISES

1. Find the domain and range of the given relations.
   (a) $\{(0, 1), (2, 3), (4, 5), (6, 7)\}$
   (b) $\{((a, b), 1), ((a, c), 2), ((a, d), 3)\}$
   (c) $\mathbb{R} \times \mathbb{Z}$
   (d) $\varnothing \times \varnothing$
   (e) $\mathbb{Q} \times \varnothing$
   (f) $\{(x, y) : x, y \in [0, 1] \text{ and } x < y\}$
   (g) $\{(x, y) \in \mathbb{R}^2 : y = 2x + 3\}$
   (h) $\{(x, y) \in \mathbb{R}^2 : y = |x|\}$
   (i) $\{(x, y) \in \mathbb{R}^2 : x^2 + y^2 = 1\}$
   (j) $\{(x, y) \in \mathbb{R}^2 : y \leq \sqrt{x}, x \geq 0\}$
   (k) $\{(f, g) : f(x) = e^x \text{ and } g(x) = ax \text{ for some } a \in \mathbb{R}\}$
   (l) $\{((a, b), a + b) : a, b \in \mathbb{Z}\}$
2. Write $R \circ S$ as a roster.
   (a) $R = \{(1, 0), (2, 3), (4, 6)\}$
   $S = \{(1, 2), (2, 3), (3, 4)\}$
   (b) $R = \{(1, 3), (2, 5), (3, 1)\}$
   $S = \{(1, 3), (3, 1), (5, 2)\}$
   (c) $R = \{(1, 2), (3, 4), (5, 6)\}$
   $S = \{(1, 2), (3, 4), (5, 6)\}$
   (d) $R = \{(1, 2), (3, 4), (5, 6)\}$
   $S = \{(2, 1), (3, 5), (5, 7)\}$
3. **3.** For each composition in the previous problem, find both $(R \circ S)^{-1}$ and $S^{-1} \circ R^{-1}$ to confirm Theorem 6.1.9.

4. Write $R \circ S$ using set-builder notation where appropriate.
   (a) $R = \{(x, y) \in \mathbb{R}^2 : x^2 + y^2 = 1\}$
   $S = \mathbb{R}^2$
   (b) $R = \{(x, y) \in \mathbb{R}^2 : x^2 + y^2 = 1\}$
   $S = \mathbb{Z}^2$
   (c) $R = \{(x, y) \in \mathbb{R}^2 : x^2 + y^2 = 1\}$
   $S = \{(x, y) \in \mathbb{R}^2 : (x - 2)^2 + y^2 = 1\}$
   (d) $R = \{(x, y) \in \mathbb{R}^2 : x^2 + y^2 = 1\}$
   $S = \{(x, y) \in \mathbb{R}^2 : y = 2x - 1\}$
5. Write the inverse of each relation using set-builder notation where appropriate.
   (a) $\varnothing$
   (b) $I_{\mathbb{Z}}$
   (c) $\{(1, 0), (2, 3), (4, 6)\}$
   (d) $\{(1, 0), (1, 1), (2, 1)\}$
   (e) $\mathbb{Z} \times \mathbb{R}$
   (f) $\{(x, \sin x) : x \in \mathbb{R}\}$
   (g) $\{(x, y) \in \mathbb{R}^2 : x^2 + y^2 = 1\}$
   (h) $\{(x, y) \in \mathbb{R}^2 : x + y = 1\}$
6. Let R be a relation such that $R \subseteq A \times B$.
   (a) Prove $R \circ I_A = R$ and $I_B \circ R = R$.
   (b) Show that if there exists a set C such that A and B are subsets of C, then $R \circ I_C = I_C \circ R = R$.
7. Let $R \subseteq A \times B$ and $S \subseteq B \times C$. Show that $S \circ R = \varnothing$ if and only if $\text{dom}(S) \cap \text{ran}(R) = \varnothing$.
8. For any relation R, prove $I_{\text{ran}(R)} \subseteq R \circ R^{-1}$.
9. Let $L \subseteq \mathbb{Z} \times \mathbb{Z}$ be the less-than relation as defined on page 208. Prove that

$$L = \{(a, b) \in \mathbb{Z} \times \mathbb{Z} : a - b \in \mathbb{Z}^-\}.$$

10. Let $R \subseteq A \times B$ and $S \subseteq B \times C$. Show:
    (a) $R^{-1} \subseteq B \times A$
    (b) $S \circ R \subseteq A \times C$
    (c) $R^{-1} \circ S^{-1} \subseteq C \times A$
    (d) $\text{dom}(R) = \text{ran}(R^{-1})$
    (e) $\text{ran}(R) = \text{dom}(R^{-1})$
11. Prove that if R is a relation on A, then $(R^{-1})^{-1} = R$.
12. Let $R, S \subseteq A \times B$ be two relations. Prove the following:
    (a) If $R \subseteq S$, then $R^{-1} \subseteq S^{-1}$.
    (b) $(R \cup S)^{-1} = R^{-1} \cup S^{-1}$.
    (c) $(R \cap S)^{-1} = R^{-1} \cap S^{-1}$.
13. Let A be a set. Prove that composition of relations is an operation on the set of all relations on A.

6.2. EQUIVALENCE RELATIONS

Thinking of relations as ordered pairs may seem odd, for in practice we do not usually write them in this way. We instead write statements like $4 = 4$ or $3 < 9$. To mimic this, we will introduce an alternate notation. Let R be a relation on a set A. For all $(a, b) \in R$ define:

$$a \sim_R b \text{ if and only if } (a, b) \in R.^*$$

This is read as "a is related to b (under R)." If the relation is clear from context, just the tilde ($\sim$) is employed. Use the symbol $\nsim_R$ or $\nsim$ to mean that two elements are not related.

Example. Define a relation R on $\mathbb{Z}$ by

$$R = \{(a, b) \in \mathbb{Z} \times \mathbb{Z} : a \mid b\}.$$

Therefore, for all $a, b \in \mathbb{Z}$, $a \sim b$ if and only if a divides b. Therefore, $4 \sim 8$ but $4 \nsim 9$.

Sometimes we will need to study relations that behave like $<$ or $\leq$. The following definitions name important properties of these relations.

6.2.1. Definition

Let $\sim$ be a relation on A.

- The relation is ***irreflexive*** when for all $a \in A$,

$$a \nsim a.$$

- The relation is ***asymmetric*** if for all $a, b \in A$,

$$a \sim b \text{ implies } b \nsim a.$$

- The relation is ***antisymmetric*** when for all $a, b \in A$,

$$\text{if } a \sim b \text{ and } b \sim a, \text{ then } a = b.$$

Example. As a relation on $\mathbb{Z}$, the less-than relation is irreflexive and asymmetric. It is also antisymmetric. To see this, let $a, b \in \mathbb{Z}$. By the Trichotomy Axiom, the conjunction $a < b$ *and* $b < a$ is false. Hence, the implication

$$\text{if } a < b \text{ and } b < a, \text{ then } a = b$$

is true. The $\leq$ relation is also antisymmetric. However, it is neither irreflexive nor asymmetric since $3 \leq 3$.

*Some texts write $a \sim_R b$ as $a \, R \, b$.

Example. Let $R = \{(1, 2)\}$ and $S = \{(1, 2), (2, 1)\}$. Both are relations on $\{1, 2\}$. The first relation is asymmetric since $1 \sim_R 2$ but $2 \nsim_R 1$. It is also antisymmetric, but S is not because $2 \sim_S 1$ and $1 \sim_S 2$ but $1 \neq 2$. Both are irreflexive.

Example. Let R be a relation on a set A. We will prove that

$$R \text{ is antisymmetric if and only if } R \cap R^{-1} \subseteq I_A.$$

($\Leftarrow$) Assume that R is antisymmetric and take $(a, b) \in R \cap R^{-1}$. This means $(a, b) \in R$ and $(a, b) \in R^{-1}$. Because of the second conjunct, $(b, a) \in R$. Therefore, since R is antisymmetric, $a = b$, which means $(a, b) \in I_A$.

($\Rightarrow$) Now suppose $R \cap R^{-1} \subseteq I_A$. Let $(a, b) \in R$ and $(b, a) \in R$. Thus, $(a, b) \in R^{-1}$, which gives $(a, b) \in R \cap R^{-1}$. By hypothesis, $(a, b) \in I_A$. Hence, $a = b$.

Although relations like the above play an important role in mathematics, the most important relation is equality. Its key properties are generalized in the next definition. Compare this definition with Axiom 5.1.1.

6.2.2. Definition

Let $\sim$ be a relation on a set A. We call $\sim$ an ***equivalence relation*** when for all $a, b, c \in A$, the following hold:

1. $a \sim a$ [***Reflexive***];
2. if $a \sim b$; then $b \sim a$ [***Symmetric***];
3. if $a \sim b$ and $b \sim c$, then $a \sim c$ [***Transitive***].

Notice that a relation on a nonempty set cannot be both reflexive and irreflexive at the same time. However, many relations have neither property. For example, consider the relation $R = \{(1, 1)\}$ on $\{1, 2\}$. Since $1 \sim 1$, R is not irreflexive, and R is not reflexive because $2 \nsim 2$. The same can be said about symmetric and asymmetric relations.

Example. Let $m \in \mathbb{Z}^+$. Theorem 5.4.3 has already shown us that congruence modulo m is reflexive, symmetric, and transitive. This means that it is an equivalence relation.

The next example uses ordered pairs as an alternate way to write fractions. In the definition, notice that (a, b) corresponds to a/b, and the relation is equivalent to cross-multiplication.

Example. Define the following relation on $\mathbb{Z} \times \mathbb{Z}^+$: for all $a, c \in \mathbb{Z}$ and $b, d \in \mathbb{Z}^+$,

$$(a, b) \sim (c, d) \text{ if and only if } ad = bc.$$

To see that this is an equivalence relation, we must show that $\sim$ is reflexive, symmetric, and transitive. Let (a, b), (c, d), and (e, f) be elements of $\mathbb{Z} \times \mathbb{Z}^+$.

1. Since $ab = ab$, $(a, b) \sim (a, b)$.
2. Assume $(a, b) \sim (c, d)$. Then, $ad = bc$. This implies $cb = da$, and so $(c, d) \sim (a, b)$.
3. Let $(a, b) \sim (c, d)$ and $(c, d) \sim (e, f)$. This gives $ad = bc$ and $cf = de$. Therefore, $(a, b) \sim (e, f)$ because

$$af = \frac{bcf}{d} = \frac{bdef}{df} = be.$$

An equivalence relation takes all of the elements of a set and groups those that are related together. These groupings are called equivalence classes.

6.2.3. Definition

Let $\sim$ be an equivalence relation on A. If $a \in A$, then the ***equivalence class*** of a is denoted by $[a]_\sim$ and defined as the set

$$\{x \in A : x \sim a\}.$$

Write $[a]$ when the relation is clear from context.

Example. The equivalence classes of congruence modulo 5 are simply the congruence classes:

$$[0] = \{\ldots, -10, -5, 0, 5, 10, \ldots\}$$
$$[1] = \{\ldots, -9, -4, 1, 6, 11, \ldots\}$$
$$[2] = \{\ldots, -8, -3, 2, 7, 12, \ldots\}$$
$$[3] = \{\ldots, -7, -2, 3, 8, 13, \ldots\}$$
$$[4] = \{\ldots, -6, -1, 4, 9, 14, \ldots\}$$

If C is an equivalence class for some equivalence relation and $a \in C$, then $C = [a]$. (See Exercise 6.) The element a is called a ***representative*** of C. Notice that in the last example $[5] = [0]$, so 5 is another representative of $[0]$.

Example. Let $\sim$ be the equivalence relation on $\mathbb{Z} \times \mathbb{Z}^+$ defined by

$$(a, b) \sim (c, d) \text{ if and only if } ad = bc.$$

The equivalence class of $(1, 3)$ is

$$S = \{(n, 3n) : n \in \mathbb{Z} \setminus \{0\}\}.$$

To show this, we must prove that

$$\{(a, b) \in \mathbb{Z} \times \mathbb{Z}^+ : (a, b) \sim (1, 3)\}$$

equals S. The proof is as follows:

($\subseteq$) If $(a, b) \sim (1, 3)$ with $a \in \mathbb{Z}$ and $b \in \mathbb{Z}^+$, then $3a = b$. Since $b \neq 0$, $a \neq 0$. Hence, $(a, b) = (a, 3a) \in S$.

($\supseteq$) Let $(n, 3n) \in S$. Because $n \cdot 3 = 3n \cdot 1$, $(n, 3n) \sim (1, 3)$. Thus, $(n, 3n) \in [(1, 3)]$.

If we take all equivalence classes of an equivalence relation $\sim$ on A and put them in a set, we have a ***quotient set***. It is denoted by $A/\sim$, and in set-builder form it looks like

$$A/\sim = \{[a]_\sim : a \in A\}.$$

Read this as "A modulo tilde." By Exercise 5, it is always the case that

$$A = \bigcup_{a \in A} [a]_\sim.$$

Example. Define an equivalence relation on $\mathbb{R}^2$ by

$$(a, b) \sim (c, d) \text{ if and only if } b - a = d - c.$$

(See Exercise 4.) For any $(a, b) \in \mathbb{R}^2$,

$$\begin{aligned}[(a, b)] &= \{(x, y) : (x, y) \sim (a, b)\} \\ &= \{(x, y) : y - x = b - a\} \\ &= \{(x, y) : y = x + (b - a)\}.\end{aligned}$$

Therefore, the equivalence class of (a, b) is the line with a slope of 1 and a y-intercept equal to $(0, b - a)$. The equivalence classes of $(0, 1.5)$ and $(0, -1)$ are illustrated in the graph in Figure 6.2.4. The quotient set $\mathbb{R}^2/\sim$ is the collection of all such lines. Notice that

$$\mathbb{R}^2 = \bigcup \mathbb{R}^2/\sim.$$

In the previous example, $\mathbb{R}^2$ is the union of all the lines, and since the lines are parallel, they form a pairwise disjoint set. These properties can be observed in the other equivalence relations that we have seen. Each set is equal

6.2.4. Figure

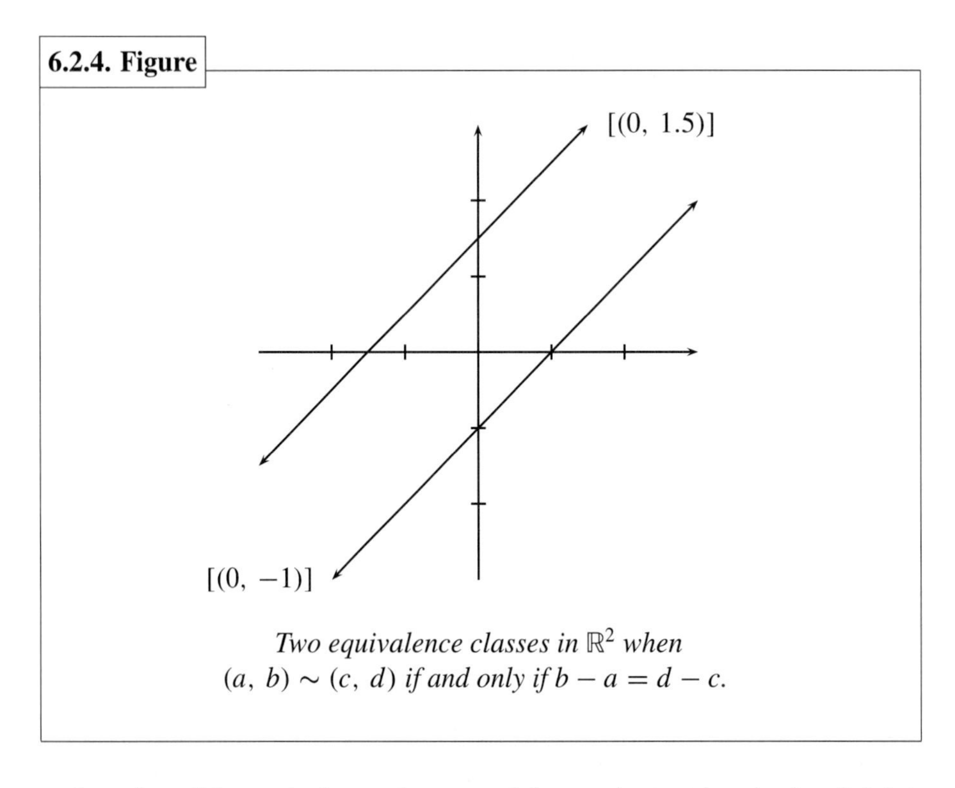

Two equivalence classes in $\mathbb{R}^2$ *when*
$(a, b) \sim (c, d)$ *if and only if* $b - a = d - c$.

to the union of the equivalence classes, and the quotient set is pairwise disjoint. Generalizing these two properties leads to the next definition.

6.2.5. Definition

Let $A \neq \varnothing$ be a set and $\mathcal{P} \subseteq \mathbf{P}(A)$. The set $\mathcal{P}$ is a ***partition*** of A if and only if

1. $\bigcup \mathcal{P} = A$, and
2. $\mathcal{P}$ is pairwise disjoint.

To illustrate the definition, let $A = \{1, 2, 3, 4, 5, 6, 7\}$ and define $A_1 = \{1, 2, 5\}$, $A_2 = \{3\}$, and $A_3 = \{4, 6, 7\}$. The family $\{A_1, A_2, A_3\}$ is pairwise disjoint, and $A = A_1 \cup A_2 \cup A_3$. This is illustrated in Figure 6.2.6.

Example. The collection of lines in the example on the previous page forms a partition of $\mathbb{R}^2$, so does the following: for each real number $r \geq 0$, define C_r to be the circle with radius r centered around the origin. Namely,

$$C_r = \{(x, y) : \sqrt{x^2 + y^2} = r\}.$$

Let $\mathcal{C} = \{C_r : r \in [0, \infty)\}$. We claim that $\mathcal{C}$ is a partition of $\mathbb{R}^2$.

6.2.6. Figure

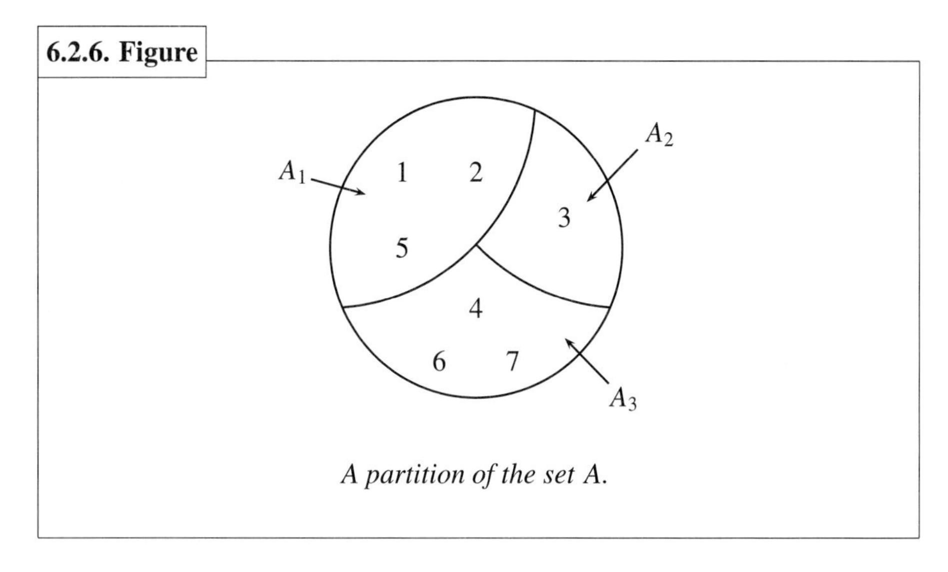

A partition of the set A.

1. Since one inclusion is clear, to prove that $\mathbb{R}^2 = \bigcup_{r\in[0,\infty)} C_r$, it suffices to show $\mathbb{R}^2 \subseteq \bigcup_{r\in[0,\infty)} C_r$. To do this, take $(a, b) \in \mathbb{R}^2$ and let $r_0 = \sqrt{a^2+b^2}$. Thus, $(a, b) \in C_{r_0}$, and we have $(a, b) \in \bigcup_{r\in[0,\infty)} C_r$.
2. To see that $\mathscr{C}$ is pairwise disjoint, let $r, s \geq 0$ such that $r \neq s$. Assume that $(a, b) \in C_r \cap C_s$. Then,

$$r = \sqrt{a^2+b^2} = s,$$

a contradiction. Therefore, $C_r \cap C_s = \varnothing$.

In the next theorem we will use an arbitrary equivalence relation on a set to build a partition for that set. In this case we say that the equivalence relation ***induces*** the partition.

6.2.7. Theorem

If $\sim$ is an equivalence relation on a A, then $A/\!\sim$ is a partition of A.

Proof. Take a set A with an equivalence relation $\sim$. We must show that $A/\!\sim$ satisfies the two parts of the definition of partition.

1. First prove $\bigcup A/\!\sim\, = A$.
   - ($\subseteq$) Suppose $x \in \bigcup A/\!\sim$. This means that there exists $a \in A$ such that $x \in [a]$. Since $[a] \subseteq A$, $x \in A$.
   - ($\subseteq$) Let $x \in A$. By definition, $x \in [x]$. This implies $x \in \bigcup A/\!\sim$, for $[x] \in A/\!\sim$.

2. We will prove the contrapositive of the pairwise disjoint test. Let $[a],\ [b] \in A/\sim$ and assume $[a] \cap [b] \neq \varnothing$. We conclude that there exists $y \in [a] \cap [b]$. In other words, $y \sim a$ and $y \sim b$. Now take $x \in [a]$. This means $x \sim a$. Since $a \sim y$ by symmetry, $x \sim y$ and then $x \sim b$ by transitivity. Thus, $x \in [b]$, which shows $[a] \subseteq [b]$. Similarly, $[b] \subseteq [a]$. Therefore, $[a] = [b]$, and we conclude that the collection is pairwise disjoint. ■

Example. The set

$$\mathbb{Z}_5 = \{[0]_5,\ [1]_5,\ [2]_5,\ [3]_5,\ [4]_5\}$$

is pairwise disjoint, and $\bigcup \mathbb{Z}_5 = \mathbb{Z}$. Hence, $\mathbb{Z}_5$ is a partition for $\mathbb{Z}$. For any $n \in \mathbb{Z}$, to determine which class n belongs, divide by 5 and find the remainder r. Then $n \in [r]_5$.

The collection of equivalence relations forms a partition of a set. Conversely, if we have a partition of a set, the partition gives rise to an equivalence relation for the set. To see this, take any set A and a partition $\mathcal{P}$ of A. For all $a,\ b \in A$, define

$$a \sim b \text{ if and only if there exists } S \in \mathcal{P} \text{ such that } a,\ b \in S.$$

Let us show that $\sim$ is an equivalence relation: Take a, b, and c in A.

1. Since $a \in A$ and $A = \bigcup \mathcal{P}$, there exists $S \in \mathcal{P}$ such that $a \in S$. Therefore, $a \sim a$.
2. Assume $a \sim b$. Then $a,\ b \in S$ for some $S \in \mathcal{P}$. This, of course, is the same as $b,\ a \in S$. Hence, $b \sim a$.
3. Suppose $a \sim b$ and $b \sim c$. Then there are sets S and T in $\mathcal{P}$ so that $a,\ b \in S$ and $b,\ c \in T$. This means that $S \cap T \neq \varnothing$. Since $\mathcal{P}$ is pairwise disjoint, $S = T$. So, a and c are in S, and we have $a \sim c$.

This equivalence relation is said to be ***induced*** from the partition.

Example. The sets

$$\{\ldots,\ -3,\ 0,\ 3,\ \ldots\},$$
$$\{\ldots,\ -2,\ 1,\ 4,\ \ldots\},$$
$$\{\ldots,\ -1,\ 2,\ 5,\ \ldots\}$$

form a collection that is a partition of $\mathbb{Z}$. The equivalence relation that arises from this partition is congruence modulo 3.

EXERCISES

1. For each relation on $\{1, 2\}$, determine if it is reflexive, irreflexive, symmetric, asymmetric, antisymmetric, or transitive.

(a) $\{(1, 2)\}$
(b) $\{(1, 2), (2, 1)\}$
(c) $\{(1, 1), (1, 2), (2, 1)\}$
(d) $\{(1, 1), (1, 2), (2, 2)\}$
(e) $\{(1, 1), (1, 2), (2, 1), (2, 2)\}$
(f) $\varnothing$

2. Give an example of an infinite set $A \subseteq \mathbb{R}$ and a relation R on A so that:

(a) R is symmetric but not asymmetric.
(b) R is asymmetric but not symmetric.
(c) R is neither symmetric nor asymmetric.

3. For all $a, b \in \mathbb{R} \setminus \{0\}$, let $a \sim b$ if and only if $ab > 0$.

(a) Show that $\sim$ is an equivalence relation on $\mathbb{R} \setminus \{0\}$.
(b) Find the following equivalence classes:

(i) $[1]$ (ii) $[-3]$

4. Define a relation on $\mathbb{R}^2$ by

$$(a, b) \sim (c, d) \text{ if and only if } b - a = d - c.$$

Prove that $\sim$ is an equivalence relation.

5. Let $\sim$ be an equivalence relation on A. Prove that $A = \bigcup_{a \in A} [a]_\sim$.

6. Prove that if C is an equivalence class for some equivalence relation and $a \in C$, then $C = [a]$.

7. For all $a, b \in \mathbb{Z}$, let $a \sim b$ if and only if $|a| = |b|$.

(a) Prove $\sim$ is an equivalence relation on $\mathbb{Z}$.
(b) Sketch the partition of $\mathbb{Z}$ induced by this equivalence relation.

8. For all $(a, b), (c, d) \in \mathbb{Z} \times \mathbb{Z}$, define $(a, b) \sim (c, d)$ if and only if $ab = cd$.

(a) Show that $\sim$ is an equivalence relation on $\mathbb{Z} \times \mathbb{Z}$.
(b) What is the equivalence class of $(1, 2)$?
(c) Sketch the partition of $\mathbb{Z}$ induced by this equivalence relation.

9. Let S be a set and $a \in S$. Show that the following are not equivalence relations on $\mathbf{P}(S)$.

(a) For all $A, B \subseteq S$, define $A \sim B$ if and only if $A \cap B \neq \varnothing$.
(b) For all $A, B \subseteq S$, define $A \sim B$ if and only if $a \in A \cap B$.

10. How many partitions into 3 sets do the following have?

(a) $\{1, 2, 3\}$
(b) $\{1, 2, 3, 4\}$
(c) $\{1, 2, 3, 4, 5\}$

11. Prove that $\{(n, n+1] : n \in \mathbb{Z}\}$ is a partition of $\mathbb{R}$.

12. Prove that the following are partitions of $\mathbb{R}^2$.
   (a) $\mathcal{P} = \{\{(a, b)\} : a, b \in \mathbb{R}\}$.
   (b) $\mathcal{V} = \{V_r : r \in \mathbb{R}\}$ where $V_r = \{(r, y) : y \in \mathbb{R}\}$ for all real numbers r.
   (c) $\mathcal{H} = \{H_n : n \in \mathbb{Z}\}$ where $H_n = \{\mathbb{R} \times (i, i+1] : i \in \mathbb{Z}\}$ for all $n \in \mathbb{Z}$.

13. Is $\{[n, n+1] \times (n, n+1) : n \in \mathbb{Z}\}$ is a partition of $\mathbb{R}^2$? Explain.

14. Let R be a relation on A and show the following:
   (a) R is reflexive if and only if R^{-1} is reflexive.
   (b) R is reflexive if and only if $(A \times A) \setminus R$ is irreflexive.
   (c) R is symmetric if and only if $R = R^{-1}$.
   (d) R is symmetric if and only if $(A \times A) \setminus R$ is symmetric.

15. Let R and S be relations on A. Prove:
   (a) If R and S are reflexive, then $R \cup S$ and $R \cap S$ are reflexive.
   (b) If R and S are symmetric, then $R \cup S$ and $R \cap S$ are symmetric.
   (c) If R and S are transitive, then $R \cup S$ and $R \cap S$ are transitive.

16. Show that a relation R on A is asymmetric if and only if $R \cap R^{-1} = \varnothing$.

17. Let R be a relation on A. For all $a \in A$, define

$$R(a) = \{a' \in A : (a, a') \in R\}.$$

   Prove the following: for all $a, b, c \in A$,
   (a) R is reflexive if and only if $a \in R(a)$.
   (b) R is symmetric if and only if

$$a \in R(b) \Leftrightarrow b \in R(a).$$

   (c) R is transitive if and only if

$$[b \in R(a) \wedge c \in R(b)] \Rightarrow c \in R(a).$$

   (d) R is an equivalence relation if and only if

$$(a, b) \in R \Leftrightarrow R(a) = R(b).$$

6.3. FUNCTIONS

From algebra and calculus we know what a function is. It is a rule or equation that assigns to each possible x-value a unique y-value. The common picture is that of a machine that when a certain button is pushed the same result always happens. Translating this into algebra, if we graph an equation that is a function, the curve will pass the ***vertical line test*** (Figure 6.3.1). Namely, every vertical line intersects the graph at most once if and only if it is a function.

6.3.1. Figure

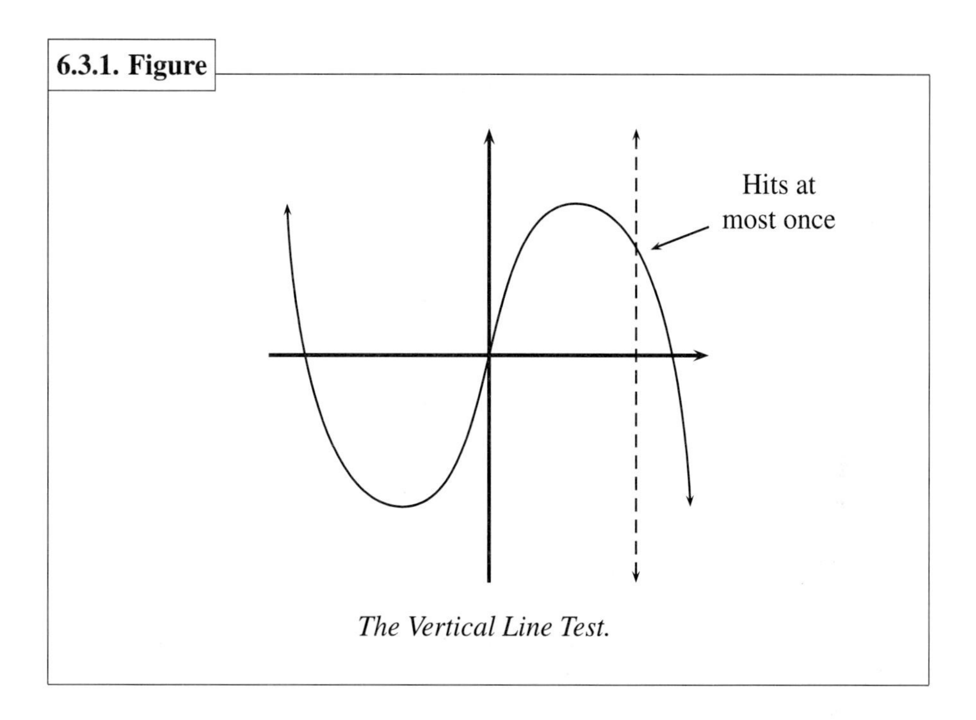

The Vertical Line Test.

Furthermore, if a graph can represent a function, then a function can be viewed as a set of ordered pairs. If it is a set of ordered pairs, then a function is a relation. Let us make this precise.

6.3.2. Definition

Let A and B be sets. A relation $f \subseteq A \times B$ with domain A is a ***function*** when for all $x \in A$ and $y,\ y' \in \operatorname{ran}(f)$,

$$\text{if } (x,\ y) \in f \text{ and } (x,\ y') \in f, \text{ then } y = y'.$$

The set B is called a ***codomain*** of f.

Notice that the definition is a general version of the vertical line test. The only way both $(x,\ y)$ and $(x,\ y')$ can be elements of the function is to have them be the same ordered pair.

Example. The set $\{(1,\ 2),\ (4,\ 5),\ (6,\ 5)\}$ is a function, but the relation given by $\{(1,\ 2),\ (1,\ 5),\ (6,\ 5)\}$ is not since it contains both $(1,\ 2)$ and $(1,\ 5)$.

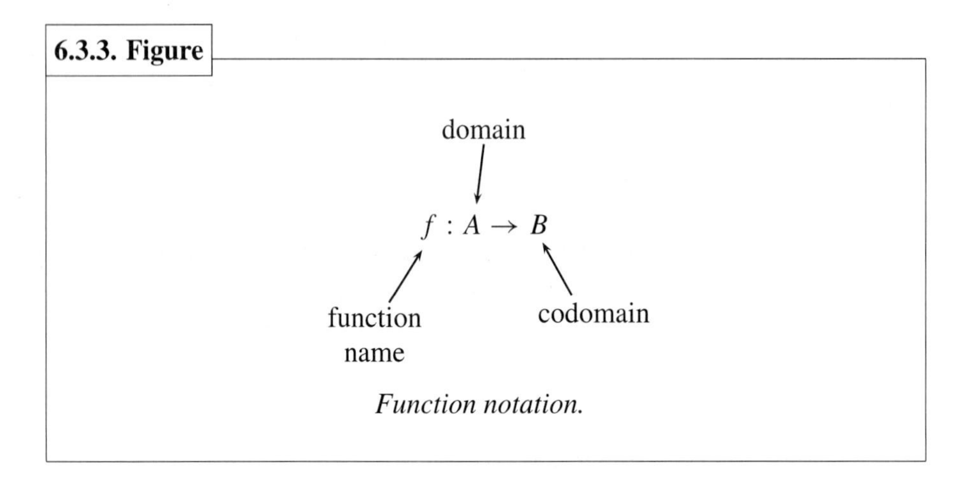

Function notation.

When f is a function, we will use standard ***function notation*** and write $y = f(x)$ for $(x, y) \in f$. If A is the domain of f and B is a codomain, then write $f : A \to B$. (See Figure 6.3.3.) Because functions are often represented by arrows that "send" one element to another, a function can be called a ***map***. If $f(x) = y$, we can say that "f maps x to y." We can also say that y is the ***image*** of x under f and x is a ***pre-image*** of y. If g is another function with domain A and codomain B, then we may use the abbreviation $f, g : A \to B$ to represent both functions. An alternate choice of notation involves referring to the functions as $A \to B$.

Example. If $f(x) = \cos x$, then we may say *f maps π to -1, -1 is the image of π*, or *π is a pre-image of -1*.

Now to show that some relations are functions.

Example. Let $f(x) = 1 + 2\cos \pi x$. The function f as a relation is the set

$$\{(x, 1 + 2\cos \pi x) : x \in \mathbb{R}\}.$$

Let us show that f is a function. Take $(x, 1 + 2\cos \pi x) \in f$. We need another coordinate in f with the property that the x-coordinate matches the original, yet the y may be different. The best way to do this is to take $(x', 1 + 2\cos \pi x') \in f$ and assume $x = x'$. Certainly the y-coordinates must match. The computation would go like this:

$$\begin{aligned} x = x' &\Rightarrow \pi x = \pi x' \\ &\Rightarrow \cos \pi x = \cos \pi x' \\ &\Rightarrow 2\cos \pi x = 2\cos \pi x' \\ &\Rightarrow 1 + 2\cos \pi x = 1 + 2\cos \pi x'. \end{aligned}$$

(This second implication holds because cosine is a function.) Therefore, f is a function, and we may write $f : \mathbb{R} \to \mathbb{R}$.

The situation in the last example is common. Sometimes an expression is written using the $y = f(x)$ notation even though it may not be a function. This may be due to some unobserved complication or an oversight. To check that f is indeed a function, take $x, x' \in \text{dom}(f)$ and show

$$\text{if } x = x', \text{ then } f(x) = f(x').$$

When this is done successfully, we say that f is ***well-defined***. (This is another way to say that f is a function.)

Before we examine another example, let us set a convention about naming functions. It is mainly for aesthetics, but it does help in identifying the type of function that is being used.

- Use Latin letters (usually f, g, and h) for naming functions that involve only numbers. Typically these will be lower case, but there are occasions when we will want them to be upper case.
- Use Greek letters (often ϕ or ψ) for other types of functions. They are also usually lower case, but capital Greek letters like Φ and Ψ are sometimes appropriate.

We will use this convention in the next examples.

Example. Let $n, m \in \mathbb{Z}^+$ such that $m \mid n$. Define $\phi([a]_n) = [a]_m$ for all $a \in \mathbb{Z}$. (Remember, $[a]_k$ is the congruence class of a modulo k.) It is not clear that ϕ is well-defined since a congruence class usually has many representatives. So, assume $[a]_n = [b]_n$ for $a, b \in \mathbb{Z}$. Therefore, $n \mid a - b$. Then by hypothesis, $m \mid a - b$, and this yields

$$\phi([a]_n) = [a]_m = [b]_m = \phi([b]_n).$$

Example. Define $\psi(x, y) = x + y$ for all $(x, y) \in \mathbb{Z} \times \mathbb{Z}$. For example, $\psi(1, 4) = 5$. Notice that for simplicity sake we do not write $\psi((1, 4)) = 5$, although it is technically correct. To see that ψ is a function, let $(a, b), (a', b') \in \mathbb{Z} \times \mathbb{Z}$ and assume $(a, b) = (a', b')$. Then $a = a'$ and $b = b'$. Therefore, $a + b = a' + b'$, which means $\psi(a, b) = \psi(a', b')$.

Example. Let A be any set. The identity relation on A is actually a function. We can denote it by

$$I_A(x) = x$$

6.3.4. Figure

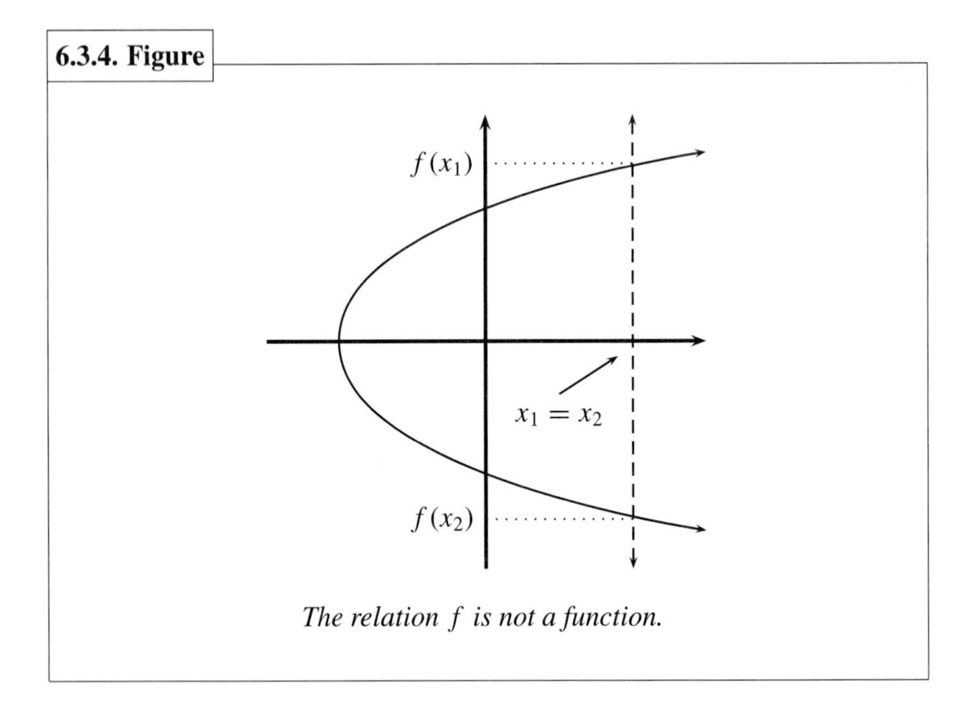

The relation f is not a function.

and call it the ***identity map***. It is an easy exercise to show that this is well-defined. (Following the convention, the I can be considered as a capital iota!)

As noted above, it is possible to write an expression using function notation, but not have the definition be a function, as with $f(x) = \pm\sqrt{x}$. We can show that such a definition is not well-defined by proving

$$\sim(\forall x_1)(\forall x_2)[x_1 = x_2 \Rightarrow f(x_1) = f(x_2)],$$

but this is logically equivalent to

$$(\exists x_1)(\exists x_2)[x_1 = x_2 \wedge f(x_1) \neq f(x_2)].$$

If both the domain and codomain are $\mathbb{R}$, this amounts to the function failing the vertical line test (Figure 6.3.4). This means that a relation $f \subseteq A \times B$ is not a function if there exist $(x, y_1), (x, y_2) \in f$ such that $y_1 \neq y_2$.

Example. The relation $f = \{(x, \pm|x|) : x \in \mathbb{R}\}$ is not a function since $(4, 4) \in f$ and $(4, -4) \in f$.

Example. Let $\phi([a]_2) = [a]_3$ for all $a \in \mathbb{Z}$. Since 3 does not divide 2, ϕ is not well-defined. This is proven by finding an example: $[0]_2 = [2]_2$, but $[0]_3 \neq [2]_3$. Hence, $\phi([0]_2) \neq \phi([2]_2)$.

There will be times when we want to examine sets of functions. If each function is to have the same domain and codomain, we use the following notation.

6.3.5. Definition

If A and B are sets, then

$$B^A = \{f : f \text{ is a function with domain } A \text{ and codomain } B\}.$$

For instance, B^A is a set of ***real-valued*** functions if $A,\ B \subseteq \mathbb{R}$. (Note: some texts use ${}^BA$ instead of B^A.)

Example. We are familiar with sequences from calculus, but we may not typically think of them as functions. If a_n is a sequence of real numbers with $n = 1, 2, 3, \ldots$, then it is an element of $\mathbb{R}^{\mathbb{Z}^+}$. For instance, if $a_n = (-1/2)^{n-1}$, then it can be viewed as a function as in the graph in Figure 6.3.6.

Example. Fix $a \in \mathbb{R}$. The ***evaluation map***,

$$\varepsilon_a : \mathbb{R}^{\mathbb{R}} \to \mathbb{R},$$

is defined as

$$\varepsilon_a(f) = f(a).$$

For example, if $g(x) = x^2$, then $\varepsilon_3(g) = 9$. Thus, the evaluation map is an element of $\mathbb{R}^{\mathbb{R}^{\mathbb{R}}}$.

6.3.6. Figure

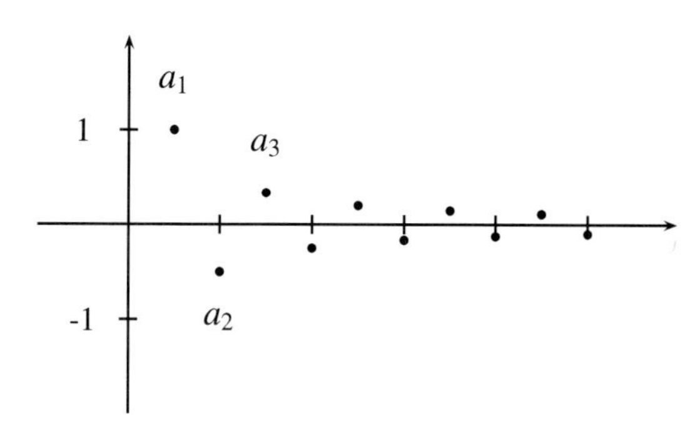

$a_n = (-1/2)^{n-1}$ *is a function.*

EXERCISES

1. Evaluate the indicated expressions.

(a) $\varepsilon_4(f)$ if $f(x) = 9x + 2$ (b) $\varepsilon_\pi(g)$ if $g(\theta) = \sin\theta$

2. Let $\phi: \mathbb{Z} \to \mathbb{Z}_7$ be defined by $\phi(a) = [a]_7$. Write the following as rosters:

(a) $\phi(0)$ (c) $\phi(3)$
(b) $\phi(7)$ (d) $\phi(-3)$

3. Let x be in the domain and y in the range of each relation. Explain why each of the given equations does not describe a function.

(a) $y = 5 \pm x$ **(c)** $x = 4y^2 - 1$
(b) $x^2 + y^2 = 1$ (d) $y^2 - x^2 = 9$

4. Indicate whether each of the following are functions. If one is not a function, find an element of its domain that is associated with two elements of its range.

(a) $\{(1, 2), (2, 3), (3, 4), (4, 5), (5, 1)\}$
(b) $\{(1, 1), (1, 2), (1, 3), (1, 4), (1, 5)\}$
(c) $f: \mathbb{R} \to \mathbb{R}$ if $f(x) = \sqrt{|x|}$
(d) $g: \mathbb{R} \to \mathbb{R}$ if $g(x) = \pm\sqrt{|x|}$
(e) $h: \mathbb{R} \to [0, \infty)$ if $h(x) = x^2$
(f) $\phi: \mathbb{Z}_5 \to \mathbb{Z}$ if $\phi([a]_5) =$ a representative of $[a]_5$
(g) $\psi: \mathbb{Z} \to \mathbb{Z}_5$ if $\psi(a) = [a]_5$

5. Give a domain and two possible codomains for each of the following functions. Write your answers in the form $\phi: A \to B$.

(a) $f(x) = |x|$ **(d)** $h(x) = |\ln x|$
(b) $g(x) = \cos x$ (e) $\psi(x, y, z) = (x, 0)$
(c) $f(x) = \lceil x \rceil$ (f) $\varepsilon_a(f) = f(a)$

6. Give an example of a function that is an element of each of the following:

(a) $\mathbb{R}^\mathbb{R}$ **(d)** $\mathbb{N}^\mathbb{R}$
(b) $\mathbb{Z}^\mathbb{R}$ (e) $[0, \infty)^\mathbb{R}$
(c) $\mathbb{R}^\mathbb{N}$ (f) $\mathbb{Z}_5^\mathbb{Z}$

7. Prove that the following famous formulas are well-defined on an appropriate domain.

(a) $f(x) = 1/x$ (c) $h(x) = |x|$
(b) $g(x) = x + 1$ (d) $f(x) = \sqrt{x}$

8. Let f be the relation $\{(x, y) \in \mathbb{R}^2 : 2x + y = 1\}$. Show that f is a function with domain and codomain equal to the set of real numbers.

9. Let $f, g: \mathbb{R} \to \mathbb{R}$ be functions. Prove that $\phi(x, y) = (f(x), g(y))$ is a function with domain and codomain equal to $\mathbb{R} \times \mathbb{R}$.

10. Let $\psi(A) = \mathbf{P}(A)$. Show that ψ is a function.
11. Define

$$f(x) = \begin{cases} x^2 & \text{if } x \geq 0 \\ 5 & \text{if } x < 0. \end{cases}$$

Show that f is well-defined with domain and codomain equal to $\mathbb{R}$.

12. Let

$$\mathbb{P}_2 = \{ax^2 + bx + c : a,\ b,\ c \in \mathbb{R}\},$$

and

$$\mathbb{P}_1 = \{ax + b : a,\ b \in \mathbb{R}\}.$$

Define $\delta \colon \mathbb{P}_2 \to \mathbb{P}_1$ by

$$\delta(ax^2 + bx + c) = 2ax + b.$$

Prove that δ is well-defined. (Note: δ should be familiar!)

13. Since functions are sets, we can perform set operations on them. Let $f(x) = x^2$ and $g(x) = -x$. Find the following:
    (a) $f \cup g$
    (b) $f \cap g$
    (c) $f \setminus g$
    (d) $g \setminus f$

6.4. FUNCTION OPERATIONS

In Section 5.1 we introduced the notion of a binary operation. Using the terminology of this chapter, the definition on page 173 means that a binary operation $*$ on a set S is a function $S \times S \to S$. For any $a,\ b \in S$, the notation $a * b$ refers to the image of $(a,\ b)$ under $*$. Therefore, standard addition and multiplication of real numbers are functions $\mathbb{R} \times \mathbb{R} \to \mathbb{R}$. For instance, $3 + 5$ is 8 and never another number. Subtraction is also an operation on $\mathbb{R}$, but it is not one on $\mathbb{Z}^+$. This is because subtraction is not a function $\mathbb{Z}^+ \times \mathbb{Z}^+ \to \mathbb{Z}^+$ since the difference $3 - 5$ is not a positive integer. This means that if $*$ is to be a binary operation on S, we want $a * b \in S$ for all $a,\ b \in S$. If this is the case, we say that S is ***closed*** under $*$. For example, $\mathbb{Z}$ is closed under addition but not under subtraction. We may now make the following definition:

6.4.1. Definition

A relation $*$ is a ***binary operation*** on S if for all $s,\ s',\ t,\ t' \in S$:

1. $s = s'$ and $t = t'$ implies $s * t = s' * t'$, and
2. S is closed under $*$.

If $*$ satisfies the first condition of the definition, we say that $*$ is ***well-defined***.

Example. Let $n \in \mathbb{Z}^+$ and define $[a]_n + [b]_n = [a+b]_n$. We will show that this is a binary operation on $\mathbb{Z}_n$.

1. Let $a, a', b, b' \in \mathbb{Z}$. Suppose $[a]_n = [a']_n$ and $[b]_n = [b']_n$. This means $a = a' + nk$ and $b = b' + n\ell$ for some $k, \ell \in \mathbb{Z}$. Hence, $a + b = a' + b' + n(k + \ell)$, and we have $[a+b]_n = [a'+b']_n$.
2. For closure, let $[a]_n, [b]_n \in \mathbb{Z}_n$ where a and b are integers. Then $[a]_n + [b]_n = [a+b]_n \in \mathbb{Z}_n$ since $a + b$ is an integer.

Example. Fix a set A. For any $X, Y \subseteq A$, define $X * Y = X \cup Y$. To show that $*$ is a binary operation on $\mathbf{P}(A)$ proceed as follows:

1. Let $X_1, X_2, Y_1, Y_2 \subseteq A$. If we assume that $X_1 = X_2$ and $Y_1 = Y_2$, then we have $X_1 \cup Y_1 = X_2 \cup Y_2$. Hence, $*$ is well-defined.
2. To show that $\mathbf{P}(A)$ is closed under $*$, let $B, C \in \mathbf{P}(A)$. Then $B * C = B \cup C$. It is an easy exercise to show $B \cup C \subseteq A$, and hence $B \cup C \in \mathbf{P}(A)$.

These examples show that operations need not only take numbers. In this section, the operations will be on sets of functions. Before we begin, however, it will be necessary to discuss what it means for two functions to be the same. This is the purpose of the next definition.

6.4.2. Definition

Let $f, g\colon A \to B$ be functions. We say that f is ***equal*** to g when for all $x \in A$,

$$f(x) = g(x).$$

If this is the case, write $f = g$.

Example. Let f and g be functions $\mathbb{R} \to \mathbb{R}$ defined by $f(x) = (x-3)^2+2$ and $g(x) = x^2 - 6x + 11$. We will show that $f = g$ by taking $x \in \mathbb{R}$ and calculating:

$$\begin{aligned} f(x) &= (x-3)^2 + 2 \\ &= (x^2 - 6x + 9) + 2 \\ &= x^2 - 6x + 11 \\ &= g(x). \end{aligned}$$

Example. Let $\phi,\ \psi\colon \mathbb{Z} \to \mathbb{Z}_6$ be functions such that

$$\phi(n) = [n]_6$$

and

$$\psi(n) = [n + 12]_6.$$

To prove that these two functions are equal take $n \in \mathbb{Z}$. We must show $[n + 12]_6 = [n]_6$. To do this, proceed as follows:

$$\begin{aligned} x \in [n + 12]_6 &\Leftrightarrow x = n + 12 + 6k, \text{ some } k \in \mathbb{Z} \\ &\Leftrightarrow x = n + 6(2 + k), \text{ some } k \in \mathbb{Z} \\ &\Leftrightarrow x \in [n]_6 \end{aligned}$$

Therefore, $\phi = \psi$.

The next example combines the notions of well-defined functions and function equality.

Example. Recall that ε_a is the evaluation map and $\mathbb{R}^{\mathbb{R}}$ is a collection of real-valued functions $\mathbb{R} \to \mathbb{R}$. Define

$$\psi\colon \mathbb{R} \to \mathbb{R}^{\mathbb{R}^{\mathbb{R}}}$$

by $\psi(x) = \varepsilon_x$, all $x \in \mathbb{R}$. We want to show that ψ is well-defined. Take $a,\ b \in \mathbb{R}$ and assume $a = b$. We must show that $\psi(a) = \psi(b)$, but this means $\varepsilon_a = \varepsilon_b$. Therefore, let $f \in \mathbb{R}^{\mathbb{R}}$. Since f is a function and $a,\ b \in \operatorname{dom}(f)$, $f(a) = f(b)$. Thus,

$$\varepsilon_a(f) = f(a) = f(b) = \varepsilon_b(f).$$

By negating the definition of function equality, we see that two functions, f and g, are not equal when $f(x) \neq g(x)$ for some x in their domain. For example, $f(x) = x^2$ and $g(x) = 2x$ are not equal because $f(3) = 9$ and $g(3) = 6$. Although these two functions differ for every $x \neq 0$, it only takes one inequality to prove that the functions are not equal. For example, if we define

$$h(x) = \begin{cases} x^2 & \text{if } x \neq 0 \\ 7 & \text{if } x = 0 \end{cases},$$

then $f \neq h$ since $f(0) = 0$ and $h(0) = 7$.

Our first function operation is an important application where understanding the domain and codomain is crucial. Let A, B, C, and D be nonempty sets. Let $f\colon A \to B$ and $g\colon C \to D$ be two functions so that $\operatorname{ran}(f) \subseteq C$. This means that as relations,

$$f \subseteq A \times C \text{ and } g \subseteq C \times D.$$

Therefore, we may find their composition:

$$\begin{aligned} g \circ f &= \{(x, z) : (x, y) \in f \text{ and } (y, z) \in g, \text{ some } y \in C\} \\ &= \{(x, z) : f(x) = y \text{ and } g(y) = z, \text{ some } y \in C\} \\ &= \{(x, z) : g(f(x)) = z\} \\ &= \{(x, g(f(x))) : x \in A\}. \end{aligned}$$

This is the motivation for the next definition.

6.4.3. Definition

Let $f: A \to B$ and $g: C \to D$ be functions so that $\text{ran}(f) \subseteq C$. We define a new function $g \circ f$ as

$$(g \circ f)(x) = g(f(x)).$$

This function is called the ***composition*** of g and f.

Although the $\circ$ can be considered an operation, the notation $g \circ f$ is the name of a new function. Furthermore, the $\text{ran}(f) \subseteq C$ condition is important. The composition of relations is always defined, even if it is empty. However, if $\text{ran}(f) \nsubseteq C$, then $g \circ f$ may be undefined. For instance, take the real-valued functions $f(x) = x$ and $g(x) = \sqrt{x}$ and try to calculate $(g \circ f)(-1)$:

$$(g \circ f)(-1) = g(f(-1)) = g(-1) = \sqrt{-1} \notin \mathbb{R}.$$

This composition is undefined because $\text{ran}(f) = \mathbb{R}$ but $\text{dom}(g) = [0, \infty)$.

Example. Define the two functions $f: \mathbb{R} \to \mathbb{Z}$ and $g: \mathbb{R} \setminus \{0\} \to \mathbb{R}$ by $f(x) = \lceil x \rceil$ and $g(x) = 1/x$. Since $\text{ran}(f) = \mathbb{Z} \nsubseteq \text{dom}(g)$, $g \circ f$ is undefined as a function. However,

$$\text{ran}(g) = \mathbb{R} \setminus \{0\} \subseteq \mathbb{R} = \text{dom}(f),$$

so $f \circ g$ is defined and is

$$(f \circ g)(x) = f(g(x)) = f(1/x) = \lceil 1/x \rceil.$$

Example. Let $\psi: \mathbb{Z}^{\mathbb{Z}} \to \mathbb{Z}$ be defined by $\psi(f) = \varepsilon_3(f)$ and also let $\phi: \mathbb{Z} \to \mathbb{Z}_7$ be $\phi(n) = [n]_7$. Since $\text{ran}(\psi) \subseteq \text{dom}(\phi)$, $\phi \circ \psi$ is defined. Thus, if g is the function $\mathbb{Z} \to \mathbb{Z}$ defined as $g(n) = 3n$, then:

$$(\phi \circ \psi)(g) = \phi(\psi(g)) = \phi(g(3)) = \phi(9) = [9]_7 = [2]_7.$$

The next set of function operations should be familiar to all former calculus students.

6.4.4. Definition

Let f and g be real-valued functions. Define for all $x,\ k \in \mathbb{R}$:

$$
\begin{aligned}
(f+g)(x) &= f(x)+g(x)\\
(f-g)(x) &= f(x)-g(x)\\
(fg)(x) &= f(x)g(x)\\
(f/g)(x) &= f(x)/g(x) \text{ if } g(x) \neq 0 \text{ for all } x \in \mathbb{R}\\
(kf)(x) &= kf(x).
\end{aligned}
$$

Remember that $f+g$ is the name of the new function. To illustrate, if $f(x) = x+3$ and $g(x) = x^2+5x-7$, then

$$(f+g)(x) = (x+3)+(x^2+5x-7) = x^2+6x-4.$$

Example. Let f be a real-valued function with domain A. We say that f is ***even*** when for all $x \in A$,

$$f(-x) = f(x),$$

and f is ***odd*** if

$$f(-x) = -f(x)$$

for all $x \in A$. Let us show that $f(x) = x^4$ is even but $g(x) = x+1$ is neither even nor odd.

1. To see that f is even, let $x \in \mathbb{R}$ and use a chain of equal signs:

$$f(-x) = (-x)^4 = x^4 = f(x).$$

2. To see that g is not even, we must find one real at which the formula does not hold:

$$
\begin{aligned}
g(-2) &= -1\\
g(2) &= 3.
\end{aligned}
$$

In fact, g is also not odd, for $-g(2) = -3 \neq -1$. Therefore, in order to show that a function is odd, it is not enough to show that it is not even.

Example. Suppose f and g are odd functions. We claim that $f+g$ is odd. To see this, take $x \in \mathbb{R}$ and calculate:

$$\begin{aligned}(f+g)(-x) &= f(-x)+g(-x)\\ &= -f(x)-g(x)\\ &= -[f(x)+g(x)]\\ &= -(f+g)(x).\end{aligned}$$

The next definition will play a role in the next section. Although it is not a binary operation, it is a function

$$B^A \times C \to B^C.$$

6.4.5. Definition

Let $f: A \to B$ be a function and $C \subseteq A$. The ***restriction*** of f to C is a function $g: C \to B$ so that

$$g(x) = f(x) \text{ for all } x \in C.$$

We will use the notation $f \restriction C$ for the restriction of f to C.

(Note: some texts will use a notation like $f|_A$ to represent $f \restriction A$.)

Example. Let $f = \{(1, 2), (2, 3), (3, 4), (4, 1)\}$. Then,

$$f \restriction \{1, 2\} = \{(1, 2), (2, 3)\}.$$

Example. Let $\phi: C^A \to C^B$ be defined by $\phi(f) = f \restriction B$. This is a function whose domain and codomain are sets of functions.

Example. Let $f: U \to V$ be a function and $A, B \subseteq U$. Let us prove

$$f \restriction (A \cup B) = (f \restriction A) \cup (f \restriction B).$$

Viewing f as a relation we have the following:

$$\begin{aligned}(x, y) \in f \restriction (A \cup B) &\Leftrightarrow y = f(x) \text{ and } x \in A \cup B\\ &\Leftrightarrow y = f(x) \text{ and } (x \in A \text{ or } x \in B)\\ &\Leftrightarrow (x, y) \in f \restriction A \text{ or } (x, y) \in f \restriction B\\ &\Leftrightarrow (x, y) \in (f \restriction A) \cup (f \restriction B).\end{aligned}$$

EXERCISES

1. Let $f, g\colon \mathbb{R} \to \mathbb{R}$ be functions. Prove that each of the following are well-defined.
   (a) $f \circ g$
   (b) $f + g$
   (c) $f - g$
   (d) fg
2. Prove that the following pairs of functions are equal.
   (a) $f(x) = (x-1)(x-2)(x+3)$ and $g(x) = x^3 - 7x + 6$, where $f, g\colon \mathbb{R} \to \mathbb{R}$
   (b) $\phi(a, b) = a + b$ and $\psi(a, b) = b + a$, where $\phi, \psi\colon \mathbb{Z} \times \mathbb{Z} \to \mathbb{Z}$
   (c) $\phi(a, b) = ([a]_5, [b+7]_5)$ and $\phi(a, b) = ([a+5]_5, [b-3]_5)$, where $\phi, \psi\colon \mathbb{Z} \times \mathbb{Z} \to \mathbb{Z}_5 \times \mathbb{Z}_5$
   (d) $\phi(f) = f \restriction \mathbb{Z}$ and $\psi(f) = \{(n, f(n)) : n \in \mathbb{Z}\}$, where $\phi, \psi\colon \mathbb{R}^{\mathbb{R}} \to \mathbb{R}^{\mathbb{Z}}$
3. Let $f, g\colon \mathbb{R} \to \mathbb{R}$ be functions defined by $f(x) = 2|x| + 1$ and

$$g(x) = \begin{cases} 2x+1 & \text{if } x \geq 0 \\ -2x+1 & \text{if } x < 0. \end{cases}$$

   Prove $f = g$.
4. Show that the following pairs of functions are not equal.
   (a) $f(x) = x$ and $g(x) = 2x$ where $f, g\colon \mathbb{R} \to \mathbb{R}$
   (b) $f(x) = x - 3$ and $g(x) = x + 3$ where $f, g\colon \mathbb{R} \to \mathbb{R}$
   (c) $\phi(a) = [a]_5$ and $\psi(a) = [a]_4$ where $\phi, \psi\colon \mathbb{Z} \to \mathbb{Z}_4 \cup \mathbb{Z}_5$
   (d) $\phi(A) = A \setminus \{0\}$ and $\psi(A) = A \cap \{1, 2, 3\}$ where $\phi, \psi\colon \mathbf{P}(\mathbb{Z}) \to \mathbf{P}(\mathbb{Z})$
5. Let $\psi\colon \mathbb{R} \to \mathbb{R}^{\mathbb{R}}$ be defined by $\psi(a) = f_a$ where f_a is the function $f_a\colon \mathbb{R} \to \mathbb{R}$ with $f_a(x) = ax$. Prove that ψ is well-defined.
6. For each pair of functions, find the indicated values when possible.
   (a) $f\colon \mathbb{R} \to \mathbb{R}$ and $f(x) = 2x^3$, $g\colon \mathbb{R} \to \mathbb{R}$ and $g(x) = x + 1$
      (i) $(f \circ g)(2)$
      (ii) $(g \circ f)(0)$
   (b) $f\colon [0, \infty) \to \mathbb{R}$ and $f(x) = \sqrt{x}$, $g\colon \mathbb{R} \to \mathbb{R}$ and $g(x) = |x| - 1$
      (i) $(f \circ g)(0)$
      (ii) $(g \circ f)(4)$

(c) $\phi\colon \mathbb{Z} \to \mathbb{Z}_5$ and $\phi(a) = [a]_5$,
$\psi : \mathbb{R}^{\mathbb{R}} \to \mathbb{R}$ and $\psi(f) = f(0)$
(i) $(\phi \circ \psi)(.5x + 1)$
(ii) $(\psi \circ \phi)(2)$

7. For each of the following functions, find the composition of the function with itself. For example, find $f \circ f$ for part (a).
   (a) $f : \mathbb{R} \to \mathbb{R}$ with $f(x) = x^2$
   (b) $g : \mathbb{R} \to \mathbb{R}$ with $f(x) = 3x + 1$
   (c) $\phi\colon \mathbb{Z} \times \mathbb{Z} \to \mathbb{Z} \times \mathbb{Z}$ with $\phi(x, y) = (2y, 5x - y)$
   (d) $\psi : \mathbb{Z}_m \to \mathbb{Z}_m$ with $\psi([n]_m) = [n + 2]_m$
8. Let $f, g : \mathbb{R} \to \mathbb{R}$ be functions. For each of the following either prove the proposition true or show that it is false by finding a counterexample.
   (a) If f and g are even, then $f + g$ is even.
   (b) If f and g are odd, then $f - g$ is odd.
   (c) If f and g are even, then fg is even.
   (d) If f and g are odd, then fg is odd.
   (e) If f and g are odd, then fg is even.
9. Show the following:
   (a) $f(x) = \sqrt{x^2 + x^4}$ is even.
   (b) $g(x) = -|x|$ is odd.
   (c) $h(x) = x^4 + x^2 + 1$ is even and not odd.
   (d) $f(x) = x^3$ is odd and not even.
10. Can a function be both even and odd? Explain.
11. Write the following restrictions as rosters:
    (a) $\{(1, 2), (2, 2), (3, 4), (4, 7)\}\restriction\{1, 3\}$
    (b) $f\restriction\{0, 1, 2, 3\}$ where $f(x) = 7x - 1$ and $\operatorname{dom}(f) = \mathbb{R}$.
    (c) $(g + h)\restriction\{-3.3, 1.2, 7\}$ where $g(x) = \lceil x \rceil$, $h(x) = x + 1$, and $\operatorname{dom}(g) = \operatorname{dom}(h) = \mathbb{R}$.
12. For any function f such that $A, B \subseteq \operatorname{dom}(f)$, prove:
    (a) $f\restriction A = f \cap [A \times \operatorname{ran}(f)]$
    (b) $f\restriction(A \cap B) = (f\restriction A) \cap (f\restriction B)$
    (c) $f\restriction(A \setminus B) = (f\restriction A) \setminus (f\restriction B)$
    (d) $(f \circ g)\restriction A = f \circ (g\restriction A)$
13. Let $f : U \to V$ be a function. Prove that if $A \subseteq U$, $f\restriction A = f \circ I_A$.
14. Let $\phi\colon C^A \to C^B$ be defined by $\phi(f) = f\restriction B$. Prove that ϕ is well-defined.
15. A real-valued function f is ***periodic*** with ***period*** $k \in \mathbb{R}$ if for all $x \in \operatorname{dom}(f)$, $f(x) = f(x + k)$. Let $g, h : \mathbb{R} \to \mathbb{R}$ be functions with period k. Prove that $g + h$, gh, and $g \circ h$ are periodic with period k.

16. A real valued function f with domain $\mathbb{R}$ is ***increasing*** means for all $x, y \in \mathbb{R}$, if $x \leq y$, then $f(x) \leq f(y)$. A ***decreasing*** function is defined similarly. Suppose that g and h are increasing. Prove that $g + h$ and $g \circ h$ are increasing.

17. Find two increasing functions whose product is not increasing.

6.5. ONE-TO-ONE AND ONTO

While looking at relations, we studied the notation of an inverse relation. Given R, obtain R^{-1} by exchanging the x- and y-coordinates. The same can be done with functions, but the inverse may not be a function. For example, given

$$f = \{(1, 2), (2, 3), (3, 2)\}$$

its inverse is

$$f^{-1} = \{(2, 1), (3, 2), (2, 3)\}.$$

However, if the original relation is a function, we often want the inverse also to be a function. This leads to the next definition.

6.5.1. Definition

Let $f : A \to B$ be a function. We say that f is ***invertible*** if and only if its inverse relation f^{-1} is a function such that $B \to A$.

A consequence of the definition is that when f is invertible,

$$f(x) = y \text{ if and only if } f^{-1}(y) = x$$

for all $x \in \text{dom}(f)$. We will use this in the proof of the next theorem.

6.5.2. Theorem

Let $f : A \to B$ be a function. Its inverse relation f^{-1} is a function with domain B if and only if $f^{-1} \circ f = I_A$ and $f \circ f^{-1} = I_B$.

Proof. Take a function $f : A \to B$. Then, $f^{-1} \subseteq B \times A$.

($\Rightarrow$) Assume that $f^{-1} : B \to A$ is a function. Let $x \in A$ and $y \in B$ such that $y = f(x)$. This means $f^{-1}(y) = x$. Therefore,

$$(f^{-1} \circ f)(x) = f^{-1}(f(x)) = f^{-1}(y) = x,$$

and

$$(f \circ f^{-1})(y) = f(f^{-1}(y)) = f(x) = y.$$

Hence, $f^{-1} \circ f = I_A$ and $f \circ f^{-1} = I_B$.

($\Leftarrow$) Now assume $f^{-1} \circ f = I_A$ and $f \circ f^{-1} = I_B$. To show that f^{-1} is a function, take $(y, x) \in f^{-1}$ and $(y, x') \in f^{-1}$. From this we know that $(x, y) \in f$. Therefore, $(x, x') \in f^{-1} \circ f = I_A$, so $x = x'$.

We know that $\operatorname{dom}(f^{-1}) \subseteq B$, so to prove equality, let $y \in B$. Then, $(y, y) \in I_B = f \circ f^{-1}$. Thus, there exists $x \in A$ such that $(y, x) \in f^{-1}$ and $(x, y) \in f$. In particular, $f^{-1}(y) = x$ so that $y \in \operatorname{dom}(f^{-1})$. ■

Example. The following functions are invertible.

1. Let $f : \mathbb{R} \to \mathbb{R}$ be the function given by $f(x) = x + 2$. Its inverse is $g(x) = x - 2$. This is because:

$$(g \circ f)(x) = g(x + 2) = (x + 2) - 2 = x,$$

and

$$(f \circ g)(x) = f(x - 2) = (x - 2) + 2 = x.$$

Hence, we may write $g = f^{-1}$.

2. Define $g : \mathbb{R} \to [0, \infty)$ by $g(x) = x^2$. Then g^{-1} is a function $[0, \infty) \to \mathbb{R}$ and is defined by $g^{-1}(x) = \sqrt{x}$.
3. Let $h : \mathbb{R} \to (0, \infty)$ be defined as $h(x) = e^x$. From calculus we know that $h^{-1}(x) = \ln x$. This is because

$$e^{\ln x} = x \text{ and } \ln e^x = x.$$

Theorem 6.5.2 can be improved by finding a condition for the invertibility of a function that examines only the function. Consider the following. In order for a relation to be a function, it cannot look like Figure 6.3.4. Namely, it must pass the vertical line test (see Figure 6.3.1). An inverse exchanges the roles of the two coordinates. Hence, for an inverse to be a function, the original function cannot look like the graph in Figure 6.5.4. Generalizing, we do not want there to exist x_1 and x_2 so that $x_1 \neq x_2$ and $f(x_1) = f(x_2)$. But then,

$$\sim(\exists x_1)(\exists x_2)[x_1 \neq x_2 \wedge f(x_1) = f(x_2)]$$

is equivalent to

$$(\forall x_1)(\forall x_2)[x_1 = x_2 \vee f(x_1) \neq f(x_2)],$$

which in turn is equivalent to

$$(\forall x_1)(\forall x_2)[f(x_1) = f(x_2) \Rightarrow x_1 = x_2].$$

This condition is necessary for the inverse to be a function. Hence, we make the next definition.

6.5.3. Definition

The function $f : A \to B$ is ***one-to-one*** if and only if for all $x_1,\ x_2 \in A$,

$$\text{if } f(x_1) = f(x_2), \text{ then } x_1 = x_2.$$

A one-to-one function is sometimes called an ***injection***.

Example. Define $f\colon \mathbb{R} \to \mathbb{R}$ by $f(x) = 5x + 1$. To show that f is one-to-one, let $x_1,\ x_2 \in \mathbb{R}$ and assume $f(x_1) = f(x_2)$. Then,

$$\begin{aligned} 5x_1 + 1 &= 5x_2 + 1 \\ 5x_1 &= 5x_2 \\ x_1 &= x_2. \end{aligned}$$

Example. We will show that the following function is an injection. Let $\phi\colon \mathbb{Z} \times \mathbb{Z} \to \mathbb{Z} \times \mathbb{Z} \times \mathbb{Z}$ be the function

$$\phi(a,\ b) = (a,\ b,\ 0).$$

For any $(a_1,\ b_1),\ (a_2,\ b_2) \in \mathbb{Z} \times \mathbb{Z}$, assume

$$\phi(a_1,\ b_1) = \phi(a_2,\ b_2).$$

6.5.4. Figure

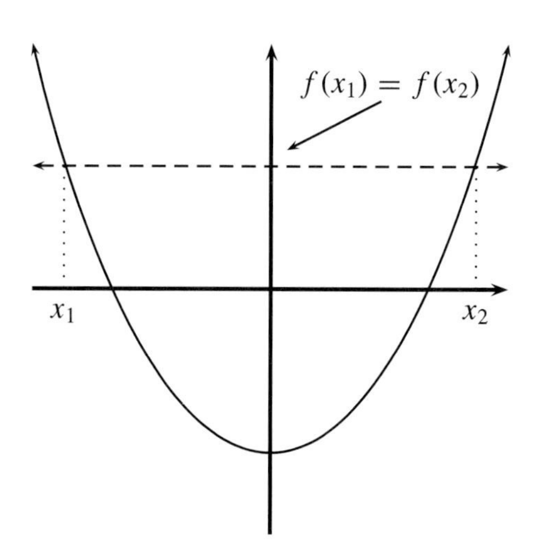

The inverse of f is not a function.

This means

$$(a_1, b_1, 0) = (a_2, b_2, 0).$$

Hence, $a_1 = a_2$ and $b_1 = b_2$, and this yields $(a_1, b_1) = (a_2, b_2)$.

To clarify the concept of a one-to-one function, consider the following circle diagrams. If a relation is a function, every element of the domain must be mapped to exactly one element of the codomain. So, if the inverse is also to be a function, every element of the original function's range must be mapped from exactly one element of the domain. (See Figure 6.5.5.)

If a function is not one-to-one, then there must be an element of the range that has at least two pre-images. This is represented by the circle diagram found in Figure 6.5.6. An example of a function that is not one-to-one is $f(x) = x^2$ where both the domain and codomain of f are $\mathbb{R}$. This is because $f(2) = 4$ and $f(-2) = 4$. Another example is the following.

Example. The function $g(\theta) = \cos\theta$ is not one-to-one. To see this we must show that there exist $\theta_1, \theta_2 \in \mathbb{R}$ such that $g(\theta_1) = g(\theta_2)$ but $\theta_1 \neq \theta_2$. To do this let $\theta_1 = 0$ and $\theta_2 = 2\pi$. Then

$$g(\theta_1) = \cos 0 = 1$$

and

$$g(\theta_2) = \cos 2\pi = 1.$$

Although the original function may not be one-to-one, we can always restrict the function to a subset of its domain so that the resulting function is one-to-one. This is illustrated in the next two examples.

6.5.5. Figure

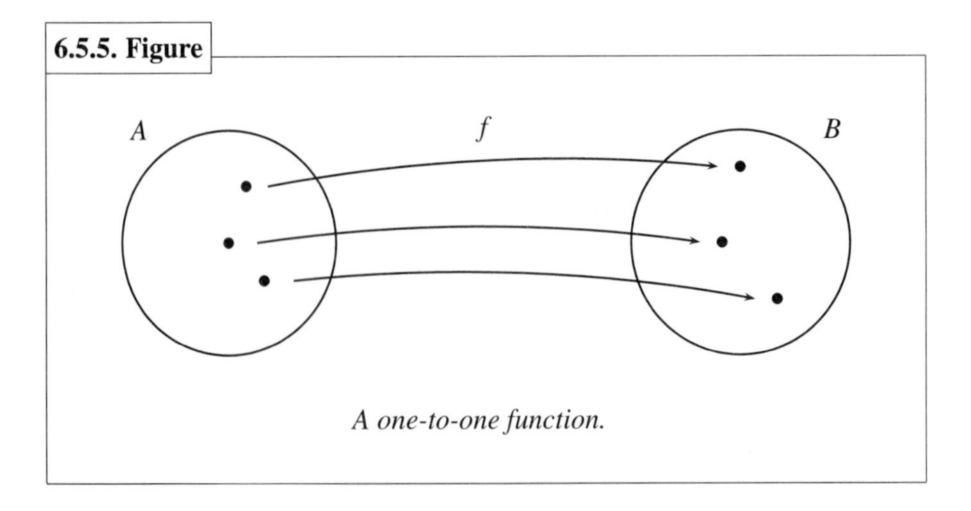

A one-to-one function.

6.5.6. Figure

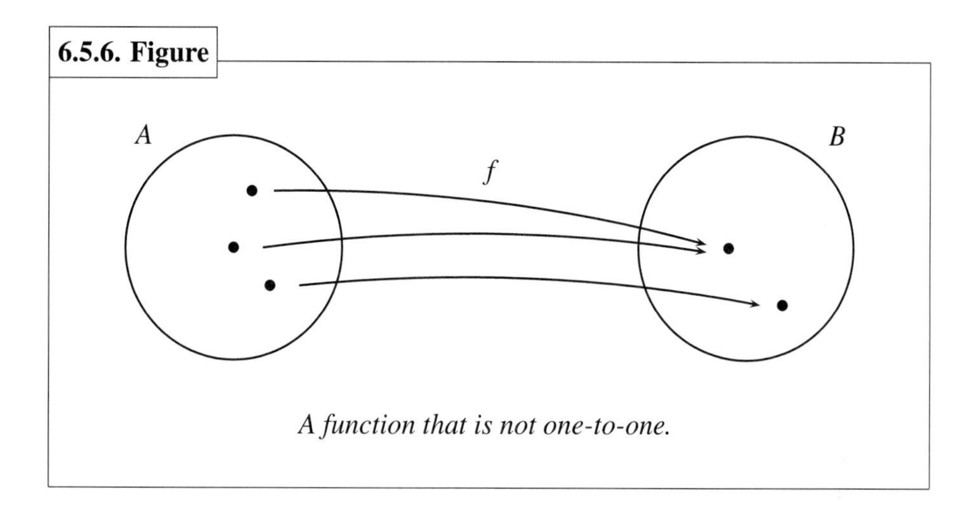

A function that is not one-to-one.

Example. Let f be the function $\{(1, 5), (2, 8), (3, 8), (4, 6)\}$. We observe that f is not one-to-one, but if we let $A = \{1, 2\}$ and $B = \{3, 4\}$, then $f \restriction A = \{(1, 5), (2, 8)\}$ and $f \restriction B = \{(3, 8\}, (4, 6)\}$ are both one-to-one as in Figure 6.5.7.

Example. Let $g\colon \mathbb{R} \to \mathbb{R}$ be the function $g(x) = x^2$. This function is not one-to-one, but if we restrict the function to another domain we can form a function that is. As examples, $g \restriction [0, \infty)$ and $g \restriction (-10, -5)$ are one-to-one.

6.5.7. Figure

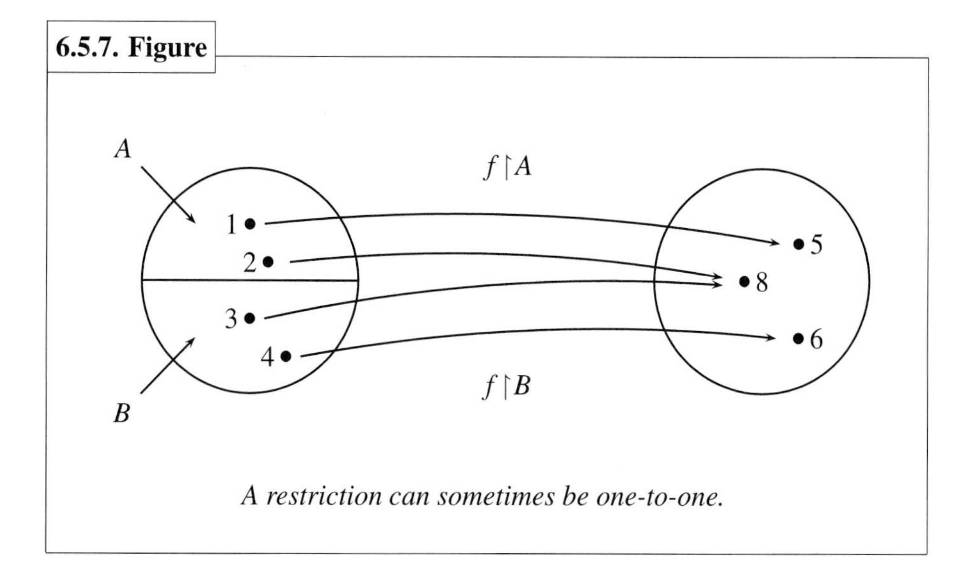

A restriction can sometimes be one-to-one.

The function being an injection is not sufficient for it to be invertible since not every element of the codomain may have a pre-image. This will not allow the codomain to be the domain of the inverse. To prevent this situation, we will need the function to satisfy the next definition.

6.5.8. Definition

A function $f\colon A \to B$ is ***onto*** if and only if

$$\text{for every } y \in B, \text{ there exists } x \in A \text{ such that } f(x) = y.$$

An onto function can be called a ***surjection***.

This definition is related to the ***range*** (or ***image***) of a function. The range of $f\colon A \to B$ as a relation is the set

$$\begin{aligned}\operatorname{ran}(f) &= \{y : (x,\ y) \in f \text{ for some } x \in A\}\\ &= \{y : f(x) = y \text{ for some } x \in \operatorname{dom}(f)\}\\ &= \{f(x) : x \in \operatorname{dom}(f)\}.\end{aligned}$$

(See Figure 6.5.9.) Thus, f is onto if and only if $\operatorname{ran}(f) = B$.

Example. The equation $f(x) = \sqrt{x}$ is a function $[0,\ \infty) \to [0,\ \infty)$. Its range is also $[0,\ \infty)$, so it is onto.

Example. The ranges of the following functions are different than their codomains.

- Let $g\colon \mathbb{R} \to \mathbb{R}$ be defined by $g(x) = |x|$. Then $\operatorname{ran}(g) = [0,\ \infty)$.
- Define $h\colon \mathbb{Z} \to \mathbb{Z}$ by $h(n) = 2n$. Here, $\operatorname{ran}(h) = \{2n : n \in \mathbb{Z}\}$.

These functions are not onto.

The functions illustrated in Figures 6.5.5 and 6.5.6 are onto functions as are those in the next examples.

Example. Any linear function $f\colon \mathbb{R} \to \mathbb{R}$ that is not a horizontal line is onto. To see this, let $f(x) = ax + b$ for some $a \neq 0$. Take $y \in \mathbb{R}$. We need to find $x \in \mathbb{R}$ so that $ax + b = y$. Choose

$$x = \frac{y - b}{a}.$$

Then

$$f(x) = a\left(\frac{y-b}{a}\right) + b = y - b + b = y.$$

6.5.9. Figure

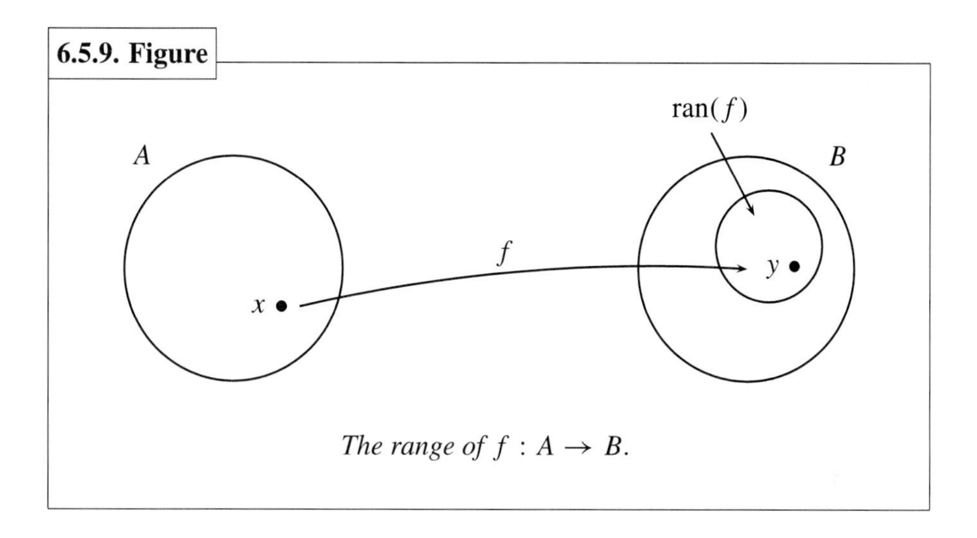

The range of $f : A \to B$.

The approach in the example is typical. To show that a function is onto, take an arbitrary element of the codomain and search for a candidate to serve as its pre-image. When found, check it.

Example. Take a positive integer n and let $\phi\colon \mathbb{Z} \to \mathbb{Z}_n$ be defined as $\phi(k) = [k]_n$. To see that ϕ is onto, take $[\ell]_n \in \mathbb{Z}_n$ for some $\ell \in \mathbb{Z}$. We then find that $\phi(\ell) = [\ell]_n$.

Example. Define the function $\pi : \mathbb{R} \times \mathbb{R} \times \mathbb{R} \to \mathbb{R} \times \mathbb{R}$ by

$$\pi(x,\ y,\ z) = (x,\ y).$$

Such functions are called ***projections***. Notice that π is not one-to-one, but it is onto. To show this, let $(a,\ b) \in \mathbb{R} \times \mathbb{R}$. Then

$$\pi(a,\ b,\ 0) = (a,\ b).$$

If a function is not onto, it has a circle diagram like that found in Figure 6.5.10. Therefore, to show that a function is not a surjection, we must find an element of the codomain that does not have a pre-image.

Example. Define $f\colon \mathbb{Z} \to \mathbb{Z}$ by $f(n) = 3n$. This function is not onto because 5 does not have a pre-image in $\text{dom}(f)$.

Example. The function $\phi\colon \mathbb{Z} \times \mathbb{Z} \to \mathbb{Z} \times \mathbb{Z} \times \mathbb{Z}$ defined as

$$\phi(a,\ b) = (a,\ b,\ 0)$$

is also not onto because $(1,\ 1,\ 1)$ does not have a pre-image.

6.5.10. Figure

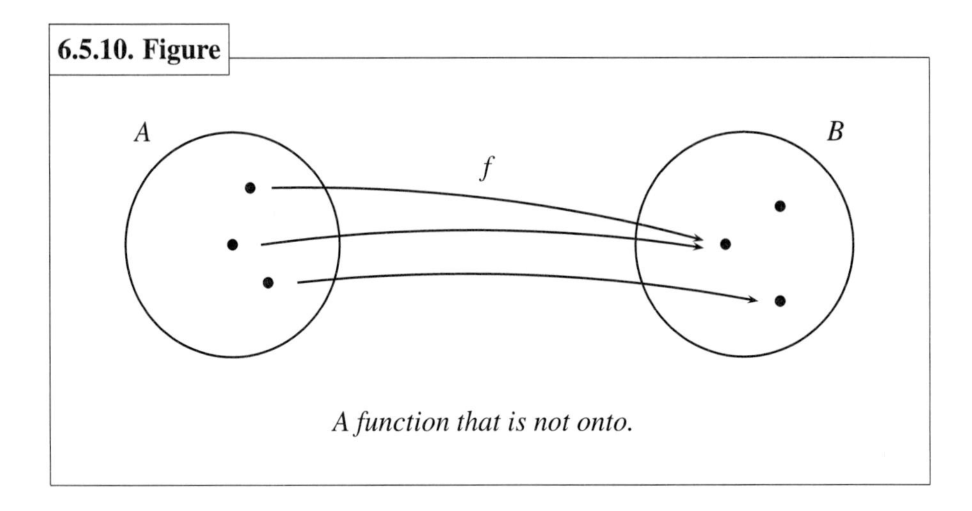

A function that is not onto.

We are now ready to prove the main theorem. Compare its proof to that of Theorem 6.5.2. The benefit this theorem has is that we only need to examine f to determine if it has an inverse.

6.5.11. Theorem

If $f\colon A \to B$ is a function, then f is invertible if and only if f is one-to-one and onto.

Proof. Let $f : A \to B$ be a function.

($\Rightarrow$) Suppose that f is invertible. To show that f is one-to-one, take $x_1, x_2 \in A$ and let $f(x_1) = f(x_2)$. Then by Theorem 6.5.2,

$$x_1 = (f^{-1} \circ f)(x_1) = (f^{-1} \circ f)(x_2) = x_2.$$

To see that f is onto, take $y \in B$. Then there exists $x \in A$ such that $f^{-1}(y) = x$. Hence, $f(x) = y$.

($\Leftarrow$) Assume that f is both one-to-one and onto. To show that f^{-1} is a function, let $(y, x), (y, x') \in f^{-1}$. This means $(x, y), (x', y) \in f$. In other words, $f(x) = y = f(x')$. Since f is one-to-one, $x = x'$.

To prove that the domain of f^{-1} is B, it suffices to show that B is a subset of $\text{dom}(f^{-1})$. Let $y \in B$. Since f is onto, there exists $x \in A$ such that $f(x) = y$. This means $f^{-1}(y) = x$ and $y \in \text{dom}(f^{-1})$. ■

If we have a function that is both one-to-one and onto, then it is called a ***bijection*** or a ***one-to-one correspondence***. As illustrated in Figure 6.5.5, if there is a bijection between two sets, the sets must be the same size. (See Section 6.7.)

Example. The following are bijections. We will prove the first one.

1. $f: \mathbb{R} \to \mathbb{R}$ where $f(x) = 9x - 6$.

   Proof. We must show that f is both one-to-one and onto.

   - Let $x_1, x_2 \in \mathbb{R}$. Suppose $f(x_1) = f(x_2)$, which means
     $$9x_1 - 6 = 9x_2 - 6.$$
     After adding 6 to both sides and dividing by 9, we find that $x_1 = x_2$. Therefore, f is one-to-one.
   - For onto, take $y \in \mathbb{R}$. We must find $x \in \mathbb{R}$ such that $f(x) = y$. After a little algebra, our candidate is $x = [y + 6]/9$. Let us check:
     $$f([y+6]/9) = 9([y+6]/9) - 6 = y. \blacksquare$$

2. $g: (-\pi/2, \pi/2) \to \mathbb{R}$ such that $g(\theta) = \tan\theta$.
3. $h: \mathbb{R} \to (0, \infty)$ where $h(x) = e^x$.

We close the section with an important theorem that will be needed later.

6.5.12. Theorem

If $\phi: A \to B$ and $\psi: B \to C$ are bijections, then $\psi \circ \phi$ is a bijection.

Proof. Assume that $\phi: A \to B$ and $\psi: B \to C$ are bijections.

1. Let $a_1, a_2 \in A$ and assume $(\psi \circ \phi)(a_1) = (\psi \circ \phi)(a_2)$. Then,
   $$\psi(\phi(a_1)) = \psi(\phi(a_2)).$$
   Since ψ is one-to-one,
   $$\phi(a_1) = \phi(a_2),$$
   and since ϕ is one-to-one, $a_1 = a_2$.
2. Take $c \in C$. Then there exists $b \in B$ so that $\psi(b) = c$. But then there is an $a \in A$ such that $\phi(a) = b$. Therefore,
   $$(\psi \circ \phi)(a) = \psi(\phi(a)) = \psi(b) = c.$$
   This means that $\psi \circ \phi$ is onto. $\blacksquare$

Using the functions g and h from the last example, we conclude from the theorem that $h \circ g$ is a bijection with domain $(-\pi/2, \pi/2)$ and codomain $(0, \infty)$.

EXERCISES

1. Show that the following pairs of functions are inverses.
   (a) $f(x) = 3x + 2$ and $g(x) = \frac{1}{3}x - \frac{2}{3}$
   (b) $\phi(a, b) = (2a, b + 2)$ and $\psi(a, b) = (\frac{1}{2}a, b - 2)$
   (c) $f(x) = a^x$ and $g(x) = \log_a x, a > 0$
2. Find the inverses of the following real-valued functions.
   (a) $f(x) = 7x + 3$
   (b) $g(x) = \sqrt{x} + 1$
   (c) $h(x) = 2e^{x+1}$
   (d) $f(x) = \sin e^x$
3. Show that when f is invertible,
$$f(x) = y \text{ if and only if } f^{-1}(y) = x$$
for all $x \in \text{dom}(f)$.
4. Let f be an invertible function. Prove f^{-1} is invertible with inverse equal to f.
5. Draw a circle diagram that represents a one-to-one function that is not onto.
6. For each function, graph the indicated restriction. This exercise will illustrate how a restriction can be one-to-one even if the original function is not.
   (a) $f\restriction(0, \infty), f(x) = x^2$
   (b) $g\restriction[-5, -2], g(x) = |x|$
   (c) $h\restriction(0, 2\pi), h(x) = \cos x$
   (d) $f\restriction\{5\}, f(x) = 2$
7. Prove that the following are one-to-one.
   (a) $f: \mathbb{R} \to \mathbb{R}, f(x) = 2x + 1$
   (b) $g: \mathbb{R}^2 \to \mathbb{R}^2, g(x, y) = (3y, 2x)$
   (c) $h: \mathbb{R} \setminus \{9\} \to \mathbb{R} \setminus \{0\}, h(x) = 1/(x - 9)$
   (d) $\phi: \mathbb{Z} \times \mathbb{R} \to \mathbb{Z} \times (0, \infty), \phi(n, x) = (3n, e^x)$
   (e) $\psi: \mathbf{P}(A) \to \mathbf{P}(B), \psi(C) = C \cup \{b\}$ where $A \subseteq B$ and $b \in B \setminus A$
8. Let $f: (a, b) \to (c, d)$ be defined by
$$f(x) = \frac{d - c}{b - a}(x - a) + c.$$
Graph f and then show that it is one-to-one and onto.
9. Let f and g be functions such that $\text{ran}(g) \subseteq \text{dom}(f)$. Prove:
   (a) If $f \circ g$ is one-to-one, then g is one-to-one.
   (b) Give an example of functions f and g such that $f \circ g$ is one-to-one, but f is not one-to-one.
10. Define $\phi: \mathbb{Z} \to \mathbb{Z}_n$ to be the function $\phi(k) = [k]_n$. Prove that ϕ is not one-to-one.

11. Show that the following functions are not one-to-one.
    (a) $f\colon \mathbb{R} \to \mathbb{R}$, $f(x) = x^4 + 3$
    (b) $g\colon \mathbb{R} \to \mathbb{R}$, $g(x) = |x - 2| + 4$
    (c) $\phi\colon \mathbf{P}(A) \to \{\{a\}, \varnothing\}$, $\phi(B) = B \cap \{a\}$, where $\{a, b\} \subseteq A$ and $a \neq b$.
    (d) $\varepsilon_5\colon \mathbb{R}^{\mathbb{R}} \to \mathbb{R}$, $\varepsilon_5(f) = f(5)$
12. Show that the following functions are onto.
    (a) $f\colon \mathbb{R} \to \mathbb{R}$, $f(x) = 2x + 1$
    (b) $g\colon \mathbb{R} \to (0, \infty)$, $g(x) = e^x$
    (c) $h\colon \mathbb{R} \setminus \{0\} \to \mathbb{R} \setminus \{0\}$, $h(x) = 1/x$
    (d) $\phi\colon \mathbb{Z} \times \mathbb{Z} \to \mathbb{Z}$, $\phi(a, b) = a + b$
    (e) $\varepsilon_5\colon \mathbb{R}^{\mathbb{R}} \to \mathbb{R}$, $\varepsilon_5(f) = f(5)$
13. Show that the following functions are not onto.
    (a) $f\colon \mathbb{R} \to \mathbb{R}$, $f(x) = e^x$
    (b) $g\colon \mathbb{R} \to \mathbb{R}$, $g(x) = |x|$
    (c) $\phi\colon \mathbb{Z} \times \mathbb{Z} \to \mathbb{Z} \times \mathbb{Z}$, $\phi(a, b) = (3a, b^2)$
    (d) $\psi\colon \mathbb{R} \to \mathbb{R}^{\mathbb{R}}$, $\psi(a) = f$, where $f(x) = a$ for all $x \in \mathbb{R}$
14. Let f and g be functions such that $\operatorname{ran}(g) \subseteq \operatorname{dom}(f)$. Prove:
    (a) If $f \circ g$ is onto, then f is onto.
    (b) Give an example of functions f and g such that $f \circ g$ is onto, but g is not onto.
15. Let $\phi\colon \mathbb{Q} \times \mathbb{Z} \to \mathbb{Z} \times \mathbb{Q}$ be defined by $\phi(x, y) = (y, x)$. Show that ϕ is a bijection.
16. **16.** Show that function $\gamma\colon A \times B \to C \times D$ defined by

$$\gamma(a, b) = (\phi(a), \psi(b))$$

    is a bijection if both $\phi\colon A \to C$ and $\psi\colon B \to D$ are bijections.
17. Define $\gamma\colon A \times (B \times C) \to (A \times B) \times C$ by

$$\gamma(a, (b, c)) = ((a, b), c).$$

    Prove γ is a bijection.
18. Demonstrate that the inverse of a bijection is a bijection.
19. Let $f, g\colon A \to A$ be two bijections. Show that $f + g$ and fg may not be bijections.
20. **20.** Let $A \subseteq \mathbb{R}$ and define $\phi\colon \mathbb{R}^{\mathbb{R}} \to \mathbb{R}^A$ by $\phi(f) = f \restriction A$. Is ϕ always one-to-one? Is it always onto?
21. A function f has a ***left inverse*** if there exists g such that $g \circ f = I$. Prove that a function is one-to-one if and only if it has a left inverse.
22. A function $f\colon A \to B$ has a ***right inverse*** if there exists $g\colon B \to A$ so that $f \circ g = I_B$. Prove f is onto if and only if f has a right inverse.

23. A real-valued function f with domain D is ***strictly increasing*** if for all $x,\ y \in D$,
$$x < y \Rightarrow f(x) < f(y).$$
A ***strictly decreasing*** function satisfies
$$x < y \Rightarrow f(x) > f(y)$$
for all $x,\ y \in D$. Show that if a real-valued function is strictly increasing or strictly decreasing, then it is one-to-one. (Compare this with Exercise 6.4.16.)
24. Find examples of two functions, one strictly increasing and the other strictly decreasing, such that neither function is onto.
25. Prove or show false this modification of Theorem 6.5.12: If $\phi\colon A \to B$ and $\psi\colon C \to D$ are two bijections with $\operatorname{ran}(\phi) \subseteq C$, then $\psi \circ \phi$ is a bijection.

6.6. IMAGES AND INVERSE IMAGES

So far we have focused on the images of single elements in the domain of a function. Sometimes we will need to look at the image of each element that belongs to a given subset of the domain. Let $\phi\colon A \to B$ be a function and take $C \subseteq A$. Define

$$\begin{aligned}\phi[C] &= \{\phi(x) : x \in C\} \\ &= \{y \in B : \phi(x) = y \text{ for some } x \in C\}.\end{aligned}$$

The set $\phi[C]$ is called the ***image*** of C under ϕ. If the function is clear from context, it is simply called the image of C. Its circle diagram can be found in Figure 6.6.1. Notice that $\phi[C] \subseteq \operatorname{ran}(\phi)$ and $\phi[A] = \operatorname{ran}(\phi)$ (see the exercises). Another way to state the definition is to note

$$y \in \phi[C] \text{ if and only if } (\exists x \in C)(\phi(x) = y).$$

Example. The set $f = \{(1,\ 2),\ (2,\ 4),\ (3,\ 5),\ (4,\ 5)\}$ is a function with domain $\{1,\ 2,\ 3,\ 4\}$ and range $\{2,\ 4,\ 5\}$. With this definition, $f[\{1,\ 3\}] = \{2,\ 5\}$.

A similar definition can be made with subsets of the codomain. If $D \subseteq B$, then

$$\begin{aligned}\phi^{-1}[D] &= \{x \in A : \phi(x) \in D\} \\ &= \{x \in A : \phi(x) = y \text{ for some } y \in D\}.\end{aligned}$$

6.6.1. Figure

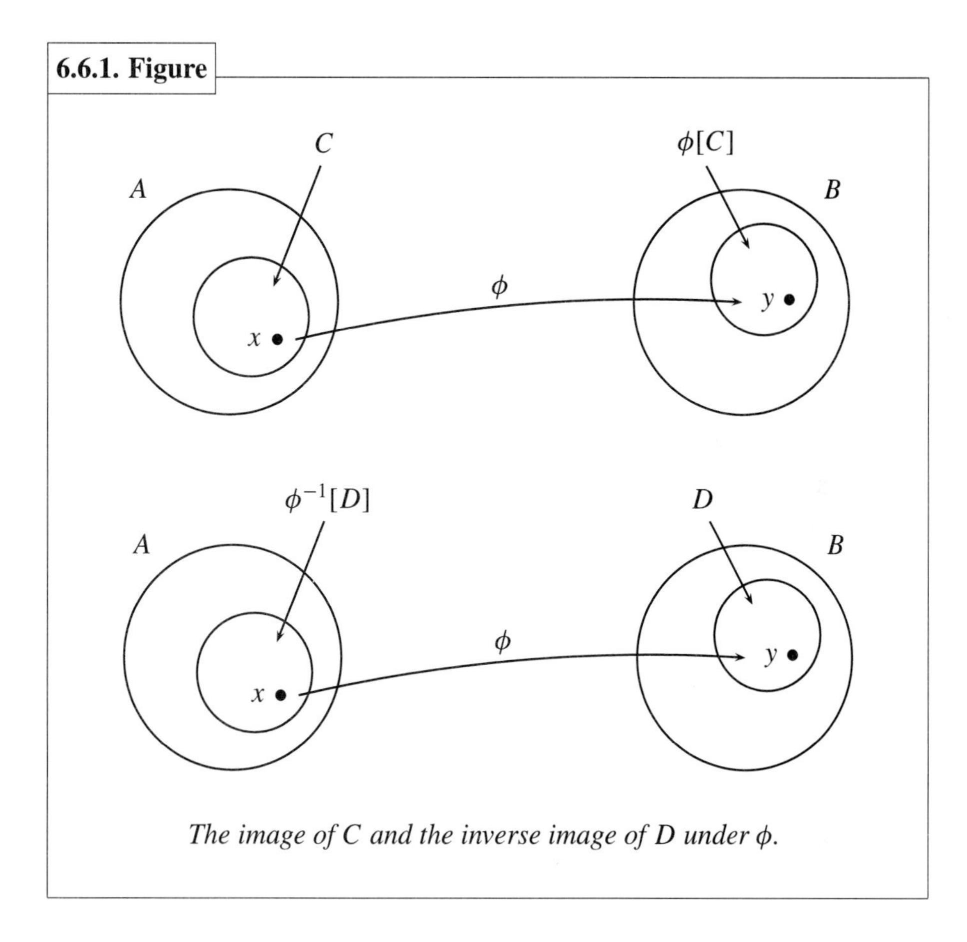

The image of C and the inverse image of D under ϕ.

This set is called the ***inverse image*** of D under ϕ. See Figure 6.6.1 for the circle diagram of an inverse image. Stated another way:

$$x \in \phi^{-1}[D] \text{ if and only if } \phi(x) \in D.$$

Example. If f is defined as in the last example, $f^{-1}[\{5\}] = \{3, 4\}$.

In the next example we will follow the convention that elements of the domain are represented by x and elements of the codomain are represented by y. This will help keep clear which elements belong to which sets.

Example. Define $f: \mathbb{R} \to \mathbb{R}$ by $f(x) = x^2 + 1$.

- To prove $f[(1, 2)] = (2, 5)$, we must show both inclusions.
  ($\subseteq$) Let $y \in f[(1, 2)]$. Then $f(x) = y$ for some $x \in (1, 2)$. This means $y = x^2 + 1$. By a little algebra we see $2 < x^2 + 1 < 5$. Hence, $y \in (2, 5)$.

6.6.2. Figure

$$f(x) = x^2 + 1$$

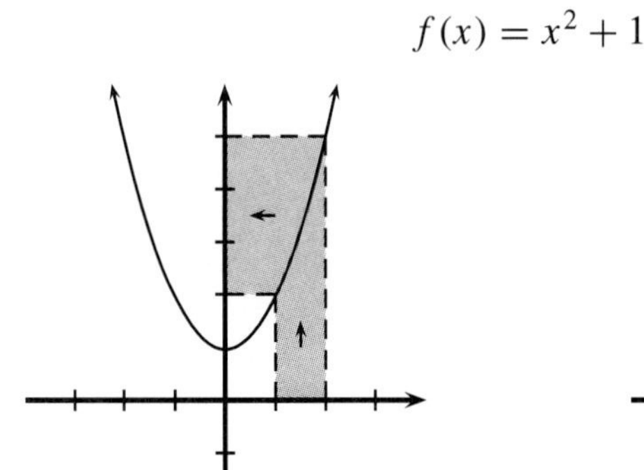

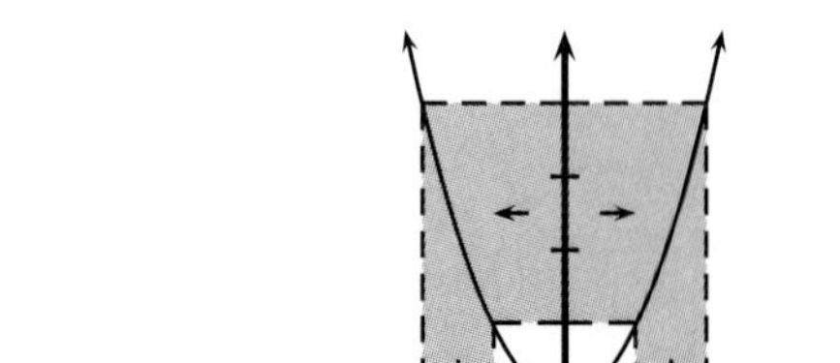

$f[(1, 2)] = (2, 5)$ $\quad$ $f^{-1}[(2, 5)] = (-2, -1) \cup (1, 2)$

($\supseteq$) Let $y \in (2, 5)$. We must find $x \in (1, 2)$ so that $f(x) = y$. Now,

$$\begin{aligned} 2 < y < 5 &\Leftrightarrow 1 < y - 1 < 4 \\ &\Leftrightarrow 1 < \sqrt{y-1} < 2. \end{aligned}$$

So take $x = \sqrt{y-1}$. Then $x \in (1, 2)$ and

$$\begin{aligned} f(x) &= f(\sqrt{y-1}) \\ &= (\sqrt{y-1})^2 + 1 \\ &= y - 1 + 1 \\ &= y. \end{aligned}$$

Therefore, $y \in f[(1, 2)]$. (See the first graph of Figure 6.6.2.)

- Simply because $f[(1, 2)] = (2, 5)$, we cannot conclude $f^{-1}[(2, 5)]$ equals $(1, 2)$. Instead, $f^{-1}[(2, 5)] = (-2, 1) \cup (1, 2)$, as seen in the second graph of Figure 6.6.2. The proof is as follows:

$$\begin{aligned} x \in (-2, -1) \cup (1, 2) &\Leftrightarrow -2 < x < -1 \text{ or } 1 < x < 2 \\ &\Leftrightarrow 1 < x^2 < 4 \\ &\Leftrightarrow 2 < x^2 + 1 < 5 \\ &\Leftrightarrow f(x) \in (2, 5) \\ &\Leftrightarrow x \in f^{-1}[(2, 5)]. \end{aligned}$$

- We shall show that $f^{-1}[(-2, -1)]$ is empty. To do this take $x \in f^{-1}[(-2, -1)]$. This means that $f(x) \in (-2, -1)$. Therefore, $-2 < f(x) < -1$, but this is impossible since $f(x) = x^2 + 1 \geq 1$.

Our next theorem shows how we can find the image and pre-image of unions and intersections.

6.6.3. Theorem

Let $\phi\colon A \to B$ be a function with $C,\ D \subseteq A$ and $E,\ F \subseteq B$.

1. $\phi[C \cup D] = \phi[C] \cup \phi[D]$.
2. $\phi[C \cap D] \subseteq \phi[C] \cap \phi[D]$.
3. $\phi^{-1}[E \cup F] = \phi^{-1}[E] \cup \phi^{-1}[F]$.
4. $\phi^{-1}[E \cap F] = \phi^{-1}[E] \cap \phi^{-1}[F]$.

Proof. We will prove parts 1 and 3, leaving the others as exercises.

1. If $y \in B$, then:

$$\begin{aligned} &y \in \phi[C \cup D] \\ &\qquad \Leftrightarrow (\exists x \in C \cup D)(\phi(x) = y) \\ &\qquad \Leftrightarrow (\exists x \in C)(\phi(x) = y) \text{ or } (\exists x \in D)(\phi(x) = y) \\ &\qquad \Leftrightarrow y \in \phi[C] \text{ or } y \in \phi[D] \\ &\qquad \Leftrightarrow y \in \phi[C] \cup \phi[D]. \end{aligned}$$

The justification of the second biconditional is found in Chapter Exercise 3.20d.

3. For every $x \in A$:

$$\begin{aligned} x \in \phi^{-1}[E \cup F] &\Leftrightarrow \phi(x) \in E \cup F \\ &\Leftrightarrow \phi(x) \in E \text{ or } \phi(x) \in F \\ &\Leftrightarrow x \in \phi^{-1}[E] \text{ or } x \in \phi^{-1}[F] \\ &\Leftrightarrow x \in \phi^{-1}[E] \cup \phi^{-1}[F]. \ \blacksquare \end{aligned}$$

It may seem odd that we only have an inclusion in part 2 of the theorem. To see that the other inclusion is false, let $f = \{(1, 3), (2, 3)\}$. Then,

$$f[\{1\} \cap \{2\}] = f[\varnothing] = \varnothing,$$

but

$$f[\{1\}] \cap f[\{2\}] = \{3\} \cap \{3\} = \{3\}.$$

Hence, $f[\{1\}] \cap f[\{2\}] \nsubseteq f[\{1\} \cap \{2\}]$. However, if f had been a bijection, then the inclusion would hold. (See Exercise 13.)

Example. Let $f(x) = x^2 + 1$. We will check the union results of the last theorem.

- We have already seen that $f[(1, 2)] = (2, 5)$. Since $(1, 2) = (1, 1.5] \cup [1.5, 2)$, apply f to both of these intervals. When we do this we find:
$$f[(1, 1.5]] = (2, 3.25],$$
$$f[[1.5, 2)] = [3.25, 5).$$
Therefore, $f[(1, 2)] = f[(1, 1.5]] \cup f[[1.5, 2)]$.
- We have also seen that $f^{-1}[(2, 5)] = (-2, -1) \cup (1, 2)$. We may write $(2, 5)$ as the union of $(2, 4)$ and $(3, 5)$. Let us find the inverse image of both of these sets:
$$f^{-1}[(2, 4)] = (-\sqrt{3}, -1) \cup (1, \sqrt{3}),$$
$$f^{-1}[(3, 5)] = (-2, -\sqrt{2}) \cup (\sqrt{2}, 2).$$
Hence,
$$f^{-1}[(2, 5)] = f^{-1}[(2, 4)] \cup f^{-1}[(3, 5)].$$

The last two results of the section show a relationship between images and inverse images and functions that are either one-to-one or onto.

6.6.4. Theorem

Let $\phi: A \to B$ be a function. Suppose $C \subseteq A$ and $D \subseteq B$.

1. If ϕ is one-to-one, then $\phi^{-1}[\phi[C]] = C$.
2. If ϕ is onto, then $\phi[\phi^{-1}[D]] = D$.

Proof. We will prove part one and leave the second as an exercise. Suppose that ϕ is an injection. We must show $\phi^{-1}[\phi[C]] = C$.

($\subseteq$) Let $x \in \phi^{-1}[\phi[C]]$. This means that there exists $y \in \phi[C]$ such that $\phi(x) = y$. Furthermore, there is a $z \in C$ so that $\phi(z) = y$. Therefore, $\phi(x) = \phi(z)$. Since ϕ is one-to-one, $x = z$, and this means $x \in C$.

($\supseteq$) This step will work for any function. Take $x \in C$. From here we can conclude that $\phi(x) \in \phi[C]$. Now by definition,
$$\phi^{-1}[\phi[C]] = \{z \in A : \phi(z) \in \phi[C]\}.$$
Since x is also a member of A, we may conclude $x \in \phi^{-1}[\phi[C]]$. ■

Because we only used the one-to-one condition to prove $\phi^{-1}[\phi[C]] \subseteq C$, we suspect that this inclusion is false if the function is not one-to-one. To confirm this, we need an example. Let $f\colon \mathbb{R} \to \mathbb{R}$ be defined as $f(x) = x^2$. We know that f is not one-to-one. Now choose $C = \{1\}$. Then,

$$f^{-1}[f[C]] = f^{-1}[f[\{1\}]] = f^{-1}[\{1\}] = \{-1, 1\}.$$

Therefore, $f^{-1}[f[C]] \nsubseteq C$.

When examining the example carefully, we may conjecture that the function being one-to-one is necessary for equality. Indeed, this is the case as the following theorem shows.

6.6.5. Theorem

Let $\phi\colon A \to B$ be a function.

1. If $\phi^{-1}[\phi[C]] = C$ for all $C \subseteq A$, then ϕ is one-to-one.
2. If $\phi[\phi^{-1}[D]] = D$ for all $D \subseteq B$, then ϕ is onto.

Proof. As with the previous theorem, we will prove only the first part. Assume

$$\phi^{-1}[\phi[C]] = C \text{ for all } C \subseteq A.$$

To show that ϕ is one-to-one, take $x_1, x_2 \in A$ and let $\phi(x_1) = \phi(x_2)$. Now,

$$\begin{aligned}\{x_1\} &= \phi^{-1}[\phi[\{x_1\}]] \\ &= \{z \in A : \phi(z) \in \phi[\{x_1\}]\} \\ &= \{z \in A : \phi(z) \in \{\phi(x_1)\}\} \\ &= \{z \in A : \phi(z) = \phi(x_1)\} \\ &\supseteq \{x_1, x_2\}.\end{aligned}$$

The hypothesis gives the first equality, and the last step is true because $\phi(x_1) = \phi(x_2)$. Since $\{x_1, x_2\} \subseteq \{x_1\}$, we must have $x_1 = x_2$. ■

EXERCISES

1. Let $f : \mathbb{R} \to \mathbb{R}$ be defined by $f(x) = 2x + 1$. Find the following:
   (a) $f[(1, 3]]$
   (b) $f[(-\infty, 0)]$
   (c) $f^{-1}[(-1, 1)]$
   (d) $f^{-1}[(0, 2) \bigcup (5, 8)]$

2. Let $g : \mathbb{R} \to \mathbb{R}$ be the function $g(x) = x^4 - 1$. Find:
   (a) $g[\{0\}]$
   (b) $g[\mathbb{Z}]$
   (c) $g^{-1}[\{0, 15\}]$
   (d) $g^{-1}[[-9, -5] \bigcap [0, 5]]$

3. Define $\phi\colon \mathbb{R} \times \mathbb{R} \to \mathbb{Z}$ by $\phi(a, b) = \lceil a \rceil + \lceil b \rceil$. Find the following:
   (a) $\phi[\{0\} \times \mathbb{R}]$
   (b) $\phi[(0, 1) \times (0, 1)]$
   (c) $\phi^{-1}[\{2, 4\}]$
   (d) $\phi^{-1}[\mathbb{N}]$
4. Let $\psi\colon \mathbb{Z} \to \mathbb{Z}$ be a function and define $\gamma\colon \mathbf{P}(\mathbb{Z}) \to \mathbf{P}(\mathbb{Z})$ by $\gamma(C) = \psi[C]$.
   (a) Prove that γ is well-defined.
   (b) Under what conditions is γ one-to-one?
   (c) Under what conditions is γ onto?
   (d) Find the following when $\psi(n) = 2n$:
      (i) $\gamma[\mathbb{Z}]$
      (ii) $\gamma[\{1, 2, 3\}]$
      (iii) $\gamma^{-1}[\mathbb{Z}]$
      (iv) $\gamma^{-1}[\{1, 2, 3\}]$
5. For any function ψ, show $\psi[\varnothing] = \varnothing$ and $\psi^{-1}[\varnothing] = \varnothing$.
6. Recall that I_A is the identity map on A. Prove for every $B \subseteq A$:
   (a) $I_A[B] = B$
   (b) $(I_A)^{-1}[B] = B$
7. Let $f\colon A \to B$ be a function with $C \subseteq A$. Prove:
   (a) $f[C] \subseteq \text{ran}(f)$
   (b) $f[A] = \text{ran}(f)$
   (c) $f^{-1}[B] = A$.
8. Let $\phi\colon A \to B$ be an injection and $C \subseteq A$.
   (a) Prove $\phi(x) \in \phi[C]$ if and only if $x \in C$.
   (b) Show that this is false if the function is not one-to-one.
9. Let $\phi\colon A \to B$ be a function. Prove:

$$\phi \text{ is onto if and only if for all } D \subseteq B, \phi[\phi^{-1}[D]] = D.$$

10. If f is a function, $A \subseteq B \subseteq \text{dom}(f)$, and $C \subseteq D \subseteq \text{ran}(f)$, show that $f[A] \subseteq f[B]$ and $f^{-1}[C] \subseteq f^{-1}[D]$.
11. Let $f\colon A \to B$ be a function and take two disjoint sets U and V.
   (a) Prove that the following are false.
      (i) If $U, V \subseteq A$, then $f[U] \cap f[V] = \varnothing$.
      (ii) If $U, V \subseteq B$, then $f^{-1}[U] \cap f^{-1}[V] = \varnothing$.
   (b) What additional hypothesis is needed to prove both of the implications? Prove it.
12. Assume that f and g are functions such that $\text{ran}(g) \subseteq \text{dom}(f)$. Let $A \subseteq \text{dom}(g)$. Prove $(f \circ g)[A] = f[g[A]]$.
13. Let $\phi\colon A \to B$ be a bijection. Prove the following.
   (a) $\phi[A] \cap \phi[B] \subseteq \phi[A \cap B]$.
   (b) If $C \subseteq A$ and $D \subseteq B$, then $\phi[C] = D$ if and only if $\phi^{-1}[D] = C$.
14. Let $\phi\colon A \to B$ and $C \subseteq A$.
   (a) Prove: if ϕ is a bijection, then $\phi[A \setminus C] = B \setminus \phi[C]$.
   (b) Show that the bijection condition is necessary for equality.

15. Find a function $\phi\colon A \to B$ and a set $D \subseteq B$ such that $D \not\subseteq \phi[\phi^{-1}[D]]$. Be sure to prove the result.

6.7. CARDINALITY

How can we determine whether two sets are the same size? One possibility is to count their elements. What happens, however, if the sets are infinite? We need another method. Suppose $A = \{12, 47, 84\}$ and $B = \{17, 101, 200\}$. We can see that these two sets are the same size without counting. Define a function $f\colon A \to B$ so that

$$f(12) = 17$$
$$f(47) = 101$$
$$f(84) = 200.$$

This function is a bijection. Since each element is paired with exactly one element of the opposite set, A and B must be the same size. This is the motivation behind our first definition.

6.7.1. Definition

For any two sets A and B, A is ***equinumerous*** with B if and only if there exists a bijection $\phi : A \to B$. Write $A \approx B$ if A is equinumerous with B.

If A and B are not equinumerous, write $A \not\approx B$.

Example. Take $n \in \mathbb{Z}$ and define

$$n\mathbb{Z} = \{nk : k \in \mathbb{Z}\}.$$

We will prove that if $n \neq 0$, then $\mathbb{Z} \approx n\mathbb{Z}$. To show this we must find a bijection $f\colon \mathbb{Z} \to n\mathbb{Z}$. Define $f(k) = nk$.

1. Assume $x_1, x_2 \in \mathbb{Z}$ and let $f(x_1) = f(x_2)$. Then $nx_1 = nx_2$, which yields $x_1 = x_2$ since $n \neq 0$. Thus, f is one-to-one.
2. Let $y \in n\mathbb{Z}$. This means $y = nk$, some $k \in \mathbb{Z}$. Then

$$f(k) = nk = y,$$

which shows that f is onto and, hence, a bijection.

6.7.2. Figure

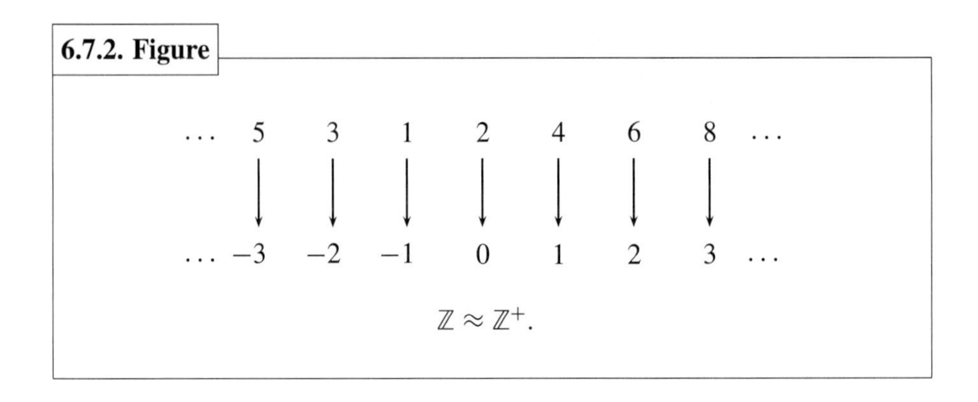

Example. To see $\mathbb{Z}^+ \approx \mathbb{Z}$, we will define a one-to-one correspondence so that each even integer will be mapped to a nonnegative one and every odd will be paired with a negative. (See Figure 6.7.2.) Let $g\colon \mathbb{Z}^+ \to \mathbb{Z}$ be defined by

$$g(n) = \begin{cases} k - 1 & \text{if } n = 2k \text{ for some } k \in \mathbb{Z}^+ \\ -k & \text{if } n = 2k - 1 \text{ for some } k \in \mathbb{Z}^+. \end{cases}$$

Notice that $g(4) = 1$ since $4 = 2(2)$, and $g(5) = -3$ because $5 = 2(3) - 1$. This function is a bijection by Exercise 7.

Equinumerosity plays a role similar to that of equality of integers. This is seen in the next theorem. As with Theorem 5.4.3, we cannot take this for granted but must carefully show each step.

6.7.3. Theorem

Let A, B, and C be sets.

1. $A \approx A$. [***Reflexive***]
2. If $A \approx B$, then $B \approx A$. [***Symmetric***]
3. If $A \approx B$ and $B \approx C$, then $A \approx C$. [***Transitive***]

Proof. Take three sets A, B, and C.

1. $A \approx A$ since the identity map is a bijection.
2. Assume $A \approx B$. Then there exists a bijection $\phi\colon A \to B$. Therefore, ϕ^{-1} exists and is a bijection. Hence, $B \approx A$.
3. By Theorem 6.5.12, the composition of two bijections is a bijection. Therefore, $A \approx B$ and $B \approx C$ implies $A \approx C$. ■

The symmetric property allows us to write $n\mathbb{Z} \approx \mathbb{Z}$ and $\mathbb{Z} \approx \mathbb{Z}^+$ by relying on the previous examples. The transitivity part of the lemma allows us to conclude from this that $n\mathbb{Z} \approx \mathbb{Z}^+$.

Example. Let us show $(0, 1) \approx \mathbb{R}$. We will do this in two parts. First, let $f: (0, 1) \to (-\pi/2, \pi/2)$ be defined by $f(x) = \pi x - \pi/2$. This function is a one-to-one correspondence since its graph is a non-vertical line (see Exercise 6.5.8). Second, define $g: (-\pi/2, \pi/2) \to \mathbb{R}$ to be the function $g(x) = \tan f(x)$. From trigonometry we know that tangent is a bijection on $(-\pi/2, \pi/2)$. Hence,

$$(0, 1) \approx (-\pi/2, \pi/2)$$

and

$$(-\pi/2, \pi/2) \approx \mathbb{R}.$$

The last theorem then allows us to conclude that $(0, 1) \approx \mathbb{R}$.

Another parallel with number theory is the next definition. It is analogous to $\leq$. For any two sets A and B, we say that B ***dominates*** A and write

$$A \preccurlyeq B$$

if there exists an injection $\phi: A \to B$. If B does not dominate A, write $A \not\preccurlyeq B$. Furthermore, define $A \prec B$ to mean $A \preccurlyeq B$ but $A \not\approx B$.

Example. If $A \subseteq B$, then $A \preccurlyeq B$. The ***inclusion map*** $\iota: A \to B$ defined by $\iota(a) = a$ shows this. For instance, $\mathbb{Z}^+ \preccurlyeq \mathbb{R}$ and $\mathbb{Q} \preccurlyeq \mathbb{C}$. However, $A \subset B$ does not imply $A \prec B$. For instance, $n\mathbb{Z} \approx \mathbb{Z}$, but $n\mathbb{Z} \subset \mathbb{Z}$ when $n \neq \pm 1$.

Another way to prove that $A \preccurlyeq B$ is by finding a surjection $B \to A$. Consider the sets $A = \{1, 2\}$ and $B = \{3, 4, 5\}$. Define $f: B \to A$ to be the surjection given by $f(3) = 1$, $f(4) = 2$, and $f(5) = 2$. This is the inverse of the relation R in Figure 6.1.3. By definition, to show that B dominates A, we must find an injection $A \to B$. We can modify f^{-1} by deleting $(2, 4)$ so that it becomes a function. Call it g. Hence, $g(1) = 3$ and $g(2) = 5$, which is an injection, and we have shown $A \preccurlyeq B$.

However, what happens if A and B are infinite sets? We may have to make infinitely many deletions to convert the inverse into a function. This is something that we cannot do if there is not a systematic way to do it. However, it appears reasonable that there is a subset of the inverse relation that is a function. This means we need an axiom. The one that we need is called the ***Axiom of Choice***. We will have to use it every time that an infinite number of arbitrary choices need to be made.

6.7.4. Axiom of Choice

Let A and B be sets. If $R \subseteq A \times B$, then there exists a function F such that $F \subseteq R$ and $\text{dom}(F) = \text{dom}(R)$.

The function F in the axiom is called a ***choice function***.

Example. Let $\mathcal{A} = \{A_i : i \in I\}$ be a family of nonempty sets such that I may be infinite. We want to define a family of singletons $\mathcal{B}$ such that

$$\{a_i\} \in \mathcal{B} \text{ if and only if } a_i \in A_i$$

for all $i \in I$. Since I may be infinite and there is no formula to guide the choices, we must use the Axiom of Choice. First define a relation $R \subseteq I \times \bigcup \mathcal{A}$ by

$$(i, a) \in R \text{ if and only if } a \in A_i.$$

(Remember that $\bigcup \mathcal{A}$ contains all the elements of the members of $\mathcal{A}$.) The axiom then gives us a choice function $F : I \to \bigcup \mathcal{A}$. Now define $\mathcal{B}$ to be the collection $\{\{F(i)\} : i \in I\}$. This is the desired family because if $(i, F(i)) \in R$, then $F(i) \in A_i$.

Example. Here is a situation that does not need the Axiom of Choice. Take the relation $R \subseteq \mathbb{R} \times \{0, 1\}$ defined by

$$(a, 0) \in R \text{ and } (a, 1) \in R \text{ if and only if } a \in \mathbb{R}.$$

Clearly R is not a function, but we can define a function

$$F : \mathbb{R} \to \{0, 1\}$$

by $F(a) = 0$ for all $a \in \mathbb{R}$. We can make infinitely many choices because there is a systematic way in which to make them.

We can now prove the generalization of our earlier argument.

6.7.5. Theorem

If there exists a surjection $\phi : A \to B$, then $B \preccurlyeq A$.

Proof. Let $\phi : A \to B$ be onto. We will use the Axiom of Choice. Define a relation $R \subseteq B \times A$ by

$$R = \{(b, a) : \phi(a) = b\}.$$

Since ϕ is onto,

$$\text{dom}(R) = \text{ran}(\phi) = B.$$

The Axiom of Choice yields a function F so that $\operatorname{dom}(F) = \operatorname{dom}(R)$ and $F \subseteq R$. We claim that F is one-to-one. Indeed, let $b_1, b_2 \in B$. Assume $F(b_1) = F(b_2)$. Let $a_1 = F(b_1)$ and $a_2 = F(b_2)$ where $a_1, a_2 \in A$. This means $a_1 = a_2$. Also, $\phi(a_1) = b_1$ and $\phi(a_2) = b_2$ because $F \subseteq R$. Since ϕ is a function, $b_1 = b_2$. ■

Example. Let $\sim$ be an equivalence relation on a set S. The map $\phi\colon S \to S/{\sim}$ defined by $\phi(a) = [a]_\sim$ is a surjection. Therefore, $S/{\sim} \preccurlyeq S$.

Example. We know that $\mathbb{Z}^+ \preccurlyeq \mathbb{R}$, since $\mathbb{Z}^+ \subseteq \mathbb{R}$. We can also prove this by using the function $f(x) = |\lceil x \rceil| + 1$ and appealing to Theorem 6.7.5.

The strict inequality $A \prec B$ is sometimes difficult to prove because we must show that there does not exist any bijection from A onto B. The next theorem is an example of this. It was first proven by Georg Cantor* using a method called ***diagonalization***.

6.7.6. Cantor's Theorem

$\mathbb{Z}^+ \prec \mathbb{R}$.

Proof. We already know $\mathbb{Z}^+ \preccurlyeq \mathbb{R}$, so we are left to prove $\mathbb{Z}^+ \not\approx \mathbb{R}$. Let $f\colon \mathbb{Z}^+ \to \mathbb{R}$ be a function. We will show that f cannot be onto. To do this enumerate the elements of $\operatorname{ran}(f)$. Suppose the list looks like:

$$f(1) = 1.\boxed{7}83477492\ldots$$
$$f(2) = 0.3\boxed{0}4934587\ldots$$
$$f(3) = 4.98\boxed{8}647655\ldots$$
$$f(4) = 3.320\boxed{9}98723\ldots$$
$$\vdots$$

We want to define a real number that differs from each of these values at the boxed digit. To do this let $i \in \mathbb{Z}^+$ and set

$$a_i = \begin{cases} 0 & \text{if the digit in the } 10^{-i} \text{ place of } f(i) \neq 0 \\ 1 & \text{otherwise.} \end{cases}$$

---

*Georg Cantor (St. Petersburg, Russia, 1845 – Halle, Germany, 1918): Cantor is most famous for his work in set theory. He showed that there were infinite sets of different magnitudes. His work was not accepted among many mathematicians, for at the time many believed that actual infinite sets did not exist.

For the above example, $a_1 = 0$ and $a_2 = 1$. Let $a \in (0, 1)$ be the number formed by letting the digit in the 10^{-i} position be a_i. In our example, $a = .0100\ldots$ Then $a \neq f(i)$ for all $i \in \mathbb{Z}^+$ because a differs from $f(i)$ in the 10^{-i} decimal place, so f is not onto. ■

(Note: there is a possible problem with the proof of Cantor's Theorem since two infinite decimals may represent the same number. See Chapter Exercise 38 for a discussion of this and why the proof is actually valid.)

Cantor's diagonalization argument can be generalized, but we first need a definition. Let A be a set and $B \subseteq A$. The function

$$\chi_B : A \to \{0, 1\}$$

is called a ***characteristic function*** and is defined by

$$\chi_B(a) = \begin{cases} 1 & \text{if } a \in B \\ 0 & \text{if } a \notin B. \end{cases}$$

For example, if $A = \mathbb{Z}$ and $B = \{0, 1, 3, 5\}$, then $\chi_B(1) = 1$ but $\chi_B(2) = 0$.

6.7.7. Theorem

If A is a set, then $A \prec \mathbf{P}(A)$.

Proof. This is clear for any finite set A. To prove the infinite case, we must show $A \preccurlyeq \mathbf{P}(A)$ and $A \not\approx \mathbf{P}(A)$.

1. Since the map $\psi(a) = \{a\}$ is an injection $A \to \mathbf{P}(A)$, the power set of A dominates A.
2. To show that A is not equinumerous with $\mathbf{P}(A)$, we will show that every function $A \to \mathbf{P}(A)$ cannot be a surjection. Define

$$X = \{\chi_B : B \in \mathbf{P}(A)\}.$$

It will be left as an exercise to prove that $\mathbf{P}(A) \approx X$. It now suffices to show that $A \not\approx X$. Let $\phi : A \to X$ be a function, and for all $a \in A$, write $\phi(a) = \chi_{B_a}$ for some $B_a \subseteq A$. It is impossible for ϕ to be onto. To see this, define χ so that

$$\chi(a) = \begin{cases} 1 & \text{if } \chi_{B_a}(a) = 0 \\ 0 & \text{if } \chi_{B_a}(a) = 1. \end{cases}$$

Therefore, $\chi \notin \operatorname{ran}(\phi)$ because $\phi(a) \neq \chi$ for all $a \in A$. This is because $\chi_{B_a}(a) \neq \chi(a)$. However, $\chi \in X$. To prove this, we must find $B \subseteq A$ such that $\chi = \chi_B$. Define

$$B = \{a \in A : \chi_{B_a}(a) = 0\}.$$

We have two cases to check:

(Case 1) If $\chi_{B_a}(a) = 0$, then $\chi(a) = 1$ and $\chi_B(a) = 1$ since $a \in B$. Hence, $\chi(a) = \chi_B(a)$.

(Case 2) If $\chi_{B_a}(a) = 1$, then $\chi(a) = 0$ and $\chi_B(a) = 0$. Again, $\chi(a) = \chi_B(a)$.

Therefore, $\chi = \chi_B$. Hence, ϕ is not onto. ■

From the theorem, we conclude that

$$\mathbb{N} \prec \mathbf{P}(\mathbb{N}) \prec \mathbf{P}(\mathbf{P}(\mathbb{N})) \prec \mathbf{P}(\mathbf{P}(\mathbf{P}(\mathbb{N}))) \prec \cdots$$

Thus, there are larger and larger magnitudes of infinity. To discuss these different sizes, we introduce the next concept. Take a set A. The ***cardinal number*** (or simply the ***cardinality***) of A refers to its size. Let $|A|$ represent the cardinal number of A. If A is finite, we understand that the size of A is represented by some natural number. Hence, the cardinal number of $\{1, -6, 28\}$ is 3 while the cardinality of $\varnothing$ is 0, and we write $|\{1, -6, 28\}| = 3$ and $|\varnothing| = 0$. For the infinite sets, Cantor denoted $|\mathbb{N}|$ by $\aleph_0$. The symbol $\aleph$ (**aleph**) is the first letter in the Hebrew alphabet. The next magnitude of infinity is $\aleph_1$. This continues and gives a strictly increasing sequence of infinite cardinals, and since natural numbers must be less than any infinite cardinal, we have

$$(*) \qquad 0 < 1 < 2 < \cdots < \aleph_0 < \aleph_1 < \aleph_2 < \cdots$$

For instance, $4 < \aleph_1$, $\aleph_0 \leq \aleph_0$, and $\aleph_3 < \aleph_7$.

Example. Although he was unable to prove it, Cantor conjectured that $|\mathbb{R}| = \aleph_1$. This conjecture is called the ***Continuum Hypothesis***. He was unable to prove it because it is possible that $|\mathbb{R}| > \aleph_1$.*

Summarizing, the results of the first half of the section can be translated in terms of cardinality.

- To prove $|A| = |B|$, it suffices to prove $A \approx B$.
- To prove $|A| \leq |B|$, we prove $A \preccurlyeq B$.
- To prove $|A| < |B|$, we must show that $A \preccurlyeq B$ but $A \not\approx B$.

This means that since $\mathbb{Z} \approx \mathbb{N}$, $|\mathbb{Z}| = |\mathbb{N}|$. However, $|\mathbb{Z}^+| < |\mathbb{R}|$ and $|A| < |\mathbf{P}(A)|$ because $\mathbb{Z}^+ \prec \mathbb{R}$ and $A \prec \mathbf{P}(A)$ for any set A. Furthermore,

*Paul C. Cohen (Long Branch, New Jersey, 1934 –): Cohen is most famous for his contributions to mathematical logic. He developed a method known as forcing to show that the Continuum Hypothesis could be false.

$$A \text{ is } \textbf{\textit{finite}} \text{ if and only if } |A| < \aleph_0,$$

and

$$A \text{ is } \textbf{\textit{infinite}} \text{ if and only if } |A| \geq \aleph_0.$$

For the rest of the section we will focus on sets of cardinality less than $\aleph_1$. Sets of this size are very important to mathematics, so we make a definition.

6.7.8. Definition

A set A is ***countable*** if $|A| \leq \aleph_0$.

Sometimes countable sets are called ***discrete*** or ***denumerable***. According to the definition, a finite set is countable. Any finite set can be written as

$$\{a_1, a_2, \ldots, a_n\}$$

for some positive integer n. A countably infinite set can be written as

$$\{a_1, a_2, a_3, \ldots\}.$$

(Warning: The sequence of cardinals $(*)$ is written as if there were countably many cardinals. This is not the case. The terms shown are simply the beginning of the sequence.)

Since by definition $\aleph_0 = |\mathbb{N}|$, to prove that an infinite set A is countable, it suffices to show that $A \approx \mathbb{N}$. For example, the bijection $f\colon \mathbb{Z}^+ \to \mathbb{N}$ defined by $f(n) = n - 1$ shows that $\mathbb{Z}^+$ is countable.

Example. The set of rational numbers is countable. To prove this, define a bijection $f\colon \mathbb{N} \to \mathbb{Q}$ by first mapping the even natural numbers to the nonnegative rationals as indicated:

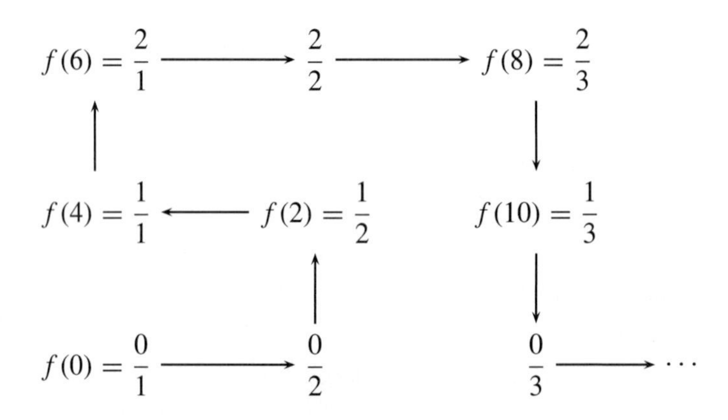

The function is defined along the indicated path. When a rational number is encountered that has previously been used, it is skipped. To complete the definition, associate the odd naturals with the negative rational numbers using a path as in the diagram. This function is a bijection, so we may conclude that $|\mathbb{Q}| = \aleph_0$.

We have defined countability in terms of one-to-one functions. Now let us define it using surjections. The proof of the next theorem relies on Theorem 6.7.5.

6.7.9. Theorem

The set A is countable if and only if there exists a function from $\mathbb{N}$ onto A.

Proof.

($\Rightarrow$) Suppose A is countable. Then, $|A| = \aleph_0$ or there exists $n \in \mathbb{N}$ such that $|A| = n$. Let us examine both cases:

(Case 1) When $|A| = \aleph_0$, $A \approx \mathbb{N}$. Hence, there is a surjection from the set of natural numbers to A.

(Case 2) If $|A| = n$, then there is no bijection between $\mathbb{N}$ and A, but a surjection can still be defined. Write A as $\{a_1, \ldots, a_n\}$ and fix any element $a \in A$. Define $\psi: \mathbb{N} \to A$ by

$$\psi(i) = \begin{cases} a_{i+1} & \text{if } i = 0, \ldots, n-1 \\ a & \text{otherwise.} \end{cases}$$

This function is certainly onto.

($\Leftarrow$) Conversely, let $\phi: \mathbb{N} \to A$ be a surjection. By Theorem 6.7.5, $A \preccurlyeq \mathbb{N}$, which means $|A| \leq \aleph_0$. ■

Example. As sets go, countable sets are relatively small, even the infinite ones. In fact, every set has a countable subset. To prove this let A be a set. If A itself is countable, then we are done. So, suppose that $|A| > \aleph_0$. By definition there exists an injection $\phi: \mathbb{N} \to A$. Since ϕ is onto its range, its range is countable and the desired subset.

If $|A_i| \leq \aleph_0$ for all $i = 0, \ldots, n$, then $A_1 \times A_2 \times \cdots \times A_n$ is countable (Exercise 16). In particular, $\mathbb{N} \times \mathbb{N}$ and $\mathbb{Z} \times \mathbb{Z}$ are countable. We also have the next theorem.

6.7.10. Theorem

A countable union of countable sets is countable.

Proof. Let $\{A_i : i \in I\}$ be a family of countable sets with $|I| \leq \aleph_0$. Then for each $i \in I$, there exists an onto function $\psi_i : \mathbb{N} \to A_i$. We also have a surjection $\phi : \mathbb{N} \to I$. Therefore, define

$$\Phi : \mathbb{N} \times \mathbb{N} \to \bigcup_{i \in I} A_i$$

by $\Phi(n, m) = \psi_{\phi(m)}(n)$. To check that Φ is onto, let $a \in \bigcup_{i \in I} A_i$. Thus, $a \in A_i$, some $i \in I$. Since ϕ is onto, there exists $m_0 \in \mathbb{N}$ so that $\phi(m_0) = i$. Furthermore, since $\psi_{\phi(m_0)}$ is onto, we have $n_0 \in \mathbb{N}$ such that

$$\psi_{\phi(m_0)}(n_0) = a.$$

This means $\Phi(n_0, m_0) = a$ and Φ is onto. Thus, the union is countable. ■

EXERCISES

1. Indicate whether each of the following are countable or uncountable.
 (a) $\mathbb{Z}^+$
 (b) $\mathbb{Z} \cup \mathbb{Z}$
 (c) $\mathbb{Z}^+ \setminus \{5, 6, 7, \ldots\}$
 (d) $\mathbb{Q} \setminus \mathbb{R}$
 (e) $\mathbb{R} \setminus \mathbb{Q}$
 (f) $[0, 10] \cup \mathbb{Z}$
 (g) $[0, 10] \cap \mathbb{Z}$
 (h) The set of all propositional forms
 (i) The set of all possible words in a dictionary
 (j) Possible actual room temperatures
 (k) Possible measured room temperatures

2. Given $\chi_B : \mathbb{Z} \to \{0, 1\}$ with $B = \{2, 5, 19, 23\}$, find
 (a) $\chi_B(1)$
 (b) $\chi_B(2)$
 (c) $\chi_B(-10)$
 (d) $\chi_B(19)$

3. Let χ_A and χ_B be characteristic functions. Prove:
 (a) $\chi_{A \cup B} = \chi_A + \chi_B - \chi_A \chi_B$
 (b) $\chi_{A \cap B} = \chi_A \chi_B$
 (c) $\chi_{A \Delta B} = \chi_A + \chi_B - 2\chi_A \chi_B$

4. Prove the following by finding a bijection between the two sets.
 (a) $[0, \pi] \approx [-1, 1]$
 (b) $[-\pi/2, \pi/2] \approx [-1, 1]$

(c) $(0, \infty) \approx \mathbb{R}$
(d) $\mathbb{N} \approx \mathbb{Z}$
(e) $|\mathbb{Z}^+| = |\mathbb{Z}^-|$
(f) $|\{(x, 0) : x \in \mathbb{R}\}| = |\mathbb{R}|$
(g) $|\mathbb{Z}| = |\mathbb{Z} \times \mathbb{Z}|$
(h) $|\{(x, y) \in \mathbb{R} : y = 2x + 4\}| = |\mathbb{R}|$

5. Show the following:
   (a) $(a, b) \approx (c, d)$
   (b) $[a, b] \approx [c, d]$
   (c) $(a, b) \approx [c, d]$
   (d) $(a, b) \approx (c, d]$
6. Prove $|\mathbf{P}(A)| = |\{\chi_B : B \in \mathbf{P}(A)\}|$ for any set A.
7. Prove $|\mathbb{Z}^+| = |\mathbb{Z}|$ by showing that the function $g : \mathbb{Z}^+ \to \mathbb{Z}$ defined by

$$g(n) = \begin{cases} k & \text{if } n = 2k \text{ for some } k \in \mathbb{N} \\ -k & \text{if } n = 2k + 1 \text{ for some } k \in \mathbb{N} \end{cases}$$

   is a bijection.
8. Let A be an infinite set and assume $a_1, a_2, \ldots$ is a sequence of distinct elements from A. Prove that

$$\phi(x) = \begin{cases} a_{n+1} & \text{if } x = a_n \\ x & \text{otherwise} \end{cases}$$

   is a bijection $A \to A \setminus \{a_1\}$.
9. Assuming that A, B, C, and D are nonempty sets, prove the following by finding an appropriate injection.
   (a) $\mathbb{N} \preccurlyeq \mathbb{Z}^-$
   (b) $A \cap B \preccurlyeq \mathbf{P}(A)$
   (c) $A \preccurlyeq A \times B$
   (d) $[0, 2] \preccurlyeq [5, 7]$
   (e) $A \cap B \preccurlyeq A$
   (f) $|A \times B| \leq |B^A|$
   (g) $|A \times \{0\}| \leq |(A \cup B) \times \{1\}|$
   (h) $|A \times B \times C| \leq |A \times B \times C \times D|$
   (i) $|A \setminus B| \leq |C \times A|$
10. Prove the following:
   (a) If $A \preccurlyeq B$ and $B \approx C$, then $A \preccurlyeq C$.
   (b) If $A \approx B$ and $B \preccurlyeq C$, then $A \preccurlyeq C$.
   (c) If $A \preccurlyeq B$ and $B \preccurlyeq C$, then $A \preccurlyeq C$.
11. If A is finite, why is it the case that $|A| < |\mathbf{P}(A)|$?
12. Let ϕ be a function with codomain B. Show the following:
   (a) $\text{ran}(\phi) \preccurlyeq B$
   (b) There exists $C \subseteq B$ such that $\text{dom}(\phi) \approx C$ if ϕ is one-to-one.

13. Show that $|\mathbb{N}| \leq |\mathbb{N} \times \mathbb{N}|$ by finding a surjection $\phi \colon \mathbb{N} \times \mathbb{N} \to \mathbb{N}$.
14. If A and B are sets and A is countable, prove that B is countable when $A \approx B$.
15. Let A and B be countable sets. Prove that each of the following are countable:
   (a) $A \cup B$
   (b) $A \cap B$
   (c) $A \times B$
   (d) $A \setminus B$
16. Let $\{A_1, \ldots A_n\}$ be a family of countable sets. Prove that each of the following are countable:
   (a) $A_1 \cup \cdots \cup A_n$
   (b) $A_1 \cap \cdots \cap A_n$
   (c) $A_1 \times \cdots \times A_n$
17. Prove the following using the Axiom of Choice: Let $f \colon A \to B$ be a surjection. Prove that there exists a function $g \colon B \to A$ such that $g \circ f = I_A$.

CHAPTER EXERCISES

1. Let $\sim$ be a relation on A with the property that if $a \sim b$ and $b \sim c$, then $c \sim a$. This is called a ***circular relation***. Prove that a reflexive circular relation is an equivalence relation.
2. Define a relation on $\mathbb{C}$: $a + bi \sim c + di$ if and only if

$$\sqrt{a^2 + b^2} = \sqrt{c^2 + d^2}.$$

   (a) Prove $\sim$ is an equivalence relation on $\mathbb{C}$.
   (b) Graph $[1 + i]_\sim$.
   (c) How does $\sim$ partition $\mathbb{C}$?
3. If $f \colon A \to B$ is a function, then define a relation on A by

$$a \sim b \text{ if and only if } f(a) = f(b).$$

   Answer the following:
   (a) Show $\sim$ is an equivalence relation.
   (b) Find $[3]_\sim$ if $f \colon \mathbb{Z} \to 2\mathbb{Z}$ is defined by $f(n) = 2n$.
   (c) Find $[2]_\sim$ if $f \colon \mathbb{Z} \to \mathbb{Z}_5$ is given by $f(n) = [n]_5$.
4. Let R_1 and R_2 be equivalence relations on A. For each of the following, either prove that the set is an equivalence relation or disprove it by finding a counterexample.
   (a) $R_1 \cup R_2$
   (b) $R_1 \cap R_2$
   (c) $(A \times A) \setminus R_1$

5. Let $\sim$ be a relation on a set A. We say that $\sim$ is a ***partial ordering*** if $\sim$ is reflexive, antisymmetric, and transitive. The pair $(A, \sim)$ is called a ***partially ordered set*** or a ***poset***. (Note: it is traditional to represent partial orderings by other symbols such as $\preccurlyeq$ or $\leq$.) Use the definitions on pages 176 and 177 to prove that $\leq$ is a partial ordering on $\mathbb{Z}$.
6. Fix a set A. For all $B, C \in \mathbf{P}(A)$, define $B \sim C$ if and only if $B \subseteq C$. Prove that $(\mathbf{P}(A), \sim)$ is a poset.
7. Let $(A, \preccurlyeq)$ and $(B, \leq)$ be posets. Define $\sim$ on $A \times B$ by $(a, b) \sim (a', b')$ if and only if

$$a \preccurlyeq a' \text{ and } b \leq b'.$$

Show that this is a partial ordering on $A \times B$.

8. Let $\mathbb{A}$ be a finite, nonempty set of symbols. We will call this set an ***alphabet***. A ***string*** from $\mathbb{A}$ is a finite sequence of elements of $\mathbb{A}$ written without punctuation or spaces. Let $\mathbb{A}^*$ denote the set of all strings from $\mathbb{A}$. For example, if $\mathbb{A} = \{a, b, c\}$, then abc, $aaabbb$, c, and $\varnothing$ are elements of $\mathbb{A}^*$. The symbol $\varnothing$ denotes the ***empty string***, the string of length zero.
   (a) Find five elements from each of the following:
      (i) $\{0, 1\}^*$
      (ii) $\{a, b, c, \ldots, y, z\}^*$
      (iii) $\mathbb{N}^*$
      (iv) $\{\%, \#, +, @\}^*$
   (b) Let $\sigma, \tau \in \mathbb{A}^*$. The ***concatenation*** of σ and τ is denoted by $\sigma^\frown\tau$ and is formed by combining σ and τ into one string. For example, if $\mathbb{A} = \{0, 1\}$, $\sigma = 011$, and $\tau = 1010$, then $\sigma^\frown\tau = 0111010$. For any alphabet $\mathbb{A}$, prove the following:
      (i) Concatenation is associative. In other words, fix $\mathbb{A}$ and show that for all $\sigma, \tau, \nu \in \mathbb{A}^*$,

$$\sigma^\frown(\tau^\frown\nu) = (\sigma^\frown\tau)^\frown\nu.$$

      (ii) Concatenation is not commutative.
      (iii) $\mathbb{A}^*$ has an identity with respect to $\frown$, but for all $\sigma \in \mathbb{A}^*$, there is no inverse for σ if $\sigma \neq \varnothing$.
9. Let $\mathbb{A}$ be an alphabet. Define for all $\sigma, \tau \in \mathbb{A}^*$,

$$\sigma \preccurlyeq \tau \text{ if and only if there exists } \nu \in \mathbb{A}^* \text{ such that } \tau = \sigma^\frown\nu.$$

Prove that $(\mathbb{A}^*, \preccurlyeq)$ is a poset.

10. Let A be a set.
   (a) If $(A, \preccurlyeq)$ is a poset such that for every $a, b \in A$ either $a \preccurlyeq b$ or $b \preccurlyeq a$, then $\preccurlyeq$ is called a ***linear ordering***.
      (i) Prove that $\leq$ is a linear ordering on $\mathbb{Z}$.

(ii) Let $(A, \preccurlyeq)$ and $(B, \leq)$ be two linear orderings. Define $\sim$ on $A \times B$ by $(a, b) \sim (a', b')$ if and only if

$$a \preccurlyeq a' \text{ or } a = a' \text{ and } b \leq b'.$$

Prove that $(A \times B, \sim)$ is a linear ordering. (Note: this relation is called a ***lexicographical ordering*** since it mimics the ordering of a dictionary.)

11. Prove that $(\mathbf{P}(A), \subseteq)$ is not a linear ordering if $|A| > 2$.
12. Using the definitions of Chapter Exercise 9, show that $(\mathbb{A}^*, \preccurlyeq)$ is not a linear ordering if $|\mathbb{A}| > 1$.
13. Let $\sim$ and $\sim'$ be relations on A and B, respectively. A function $\phi: A \to B$ is called ***order-preserving*** when for all $x, y \in A$,

$$\text{if } x \sim y, \text{ then } \phi(x) \sim' \phi(y).$$

Show that the following are order-preserving.

(a) Define $f: \mathbb{R} \to \mathbb{R}$ by $f(x) = 2x + 1$ and take both relations to be regular less-than.

(b) Let $A \subseteq B$ and take $b \in B$. Order both $\mathbf{P}(A)$ and $\mathbf{P}(B)$ by inclusion. Define $\psi_b: \mathbf{P}(A) \to \mathbf{P}(B)$ such that $\psi_b(D) = D \cup \{b\}$.

14. Let $A, B \subseteq \mathbb{R}$ be two sets ordered by $\leq$. Suppose that A is well-ordered. Prove that if $f: A \to B$ is an order-preserving surjection, then B is well-ordered.
15. Let $A \subseteq \mathbb{R}$ be a well-ordered, infinite set (using $\leq$). For all $x \in A$, define x^+ to be the first element after x. In other words, x^+ has the property that $x < x^+$ and

$$x^+ \leq y \text{ for all } y > x.$$

This is called the ***successor*** of x. Prove that x^+ exists.

16. Let A and B be well-ordered, infinite sets of real numbers. A function $f: A \to B$ is called ***successor-preserving*** when for all $x \in A$,

$$f(x^+) = f(x)^+.$$

Prove the following.

(a) There is at most one successor-preserving surjection $f: A \to B$.

(b) If f is successor-preserving:

(i) f is one-to-one.

(ii) f is order-preserving.

(iii) f^{-1} is order-preserving.

(c) Let $\phi: A \to B$ be a successor-preserving function. Fix $a \in A$ and define $S = \{x \in A : x \leq a\}$. Prove that $f \restriction A$ is order-preserving.

17. A relation S on A is the ***reflexive closure*** of R if it is the smallest reflexive relation on A containing R. This means that

$$\text{if } T \subseteq A \times A \text{ is reflexive and } R \subseteq T, \text{ then } S \subseteq T.$$

Identify the reflexive closure of R. Be sure to prove the result.

18. Let R be a relation on A. The ***symmetric closure*** of R is the smallest symmetric relation on A containing R. Prove the following:
    (a) $R \cup R^{-1}$ is the symmetric closure of R.
    (b) A symmetric closure is unique.
19. Define $\phi([a]_n) = [a]_m$ for all $a \in \mathbb{Z}$. If ϕ is a function, then does this mean that $m \mid n$? Explain.
20. Let $\mathscr{C}$ be a chain of functions. Demonstrate that $\bigcup \mathscr{C}$ is a function.
21. Let f and g be functions. Prove the following:
    (a) If f and g are functions, then $f \cap g$ is a function.
    (b) The union $f \cup g$ is a function if and only if $f(x) = g(x)$ for all $x \in \text{dom}(f) \cap \text{dom}(g)$.
22. Prove that the empty set is a bijection with domain and range equal to $\varnothing$.
23. For both of the following, either prove that the result is true or find a counterexample:
    (a) For all $\psi, \gamma \in B^A$ and $\phi \in C^B$, if $\phi \circ \psi = \phi \circ \gamma$, then $\psi = \gamma$.
    (b) For all $\phi \in B^A$ and $\psi, \gamma \in C^B$, if $\psi \circ \phi = \gamma \circ \phi$, then $\psi = \gamma$.
24. Let $f : \mathbb{R} \to \mathbb{R}$ be periodic. (See Exercise 6.4.15.) Prove that f is not one-to-one.
25. Let $\phi : A \to B$ be a function and $\psi = \phi \restriction C$ where $C \subseteq A$. Prove that if ι is the inclusion map $C \to A$, then $\psi = \phi \circ \iota$.
26. Let f be an injection with $A \subseteq \text{dom}(f)$ and $B \subseteq \text{ran}(f)$. Prove:

$$B \subseteq f[A] \text{ if and only if } f^{-1}[B] \subseteq A.$$

27. Let $\{A_i : i \in I\}$ be a family of sets. Prove these generalizations of results from Section 6.6.
    (a) $f\left[\bigcup_{i \in I} A_i\right] = \bigcup_{i \in I} f[A_i]$
    (b) $f\left[\bigcap_{i \in I} A_i\right] \subseteq \bigcap_{i \in I} f[A_i]$
    (c) $f^{-1}\left[\bigcup_{i \in I} A_i\right] = \bigcup_{i \in I} f^{-1}[A_i]$
    (d) $f^{-1}\left[\bigcap_{i \in I} A_i\right] = \bigcap_{i \in I} f^{-1}[A_i]$

28. We will say that a function $f\colon \mathbb{Z}^+ \to \mathbb{Z}$ is ***multiplicative*** if for all $m, n \in \mathbb{Z}^+$, if m and n are relatively prime, then $f(mn) = f(n)f(m)$. If $n = p_1^{r_1} \cdots p_n^{r_n}$ is a prime power decomposition of n, prove
$$f(n) = f(p_1^{r_1}) \cdots f(p_n^{r_n}).$$
29. Prove that Euler's phi-function is multiplicative. (See Chapter Exercise 5.26.)
30. Let R be a relation on A. For any $a \in A$, define:
$$R_a = R \cap (\{a\} \times A).$$
The relation R_a is called a ***vertical section***. Prove $\bigcup_{a\in A} R_a = R$.
31. Let R be a relation on A and $a, b \in A$. Show:
$$R_a = R_b \text{ if and only if } a = b.$$
32. Prove or give a counterexample:
    (a) A vertical section can be a function.
    (b) A vertical section can be an injection.
    (c) A vertical section can be a surjection.
33. Let $\mathscr{A} = \{A_1, \ldots, A_k\}$ be a pairwise disjoint family of sets. Assuming that the sets are distinct, prove that the cardinality of $\bigcup \mathscr{A}$ is equal to the sum
$$|A_1| + \cdots + |A_k|.$$
34. Use the Second Principle of Mathematical Induction to prove that every infinite set is equinumerous to a proper subset of itself.
35. Let $\mathscr{C} \subseteq \mathbf{P}(A)$ be a chain. Prove that $\bigcup \mathscr{C}$ is an upper bound for $\mathscr{C}$ with respect to $\subseteq$. In other words, for all $A \in \mathscr{C}$, $A \subseteq \bigcup \mathscr{C}$.
36. The following is known as Zorn's Lemma:*

    Let $(A, \preccurlyeq)$ be a poset. With respect to $\preccurlyeq$, if every linear ordering $(C, \preccurlyeq)$ such that $C \subseteq A$ has an upper bound in A, then A has a maximal element.

    Prove that Zorn's Lemma implies the Axiom of Choice. (Actually, these two propositions are equivalent, but the converse is beyond the scope of this text.)
37. Use Zorn's Lemma to prove the following:
    (a) For every function $f : A \to B$ there exists a maximal $C \subseteq A$ such that $f \restriction C$ is one-to-one.
    (b) If $\mathscr{A}$ is a collection of sets such that for every chain $\mathscr{C} \subseteq \mathscr{A}$, $\bigcup \mathscr{C} \in \mathscr{A}$, then $\mathscr{A}$ contains a maximal element.

*Max Zorn (Germany, 1906 – Bloomington, Indiana, 1993): Zorn studied set theory, topology, and algebra. He first began working with his lemma in the mid 1930s while he was at Yale. Although the proposition is known as Zorn's Lemma, it is often assumed as an axiom.

38. The proof of Cantor's Theorem (6.7.6) contains a potential problem since two different decimal expansions may represent the same real number. For example, $.1\overline{9} = .2$.
    (a) Show $.1\overline{9} = .2$.
    (b) Explain why the given proof of Cantor's Theorem escapes this problem.

III Coming Attractions

The last part covers two topics that will provide a brief preview of junior- and senior-level mathematics. They will also show how logic and set theory interact with both discrete and continuous mathematics. The topics are ring theory (a generalization of number theory) and topology (the study of continuous functions).

7 Ring Theory

Consider the simple equation $2x + 1 = 0$. The exact steps needed to find its solution are:

$$
\begin{aligned}
(2x+1) + -1 &= 0 + -1 \\
2x + (1 + -1) &= 0 + -1 \\
2x + 0 &= 0 + -1 \\
2x &= -1 \\
1/2(2x) &= 1/2(-1) \\
(1/2 \cdot 2)x &= -1/2 \\
1x &= -1/2 \\
x &= -1/2
\end{aligned}
$$

Now examine the steps. There are two operations, addition and multiplication. We used inverses and identities. The associative law was also used. When studying these properties, mathematicians will often define an object that has only these properties and then examine it looking for theorems that result from the definition. The object that is defined using what is needed to solve the above equation is modeled on the integers and called a ring.

7.1. TYPES OF RINGS

When we studied the integers, we mainly limited ourselves to the two operations of addition and multiplication. This is what we will do here. The main definition is a generalization of the properties of $\mathbb{Z}$ and its operations of addition and multiplication.

7.1.1. Definition

Let $\oplus$ and $\otimes$ be two binary operations on a nonempty set R. Call $\oplus$ ***addition*** and $\otimes$ ***multiplication***. The triple $(R, \oplus, \otimes)$ is called a ***ring*** if the following hold for all $a, b, c \in R$:

Associative Laws $a \oplus (b \oplus c) = (a \oplus b) \oplus c$
$a \otimes (b \otimes c) = (a \otimes b) \otimes c$

Commutative Law $a \oplus b = b \oplus a$

Additive Identity There exist $0_R \in R$ such that for all $a \in R, a \oplus 0_R = a$.

Additive Inverse For every $a \in R$, there exists $b \in R$ such that b is the additive inverse of a.

Distributive Laws $a \otimes (b \oplus c) = (a \otimes b) \oplus (a \otimes c)$
$(a \oplus b) \otimes c = (a \otimes c) \oplus (b \otimes c)$

Notes:

- Since $\oplus$ is commutative, we do not have to write $a \oplus 0_R = 0_R \oplus a = a$.
- If b is the additive inverse of a, then we write $-a = b$.
- To minimize the use of parentheses, we assume that multiplication has precedence over addition.

Example. $(\mathbb{Z}, +, \cdot)$, $(\mathbb{R}, +, \cdot)$, $(\mathbb{Q}, +, \cdot)$,$(\mathbb{C}, +, \cdot)$ are all rings where both $+$ and $\cdot$ denote standard addition and multiplication. We can also make $\mathbb{Z}_n$ into a ring for $n \in \mathbb{Z}^+$. Take $[a]_n, [b]_n \in \mathbb{Z}_n$ and define

$$[a]_n + [b]_n = [a + b]_n$$

and

$$[a]_n \cdot [b]_n = [ab]_n.$$

Then $(\mathbb{Z}_n, +, \cdot)$ is a ring. (See Exercise 8.)

Example. A ***polynomial with real coefficients*** is a function $f(X)$ so that

$$f(X) = a_n X^n + a_{n-1} X^{n-1} + \cdots + a_0,$$

where $a_i \in \mathbb{R}$. (For reasons that will be explained in a later section, we use a capital X in the definition instead of x.) Denote the set of polynomials with real coefficients by $\mathbb{R}[X]$. If the operations of addition and multiplication of polynomials are defined as usual, then $(\mathbb{R}[X], +, \cdot)$ is a ring.

We next examine a more complicated ring. After we define the set and the two binary operations, we will outline the proof that it is a ring. For all $m, n \in \mathbb{Z}^+$, define $\mathrm{M}_{m,n}(\mathbb{R})$ as the set of $m \times n$ matrices with real entries. In other words, each matrix has m rows, n columns, and looks like

$$\begin{bmatrix} a_{1,1} & \cdots & a_{1,n} \\ \vdots & \ddots & \vdots \\ a_{m,1} & \cdots & a_{m,n} \end{bmatrix}$$

where $a_{i,j} \in \mathbb{R}$ for $i = 1, \ldots, m$ and $j = 1, \ldots, n$. As an example:

$$\begin{bmatrix} 1 & 2 \\ 3 & 4 \\ 5 & 6 \end{bmatrix} \in \mathrm{M}_{3,2}(\mathbb{R}).$$

Now define ***matrix addition*** entrywise. For instance,

$$\begin{bmatrix} 1 & 2 \\ 3 & 4 \\ 5 & 6 \end{bmatrix} + \begin{bmatrix} 1 & 0 \\ 2 & -5 \\ 0 & -2 \end{bmatrix} = \begin{bmatrix} 2 & 2 \\ 5 & -1 \\ 5 & 4 \end{bmatrix}.$$

Next define ***matrix multiplication*** for $n \times n$ matrices.* Such matrices are called ***square matrices***. For 2×2 matrices, the product

$$\begin{bmatrix} a_{1,1} & a_{1,2} \\ a_{2,1} & a_{2,2} \end{bmatrix} \cdot \begin{bmatrix} b_{1,1} & b_{1,2} \\ b_{2,1} & b_{2,2} \end{bmatrix}$$

is equal to

$$\begin{bmatrix} a_{1,1}b_{1,1} + a_{1,2}b_{2,1} & a_{1,1}b_{1,2} + a_{1,2}b_{2,2} \\ a_{2,1}b_{1,1} + a_{2,2}b_{2,1} & a_{2,1}b_{1,2} + a_{2,2}b_{2,2} \end{bmatrix}.$$

For example,

$$\begin{bmatrix} 1 & 2 \\ 3 & 4 \end{bmatrix} \cdot \begin{bmatrix} 0 & -1 \\ 2 & 1 \end{bmatrix} = \begin{bmatrix} 4 & 1 \\ 8 & 1 \end{bmatrix}.$$

This can be generalized to any $n \times n$ matrix.

To prove that $\mathrm{M}_{m,n}(\mathbb{R})$ is a ring with these two operations, we will have to show that they satisfy the five parts of the definition.

---

*It is possible to define matrix multiplication for certain non-square matrices, but we do not need them here.

1. To see that matrix addition is associative, we rely on the fact that regular addition is associative. Take three matrices from $M_{2,2}(\mathbb{R})$ and add:

$$\begin{bmatrix} a_{1,1} & a_{1,2} \\ a_{2,1} & a_{2,2} \end{bmatrix} + \left(\begin{bmatrix} b_{1,1} & b_{1,2} \\ b_{2,1} & b_{2,2} \end{bmatrix} + \begin{bmatrix} c_{1,1} & c_{1,2} \\ c_{2,1} & c_{2,2} \end{bmatrix} \right)$$
$$= \begin{bmatrix} a_{1,1} & a_{1,2} \\ a_{2,1} & a_{2,2} \end{bmatrix} + \begin{bmatrix} b_{1,1}+c_{1,1} & b_{1,2}+c_{1,2} \\ b_{2,1}+c_{2,1} & b_{2,2}+c_{2,2} \end{bmatrix}$$
$$= \begin{bmatrix} a_{1,1}+(b_{1,1}+c_{1,1}) & a_{1,2}+(b_{1,2}+c_{1,2}) \\ a_{2,1}+(b_{2,1}+c_{2,1}) & a_{2,2}+(b_{2,2}+c_{2,2}) \end{bmatrix}$$
$$= \begin{bmatrix} (a_{1,1}+b_{1,1})+c_{1,1} & (a_{1,2}+b_{1,2})+c_{1,2} \\ (a_{2,1}+b_{2,1})+c_{2,1} & (a_{2,2}+b_{2,2})+c_{2,2} \end{bmatrix}$$
$$= \begin{bmatrix} a_{1,1}+b_{1,1} & a_{1,2}+b_{1,2} \\ a_{2,1}+b_{2,1} & a_{2,2}+b_{2,2} \end{bmatrix} + \begin{bmatrix} c_{1,1} & c_{1,2} \\ c_{2,1} & c_{2,2} \end{bmatrix}$$
$$= \left(\begin{bmatrix} a_{1,1} & a_{1,2} \\ a_{2,1} & a_{2,2} \end{bmatrix} + \begin{bmatrix} b_{1,1} & b_{1,2} \\ b_{2,1} & b_{2,2} \end{bmatrix} \right) + \begin{bmatrix} c_{1,1} & c_{1,2} \\ c_{2,1} & c_{2,2} \end{bmatrix}.$$

2. Similarly, matrix multiplication is associative.
3. Matrix addition is commutative because regular addition is commutative.
4. Next, we show that $M_{2,2}(\mathbb{R})$ contains an identity element. This element is the ***zero matrix***,

$$\begin{bmatrix} 0 & 0 \\ 0 & 0 \end{bmatrix}.$$

   It is the additive identity because for any matrix in $M_{2,2}(\mathbb{R})$,

$$\begin{bmatrix} 0 & 0 \\ 0 & 0 \end{bmatrix} + \begin{bmatrix} a_{1,1} & a_{1,2} \\ a_{2,1} & a_{2,2} \end{bmatrix} = \begin{bmatrix} 0+a_{1,1} & 0+a_{1,2} \\ 0+a_{2,1} & 0+a_{2,2} \end{bmatrix}$$
$$= \begin{bmatrix} a_{1,1} & a_{1,2} \\ a_{2,1} & a_{2,2} \end{bmatrix}.$$

5. To demonstrate that every element of $M_{2,2}(\mathbb{R})$ has an additive inverse, take

$$A = \begin{bmatrix} a_{1,1} & a_{1,2} \\ a_{2,1} & a_{2,2} \end{bmatrix} \in M_{2,2}(\mathbb{R}).$$

   Then $-A$ must equal

$$\begin{bmatrix} -a_{1,1} & -a_{1,2} \\ -a_{2,1} & -a_{2,2} \end{bmatrix}.$$

   To see this, show that $A + -A$ yields the zero matrix. This will be left as an exercise.

6. Lastly, to show that the operations are distributive, we must show for all $A,\ B,\ C \in \mathrm{M}_{2,2}(\mathbb{Z})$,

$$A \cdot (B + C) = A \cdot B + A \cdot C$$

and

$$(A + B) \cdot C = A \cdot C + B \cdot C.$$

This is also left as an exercise. (Note: for matrices, $A \cdot B$ is usually written as AB.)

It turns out that if $(R,\ \oplus,\ \otimes)$ is a ring, then the set of all $n \times n$ matrices with entries in R, denoted by $\mathrm{M}_{n,n}(R)$, is a ring with the definitions of matrix addition and multiplication based on $\oplus$ and $\otimes$.

Notice that the definition of a ring does not demand that its multiplication is commutative. If it is, then we say that the ring is ***commutative***. We do not even demand that it has a multiplicative identity, namely an element 1_R such that

$$a \otimes 1_R = 1_R \otimes a = a$$

for all $a \in R$. If it does, it is a ***ring with unity***.

Example.

1. From our work with number theory, we know $(\mathbb{Z},\ +,\ \cdot)$ is a commutative ring with unity.
2. The ring $(\mathrm{M}_{n,n}(\mathbb{Z}),\ +,\ \cdot)$ is a noncommutative ring with unity. The multiplicative identity is

$$\mathbf{I} = \begin{bmatrix} 1 & 0 & \cdots & 0 \\ 0 & 1 & \cdots & 0 \\ \vdots & \vdots & \ddots & \vdots \\ 0 & 0 & \cdots & 1 \end{bmatrix}.$$

This is called the ***identity matrix***. To see that matrix multiplication is not commutative, it suffices to give an example. Ours will be for $n = 2$:

$$\begin{bmatrix} 1 & 2 \\ 3 & 4 \end{bmatrix} \cdot \begin{bmatrix} 1 & 0 \\ 2 & 1 \end{bmatrix} = \begin{bmatrix} 5 & 2 \\ 11 & 4 \end{bmatrix},$$

but

$$\begin{bmatrix} 1 & 0 \\ 2 & 1 \end{bmatrix} \cdot \begin{bmatrix} 1 & 2 \\ 3 & 4 \end{bmatrix} = \begin{bmatrix} 1 & 2 \\ 5 & 8 \end{bmatrix}.$$

3. Let $n\mathbb{Z} = \{nk : k \in \mathbb{Z}\}$. If $+$ and $\cdot$ denote regular addition and multiplication, then $(n\mathbb{Z},\ +,\ \cdot)$ is a commutative ring that does not have unity if $n \neq \pm 1$.

Now to illustrate the purpose of a ring. If we have an equation written with elements from a ring, we can sometimes find a solution to the equation. The next theorem provides a starting point.

7.1.2. Theorem

Let $(R, \oplus, \otimes)$ be a ring and take $a, b \in R$. The element $b \oplus -a$ is the unique solution to the equation $x \oplus a = b$.

Proof. To prove that $b \oplus -a$ is a solution, calculate:

$$(b \oplus -a) \oplus a = b \oplus (-a \oplus a) = b \oplus 0_R = b.$$

To show that this is the only solution, suppose that both x_1 and x_2 are solutions to $x \oplus a = b$. Hence, $x_1 \oplus a = b$ and $x_2 \oplus a = b$. Then,

$$\begin{aligned} x_1 \oplus a &= x_2 \oplus a \\ (x_1 \oplus a) \oplus -a &= (x_2 \oplus a) \oplus -a \\ x_1 \oplus (a \oplus -a) &= x_2 \oplus (a \oplus -a) \\ x_1 \oplus 0_R &= x_2 \oplus 0_R \\ x_1 &= x_2. \blacksquare \end{aligned}$$

Example. In the ring $(\mathrm{M}_{2,2}(\mathbb{Z}), +, \cdot)$, the solution to the equation,

$$x + \begin{bmatrix} 1 & 5 \\ 0 & -3 \end{bmatrix} = \begin{bmatrix} -1 & 3 \\ 6 & 19 \end{bmatrix}$$

is

$$x = \begin{bmatrix} -2 & -2 \\ 6 & 22 \end{bmatrix}.$$

The proof of the theorem shows that in a ring we have a ***cancellation law*** with addition:

$$\text{if } a \oplus b = a \oplus c, \text{ then } b = c.$$

Multiplication is a different story. We need another assumption. Let R be a ring. Two nonzero elements a and b of R are ***zero divisors*** if $a \otimes b = 0_R$. The ring $\mathbb{Z} \times \mathbb{Z}$, where addition and multiplication is defined coordinatewise (see Exercise 20), has zero divisors as evidenced by

$$(1, 0) \cdot (0, 1) = (0, 0).$$

Other examples can be found in $\mathrm{M}_{2,2}(\mathbb{R})$ where

$$\begin{bmatrix} 1 & 1 \\ 0 & 0 \end{bmatrix} \cdot \begin{bmatrix} 1 & 1 \\ -1 & -1 \end{bmatrix} = \begin{bmatrix} 0 & 0 \\ 0 & 0 \end{bmatrix}.$$

However,

$$\begin{bmatrix} 1 & 1 \\ -1 & -1 \end{bmatrix} \cdot \begin{bmatrix} 1 & 1 \\ 0 & 0 \end{bmatrix} = \begin{bmatrix} 1 & 1 \\ -1 & -1 \end{bmatrix},$$

showing that an element can be a ***left zero divisor*** but not a ***right zero divisor***. This situation where the result is dependent on which side the element is multiplied is common for rings since multiplication need not be commutative.

We do, however, have many rings that do not have zero divisors.

7.1.3. Definition

An ***integral domain*** is a commutative ring with unity that does not have zero divisors.

Example. The rings $(\mathbb{Z}, +, \cdot)$, $(\mathbb{Q}, +, \cdot)$, $(\mathbb{R}, +, \cdot)$, and $(\mathbb{C}, +, \cdot)$ are integral domains.

The equation $2x + 1 = 0$ is written with elements of $\mathbb{Z}$ and the operations of regular addition and multiplication. Although $\mathbb{Z}$ has no zero divisors, there is no integer that is a solution to this equation. For integral domains, the best that we can do is the following theorem.

7.1.4. Theorem

If $(R, \oplus, \otimes)$ is a ring with no zero divisors and $a, b, c \in R$ with $a \neq 0_R$, then the equation $a \otimes x \oplus b = c$ has at most one solution in R.

Proof. Let $a, b, c \in R$ with $a \neq 0_R$. Suppose that both x_1 and x_2 are elements of R that are solutions to the equation. This means $a \otimes x_1 \oplus b = c$ and $a \otimes x_2 \oplus b = c$. Hence,

$$a \otimes x_1 \oplus b = a \otimes x_2 \oplus b.$$

After adding $-b$ on the right to both sides, we obtain $a \otimes x_1 = a \otimes x_2$. Adding $-(a \otimes x_2)$ to both sides of this equation yields

$$(a \otimes x_1) \oplus -(a \otimes x_2) = 0_R.$$

Since $-(a \otimes x_2) = a \otimes -x_2$ by Exercise 9c,

$$(a \otimes x_1) \oplus (a \otimes -x_2) = 0_R$$

from which $a \otimes (x_1 \oplus -x_2) = 0_R$ follows by Distribution. It then must be the case that $x_1 \oplus -x_2 = 0_R$ since R has no zero divisors and $a \neq 0_R$. Hence, $x_1 = x_2$. ■

The proof shows that in an integral domain D, we have a ***cancellation law*** with multiplication: for all $a, b, c \in D$ with $a \neq 0_D$,

$$\text{if } a \otimes b = a \otimes c, \text{ then } b = c.$$

This is not enough to guarantee a solution. What we need is the existence of multiplicative inverses. Let $(R, \oplus, \otimes)$ be a ring with unity. If $u \in R$ has the property that there exists $v \in R$ such that $u \otimes v = v \otimes u = 1_R$, then u is called a ***unit*** and v is the ***multiplicative inverse*** of u. In this case we write $v = u^{-1}$. (Be careful not to confuse a *unit* with *unity*.) With this terminology, we may make the next definition.

7.1.5. Definition

Let $(R, \oplus, \otimes)$ be a ring with unity.

- If all nonzero elements of R are units, then R is called a ***division ring*** or sometimes a ***skew field***.
- A commutative division ring is called a ***field***.

Example. The standard example of a noncommutative division ring is due to William R. Hamilton.* Let $\mathbf{I}$ be the identity matrix in $M_{2,2}(\mathbb{C})$ and write:

$$\mathbf{i} = \begin{bmatrix} i & 0 \\ 0 & -i \end{bmatrix} \qquad \mathbf{j} = \begin{bmatrix} 0 & 1 \\ -1 & 0 \end{bmatrix} \qquad \mathbf{k} = \begin{bmatrix} 0 & i \\ i & 0 \end{bmatrix}$$

where $i^2 = -1$. Define ***scalar multiplication*** of matrices by

$$k \begin{bmatrix} x & y \\ z & w \end{bmatrix} = \begin{bmatrix} kx & ky \\ kz & kw \end{bmatrix}$$

for all $k \in \mathbb{R}$. The set of ***quaternions*** is

$$\mathcal{Q} = \{a\mathbf{I} + b\mathbf{i} + c\mathbf{j} + d\mathbf{k} : a, b, c, d \in \mathbb{R}\}.$$

It is left as an exercise to show that $(\mathcal{Q}, +, \cdot)$ is a noncommutative division ring, where $+$ is matrix addition and $\cdot$ is matrix multiplication. (See Exercise 21.)

---

*William R. Hamilton (Dublin, Ireland, 1805 – Dunsink Observatory, Ireland, 1865): Hamilton represented complex numbers as ordered pairs and developed the quaternions as a generalization of $\mathbb{C}$. The quaternions have applications in both pure and applied mathematics.

Example. While $\mathbb{Z}$ is not a field with standard addition and multiplication, $\mathbb{Q}$, $\mathbb{R}$, and $\mathbb{C}$ are. A more interesting field is $\mathbb{Z}_p$ when p is a prime. To prove that it is a field, let $[a]_p \in \mathbb{Z}_p$ so that $[a]_p \neq [0]_p$. We must find an element of $\mathbb{Z}_p$ so that when it is multiplied with $[a]_p$ the result is $[1]_p$. Since $[a]_p \neq [0]_p$, p does not divide a. Hence, p and a are relatively prime, so there are integers u and v such that $ua + vp = 1$. We are then able to calculate:

$$\begin{aligned} [u]_p \cdot [a]_p &= [ua]_p \\ &= [1 - vp]_p \\ &= [1]_p + [-vp]_p \\ &= [1]_p + [0]_p \\ &= [1]_p. \end{aligned}$$

(See Chapter Exercise 5.15a.)

Before we close the section with the general result, we make the following observation: Let R be a division ring and take $u, v \in R$. Assume $u \otimes v = 0_R$ and $u \neq 0_R$. Then, u^{-1} exists, and we may calculate:

$$\begin{aligned} u^{-1} \otimes (u \otimes v) &= u^{-1} \otimes 0_R \\ (u^{-1} \otimes u) \otimes v &= 0_R \\ 1_R \otimes v &= 0_R \\ v &= 0_R. \end{aligned}$$

Therefore, R has no zero divisors. This fact plays a role in the next proof.

7.1.6. Theorem

Let $(R, \oplus, \otimes)$ be a division ring. If $a, b \in R$ and $a \neq 0_R$, then there exists a unique solution to the equation $a \otimes x \oplus b = 0_R$.

Proof. Since $a \neq 0_R$, a^{-1} exists. We can then check that $a^{-1} \otimes (-b)$ is a solution:

$$\begin{aligned} a \otimes [a^{-1} \otimes (-b)] \oplus b &= [(a \otimes a^{-1}) \otimes (-b)] \oplus b \\ &= [1_R \otimes (-b)] \oplus b \\ &= -b \oplus b \\ &= 0_R. \end{aligned}$$

Since uniqueness follows by Theorem 7.1.4, the theorem is proven. ■

EXERCISES

1. Assuming the addition and subtract that is typical for each set, indicate which of the following are rings. If it is not a ring, list the ring properties that fail.
   (a) $\mathbb{N}$
   (b) $\mathbb{Z}^+$
   (c) $\mathbb{Q} \setminus \{0\}$
   (d) $\{a + b\sqrt{2} : a,\ b \in \mathbb{Z}\}$
   (e) $\{a + b\sqrt{2} : a,\ b \in \mathbb{Q}\}$
   (f) $\{a + bi : a,\ b \in \mathbb{Z}\}$
   (g) $\{bi : b \in \mathbb{Z}\}$
   (h) $\{bi : b \in \mathbb{R}\}$
   (i) $\{ax^2 + bx + c : a,\ b,\ c \in \mathbb{N}\}$
   (j) $\{ax^2 + bx + c : a,\ b,\ c \in \mathbb{R}\}$
2. Let n be an integer. Prove that $(n\mathbb{Z},\ +,\ \cdot)$ is a commutative ring.
3. Why is

$$\left\{ \begin{bmatrix} a & b \\ c & d \end{bmatrix} : a,\ b,\ c,\ d, \in \mathbb{Z}^+ \right\}$$

   not a ring under the standard matrix operations?
4. Prove that the set

$$\left\{ \begin{bmatrix} a & 0 \\ 0 & b \end{bmatrix} : a,\ b \in \mathbb{R} \right\}$$

   is a ring with the standard matrix operations.
5. Show that $(\mathbb{R}[X],\ +,\ \cdot)$ is a ring.
6. Both $+$ and $\circ$ are binary operations on $\mathbb{R}^{\mathbb{R}}$, but $(\mathbb{R}^{\mathbb{R}},\ +,\ \circ)$ is not a ring. Identify which ring properties fail.
7. To what does -1_R refer?
8. Answer the following about $(\mathbb{Z}_n,\ +,\ \cdot)$:
   (a) Prove that addition and multiplication of congruence classes is well-defined.
   (b) Show that the additive identity is $[0]_n$.
   (c) For all integers a, show that the additive inverse of $[a]_n$ is $[n-a]_n$.
   (d) Show that $[1]_n$ is the multiplicative identity.
   (e) Prove that $(\mathbb{Z}_n,\ +,\ \cdot)$ is a commutative ring.
   (f) Prove that the ring contains zero divisors when n is not prime.
9. Let a and b be elements of a ring R. Prove:
   (a) $0_R \otimes a = 0_R$
   (b) $-a = -1_R \otimes a$
   (c) $-(a \otimes b) = -a \otimes b = a \otimes -b$
   (d) $-a \otimes -b = a \otimes b$
   (e) $-(a \oplus b) = -a \oplus -b$
   (f) $-0_R = 0_R$

10. Perform the indicated calculations:

(a) $\begin{bmatrix} 1 & 3 \\ 0 & -3 \end{bmatrix} + \begin{bmatrix} 5 & -2 \\ 8 & 2 \end{bmatrix}$ (c) $\begin{bmatrix} 0 & 1 \\ 1 & 0 \end{bmatrix} \cdot \begin{bmatrix} 5 & 4 \\ -9 & -1 \end{bmatrix}$

(b) $\begin{bmatrix} 1 & 3 \\ 0 & -3 \end{bmatrix} \cdot \begin{bmatrix} 5 & -2 \\ 8 & 2 \end{bmatrix}$ (d) $\begin{bmatrix} 4 & 0 \\ 0 & 4 \end{bmatrix} \cdot \begin{bmatrix} 7 & 3 \\ -8 & 6 \end{bmatrix}$

11. Let $\mathbf{I} \in M_{2,2}(\mathbb{R})$ be the identity matrix. Prove that the following hold for all matrices $A, B, C \in M_{2,2}(\mathbb{R})$:
   (a) $\mathbf{I}A = A\mathbf{I} = A$
   (b) $-A = (-1)A$ (see page 284)
   (c) $A(B + C) = AB + AC$
   (d) $(A + B)C = AC + BC$
12. We may perform the following matrix multiplication:

$$\begin{bmatrix} 1 & 0 \\ 0 & 1 \end{bmatrix} \cdot \begin{bmatrix} 1 & 1 \\ 1 & 1 \end{bmatrix} = \begin{bmatrix} 1 & 1 \\ 1 & 1 \end{bmatrix}.$$

and

$$\begin{bmatrix} 1 & 0 \\ 1 & 0 \end{bmatrix} \cdot \begin{bmatrix} 1 & 1 \\ 1 & 1 \end{bmatrix} = \begin{bmatrix} 1 & 1 \\ 1 & 1 \end{bmatrix}.$$

Therefore,

$$\begin{bmatrix} 1 & 0 \\ 0 & 1 \end{bmatrix} \cdot \begin{bmatrix} 1 & 1 \\ 1 & 1 \end{bmatrix} = \begin{bmatrix} 1 & 0 \\ 1 & 0 \end{bmatrix} \cdot \begin{bmatrix} 1 & 1 \\ 1 & 1 \end{bmatrix}$$

but

$$\begin{bmatrix} 1 & 0 \\ 1 & 0 \end{bmatrix} \neq \begin{bmatrix} 1 & 0 \\ 0 & 1 \end{bmatrix}.$$

How is this possible?

13. Let a and b be elements of an integral domain $(R, \oplus, \otimes)$. Prove that if $a \otimes b = 0_R$, then $a = 0_R$ or $b = 0_R$.

14. Prove the following cancellation laws:
   (a) If $(R, \oplus, \otimes)$ is a ring, then $a \oplus b = a \oplus c$ implies $b = c$.
   (b) Let $(R, \oplus, \otimes)$ be an integral domain. If $a \otimes b = a \otimes c$ and $a \neq 0_R$, then $b = c$.
15. Find all pairs of zero divisors in the following:

(a) $\mathbb{Z}_4$ (d) $M_{2,2}(\mathbb{Z}_3)$
(b) $\mathbb{Z}_{12}$ (e) $\mathbb{Z}_2 \times \mathbb{Z}_3$
(c) $M_{2,2}(\mathbb{Z}_2)$ (f) $\mathbb{Z} \times \mathbb{Z}$

16. Let $(R, \oplus, \otimes)$ be a commutative ring with unity. Show that if this ring has the property that every equation of the form $a \otimes x \oplus b = 0_R$ $(a, b \in R, a \neq 0_R)$ has a solution in R, then it is a field.

17. Is $(\{0\}, +, \cdot)$ a field? Explain.

18. Find the units of the given rings.

    (a) $\mathbb{R}$
    (b) $\mathbb{R} \times \mathbb{R}$
    (c) $\mathbb{Z}_3$
    (d) $\mathbb{Z}_4$
    (e) $\mathbb{Z}_8$
    (f) $\mathrm{M}_{2,2}(\mathbb{R})$

19. Let $a \in R$ where $(R, \oplus, \otimes)$ is a ring. We say that a is ***nilpotent*** if there exists $n \in \mathbb{Z}^+$ such that $a^n = 0_R$ where

$$a^n = \underbrace{a \otimes \cdots \otimes a}_{n \text{ times}}.$$

    Prove the following for a commutative ring:

    (a) If a is nilpotent, then $1_R \oplus a$ is a unit.
    (b) If a and b are nilpotent, then so are $a \oplus b$ and $a \otimes b$.

20. Let $(R, \oplus, \otimes)$ and $(R', \oplus', \otimes')$ be rings. (Note: $\oplus$ and $\oplus'$ can be two different additions.) Define an addition and a multiplication on $R \times R'$: for all $(a, b), (c, d) \in R \times R'$,

$$(a, b) \boxplus (c, d) = (a \oplus c, b \oplus' d),$$

and

$$(a, b) \boxtimes (c, d) = (a \otimes c, b \otimes' d).$$

Prove that $(R \times R', \boxplus, \boxtimes)$ is a ring.

21. Let $\mathfrak{Q}$ be the set of quaternions and let $+$ and $\cdot$ be matrix addition and multiplication. Prove:

    (a) The set $\mathfrak{Q}$ equals

$$\left\{ \begin{bmatrix} a + bi & c + di \\ -c + di & a - bi \end{bmatrix} : a, b, c, d \in \mathbb{R} \right\}.$$

    (b) The following relations hold:
        (i) $\mathbf{i}^2 = \mathbf{j}^2 = \mathbf{k}^2 = -\mathbf{I}$
        (ii) $\mathbf{jk} = -\mathbf{kj} = \mathbf{i}$
        (iii) $\mathbf{ij} = -\mathbf{ji} = \mathbf{k}$
        (iv) $\mathbf{ki} = -\mathbf{ik} = \mathbf{j}$
    (c) $(\mathfrak{Q}, +, \cdot)$ is a noncommutative ring.
    (d) $\mathfrak{Q}$ is a division ring by showing that if $A = a\mathbf{I} + b\mathbf{i} + c\mathbf{j} + d\mathbf{k}$ is a nonzero element of $\mathfrak{Q}$, then its multiplicative inverse is

$$\frac{1}{(a^2 + b^2 + c^2 + d^2)} A.$$

7.2. SUBRINGS AND IDEALS

We will now follow the traditional convention regarding rings. Given a ring $(R, \oplus, \otimes)$, we will simply use standard addition and multiplication notation for the operations $\oplus$ and $\otimes$. Likewise, the additive identity will be denoted by 0 and the multiplicative identity by 1. Furthermore, the ring $(R, +, \cdot)$ will just be named by R. Therefore, the phrase

let R be a ring

assumes the existence of an addition and a multiplication on R. If the set is a set like $\mathbb{Z}$ or $\mathrm{M}_{2,2}(\mathbb{R})$, then the operations are the standard ones for that set unless noted otherwise. Finally, we will write $a - b$ instead of $a + -b$.

Some of the examples of rings have been subsets of other rings. For example, $n\mathbb{Z} \subseteq \mathbb{Z}$. This idea leads to the next definition.

7.2.1. Definition

Let R be a ring. If $S \subseteq R$ and S is a ring using the same operations as R, then S is called a ***subring*** of R.

A subring of R that is a proper subset of R is called a ***proper subring***. The ring itself is called the ***improper subring***. The subring $\{0\}$ is the ***trivial subring***.

Example. $\{[0]_9, [3]_9, [6]_9\}$ is a subring of $\mathbb{Z}_9$, $\mathbb{Z}$ is a subring of $\mathbb{R}$, and $\mathrm{M}_{2,2}[\mathbb{R}]$ is a subring of $\mathrm{M}_{2,2}[\mathbb{C}]$.

To show that a subset is actually a subring, use the next theorem. It is stated without proof, for it follows quickly from the definition.

7.2.2. Theorem

Let R be a ring. A subset S of R is a subring of R if and only if

1. S is closed under addition $(+)$ and multiplication $(\cdot)$,
2. $0 \in S$, and
3. for all $a \in R$, $a \in S$ implies $-a \in S$.

Notice that the theorem does not involve a check of the commutative, associative, and distributive laws. This is because S is a subset of R and uses the same operations. For instance, if $+$ is commutative on R, then it must be commutative on any subset. We do have to check closure, however. This ensures that the addition and multiplication are binary operations on S not just on R.

Example. Let us use the theorem to show

$$S = \left\{ \begin{bmatrix} a & 0 \\ 0 & b \end{bmatrix} : a,\ b \in \mathbb{R} \right\}$$

is a subring of $M_{2,2}(\mathbb{R})$. This is called the set of diagonal matrices (with real entries).

1. Let $A,\ B \in S$. Then

$$A = \begin{bmatrix} a & 0 \\ 0 & b \end{bmatrix} \text{ and } B = \begin{bmatrix} a' & 0 \\ 0 & b' \end{bmatrix},$$

$a,\ b,\ a',\ b' \in \mathbb{R}$. When these matrices are added and multiplied together we get

$$A + B = \begin{bmatrix} a + a' & 0 \\ 0 & b + b' \end{bmatrix}$$

and

$$AB = \begin{bmatrix} aa' & 0 \\ 0 & bb' \end{bmatrix}.$$

These are both elements of S.

2. Clearly the zero matrix is in S. (Let $a = b = 0$.)
3. Suppose $A \in S$. Then,

$$A = \begin{bmatrix} a & 0 \\ 0 & b \end{bmatrix}$$

for some $a,\ b \in \mathbb{R}$. Hence,

$$-A = \begin{bmatrix} -a & 0 \\ 0 & -b \end{bmatrix}$$

is an element of S.

Example. Let S and T be subrings of a ring R. We wish to prove that $S \cap T$ is a subring of R.

1. To prove closure, let $x,\ y \in S \cap T$. This means $x + y \in S$ and $x + y \in T$. Hence, $x + y \in S \cap T$. Similarly, $xy \in S \cap T$.
2. Since $0 \in S$ and $0 \in T$, $0 \in S \cap T$.
3. Suppose $x \in S \cap T$. Then $x \in S$ and $x \in T$. Since these are subrings, $-x \in S$ and $-x \in T$. Thus, $-x \in S \cap T$.

The union of two subrings may not be a subring. Let S and T be subrings of R. Let us see which conditions might fail. The additive identity of R is an element of $S \cup T$ since it is in S (and T). Also, if $x \in S \cup T$, then $-x$ is in $S \cup T$ because x is in at least one of them. Therefore, if a condition is going to fail, it will be closure. To see this, take x and y in $S \cup T$. This means that $x \in S$ or $x \in T$. It also means that $y \in S$ or $y \in T$. If both are in S or both are in T, then there is no problem, and both the sum and product are elements of $S \cup T$. However, if $x \in S$ and $y \in T$ (or vice versa), there is no reason to conclude that the result is in $S \cup T$. We need an extra hypothesis. This is illustrated next.

Example. Take a ring R. Let $\mathscr{C} = \{S_\alpha : \alpha \in \Lambda\}$ be a chain of subrings of R. To prove that $\bigcup_{\alpha\in\Lambda} S_\alpha$ is a subring of R, the previous paragraph shows that we just need to prove closure. Let $x, y \in \bigcup_{\alpha\in\Lambda} S_\alpha$. By definition, there exist $\alpha, \beta \in \Lambda$ so that $x \in S_\alpha$ and $y \in S_\beta$. Since $\mathscr{C}$ is a chain, either $S_\alpha \subseteq S_\beta$ or $S_\beta \subseteq S_\alpha$.

(Case 1) If $S_\alpha \subseteq S_\beta$, then $x \in S_\beta$. Hence, since S_β is a subring, $x+y \in S_\beta$ and $xy \in S_\beta$. This means that

$$x + y,\ xy \in \bigcup_{\alpha\in\Lambda} S_\alpha.$$

(Case 2) When $S_\beta \subseteq S_\alpha$, the proof proceeds as in Case 1.

Subrings are important structures within a ring, but we will need a stronger definition for some later work.

7.2.3. Definition

Let R be a ring and $I \subseteq R$. If I is a subring of R and for all $r \in R$ and $a \in I$,

1. $ra \in I$, and
2. $ar \in I$,

then I is called a ***(two-sided) ideal*** of R. If I satisfies condition 1, it is a ***left ideal***. If I satisfies part 2, then I is a ***right ideal***.

Example. The subring $I = \{[0]_4, [2]_4\}$ of $\mathbb{Z}_4$ is also an ideal. To see this, check the calculations:

$$\begin{array}{ll}
[0]_4 \cdot [0]_4 = [0]_4 & [0]_4 \cdot [0]_4 = [0]_4 \\
[1]_4 \cdot [0]_4 = [0]_4 & [0]_4 \cdot [1]_4 = [0]_4 \\
[2]_4 \cdot [0]_4 = [0]_4 & [0]_4 \cdot [2]_4 = [0]_4 \\
[3]_4 \cdot [0]_4 = [0]_4 & [0]_4 \cdot [3]_4 = [0]_4 \\
[0]_4 \cdot [2]_4 = [0]_4 & [2]_4 \cdot [0]_4 = [0]_4 \\
[1]_4 \cdot [2]_4 = [2]_4 & [2]_4 \cdot [1]_4 = [2]_4 \\
[2]_4 \cdot [2]_4 = [0]_4 & [2]_4 \cdot [2]_4 = [0]_4 \\
[3]_4 \cdot [2]_4 = [2]_4 & [2]_4 \cdot [3]_4 = [2]_4
\end{array}$$

When we multiply any element of $\mathbb{Z}_4$ by an element of I on either side, the result is an element of I.

Example. Let R be a ring. It is left as an exercise to show that $R \times \{0\}$ is a subring of $R \times R$. To show that it is an ideal, let $(r, s) \in R \times R$ and $(a, 0) \in R \times \{0\}$. We then calculate:

$$(r, s) \cdot (a, 0) = (ra, 0) \in R \times \{0\},$$

and

$$(a, 0) \cdot (r, s) = (ar, 0) \in R \times \{0\}.$$

Example. Let S and T be ideals of a ring R. Define

$$S + T = \{s + t : s \in S \text{ and } t \in T\}.$$

To prove $S + T$ is an ideal of R, we will use Theorem 7.2.2 to show that it is a subring and then prove the conditions of the definition of an ideal:

1. Take $x, y \in S + T$. This means that $x = s + t$ and $y = s' + t'$ for some $s, s' \in S$ and $t, t' \in T$. We must now show that $x + y$ looks like an element of S operated with an element of T. Indeed:

$$\begin{aligned} x + y &= (s + t) + (s' + t') \\ &= s + (t + s') + t' \\ &= s + (s' + t) + t' \\ &= (s + s') + (t + t'). \end{aligned}$$

   Thus, $x + y \in S + T$ since $s + s' \in S$ and $t + t' \in T$. Also,

$$xy = (s + t)(s' + t') = (s + t)s' + (s + t)t'.$$

   Since $s + t \in R$ and S is an ideal, $(s + t)s' \in S$. Likewise, $(s + t)t'$ is an element of T. Hence, $xy \in S + T$.
2. We know that $0 \in S + T$ because $0 = 0 + 0$ and $0 \in S$ and $0 \in T$.
3. Let $x \in S + T$. Again, this means that $x = s + t$ for some $s \in S$ and $t \in T$. Then since $-s \in S$ and $-t \in t$,

$$-(s + t) = -s + -t \in S + T.$$

4. Let $r \in R$ and $s + t \in S + T$. Since $rs \in S$ and $rt \in T$,

$$r(s + t) = rs + rt \in S + T,$$

   and since $sr \in S$ and $tr \in T$,

$$(s + t)r = sr + tr \in S + T.$$

For any ring R, both R and $\{0\}$ are left and right ideals of R. In a commutative ring there is no difference between a left and right ideal. However, if the ring is not commutative, a left ideal may not be a right ideal.

Example. Define I to be the following set of matrices:

$$\left\{\begin{bmatrix} a & 0 \\ b & 0 \end{bmatrix} : a,\ b \in \mathbb{R}\right\}.$$

This is a subring of $\mathrm{M}_{2,2}(\mathbb{R})$. To check that I is a left ideal, we perform the multiplication:

$$\begin{bmatrix} x & y \\ z & w \end{bmatrix} \cdot \begin{bmatrix} a & 0 \\ b & 0 \end{bmatrix} = \begin{bmatrix} xa + yb & 0 \\ za + wb & 0 \end{bmatrix} \in I.$$

Yet, I is not a right ideal because

$$\begin{bmatrix} 1 & 0 \\ 1 & 0 \end{bmatrix} \cdot \begin{bmatrix} 1 & 1 \\ 1 & 1 \end{bmatrix} = \begin{bmatrix} 1 & 1 \\ 1 & 1 \end{bmatrix} \notin I.$$

A lot of what we will do in ring theory is motivated by our experience with number theory. The study of ideals is a good example of this. The following is needed to understand the structure of $\mathbb{Z}$ as a ring.

7.2.4. Definition

Let R be a ring. For every $a \in R$, define

$$\langle a \rangle = \{ra : r \in R\}.$$

If I is a left ideal of R such that $I = \langle a \rangle$ for some $a \in R$, then I is called a ***principal left ideal***, and a is a ***generator*** of I.

Notice that $\langle a \rangle$ may not be a two-sided ideal. The previous example combined with the next illustrates this fact.

Example. Let us show that

$$I = \left\{\begin{bmatrix} a & 0 \\ b & 0 \end{bmatrix} : a,\ b \in \mathbb{R}\right\}$$

is a principal left ideal of $\mathrm{M}_{2,2}(\mathbb{R})$. To do this, we will show that

$$I = \left\langle \begin{bmatrix} 1 & 0 \\ 0 & 0 \end{bmatrix} \right\rangle.$$

($\subseteq$) Let $a,\ b \in \mathbb{R}$. Then,

$$\begin{bmatrix} a & 0 \\ b & 0 \end{bmatrix} = \begin{bmatrix} a & 0 \\ b & 0 \end{bmatrix} \cdot \begin{bmatrix} 1 & 0 \\ 0 & 0 \end{bmatrix}.$$

Hence,

$$\begin{bmatrix} a & 0 \\ b & 0 \end{bmatrix} \in \left\langle \begin{bmatrix} 1 & 0 \\ 0 & 0 \end{bmatrix} \right\rangle.$$

(Observe that there are many matrices that would have worked in the above equation.)

($\supseteq$) Any element of

$$\left\langle \begin{bmatrix} 1 & 0 \\ 0 & 0 \end{bmatrix} \right\rangle$$

looks like

$$\begin{bmatrix} a & b \\ c & d \end{bmatrix} \cdot \begin{bmatrix} 1 & 0 \\ 0 & 0 \end{bmatrix} = \begin{bmatrix} a & 0 \\ c & 0 \end{bmatrix},$$

which is an element of I.

(Note: the ideal I has many generators. See Exercise 1.)

If R is commutative, then $\langle a \rangle$ is a two-sided ideal of R called a ***principal ideal***.

Example. The set $n\mathbb{Z}$ is an ideal of $\mathbb{Z}$. It is a principal ideal because $n\mathbb{Z} = \langle n \rangle$. (See Exercises 9 and 24.)

It turns out that every ideal in $\mathbb{Z}$ is principal. To see this, let I be an ideal of $\mathbb{Z}$. We must find an element of $\mathbb{Z}$ that generates I. We have two cases to consider:

(Case 1) If $I = \{0\}$, then $I = \langle 0 \rangle$.

(Case 2) Suppose $I \neq \{0\}$. This means $I \cap \mathbb{Z}^+ \neq \varnothing$. By the Well-Ordering Principle, I must contain a minimal positive integer. Call it m. We claim that $I = \langle m \rangle$. Since it is clear that $\langle m \rangle \subseteq I$, we must show the converse. So, take $a \in I$ and divide it by m. The Division Algorithm gives us $q,\ r \in \mathbb{Z}$ so that

$$a = mq + r$$

with $0 \le r < m$. Then, $r \in I$ because $r = a - mq$ and $a,\ mq \in I$. If $r > 0$, then we have a contradiction of the fact that m is the smallest positive integer in I. Hence, $r = 0$ and $a = mq$. This means $a \in \langle m \rangle$.

An integral domain in which every ideal is principal is called a ***principal ideal domain***. These rings are sometimes referred to as ***PID***s. The example shows that $\mathbb{Z}$ is a PID. We will see another PID when we study polynomials.

Another important type of ideal that will shed light on the structure of $\mathbb{Z}$ is the following:

7.2.5. Definition

An ideal P of a commutative ring R is ***prime*** if for all $a, b \in R$, $ab \in P$ implies $a \in P$ or $b \in P$.

Example. Let p be prime. We already know that $p\mathbb{Z}$ is an ideal of $\mathbb{Z}$. To see that it is a prime ideal of $\mathbb{Z}$, let $a, b \in \mathbb{Z}$. Assume $ab \in p\mathbb{Z}$. By definition, $ab = pk$ for some $k \in \mathbb{Z}$. In other words, $p \mid ab$. By Euclid's Lemma (5.3.2), $p \mid a$ or $p \mid b$. This implies that we have $a \in p\mathbb{Z}$ or $b \in p\mathbb{Z}$. Hence, $p\mathbb{Z}$ is prime.

In a sense, the ideals of the form $p\mathbb{Z}$ are the largest ideals of $\mathbb{Z}$. This is formalized in the last definition of the section.

7.2.6. Definition

Let R be a commutative ring and M an ideal of R. If $M \neq R$ and no proper ideal of R properly contains M, then M is called a ***maximal ideal***.

Example. Assume that p is prime. In the ring $\mathbb{Z}$, $p\mathbb{Z}$ is a maximal ideal. To see this, let I be an ideal of $\mathbb{Z}$ properly containing $p\mathbb{Z}$. Thus, there exists $m \in I \setminus p\mathbb{Z}$. This means that $p \nmid m$. Therefore, $\gcd(p, m) = 1$, so there exist $u, v \in \mathbb{Z}$ such that $up + vm = 1$. Since $up + vm \in I$, we have $1 \in I$. Hence, $I = \mathbb{Z}$ by Exercise 21, and we conclude that $p\mathbb{Z}$ is maximal.

In the next section we will see that for a commutative ring with unity, every maximal ideal is prime.

EXERCISES

1. Fix $a, b \in \mathbb{R}$. Find all values of $x_1, x_2, x_3, x_4 \in \mathbb{R}$ that satisfy the following matrix equation:

$$\begin{bmatrix} a & 0 \\ b & 0 \end{bmatrix} = \begin{bmatrix} x_1 & x_2 \\ x_3 & x_4 \end{bmatrix} \begin{bmatrix} 1 & 0 \\ 0 & 0 \end{bmatrix}.$$

2. For any ring R, show that R and $\{0\}$ are ideals of R.

3. Which of the following are subrings of the given ring? If it is not a subring, then which of the conditions fail?
   (a) $\{ai : a :\in \mathbb{R}\}$ of $\mathbb{C}$.
   (b) $\{aX^2 + bX + c : a, b, c \in \mathbb{R}\}$ of $\mathbb{R}[X]$.
   (c) $\{(a, a) : a \in \mathbb{Z}\}$ of $\mathbb{Z} \times \mathbb{Z}$.
   (d) $\left\{\begin{bmatrix} a & 0 \\ 0 & b \end{bmatrix} : a, b \in \mathbb{R}\right\}$ of $\mathrm{M}_{2,2}(\mathbb{R})$.
   (e) $\left\{\begin{bmatrix} a & b \\ c & d \end{bmatrix} : a + d = b + d = 0 \text{ and } a, b, c, d \in \mathbb{R}\right\}$ of $\mathrm{M}_{2,2}(\mathbb{R})$.
4. For those sets that are subrings in the previous problem, prove that they are using Theorem 7.2.2.
5. Prove Theorem 7.2.2.
6. **6.** Prove that S is a subring of R if and only if for all $a, b \in S$,
   - $0 \in S$,
   - $(a - b) \in S$, and
   - $ab \in S$.
7. Which of the subrings in Exercise 3 are ideals of the given ring?
8. Find all ideals of the following:
   (a) $\mathbb{Z}_2$
   (b) $\mathbb{Z}_6$
   (c) $\mathbb{Z}_7$
   (d) $\mathbb{Z}_{12}$
9. Let $n \in \mathbb{Z}$. Prove that $n\mathbb{Z}$ is an ideal of $\mathbb{Z}$.
10. **10.** Prove that $\{(2m, 2n) : m, n \in \mathbb{Z}\}$ is an ideal of $\mathbb{Z} \times \mathbb{Z}$.
11. **11.** Prove that $\{(2n, 2n) : n \in \mathbb{Z}\}$ is not an ideal of $\mathbb{Z} \times \mathbb{Z}$.
12. Suppose R_1 and R_2 are two rings with I_1 an ideal of R_1 and I_2 an ideal of R_2. Prove that $R_1 \times \{0\}$ and $I_1 \times I_2$ are ideals of $R_1 \times R_2$.
13. Let I and J be ideals of a ring R. Prove that $I \cap J$ is an ideal of R.
14. **14.** Let $\{J_\alpha : \alpha \in \Lambda\}$ be a chain of ideals in R. Prove $\bigcup_{\alpha \in \Lambda} J_\alpha$ is an ideal of R.
15. Let R be a ring and $I \subseteq R$ such that $I \neq \varnothing$. Prove that I is an ideal of R if and only if for all $a, b \in R$,
   - if $a, b \in I$, then $a - b \in I$, and
   - $ra \in I$ and $ar \in I$, for all $r \in R$.
16. Given a ring, write each of the indicated principal ideals as rosters.
   (a) $\langle [2]_{10} \rangle$ in $\mathbb{Z}_{10}$
   (b) $\langle [3]_{10} \rangle$ in $\mathbb{Z}_{10}$
   (c) $\langle [5]_{10} \rangle$ in $\mathbb{Z}_{10}$
   (d) $\langle ([1]_6, [1]_4) \rangle$ in $\mathbb{Z}_6 \times \mathbb{Z}_4$
   (e) $\langle ([0]_6, [1]_4) \rangle$ in $\mathbb{Z}_6 \times \mathbb{Z}_4$
   (f) $\langle ([2]_6, [3]_4) \rangle$ in $\mathbb{Z}_6 \times \mathbb{Z}_4$
17. Is $\{(2m, 2n) : m, n \in \mathbb{Z}\}$ a principal ideal of $\mathbb{Z} \times \mathbb{Z}$? If so, find a generator.

18. Prove that $\mathbb{Z} \times 3\mathbb{Z}$ is a principal ideal of $\mathbb{Z} \times \mathbb{Z}$.

19. Find all generators of

$$\left\{ \begin{bmatrix} a & 0 \\ b & 0 \end{bmatrix} : a,\ b \in \mathbb{R} \right\}$$

as a left ideal of $\mathrm{M}_{2,\,2}(\mathbb{R})$.

20. Show that

$$\left\{ \begin{bmatrix} 0 & a \\ 0 & b \end{bmatrix} : a,\ b \in \mathbb{Z} \right\}$$

is a principal left ideal of $\mathrm{M}_{2,\,2}(\mathbb{Z})$ and list its generators.

21. Let R be a ring with unity and I an ideal of R. Prove if u is a unit and $u \in I$, then $I = R$.

22. Prove that a field has no proper, nontrivial ideals.

23. Let $a,\ n \in \mathbb{Z}$. Show $a \in n\mathbb{Z}$ if and only if $n \mid a$.

24. Prove that $n\mathbb{Z}$ is a principal ideal of $\mathbb{Z}$ for any integer n.

25. Show that for any ideal I, if $a \in I$, then $\langle a \rangle \subseteq I$.

26. Take a ring R and let $a \in R$. Show $\langle a \rangle$ is a left ideal of R but not necessarily a right ideal if R is not commutative.

27. Let u be a unit of a ring R. Show for all $a \in R$, $\langle a \rangle = \langle ua \rangle$.

28. Let R be a ring. Define

$$\langle a,\ b \rangle = \{ra + sb : r,\ s \in R\}$$

for any $a,\ b \in R$.

(a) Show that $\langle a,\ b \rangle$ is a left ideal of R.

(b) Give an example of a ring R and elements a and b that shows that $\langle a,\ b \rangle$ may not be a right ideal.

(c) Prove if a and b are relatively prime integers, then $\langle a,\ b \rangle = \mathbb{Z}$.

29. Let R be an integral domain. Prove that $\{0\}$ is a prime ideal of R.

30. Let R be a principal ideal domain. Let $a \in R$ and $a \neq 0$. Prove that if $\langle a \rangle$ is prime and $a = uv$ for some $u,\ v \in R$, then u or v is a unit.

31. Let R be a principal ideal domain and take $p \in R$. We call p ***irreducible*** if for all $u,\ v \in R$, $p = uv$ implies that u or v is a unit. Prove that if p is irreducible, then $\langle p \rangle$ is prime.

32. Find a maximal ideal in each of the following:

(a) $\mathbb{Z}_7$ (b) $\mathbb{Z}_{10}$ **(c)** $\mathbb{Z}_6 \times \mathbb{Z}_5$ (d) $6\mathbb{Z}$ (e) $\mathrm{M}_{2,\,2}(\mathbb{R})$ (f) $\mathrm{M}_{2,\,2}(\mathbb{Z}_6)$

33. Let F be a field. Prove that $\{0\}$ is the maximal ideal of F.

34. Let p be a prime. Prove that $\{(pa,\ b) : a,\ b \in \mathbb{Z}\}$ is a maximal ideal of $\mathbb{Z} \times \mathbb{Z}$.

7.3. FACTOR RINGS

We have noted that ring theory is a generalization of number theory. (Or, number theory is a special case of ring theory. It all depends on our perspective!) Congruences play an important role in number theory, so we would like to generalize that notion. We begin with a definition.

7.3.1. Definition

Let R be a ring and I an ideal of R. For every $a \in R$, we may define a ***coset*** of I as the set

$$\{a + n : n \in I\}.$$

Denote this by $a + I$.

As with quotient sets, the set of all cosets of I in R is denoted by R/I. Read this as "R modulo I" or simply "R mod I." In other words,

$$R/I = \{a + I : a \in R\}.$$

We can make this family of sets into a ring in a natural way. Define addition and subtraction of cosets as follows: for all $a,\ b \in R$,

$$(a + I) + (b + I) = (a + b) + I,$$

and

$$(a + I) \cdot (b + I) = ab + I.$$

The proof that multiplication is well-defined is left as an exercise. We will prove that addition is well-defined by assuming $a+I = a'+I$ and $b+I = b'+I$ for $a,\ a',\ b,\ b' \in R$. Note that since $0 \in I$, $a \in a' + I$, so $a = a' + r$ for some $r \in I$. Similarly, there exists $s \in I$ such that $b = b' + s$. We are now ready to prove $ab + I = a'b' + I$:

($\subseteq$) Take $ab + n \in ab + I$ where $n \in I$. Then,

$$\begin{aligned} ab + n &= (a' + r)(b' + s) + n \\ &= a'b' + a's + rb' + rs + n. \end{aligned}$$

Because I is an ideal, $a's$, rb', and rs are all elements of I. Hence,

$$ab + n = a'b' + m$$

where $m = a's + rb' + rs + n$ and is an element of I, so we may conclude $ab + n \in a'b' + I$.

($\supseteq$) This is proven analogously.

These operations make R/I into a ring called a ***factor ring*** or a ***quotient ring***.

Example. Since $5\mathbb{Z}$ is an ideal of $\mathbb{Z}$, we may find its cosets:

$$\begin{aligned} 0+5\mathbb{Z} &= \{0+5n : n \in \mathbb{Z}\} \\ &= \{\ldots, -10, -5, 0, 5, 10, \ldots\} \\ 1+5\mathbb{Z} &= \{\ldots, -9, -4, 1, 6, 11, \ldots\} \\ 2+5\mathbb{Z} &= \{\ldots, -8, -3, 2, 7, 12, \ldots\} \\ 3+5\mathbb{Z} &= \{\ldots, -7, -2, 3, 8, 13, \ldots\} \\ 4+5\mathbb{Z} &= \{\ldots, -6, -1, 4, 9, 14, \ldots\} \end{aligned}$$

These should look familiar! They are the congruence classes modulo 5. This illustrates that cosets are generalizations of congruence classes.

Now let us try some calculations with these cosets:

$$\begin{aligned} (2+5\mathbb{Z}) + (4+5\mathbb{Z}) &= 6+5\mathbb{Z} = 1+5\mathbb{Z} \\ (0+5\mathbb{Z}) + (n+5\mathbb{Z}) &= n+5\mathbb{Z} \\ (2+5\mathbb{Z}) \cdot (4+5\mathbb{Z}) &= 8+5\mathbb{Z} = 3+5\mathbb{Z} \\ (1+5\mathbb{Z}) \cdot (n+5\mathbb{Z}) &= n+5\mathbb{Z} \end{aligned}$$

The equations show that $0+5\mathbb{Z}$ is the additive identity of $\mathbb{Z}/5\mathbb{Z}$ and $1+5\mathbb{Z}$ is its multiplicative identity.

From the last example, we can see that $6+5\mathbb{Z} = 1+5\mathbb{Z}$ by listing their elements. There is another test for equality of cosets. For this last equality, we note that $6-1 \in 5\mathbb{Z}$. Generalizing, for any ring R with ideal I,

$$a+I = b+I \text{ if and only if } a-b \in I.$$

The proof is as follows:

($\Rightarrow$) Assume $a+I = b+I$. We have already seen that $a \in a+I$. Therefore, $a \in b+I$. This means $a = b+i$ for some $i \in I$. Hence, $a-b = i$, so we have $a-b \in I$.

($\Leftarrow$) Suppose $a-b \in I$ and show $a+I = b+I$:

- ($\subseteq$) Let $x \in a+I$. This means $x = a+i$, some $i \in I$. But, $a-b \in I$, so there exists $j \in I$ such that $a-b = j$. Therefore,

  $$x = b+j+i.$$

  Since $j+i \in I$, $x \in b+I$.
- ($\supseteq$) This is proven similarly.

We will use this in the next result. Recall that the set of equivalence classes modulo n forms a partition of the set of integers. (See page 222 in Section 6.2 for an example of this.)

7.3.2. Theorem

Let R be a ring and I an ideal. R/I is a partition of R.

Proof. We must show $\bigcup R/I = R$ and that R/I is pairwise disjoint.

1. The one inclusion is clear, so let us prove $R \subseteq \bigcup R/I$. Let $a \in R$. From this we conclude that $a \in a + I$ and $a + I \in R/I$. Hence, $a \in \bigcup R/I$.
2. To prove that R/I is pairwise disjoint, we must show if $a+I \neq b+I$, then $a + I$ and $b + I$ are disjoint. The best way to prove this is by demonstrating the contrapositive, so assume

$$a + I \cap b + I \neq \varnothing.$$

This means that there exists $x \in a + I \cap b + I$. We may write $x = a + r$ and $x = b + s$ for some $r,\ s \in I$. Therefore,

$$a + r = b + s,$$

which gives

$$a - b = s - r \in I.$$

Therefore, $a + I = b + I$. ■

Example. The cosets of $I = \{[0]_6,\ [3]_6\}$ in $\mathbb{Z}_6$ are:

$$[0]_6 + I = [3]_6 + I = \{[0]_6,\ [3]_6\}$$
$$[1]_6 + I = [4]_6 + I = \{[1]_6,\ [4]_6\}$$
$$[2]_6 + I = [5]_6 + I = \{[2]_6,\ [5]_6\}.$$

Therefore,

$$\mathbb{Z}_6/I = \{[0]_6 + I,\ [1]_6 + I,\ [2]_6 + I\}$$

is a partition for $\mathbb{Z}_6$.

Notice that in the example there are three cosets, each having cardinality two. The ring has six elements. Therefore,

$$|\mathbb{Z}_6| = |\mathbb{Z}_6/I| \cdot |I|.$$

This result is generalized in the next theorem attributed to Lagrange.*

---

*Usually Lagrange's Theorem is stated for groups. See Section 8.1, page 328 for the definition of a group.

7.3.3. Lagrange's Theorem

If R is a finite ring and I and ideal of R, then

$$|R| = |R/I| \cdot |I|.$$

Proof. We begin by proving that any coset of I is equinumerous with I. Define $\phi\colon I \to a + I$ by

$$\phi(r) = a + r$$

for all $r \in I$. This function is a bijection because:

1. Take $r, s \in I$ and suppose $a + r = a + s$. By cancellation, $r = s$. Thus, ϕ is one-to-one.
2. Let $a + r \in a + I$ for some $r \in I$. Then, $\phi(r) = a + r$, so ϕ is onto.

From this we conclude $|a + I| = |I|$ for all $a \in R$.

Since R is finite, there can only be finitely many cosets of I, so list them without repetition:

$$a_1 + I,\ a_2 + I,\ \ldots,\ a_k + I,$$

where k is a positive integer and $a_i \in R$, $i = 1, \ldots, k$. From this we rely on the fact that R/I is a partition and conclude

$$|R| = |a_1 + I| + |a_2 + I| + \cdots + |a_k + I|$$

by Chapter Exercise 6.33. However, $|a_i + I| = |I|$ for all $i = 1, \ldots, k$. Therefore,

$$|R| = \underbrace{|I| + |I| + \cdots + |I|}_{k \text{ times}}.$$

Since $|R| = k|I|$ and $k = |R/I|$, we have proven the theorem. ■

To ease the notation in the next example we will simply write n to represent $[n]_m$. For instance, instead of writing $[4]_6+[5]_6 = [3]_6$, we will write $4+5 = 3$, where the modulus is understood to be 6. Which modulus we will be using will be clear from context.

Example. Consider the ring $\mathbb{Z}_6 \times \mathbb{Z}_4$. It has 24 elements. The principal ideal $\langle(1, 2)\rangle$ has 6 elements. As a roster it is

$$\{(1, 2), (2, 0), (3, 2), (4, 0), (5, 2), (0, 0)\}.$$

Therefore, by Lagrange's Theorem, $|(\mathbb{Z}_6 \times \mathbb{Z}_4)/\langle(1, 2)\rangle| = 4$. Here are the elements:

$$(0, 0) + \langle(1, 2)\rangle = \{(1, 2), (2, 0), (3, 2), (4, 0), (5, 2), (0, 0)\}$$

$$(1, 0) + \langle(1, 2)\rangle = \{(2, 2), (3, 0), (4, 2), (5, 0), (0, 2), (1, 0)\}$$
$$(0, 1) + \langle(1, 2)\rangle = \{(1, 3), (2, 1), (3, 3), (4, 1), (5, 3), (0, 1)\}$$
$$(1, 1) + \langle(1, 2)\rangle = \{(2, 3), (3, 1), (4, 3), (5, 1), (0, 3), (1, 1)\}$$

The representatives of these cosets were chosen by first picking the additive identity of $\mathbb{Z}_6 \times \mathbb{Z}_4$. The others were found by choosing an element that was not in any previous coset. This was repeated until we found the four cosets of $(\mathbb{Z}_6 \times \mathbb{Z}_4)/\langle(1, 2)\rangle$.

We close the section by examining an important relationship between prime and maximal ideals. For the following theorem we need the fact that $a + I$ is the additive identity of R/I if and only if $a \in I$ (Exercise 5).

7.3.4. Theorem

If R is a commutative ring with unity and M is a maximal ideal of R, then R/M is a field.

Proof. Let M be a maximal ideal of the commutative ring with unity R. We must show that every nonzero element of R/M has an inverse. Take $a \in R \setminus M$ so that $a + M \neq 0 + M$, which means $a \notin M$. By the example on page 292, $\langle a \rangle + M$ is an ideal of R. Since $a \notin M$ and M is maximal, it must be the case that $\langle a \rangle + M$ is all of R. Hence,

$$1 \in \langle a \rangle + M,$$

so we may write $1 = ra + m$ for $r \in R$ and $m \in M$. Therefore,

$$1 + M = ra + M = (r + M)(a + M),$$

and this means $(a + M)^{-1} = r + M$. ■

Example. In the ring $\mathbb{Z}_6 \times \mathbb{Z}_5$, $\{[0]_6, [2]_6, [4]_6\} \times \mathbb{Z}_5$ is a maximal ideal. Therefore,

$$(\mathbb{Z}_6 \times \mathbb{Z}_5)/(\{[0]_6, [2]_6, [4]_6\} \times \mathbb{Z}_5)$$

is a field. By Lagrange's Theorem, it has only two elements because

$$|\mathbb{Z}_6 \times \mathbb{Z}_5| = 30$$

and

$$|\{[0]_6, [2]_6, [4]_6\} \times \mathbb{Z}_5| = 15.$$

We can now prove the relationship between maximal and prime ideals. We note, however, the its converse is false. (See Exercise 14)

7.3.5. Theorem

Every maximal ideal of a commutative ring is prime.

Proof. Let M be a maximal ideal of R. Assume $a, b \in R$ and $ab \in M$. We must show $a \in M$ or $b \in M$. Now, $ab \in M$ means $ab + M = 0 + M$ in R/M. Since M is maximal, R/M is a field, and hence it has no zero divisors. Therefore, since

$$(a + M)(b + M) = ab + M = 0 + M,$$

$a + M = 0 + M$ or $b + M = 0 + M$. In other words, a is an element of M, or b is an element of M. ■

EXERCISES

1. Write all cosets of the following factor rings as rosters:
   (a) $\mathbb{Z}/4\mathbb{Z}$
   (b) $\mathbb{Z}/\mathbb{Z}$
   (c) $\mathbb{Z}_{10}/\langle [5]_{10} \rangle$
   (d) $\mathbb{Z}_{10}/\langle [2]_{10} \rangle$
   (e) $\mathbb{Z}_{10}/\langle [3]_{10} \rangle$
   (f) $(\mathbb{Z}_3 \times \mathbb{Z}_2)/(\{[0]_3\} \times \mathbb{Z}_2)$
   (g) $(\mathbb{Z}_4 \times \mathbb{Z}_6)/(\{[0]_4, [2]_4\} \times \{[0]_6, [3]_6\})$
   (h) $(\mathbb{Z}_6 \times \mathbb{Z}_4)/\langle([1]_6, [1]_4)\rangle$
   (i) $(\mathbb{Z}_6 \times \mathbb{Z}_4)/\langle([0]_6, [1]_4)\rangle$
   (j) $(\mathbb{Z}_6 \times \mathbb{Z}_4)/\langle([2]_6, [3]_4)\rangle$
2. Write each of the following factors of $\mathbb{Z} \times \mathbb{Z}$ using set-builder notation.
   (a) $(\mathbb{Z} \times \mathbb{Z})/\langle(0, 0)\rangle$
   (b) $(\mathbb{Z} \times \mathbb{Z})/\langle(1, 0)\rangle$
   (c) $(\mathbb{Z} \times \mathbb{Z})/\langle(1, 1)\rangle$
   (d) $(\mathbb{Z} \times \mathbb{Z})/\langle(2, 2)\rangle$
3. Find the number of elements of the following factor rings:
   (a) $\mathbb{Z}_{14}/\mathbb{Z}_2$
   (b) $\mathbb{Z}_{100}/\mathbb{Z}_4$
   (c) $\mathbb{Z}/8\mathbb{Z}$
   (d) $\mathrm{M}_{2,2}(\mathbb{Z}_4)/\mathrm{M}_{2,2}(I)$ where $I = \{[0]_4, [2]_4\}$
4. Given a ring R and an ideal I of R, prove that multiplication of cosets is well-defined and R/I is a ring.
5. For any ring R and ideal $I \subseteq R$, prove $a + I$ is the additive identity of R/I if and only if $a \in I$.

6. Let R be a ring and I an ideal. Prove that $|a + I| = |b + I|$ for all $a, b \in R$.
7. Let R be a ring and $K \subseteq I \subseteq R$ ideals of R. Prove that if I is a principal ideal of R, then I/K is a principal ideal of R/K.
8. Find a ring R and an ideal $I \subseteq R$ that satisfies each of the following:
   (a) R is an integral domain, but R/I is not an integral domain.
   (b) R is a field, but R/I is not a field.
9. Suppose that R is a ring with unity and I an ideal of R such that $1 \notin I$. Prove that R/I is a ring with unity.
10. A ***simple*** ring is a ring with no nontrivial proper ideals. Let R be a ring such that $|R| = p$ where p is a prime. Prove that R is a simple ring.
11. Demonstrate that if R is a field, then R is simple.
12. Let p be a prime. Prove that $\mathbb{Z}/p\mathbb{Z}$ is a field.
13. Show the following: if $k = |R/I|$, then $ka \in I$ for all $a \in R$. (Note: ka represents the sum $a + a + \cdots + a$ with k terms.)
14. Give an example of a prime ideal that is not maximal.

7.4. HOMOMORPHISMS

We have spent a lot of time studying functions with various domains and codomains. Often these sets could have been considered as rings. The set of integers is an example. However, these functions said nothing about the operations on the sets. We would now like to change that. Let R and R' be rings. Each ring has its own pair of operations. We could distinguish between the two with primes so that $(R, +, \cdot)$ and $(R', +', \cdot')$ are the rings. This notation is cumbersome and not needed since context will determine which operation is being used. For this reason, we will only use $+$ and $\cdot$ for the operations.

7.4.1. Definition

If R and R' are rings and $\phi\colon R \to R'$ a function, then ϕ is a ***(ring) homomorphism*** when for all $x, y \in R$,

1. $\phi(x + y) = \phi(x) + \phi(y)$, and
2. $\phi(xy) = \phi(x)\phi(y)$.

We say that the homomorphism ϕ ***preserves*** the operations.

In the expression $\phi(x + y) = \phi(x) + \phi(y)$, x and y are elements of R while $\phi(x)$, $\phi(y)$, and $\phi(x + y)$ are elements of R'. The first $+$ is the operation from R, and the second one is from R'. This is illustrated in Figure 7.4.2.

7.4.2. Figure

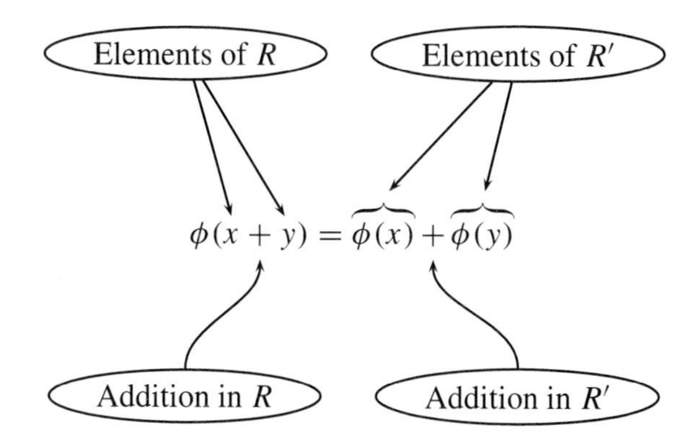

The homomorphism $\phi\colon R \to R'$ preserves addition.

Example. The function $\phi\colon \mathbb{Z} \to \mathbb{Z} \times \mathbb{Z}$ given by $\phi(n) = (n, 0)$ is a homomorphism. To demonstrate this, let $m, n \in \mathbb{Z}$. Then calculate:

$$\phi(m+n) = (m+n, 0) = (m, 0) + (n, 0) = \phi(m) + \phi(n),$$

and

$$\phi(mn) = (mn, 0) = (m, 0) \cdot (n, 0) = \phi(m) \cdot \phi(n).$$

Example. Let R and R' be rings. It is easy to show that the function $\psi\colon R \to R'$ that maps r to 0 for all $r \in R$ is a homomorphism. This function is called a ***zero map***.

For the next theorem, remember if $\phi\colon R \to R'$ is a function and $a \in R$, then $-a$ is the additive inverse of a in R, and $-\phi(a)$ is the additive inverse of $\phi(a)$ inside of R'.

7.4.3. Theorem

Let R and R' be rings and $\phi\colon R \to R'$ a ring homomorphism.

1. $\phi(0) = 0$.
2. $\phi(-a) = -\phi(a)$ for all $a \in R$.

Proof. Let $\phi\colon R \to R'$ be a ring homomorphism.

1. For this we just calculate:
$$\phi(0) + \phi(0) = \phi(0 + 0) = \phi(0).$$
Adding $-\phi(0)$ to both sides yields $\phi(0) = 0$.
2. Let $x \in R$. To show $-\phi(x) = \phi(-x)$, we add:
$$\phi(x) + \phi(-x) = \phi(x + -x) = \phi(0) = 0. \blacksquare$$

This theorem allows us to conclude $\phi(x - y) = \phi(x) - \phi(y)$ if ϕ is a homomorphism.

Example. We will check the theorem using the function ϕ as defined in the first example. First, we note that $\phi(0) = (0, 0)$, and $(0, 0)$ is the additive identity of $\mathbb{Z} \times \mathbb{Z}$. Second, 5 and -5 are additive inverses in $\mathbb{Z}$, so $\phi(5) = (5, 0)$ and $\phi(-5) = (-5, 0)$ are additive inverses in $\mathbb{Z} \times \mathbb{Z}$.

The operations of a ring create relationships among the elements. For example in $\mathbb{Z}$, $2 + 3 = 5$, so there is a relationship between 2, 3, and 5. It is often said that the operations give the set a ***structure***. For the integers, we can see the structure in the addition and multiplication tables. Another way, then, to describe the job of a homomorphism is to say that it preserves structure. If $\phi\colon R \to R'$ is a ring homomorphism, then the structure that the operations give R is similar to the structure that R' has from its operations. An example of this is the next theorem.

7.4.4. Theorem

Let R and R' be rings and $\phi\colon R \to R'$ a ring homomorphism.

1. If S is an ideal of R and ϕ is onto, then $\phi[S]$ is an ideal of R'.
2. If T is an ideal of R', then $\phi^{-1}[T]$ is an ideal of R.

Proof. The first part is left as an exercise. To prove the second, take an ideal T of R'. We will first show that $\phi^{-1}[T]$ is a subring.

1. To show closure, let $x_1, x_2 \in \phi^{-1}[T]$. This means that $\phi(x_1)$ and $\phi(x_2)$ are elements of T. Hence,
$$\phi(x_1 + x_2) = \phi(x_1) + \phi(x_2) \in T$$
and
$$\phi(x_1 x_2) = \phi(x_1)\phi(x_2) \in T.$$
Therefore, $x_1 + x_2,\ x_1 x_2 \in \phi^{-1}[T]$.
2. Since $0 \in T$ and $\phi(0) = 0$, it follows that $0 \in \phi^{-1}[T]$.

3. Let $x \in \phi^{-1}[T]$. Then, $\phi(x) \in T$, and we find that

$$\phi(-x) = -\phi(x) \in T.$$

Thus, $-x \in \phi^{-1}[T]$.

To see that $\phi^{-1}[T]$ is an ideal, take $r \in R$ and $a \in \phi^{-1}[T]$. This means $\phi(a) \in T$. We must show $ra,\ ar \in \phi^{-1}[T]$. For the first one,

$$\phi(ra) = \phi(r)\phi(a) \in T$$

since T is an ideal. Thus, $ra \in \phi^{-1}[T]$. Similarly $ar \in \phi^{-1}[T]$. ■

Example. The function $\phi \colon \mathbb{Z} \to \mathbb{Z}/6\mathbb{Z}$ defined by $\phi(n) = n + 6\mathbb{Z}$ is an onto homomorphism. (See Exercise 6.) The image of the ideal $2\mathbb{Z}$ under this map is

$$\phi[2\mathbb{Z}] = \{n + 6\mathbb{Z} : n \in 2\mathbb{Z}\} = \{0 + 6\mathbb{Z},\ 2 + 6\mathbb{Z},\ 4 + 6\mathbb{Z}\}.$$

The pre-image of $I = \{0 + 6\mathbb{Z},\ 3 + 6\mathbb{Z}\}$ is

$$\phi^{-1}[I] = \{n \in \mathbb{Z} : n \in 6\mathbb{Z} \text{ or } n \in 3\mathbb{Z}\} = 3\mathbb{Z}.$$

Notice that both $\phi[2\mathbb{Z}]$ and $\phi^{-1}[I]$ are ideals of their respective rings.

Let $\psi \colon R \to R'$ be a homomorphism where R and R' are rings. In this context, the range is usually called the ***image***. It is denoted by $\operatorname{im}(\psi)$. Another important set linked to a homomorphism is the ***kernel***. It is defined as the subset of the domain in which every element is mapped to the additive identity of the codomain. The kernel of ψ is denoted by $\ker(\psi)$. More precisely,

$$\ker(\psi) = \{x \in R : \psi(x) = 0\}.$$

Notice that this means $\ker(\psi) = \psi^{-1}[\{0\}]$. Therefore, $\ker(\psi)$ is an ideal of R.

Example. The kernel of the projection $\pi : R \times R \times R \to R \times R$ is $\{(0,\ 0,\ r) : r \in R\}$, and the image is $R \times R$ since π is onto.

Example. Let us find both the kernel and image of the homomorphism $\phi \colon \mathbb{Z} \to \mathbb{Z}_5$ defined by $\phi(n) = [n]_5$.

- To find the kernel, assume $\phi(n) = [0]_5$. Now the assumption gives $[n]_5 = [0]_5$. So, $n \in [0]_5$, which means $5 \mid n$. Hence,

$$\ker(\phi) \subseteq 5\mathbb{Z}.$$

Upon examination, however, we find that the steps are reversible, and we actually have equality.

- As for the image,

$$\text{im}(\phi) = \{[n]_5 : n \in \mathbb{Z}\},$$

but this set encompasses all congruence classes modulo 5. Therefore, ϕ is onto and $\text{im}(\phi) = \mathbb{Z}_5$.

There are times when we will need to know whether a homomorphism is one-to-one. The kernel provides a good test for this.

7.4.5. Theorem

Let $\phi\colon R \to R'$ be a ring homomorphism. Then, ϕ is one-to-one if and only if $\ker(\phi) = \{0\}$.

Proof. Let R and R' be rings and $\phi\colon R \to R'$ a homomorphism.

($\Rightarrow$) Suppose that ϕ is one-to-one. Because $\phi(0) = 0$, $\{0\} \subseteq \ker(\phi)$. To prove the other inclusion, take $x \in \ker(\phi)$. This means $\phi(x) = 0$. Since ϕ is an injection, $x = 0$. We have thus shown $\ker(\phi) = \{0\}$.

($\Leftarrow$) Now let $\ker(\phi) = \{0\}$ and assume $\phi(x_1) = \phi(x_2)$ for any $x_1, x_2 \in R$. We then have $\phi(x_1) - \phi(x_2) = 0$. Since ϕ is a homomorphism,

$$\phi(x_1 - x_2) = \phi(x_1) - \phi(x_2).$$

Hence, $x_1 - x_2 \in \ker(\phi)$, which means $x_1 - x_2 = 0$ by hypothesis. Therefore, $x_1 = x_2$. ■

Since any ring homomorphism $\phi\colon R \to R'$ maps the additive identity of R to the additive identity of R', it is always the case that $0 \in \ker(\phi)$. Therefore, to show $\ker(\phi) = \{0\}$, we only need to prove $\ker(\phi) \subseteq \{0\}$. This is used in the next example.

Example. Take a ring R. We will show that $\psi\colon R \times R \to R \times R \times R$ such that $\psi(a, b) = (a, b, 0)$ is one-to-one. Since ψ is a homomorphism, we only need to examine its kernel. Let $(r, s) \in \ker(\psi)$. This means $\psi(r, s) = (0, 0, 0)$. By definition of the function, $r = 0$ and $s = 0$. Therefore, $(r, s) = (0, 0)$, so ψ is one-to-one.

Remember, the kernel check can only be used if the function is a homomorphism.

Example. Define $\psi\colon \mathbb{Z}_5 \times \mathbb{Z}_6 \to \mathbb{Z}_6 \times \mathbb{Z}_5$ by $\psi([a]_5, [b]_6) = ([b]_6, [a]_5)$. We will prove that this function is a homomorphism and a bijection.

1. To show that ψ is a homomorphism, let $a, b, c, d \in \mathbb{Z}$. Then

$$\begin{aligned}
&\psi(([a]_5, [b]_6) + ([c]_5, [d]_6)) \\
&\quad = \psi([a]_5 + [c]_5, [b]_6 + [d]_6) \\
&\quad = \psi([a+c]_5, [b+d]_6) \\
&\quad = ([b+d]_6, [a+c]_5) \\
&\quad = ([b]_6 + [d]_6, [a]_5 + [c]_5) \\
&\quad = ([b]_6, [a]_5) + ([d]_6, [c]_5) \\
&\quad = \psi([a]_5, [b]_6) + \psi([c]_5, [d]_6).
\end{aligned}$$

   The preservation of multiplication is shown similarly.
2. Since ψ is a homomorphism, we can prove that it is one-to-one by checking its kernel. Let $([a]_5, [b]_6) \in \ker(\psi)$. In other words, $([b]_6, [a]_5) = ([0]_6, [0]_5)$. Hence, $5 \mid a$ and $6 \mid b$, but we know from our work with congruences that this means $[a]_5 = [0]_5$ and $[b]_6 = [0]_6$. Thus, ψ is an injection.
3. To see that ψ is onto, take $([c]_6, [d]_5) \in \mathbb{Z}_6 \times \mathbb{Z}_5$. Then,

$$\psi([d]_5, [c]_6) = ([c]_6, [d]_5).$$

Because there is a bijection between $\mathbb{Z}_5 \times \mathbb{Z}_6$ and $\mathbb{Z}_6 \times \mathbb{Z}_5$, they have the same cardinality. Since the bijection is also a homomorphism, the structure of the two rings are the same. Putting this together, we conclude that the two rings "look" the same as rings. The only difference is in the labeling of their elements. To formalize this notion, we make the next definition.

7.4.6. Definition

Let R and R' be rings. If $\psi\colon R \to R'$ is a ring homomorphism that is a bijection, then ψ is called a ***(ring) isomorphism***.

We say that two rings R and R' are ***isomorphic*** if there exists an isomorphism $\psi\colon R \to R'$. If two rings are isomorphic, write $R \cong R'$. The previous example shows $\mathbb{Z}_5 \times \mathbb{Z}_6 \cong \mathbb{Z}_6 \times \mathbb{Z}_5$. The next example illustrates what we mean by "looking" the same. The proofs of these facts are in the exercises.

Example. Suppose $\phi\colon R \to R'$ is a ring isomorphism.

- If R is an integral domain, then R' is an integral domain.
- If R is a division ring, then R' is a division ring.
- If R is a field, then R' is a field.

EXERCISES

1. Prove that the following are ring homomorphisms.
   (a) $\phi: \mathbb{Z} \times \mathbb{Z} \to \mathbb{Z}$ where $\phi(a, b) = a$
   (b) $\phi: \mathbb{Z}_{12} \to \mathbb{Z}_6$ where $\phi([a]_{12}) = [a]_6$
   (c) $\psi: \mathbb{Z} \times \mathbb{Z} \to \mathrm{M}_{2,2}(\mathbb{R})$ where
$$\psi(a, b) = \begin{bmatrix} a & 0 \\ 0 & b \end{bmatrix}$$
2. Prove that the zero map is a homomorphism.
3. For all $n \in \mathbb{Z}^+$, prove that $\phi: \mathbb{Z} \to \mathbb{Z}_n$ defined by $\phi(x) = [x]_n$ is a homomorphism.
4. Let $\psi: \mathbb{C} \to \mathbb{C}$ be defined by $\psi(a + bi) = a - bi$ for all $a, b \in \mathbb{R}$. (Note: $a - bi$ is called the ***complex conjugate*** of $a + bi$.) Prove:
   (a) ψ is a ring homomorphism.
   (b) ψ is one-to-one.
   (c) ψ is onto.
5. The following are not ring homomorphisms. Show which conditions of the definition fail.
   (a) $f: \mathbb{Z} \to 4\mathbb{Z}$ where $f(n) = 4n$
   (b) $g: \mathbb{R} \to \mathbb{R}$ where $f(x) = e^x$
   (c) $\phi: \mathrm{M}_{2,2}(\mathbb{R}) \to \mathrm{M}_{2,2}(\mathbb{R})$ where
$$\phi\left(\begin{bmatrix} a & b \\ c & d \end{bmatrix}\right) = \begin{bmatrix} a+b & 0 \\ 0 & c+d \end{bmatrix}$$
6. Let R be a ring and I an ideal. Let $\phi: R \to R/I$ be defined as $\phi(a) = a + I$ for all $a \in R$.
   (a) Prove that ϕ is a homomorphism. (This function is known as the ***canonical homomorphism***.)
   (b) Show that $\ker(\phi) = I$.
   (c) Prove that ϕ is onto.
7. Let R and R' be rings with unity and take an onto homomorphism $\phi: R \to R'$. Demonstrate:
   (a) $\phi(1) = 1$
   (b) Show that the previous problem may not hold if the homomorphism is not onto.
   (c) If u is a unit of R, then $\phi(u)$ is a unit of R'.
8. For any ring homomorphism $\psi: R \to R'$, show that
$$\ker(\psi) = \psi^{-1}[\{0\}].$$
9. Find the kernel and image of each homomorphism in Exercise 1.

10. Let $\phi\colon R \to R'$ be a ring homomorphism. Let I be a left ideal of R and show the following:
    (a) $\phi[I]$ is a left ideal of $\phi[R]$.
    (b) If ϕ is onto, then $\phi[I]$ is a left ideal of R'.
11. Prove directly using the definition that the kernel of a homomorphism is an ideal of the domain and the image is a subring of the codomain. Show by example that the image may not be an ideal.
12. Let R be a ring and $\phi,\ \psi\colon R \to R$ two functions. Prove:
    (a) If ϕ and ψ are homomorphisms, then $\phi \circ \psi$ is a homomorphism.
    (b) If ϕ and ψ are isomorphisms, then $\phi \circ \psi$ is an isomorphism.
13. Show if ϕ is an isomorphism, then so is ϕ^{-1}.
14. Let $\phi\colon R \to R'$ be a ring isomorphism. Define $\psi\colon R \times R \to R' \times R'$ by $\psi(a, b) = (\phi(a), \phi(b))$. Prove that ψ is an isomorphism.
15. Let R_1, R_2, and R_3 be rings. Demonstrate the following:
    (a) $R_1 \cong R_1$.
    (b) If $R_1 \cong R_2$, then $R_2 \cong R_1$.
    (c) If $R_1 \cong R_2$ and $R_2 \cong R_3$, then $R_1 \cong R_3$.
16. Let R and R' be rings. Show the following:
    (a) $R \cong R/\{0\}$
    (b) $R/R \cong \{0\}$
    (c) $R \times R'/\{0\} \cong R' \times R$
17. Suppose $\phi\colon R \to R'$ is a ring isomorphism. Prove:
    (a) If R is an integral domain, then R' is an integral domain.
    (b) If R is a division ring, then R' is a division ring.
    (c) If R is a field, then R' is a field.
18. Show that $\phi\colon \mathbb{Z}_n \to \mathbb{Z}/n\mathbb{Z}$ defined by $\phi([a]_n) = a + n\mathbb{Z}$ is an isomorphism.

7.5. POLYNOMIALS

The subject of our last section in ring theory seems to be a familiar one. We have been working with polynomials for a long time. We know what they look like. We know that $9 - 7x + x^2$ is a polynomial. Now we want to make them the object of our study, and we want the coefficients to come from an arbitrary ring. To do this, we make an observation about polynomials: the coefficients are what distinguish one polynomial from another. What symbol we use for the variable is unimportant. An x is just as good as a y. Basically, the powers of the variables serve as a placeholders. This motivates the next definition. We will identify these polynomials by what is known as a formal sum. A ***formal sum*** is a string of symbols that represents a summation of elements in a ring where one of the elements is unknown. The unknown is called an ***indeterminant*** and is represented as X. The different powers of X are placeholders only. We will not substitute into them. Since any natural number can be the highest exponent in a polynomial, the string has to be able to handle it. Therefore, the formal sum is infinite in length, but since each polynomial is finite, all but finitely many of the coefficients in the string are zero. The formal sum that represents the above polynomial is

$$9 - 7X + 1X^2 + 0X^3 + 0X^4 + 0X^5 + \cdots$$

All of this leads us to the main definition. (See Appendix B for a review of summation notation.)

7.5.1. Definition

Let R be a ring. A ***polynomial*** with coefficients in R is a formal sum

$$f(X) = \sum_{i=0}^{\infty} a_i X^i = a_0 + a_1 X + a_2 X^2 \cdots$$

where $a_i \in R$ and $a_i = 0$ for all but finitely many i. Each a_i is called a ***coefficient***, and the ***degree*** of $f(X)$ is the largest value of i so that $a_i \neq 0$. If no such i exists, we say that the degree of $f(X)$ is ***undefined***. Denote the degree of f by $\deg f(X)$.

Note: if the degree of $f(X)$ is zero, then $f(X) = r$ for some element nonzero element r of the ring.

Since it would be burdensome to have to use the definition of a polynomial verbatim each time we needed it, we introduce some shorthand. Use the expression

$$a_0 + a_1 X + a_2 X^2 + \cdots + a_n X^n$$

to refer to the polynomial

$$a_0 + a_1 X + a_2 X^2 + \cdots + a_n X^n + 0x^{n+1} + 0x^{n+2} + \cdots.$$

Since there are only finitely many nonzero coefficients, this can always be done. Furthermore, if we need to work with two (or more) polynomials, we can write them with the same number of coefficients. This is reminiscent of our work with prime power decompositions in Section 5.3. For example, suppose we have two polynomials. The first three terms of the first are nonzero, but the first four terms of the second are nonzero. This can be illustrated with the following:

$$\begin{array}{l|l} a_0 + a_1X + a_2X^2 + 0X^3 + & 0X^4 + 0X^5 + \cdots \\ b_0 + b_1X + b_2X^2 + b_3X^3 + & 0X^4 + 0X^5 + \cdots \end{array}$$

all coefficients are zero

We can then write the polynomials as

$$a_0 + a_1X + a_2X^2 + a_3X^3$$

and

$$b_0 + b_1X + b_2X^2 + b_3X^3$$

where the a_i and b_i are elements of a ring. This is sometimes needed to make our work easier.

Fix a ring R. Let us define polynomial addition and multiplication. Take two polynomials,

$$f(X) = a_0 + a_1X + \cdots + a_nX^n$$

and

$$g(X) = b_0 + b_1X + \cdots + b_nX^n.$$

The sum of the polynomials is defined by

$$f(X) + g(X) = (a_0 + b_0) + (a_1 + b_1)X + \cdots + (a_n + b_n)X^n.$$

The product is more complicated:

$$f(X)g(X) = (a_0 + a_1X + \cdots + a_nX^n)(b_0 + b_1X + \cdots + b_nX^n),$$

which equals

$$a_0b_0 + (a_0b_1 + a_1b_0)X + (a_0b_2 + a_1b_1 + a_2b_0)X^2 + \cdots .$$

Summarizing,

$$f(X)g(X) = d_0 + d_1X + \cdots + d_nX^{2n}$$

where

$$d_i = \sum_{j=0}^{i} a_jb_{i-j}.$$

If R is a ring, then the set of all polynomials with coefficients in R with these operations forms a ring called a ***polynomial ring***. This ring is denoted by $R[X]$.

Example. Let us do some arithmetic in $\mathbb{Q}[X]$:

1. $(X + \frac{1}{2}X^2) + (1 + 3X) = 1 + 4X + \frac{1}{2}X^2$
2. $(X + \frac{1}{2}X^2) \cdot (1 + 3X) = X + \frac{7}{2}X^2 + \frac{3}{2}X^3$

Notice that the product checks against the formula:

$$\begin{aligned} d_0 &= 0 \cdot 1 = 0 \\ d_1 &= 0 \cdot 3 + 1 \cdot 1 = 1 \\ d_2 &= 0 \cdot 0 + 1 \cdot 3 + \tfrac{1}{2} \cdot 1 = \tfrac{7}{2} \\ d_3 &= 0 \cdot 0 + 1 \cdot 0 + \tfrac{1}{2} \cdot 3 + 0 \cdot 1 = \tfrac{3}{2}. \end{aligned}$$

Example. Consider the polynomial ring $M_{2,2}(\mathbb{R})[X]$. The coefficients of these polynomials are matrices. To do some arithmetic in this ring, let

$$f(X) = \begin{bmatrix} 1 & 2 \\ 3 & 4 \end{bmatrix} + \begin{bmatrix} 1 & 0 \\ 1 & 2 \end{bmatrix} X, \text{ and}$$

$$g(X) = \begin{bmatrix} 1 & 1 \\ 0 & 0 \end{bmatrix} + \begin{bmatrix} 1 & 2 \\ 0 & 1 \end{bmatrix} X.$$

We will have to be careful with multiplication since $M_{2,2}(\mathbb{R})[X]$ is not commutative.

1. $f(X) + g(X) = \begin{bmatrix} 2 & 3 \\ 3 & 4 \end{bmatrix} + \begin{bmatrix} 2 & 2 \\ 1 & 3 \end{bmatrix} X.$

2. To find the product $f(X)g(X)$, first calculate the coefficients using the formula:

$$d_0 = \begin{bmatrix} 1 & 2 \\ 3 & 4 \end{bmatrix} \cdot \begin{bmatrix} 1 & 1 \\ 0 & 0 \end{bmatrix}$$

$$d_1 = \begin{bmatrix} 1 & 2 \\ 3 & 4 \end{bmatrix} \cdot \begin{bmatrix} 1 & 2 \\ 0 & 1 \end{bmatrix} + \begin{bmatrix} 1 & 0 \\ 1 & 2 \end{bmatrix} \cdot \begin{bmatrix} 1 & 1 \\ 0 & 0 \end{bmatrix}$$

$$d_2 = \begin{bmatrix} 1 & 2 \\ 3 & 4 \end{bmatrix} \cdot \begin{bmatrix} 0 & 0 \\ 0 & 0 \end{bmatrix} + \begin{bmatrix} 1 & 0 \\ 1 & 2 \end{bmatrix} \cdot \begin{bmatrix} 1 & 2 \\ 0 & 1 \end{bmatrix} + \begin{bmatrix} 0 & 0 \\ 0 & 0 \end{bmatrix} \cdot \begin{bmatrix} 1 & 1 \\ 0 & 0 \end{bmatrix}.$$

This means that $f(X)g(X)$ equals

$$\begin{bmatrix} 1 & 1 \\ 3 & 3 \end{bmatrix} + \begin{bmatrix} 2 & 5 \\ 4 & 11 \end{bmatrix} X + \begin{bmatrix} 1 & 2 \\ 1 & 4 \end{bmatrix} X^2.$$

The last example suggests that there is a relationship between the structure of the ring and the structure of the corresponding polynomial ring. This is further seen in the next theorem.

7.5.2. Theorem

Let R be a ring.

1. If R is a ring with unity, then $R[X]$ is a ring with unity.
2. If R is commutative, then $R[X]$ is commutative.
3. If R is an integral domain, then $R[X]$ is an integral domain.

Proof. We will prove the first one and leave the others to the exercises. Let R be a ring with unity. Define

$$i(X) = 1 + 0X + 0X^2 + \cdots .$$

We will show that this is the identity of $R[X]$. Let $f(X) = a_0 + a_1X + a_2X^2 + \cdots$ be an element of $R[X]$ and calculate:

$$\begin{aligned} &i(X)f(X) \\ &\qquad = (1 + 0X + 0X^2 + \cdots)(a_0 + a_1X + a_2X^2 + \cdots) \\ &\qquad = (1 \cdot a_0) + (1 \cdot a_1 + 0 \cdot a_0)X + (1 \cdot a_2 + 0 \cdot a_1 + 0 \cdot a_0)X^2 + \cdots \\ &\qquad = a_0 + a_1X + a_2X^2 + \cdots \\ &\qquad = f(X). \end{aligned}$$

Similarly, $f(X)I(X) = f(X)$. ■

Since R can be viewed as a subring of $R[X]$ (Exercise 12), the converse of each of the implications of the theorem is true.

In this section we want to look at two important results. The first states that under the right circumstances, $R[X]$ resembles $\mathbb{Z}$ in that it is a principle ideal domain. The second involves factorizations of polynomials. Both of these results require an application of a generalization of the Division Algorithm to polynomials.

7.5.3. Polynomial Division Algorithm

Let F be a field and take $f(X),\ g(X) \in F[X]$ where $g(X) \neq 0$. There exist unique polynomials $q(X),\ r(X) \in F[X]$ such that

$$f(X) = g(X)q(X) + r(X)$$

and either $r(X) = 0$ or $\deg r(X) < \deg g(X)$.

Proof. Define $S = \{f(X) - g(X)h(X) : h(X) \in F[X]\}$. Because of the Well-Ordering Principle, there exists $r(X) \in S$ such that $r(X) = 0$ or $r(X)$ has minimal degree. This means that there exists $q(X) \in F[X]$ such that

$$(*) \qquad r(X) = f(X) - g(X)q(X).$$

The polynomials $r(X)$ and $q(X)$ are the ones that we want, so suppose $r(X) \neq 0$. We are left to prove that $\deg r(X) < \deg g(X)$. Now, write

$$r(X) = a_0 + a_1 X + a_2 X^2 + \cdots + a_n X^n$$

and

$$g(X) = b_0 + b_1 X + b_2 X^2 + \cdots + b_m X^m$$

with each a_i, $b_j \in F$ and a_n, $b_m \neq 0$. In order to obtain a contradiction, assume $n \geq m$. We will use this to find a polynomial in S of degree smaller than $r(X)$. Since the coefficients come from a field and b_m is nonzero, we can multiply $g(X)$ by $(a_n b_m^{-1})X^{n-m}$ and obtain

$$(a_n b_m^{-1})b_0 X^{n-m} + (a_n b_m^{-1})b_1 X^{n-m+1} + \cdots + a_n X^n.$$

Its coefficient on X^n is the same as that for $r(X)$. Therefore, the degree of

$$r(X) - (a_n b_m^{-1})g(X)X^{n-m}$$

is less than n. But by $(*)$,

$$r(X) - (a_n b_m^{-1})g(X)X^{n-m} = f(X) - g(X)\left[q(X) + (a_n b_m^{-1})X^{n-m}\right],$$

and this is an element of S because

$$q(X) + (a_n b_m^{-1})X^{n-m} \in F[X].$$

This contradicts the minimality of the degree of $r(X)$. We then conclude that $n < m$.

To show that $q(X)$ and $r(X)$ are unique, assume that we also have $q'(X)$, $r'(X) \in F[X]$ so that

$$f(X) = q'(X)g(X) + r'(X)$$

and either $r(X) = r'(X)$ or $\deg r'(X) < \deg g(X)$. We have two cases to consider. If we have $r(X) = r'(X) = 0$, then we may calculate to find $q(X) = q'(X)$, and we are done. Otherwise, we note that

$$q(X)g(X) + r(X) = q'(X)g(X) + r'(X),$$

which yields

$$g(X)[q(X) - q'(X)] = r'(X) - r(X).$$

Since

$$\deg[r'(X) - r(X)] \leq \max[\deg r'(X), \deg r(X)]$$

by Exercise 19a, the degree of $r'(X) - r(X)$ must be less than the degree of $g(X)$. So, appealing to Exercise 19c yields $g(X) = 0$ or $q(X) - q'(X) = 0$. Since $g(X) \neq 0$, the second disjunct is true. Hence, $q(X) = q'(X)$. This gives $r'(X) - r(X) = 0$. In other words, $r'(X) = r(X)$. ■

Example. To illustrate the Polynomial Division Algorithm, let us divide $1 + 2X^2 - 5X^3$ by $4 - X^2$ in $\mathbb{Z}[X]$. We will use the standard algebra technique for dividing polynomials:

$$\begin{array}{r|l} & \quad\;\; -2 \quad +5X \\ \cline{2-2} 4 + 0X - X^2 & \;\; 1 + \;0X + 2X^2 - 5X^3 \\ & \underline{-8 + \;0X + 2X^2} \\ & \;\; 9 + \;0X + 0X^2 - 5X^3 \\ & \underline{\qquad\quad 20X + 0X^2 - 5X^3} \\ & \;\; 9 - 20X \end{array}$$

Therefore, $1 + 2X^2 - 5X^3 = (-2 + 5X)(4 - X^2) + (9 - 20X)$. Observe that the degree of $9 - 20X$ is indeed less than the degree of $4 - X^2$.

The Polynomial Division Algorithm is needed to prove the next theorem about the structure of certain polynomial rings. It is similar to the corresponding proof about $\mathbb{Z}$.

7.5.4. Theorem

If F is a field, then $F[X]$ is a principal ideal domain.

Proof. Let I be an ideal of $F[X]$. We have two cases to consider.

(Case 1) If $I = \{0\}$, then $I = \langle 0 \rangle$, and we are done.

(Case 2) Suppose $I \neq \{0\}$ and take $g(X) \in I \setminus \{0\}$ of minimal degree. We again have two cases.

- (Case 2a) Assume that the degree of $g(X)$ is 0. This means that $g(X)$ is a unit since F is a field. Hence, $I = F[X]$ by Exercise 7.2.21, so $I = \langle 1 \rangle$.
- (Case 2b) Now suppose $\deg g(X) > 0$ and show that $g(X)$ generates I. Let $f(X) \in I$. By the Polynomial Division Algorithm, there exist $q(X), r(X) \in F[X]$ such that

$$f(X) = q(X)g(X) + r(X)$$

with $r(X) = 0$ or $\deg r(X) < \deg g(X)$. If $r(X) = 0$, then $f(X) = q(X)g(X)$, and thus $I = \langle g(X) \rangle$. Otherwise, a little calculation reveals that

$$r(X) = f(X) - q(X)g(X).$$

Since I is an ideal, $q(X)g(X) \in I$. Therefore, $r(X) \in I$. Because $g(X)$ has minimal degree in $I \setminus \{0\}$, we again have $r(X) = 0$. ■

Example. Since $\mathbb{Q}$ is a field, $\mathbb{Q}[X]$ is a principal ideal domain. Define

$$I = \{f(X) \in \mathbb{Q}[X] : f(X) = a_3X^3 + a_4X^4 + a_5X^5 + \cdots\}.$$

It is left as an exercise to prove that I is an ideal of $\mathbb{Q}[X]$. To find a generator for I, examine the previous proof. It should be a nonzero polynomial in I with minimal degree. A good candidate would be X^3, so let us show that $I = \langle X^3 \rangle$. Take $f(X) \in I$. This means $f(X) = a_3X^3 + a_4X^4 + \cdots$ for some $a_i \in \mathbb{Q}$. Then,

$$f(X) = (a_3 + a_4X + a_5X^2 \cdots)X^3.$$

Since the powers of X in our polynomials serve as placeholders, it formally does not make sense to write $f(2)$. We need another way to do this common arithmetic task. We will use the ***evaluation homomorphism***. Let R be a ring. For all $r \in R$, define

$$\varepsilon_r \colon R[X] \to R$$

by $\varepsilon_r(f(X)) = a_0+a_1r+a_2r^2+\cdots$ where $f(X) = a_0+a_1X+a_2X^2 \cdots$. Notice if $f(X) = 9 - 4X + 3X^2$ in $\mathbb{R}[X]$, then $\varepsilon_0(f(X)) = 9$ and $\varepsilon_1(f(X)) = 8$. (In the past, we would write these evaluations as $f(0)$ and $f(1)$.) Let us show that ε_α is a ring homomorphism. Suppose $f(X),\ g(X) \in R[X]$ and write

$$f(X) = \sum_{i=0}^{\infty} a_iX^i$$

and

$$g(X) = \sum_{i=0}^{\infty} b_iX^i.$$

Then,

$$\varepsilon_r(f(X)) + \varepsilon_r(g(X)) = \sum_{i=0}^{\infty} a_ir^i + \sum_{i=0}^{\infty} b_ir^i = \sum_{i=0}^{\infty}(a_i + b_i)r^i,$$

and this last expression equals $\varepsilon_r(f(X) + g(X))$. It is left as an exercise to show that ε_r preserves multiplication.

The evaluation homomorphism is needed for our last theorem. Let $f(X)$ be a polynomial in $F[X]$ for some field F. An element $a \in F$ is a ***zero*** of $f(X)$ if $\varepsilon_a(f(X)) = 0$. For example, 3 is a zero for $f(X) = -6 - X - X^2$ in $\mathbb{R}[X]$. This is because

$$\varepsilon_3(f(X)) = -6 - 3 + 3^2 = 0.$$

However, there are no zeros in $\mathbb{R}$ for the polynomial $1 + X^2$. (Note: in the proof of the next theorem, we will break from convention and write $X - a$ instead of $-a + X$.)

7.5.5. Theorem

Let F be a field and $f(X) \in F[X]$. The element $a \in F$ is a zero of $f(X)$ if and only if $X - a$ is a factor of $f(X)$ in $F[X]$.

Proof. Fix a field F and take $f(X) \in F[X]$.

($\Rightarrow$) Assume $\varepsilon_a(f(X)) = 0$. To show that $X - a$ is a factor of $f(X)$, use the Polynomial Division Algorithm to divide $f(X)$ by $X - a$. The theorem gives $q(X), r(X) \in F[X]$ such that

$$f(X) = (X - a)q(X) + r(X)$$

with $r(X) = 0$ or

$$\deg r(X) < \deg(X - a) = 1.$$

From this we may conclude that $r(X)$ is a constant polynomial, and we can write it as $r(X) = k$ for some $k \in F$. Hence,

$$\begin{aligned} 0 = \varepsilon_a(f(X)) &= \varepsilon_a([X - a]q(X) + k) \\ &= \varepsilon_a(X - a)\varepsilon_a(q(X)) + \varepsilon_a(k) \\ &= (a - a)\varepsilon_a(q(X)) + k \\ &= k. \end{aligned}$$

Therefore, $r(X) = 0$, and $X - a$ divides $f(X)$.

($\Leftarrow$) Suppose $f(X) = (X - a)g(X)$ for some $g(X) \in F[X]$. Then

$$\varepsilon_a(f(X)) = (a - a)\varepsilon_a(g(X)) = 0.$$

So, a is a zero of $f(X)$. ■

EXERCISES

1. Find the coefficient of X^5 in the product of $f(X)g(X)$ where

$$f(X) = 3 + 4X - 1X^2 + 7X^3 + 3X^4 + 5X^5$$

and

$$g(X) = 3 + X - X^2 + 3X^3 - 9X^4 + X^5.$$

2. Given a polynomial $f(X)$, find $\varepsilon_a(f(X))$.
   (a) $f(X) = 1 + 2X + 3X^2, a = 0$
   (b) $f(X) = 1 + 2X + 3X^2, a = 1$
   (c) $f(X) = 4, a = 5$
   (d) $f(X) = \begin{bmatrix} 1 & 2 \\ 0 & 1 \end{bmatrix} + \begin{bmatrix} 0 & 2 \\ 3 & 1 \end{bmatrix} X + \begin{bmatrix} 2 & -1 \\ 2 & 2 \end{bmatrix} X^2, a = \begin{bmatrix} 1 & 0 \\ 0 & 1 \end{bmatrix}$
   (e) $f(X) = \begin{bmatrix} 1 & 2 \\ 0 & 1 \end{bmatrix} + \begin{bmatrix} 0 & 2 \\ 3 & 1 \end{bmatrix} X + \begin{bmatrix} 2 & -1 \\ 2 & 2 \end{bmatrix} X^2, a = \begin{bmatrix} 1 & 1 \\ 1 & 1 \end{bmatrix}$
3. Divide $f(X)$ by $g(X)$ and write the result in the form $f(X) = q(X)g(X) + r(X)$ where $r(X) = 0$ or $\deg r(X) < \deg g(X)$.
   (a) $f(X) = -15 + 2X + X^2, g(X) = 5 + X$
   (b) $f(X) = 3 - X - 2X^2 - 6X^3, g(X) = 3 - X$
   (c) $f(X) = 1 + X^5, g(X) = -1 + X$
   (d) $f(X) = X + 2X^2 + 3X^3 + 4X^4 + 5X^5, g(X) = 1 + 2X + 3X^2$
4. Use the Division Algorithm to divide $1 + X - 2X + 2X^4$ by $1 + 2X^2$ in $\mathbb{Z}_5[x]$.
5. Prove the following for any ring R:
   (a) If R is commutative, then $R[X]$ is commutative.
   (b) If R is an integral domain, then $R[X]$ is an integral domain.
6. Show that $M_{2,2}(\mathbb{R})[X]$ is not commutative.
7. Define the set

$$I = \{f(X) \in \mathbb{Q}[X] : f(X) = a_1X + a_2X^2 + a_3X^3 + \cdots\}.$$

   (a) Prove that I is an ideal of $\mathbb{R}[X]$.
   (b) Show $I = \langle X \rangle$.
8. Fix a ring R. Let

$$J = \Big\{\sum_{i=0}^{\infty} a_{2i}X^{2i} : a_{2i} \in R, \text{ all } i \in \mathbb{N}\Big\}.$$

   (a) Prove that J is an ideal of $R[X]$.
   (b) Show that J is a principal ideal.

9. Define

$$K = \{\sum_{i=0}^{\infty}(a_i, 0)X^i : a_i \in \mathbb{Z}\}.$$

Prove:

(a) K is an ideal of $(\mathbb{Z} \times \mathbb{Z})[X]$.

(b) K is a principal ideal.

10. Give an example of a nonprincipal ideal in each of the following polynomial rings if possible:

(a) $\mathbb{Z}[X]$	(c) $\mathbb{Z}_6[X]$	(e) $\mathrm{M}_{2,2}(\mathbb{R})$
(b) $\mathbb{Q}[X]$	(d) $\mathbb{Z}_7[X]$	(f) $(\mathbb{Z} \times \mathbb{Z})[X]$

11. Let R be a ring. Demonstrate the following:

(a) If S is a subring of R, then $S[X]$ is a subring of $R[X]$.

(b) If I is an ideal of R, then $I[X]$ is an ideal of $R[X]$.

12. Let R be a ring. Define

$$R_0 = \{\sum_{i=0}^{\infty} a_i X^i : a_i \in R \text{ and } a_i = 0 \text{ for all } i > 1\}.$$

(a) Prove that R_0 is a subring of $R[X]$.

(b) Show that $R \cong R_0$. (In this way R can be considered a subring of $R[X]$.)

(c) Show that R_0 is not an ideal of $R[X]$.

13. Find an example of a subring S of $R[X]$ with the property that there does not exist a subring T of R such that $S = T[X]$.

14. Repeat the previous exercise except replace the word *subring* with *ideal.*

15. Prove that the following are ring homomorphisms.

(a) $\phi\colon \mathbb{Z} \to \mathbb{Z}[X]$, $\phi(a) = a$

(b) $\psi : \mathbb{Z}[x] \to \mathbb{Z}$, $\phi(a_0 + a_1X + \cdots) = a_0$

(c) $\gamma : \mathbb{Z}[X] \to \mathbb{Z}_5[X]$, $\gamma(a_0 + a_1X + \cdots) = [a_0]_5 + [a_1]_5X + \cdots$

16. Find the kernel and the image of the homomorphisms in the previous problem.

17. Show the following:

$$\{\sum_{i=0}^{\infty} \begin{bmatrix} a_i & 0 \\ 0 & 0 \end{bmatrix} X^i : a_i \in \mathbb{R},\ \text{all } i \in \mathbb{N}\} \cong \mathbb{Z}[X].$$

18. Prove that $\mathbb{Z}_4[X]$ and $(\mathbb{Z}/4\mathbb{Z})[X]$ are isomorphic.

19. Let $f(X),\ g(X) \in R[X]$. Show:

(a) $\deg[f(X) + g(X)] \le \max[\deg f(X), \deg g(X)]$

(b) $\deg f(X)g(X) = \deg f(X) + \deg g(X)$

(c) If $\deg f(X)g(X) < \deg f(X) + \deg g(X)$, then $f(X) = 0$ or $g(X) = 0$.

20. Let F be a field. Prove that if $f(X) \in F[x] \setminus \{0\}$, then $f(X)$ has at most $\deg f(X)$ zeros.
21. Let $f(X) \in F[X]$ where F is a field. Assume $f(X)$ is of degree 2 or 3. Prove $f(X)$ is irreducible in $F[X]$ if and only if $f(X)$ has no zero in F.

CHAPTER EXERCISES

1. Let R be a ring. An element $e \in R$ is called an ***idempotent*** if $e^2 = e$. Prove the following:
   (a) Every ring has at least one idempotent.
   (b) Let e be an idempotent. Prove the following:
      (i) $1 - e$ is an idempotent.
      (ii) If R is an integral domain, then $R \cong eR \times (1 - e)R$.
2. Let R be a ring. Prove that $(R^R, +, \cdot)$ is a ring.
3. Prove the following about integral domains.
   (a) A subring of an integral domain is an integral domain.
   (b) A finite integral domain is a field.
4. Let R be a commutative ring with unity and I a proper ideal. Prove R/I is an integral domain if and only if I is prime.
5. Let $\psi : R \to R'$ be a ring isomorphism. Show that if a and b are zero divisors in R, then $\phi(a)$ and $\phi(b)$ are zero divisors in R'.
6. If R is an integral domain, then R satisfies the ***ascending chain condition*** on principal ideals (***ACC***) if for every chain of principal ideals

$$\langle a_1 \rangle \subseteq \langle a_2 \rangle \subseteq \langle a_3 \rangle \subseteq \cdots,$$

   there exists n such that $\langle a_i \rangle = \langle a_n \rangle$ for all $i \geq n$. A ring that satisfies the ACC is sometimes called a ***Noetherian*** ring.* Prove that every principal ideal domain has the ACC.
7. Let S and T be ideals of a ring R. Define

$$ST = \Big\{\sum_{i=1}^{n} s_i t_i : s_i \in S,\ t_i \in T,\ \text{and } n \in \mathbb{Z}^+\Big\}.$$

   ST is called the ***product*** of S and T. Prove:
   (a) $ST \subseteq S \cap T$.
   (b) ST is an ideal of R.
8. Use Zorn's Lemma (Chapter Exercise 6.36) to prove that every ring has a maximal ideal.

---

*These rings are named after Emmy Noether (Erlangen, Germany, 1882 – Bryn Mawr, Pennsylvania, 1935). She made contributions to abstract algebra, particularly noncommutative algebra, and to the theory of general relativity.

9. Let $\phi\colon R \to R'$ and $\psi\colon R \to R'$ be functions. Define $\gamma\colon R \times R' \to S \times S'$ by $\gamma(r, r') = (\phi(r), \psi(r'))$. Show the following:
   (a) If ϕ and ψ are isomorphisms, then γ is an isomorphism.
   (b) Find the kernel and image of γ.
10. Let $\phi\colon M_{2,2}(\mathbb{R}) \to \mathbb{R}[X]$ be defined by

$$\phi\left(\begin{bmatrix} a & b \\ c & d \end{bmatrix}\right) = (a - d) + (c - b)X.$$

   Either prove that ϕ a homomorphism or show that it is not one.
11. Let P be a prime ideal in R' and $\phi\colon R \to R'$ a ring homomorphism.
   (a) Show that $\phi^{-1}[P]$ is prime.
   (b) Give an example to show that $\phi^{-1}[P]$ may not be maximal even if P is.
12. Let $\phi\colon R \to R'$ be a ring isomorphism. Let P be a prime ideal of R.
   (a) Prove that $\phi[P]$ is a prime ideal of R'.
   (b) Is $\phi[P]$ still prime if ϕ is only a homomorphism? Explain.
13. Let R and R' be rings. If $\phi\colon R \to R'$ is an onto homomorphism, then there exists an isomorphism $\psi\colon R/\ker(\phi) \to R'$. This is known as the ***Fundamental Homomorphism Theorem***. (Hint: Use the function $\gamma\colon R \to R/\ker(\phi)$ defined by $\gamma(a) = a + \ker(\phi)$. Define ψ so that $\phi = \psi \circ \gamma$.)
14. Let R be a ring and $K \subseteq I$ ideals of R. Define $\phi\colon R/K \to R/I$ by $\phi(a+K) = a+I$ for all $a \in R$. Prove that ϕ is an onto homomorphism with kernel equal to I/K.
15. We have already seen in Exercise 7.5.7 that

$$I = \{f(X) \in \mathbb{Q}[X] : f(X) = a_1X + a_2X^2 + a_3X^3 + \cdots\}$$

   is an ideal. Prove the following:
   (a) $\varepsilon_0\colon \mathbb{Q}[X] \to \mathbb{Q}$ is a surjection.
   (b) $\ker(\varepsilon_0) = I$.
   (c) $\mathbb{Q}[X]/I \cong \mathbb{Q}$.
16. Use the Fundamental Homomorphism Theorem to prove the following:
   (a) If K and I are ideals of a ring R, then

$$(K + I)/I \cong K/K \cap I.$$

   (b) If $K \subseteq I$ are ideals of a ring R, then

$$R/I \cong R/K \big/ I/K.$$

17. Prove that if $\mathbb{Z}_n$ ($n > 1$) is a field, then n must be prime.
18. Let p be prime. Prove that a finite field of cardinality p is isomorphic to $\mathbb{Z}/p\mathbb{Z}$ and thus also to $\mathbb{Z}_p$.

19. Let $\mathcal{S}$ be a family of subrings of a ring R. Prove that $\cong$ is an equivalence relation on $\mathcal{S}$.
20. Let R be a ring.
    (a) Describe the elements of the polynomial ring $R[X][Y]$.
    (b) Prove $R[X][Y] \cong R[Y][X]$.
21. Assume F is a field. Let $p(X) \in F[X]$ be irreducible. (See Exercise 7.2.31.) Show that for all $f(X)$, $g(X) \in F[X]$, if $p(X)$ divides $f(X)g(X)$, then $p(X) \mid f(X)$ or $p(X) \mid g(X)$. (Note: $p(X) \mid f(X)$ means there exists $q(X) \in F[X]$ such that $f(X) = p(X)q(X)$.)
22. Let $p(X)$ be an irreducible polynomial in $F[X]$ where F is a field. Suppose $f_1(X), \ldots, f_k(X) \in F[X]$ for some $k \in \mathbb{Z}^+$. Prove if
$$p(X) \mid f_1(X) \cdots f_k(X),$$
then $p(X) \mid f_i(X)$ for some $i = 0, \ldots, k$.
23. Prove the Unique Factorization Theorem for Polynomials: Let F be a field. Take $f(X) \in F[X]$ such that $f(X)$ is not a constant polynomial. Then $f(X)$ can be written uniquely in the form
$$f(X) = up_1(X) \cdots p_k(X)$$
where each $p_i(X)$ is irreducible and u is a unit of F. (What should *uniquely* mean in this case?)
24. Prove that $\mathbb{R}[X]/\langle X^2 + 1\rangle$ is a field using Exercise 7.2.31.

8 Topology

From calculus we know that if a sequence a_n has a limit a, then as n increases, a_n becomes closer and closer to a. This notion is formalized and studied in analysis (advanced calculus). To describe what it means for a sequence to approach a limit, we need the notion of an open set. This is the focus of topology. In this chapter we will look at both open and closed sets. We will learn about a set's interior and its closure. We will study functions that preserve open sets by looking at isometries, and the chapter will then finish with a look at convergence.

8.1. SPACES

Before we begin, let us prove the best friend of analysis students.

8.1.1. Triangle Inequality

If $a, b \in \mathbb{R}$, then $|a + b| \leq |a| + |b|$.

Proof. Take $a, b \in \mathbb{R}$. By definition,

$$-|a| \leq a \leq |a|$$

and

$$-|b| \leq b \leq |b|.$$

Adding these inequalities yields

$$-(|a| + |b|) \leq a + b \leq |a| + |b|.$$

Therefore by Exercise 2.6.11,

$$|a + b| \leq |a| + |b|. \blacksquare$$

Often when we think of sets, we picture them as existing in a plane or in three-dimensions. These are familiar places. We choose an origin and define a set of axes. After assigning ordered pairs or ordered triples to the points, we treat the geometry algebraically. We also know how to add and multiply here. We can find distances between points and lengths of line segments. As when we generalized the properties of $\mathbb{Z}$ into the definition of a ring, we will isolate some of the important properties of $\mathbb{R}^2$ and $\mathbb{R}^3$ and study the resulting definitions. When sets are studied by placing properties on their elements so that a structure is given to the set, the set is referred to as a ***space***. In a sense, rings are spaces, although the term is typically used in analysis as opposed to algebra.

We will be looking at three types of spaces. The first one generalizes the idea of distance.

8.1.2. Definition

Take a set M and a function $d\colon M \times M \to \mathbb{R}$ such that for all $x, y, z \in M$:

1. $d(x, y) \geq 0$,
2. $d(x, y) = 0$ if and only if $x = y$,
3. $d(x, y) = d(y, x)$, and
4. $d(x, y) \leq d(x, z) + d(z, y)$ (Triangle Inequality).

The pair (M, d) is called a ***metric space***, and the function d is called a ***metric***.

There are many examples of metric spaces. The first two will be familiar.

Example. The following are well-known metric spaces.

1. $(\mathbb{R}, d)$ where $d(x, y) = |x - y|$—A routine check of the conditions using the definition of absolute value and the Triangle Inequality show that this is a metric space.
2. $(\mathbb{R}^2, d)$ where if $X = (x_1, y_1)$ and $Y = (x_2, y_2)$,

$$d(X, Y) = \sqrt{(x_2 - x_1)^2 + (y_2 - y_1)^2}$$

—Again, the conditions are easily verified. Propositions 1 and 3 follow quickly from the definition of d. The proof of 2 is left as Exercise 2. The last one is the algebraic version of Euclid's Triangle Inequality as seen in Figure 8.1.3.

Each of these distance functions is known as a ***standard metric***.

Intuitively, a ***continuous*** function on an interval has the property that it can be graphed over the interval without lifting the pencil. Define $C[a, b]$ to be the

8.1.3. Figure

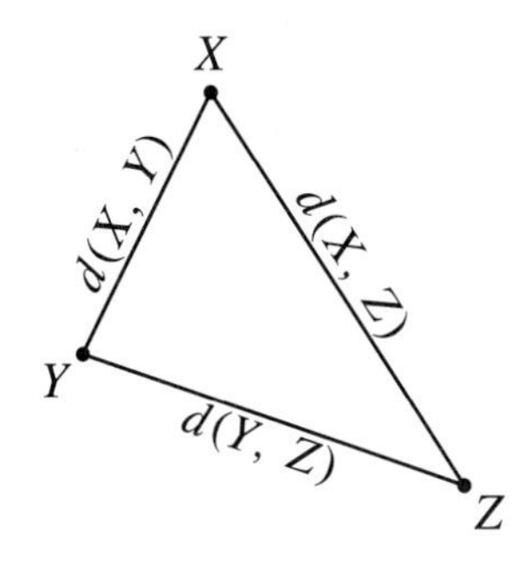

The Triangle Inequality:
$d(X, Y) + d(Y, Z) \leq d(X, Z)$.

set of all functions continuous on $[a, b]$. For example, if

$$f(x) = \begin{cases} x & \text{if } x \geq 0 \\ 2 & \text{if } x < 0, \end{cases}$$

then $f \in C[1, 2]$ despite the discontinuity at $x = 0$. (See Figure 8.1.4.) The set $C[a, b]$ will serve as the space in the next example. It may seem odd at first because the "points" of the space are functions. However, this is common in mathematics. Often the objects of study are viewed as points in a set.

Example. Let $f, g \in C[0, 1]$. Define,

$$d(f, g) = \int_0^1 |f(x) - g(x)|\, dx.$$

This is a natural definition for the distance between two functions, for this integral is the area between f and g. Take, for instance, $f(x) = x^2$ and $g(x) = x$. Then,

$$\begin{aligned} d(f, g) &= \int_0^1 |x^2 - x|\, dx \\ &= \int_0^1 (x - x^2)\, dx \\ &= (x^2/2 - x^3/3)\Big|_0^1 \\ &= 1/6. \end{aligned}$$

8.1.4. Figure

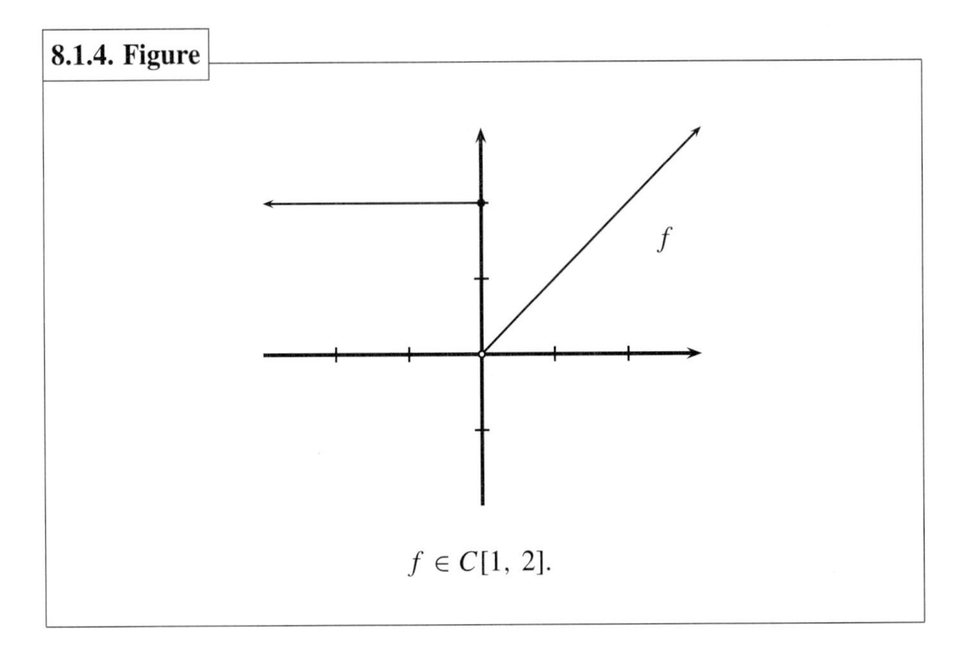

$f \in C[1, 2]$.

Now check that this is a metric space. The properties of the integral quickly satisfy conditions 1 through 3 of the definition. As for condition 4, take $h \in C[0, 1]$ and calculate:

$$\begin{aligned} d(f, g) &= \int_0^1 |f(x) - g(x)|\,dx \\ &= \int_0^1 |f(x) - h(x) + h(x) - g(x)|\,dx \\ &\leq \int_0^1 (|f(x) - h(x)| + |h(x) - g(x)|)\,dx \\ &= \int_0^1 |f(x) - h(x)|\,dx + \int_0^1 |h(x) - g(x)|\,dx \\ &= d(f, h) + d(h, g). \end{aligned}$$

The inequality holds due to the Triangle Inequality (8.1.1). Therefore, $(C[0, 1], d)$ is a metric space.

Since we now have a distance function, it is natural that we would want to calculate lengths. To do so we first need some operations. Let $(R, +, \cdot)$ be a ring. If we look at the space $(R, +)$, we have a structure called a ***group***. We can define a group as a set with one binary operation. The operation is associative, has an identity, and every element has an inverse. In the case of our ring, the operation $+$ is also commutative. Hence, $(R, +)$ is what is called an ***abelian group***. It is named after Niels Abel who studied such groups.*

> **Example.** $(\mathbb{Z}, +)$ is an example of an abelian group. If F is the set of all invertible functions $\mathbb{R} \to \mathbb{R}$ and $\circ$ is composition, then $(F, \circ)$ is an example of a nonabelian group.

Now for our definition.

8.1.5. Definition

Let F be a field. Take an abelian group $(V, +)$. On this set, define an operation

$$F \times V \to V,$$

called ***scalar multiplication***, such that for all $\alpha, \beta \in F$ and $u, v \in V$:

1. $(\alpha + \beta)u = \alpha u + \beta u$,
2. $\alpha(u + v) = \alpha u + \alpha v$,
3. $(\alpha\beta)u = \alpha(\beta u)$, and
4. $1u = u$.

The triple $(V, +, \cdot)$, where $\cdot$ represents the scalar multiplication, is called a ***vector space*** (over F). In this context, the elements of V are called ***vectors*** and F is the field of ***scalars***. The addition is called ***vector addition***.

We follow the tradition of naming scalars with Greek letters and vectors with Roman. (Note: if F is a ring with unity and not necessarily a field, then $(V, +, \cdot)$ is called a ***left F-module***.)

The motivation for this definition comes from the study of $\mathbb{R}^2$. Each ordered pair (x, y) can be considered as an arrow called a ***vector*** originating at the origin and pointing to (x, y). (See Figure 8.1.6.) The vector addition is defined by

$$(x_1, y_1) + (x_2, y_2) = (x_1 + x_2, y_1 + y_2)$$

and the scalar multiplication by

$$\alpha(x, y) = (\alpha x, \alpha y).$$

---

*Niels Abel (Finnøy, Norway, 1802 – Froland, Norway, 1829): Abel proved that an arbitrary fifth-degree (quintic) polynomial function is insolvable. This means that there is no finite sequence of steps using only addition, subtraction, multiplication, division, and taking of roots that will lead to the zeros of such a polynomial.

8.1.6. Figure

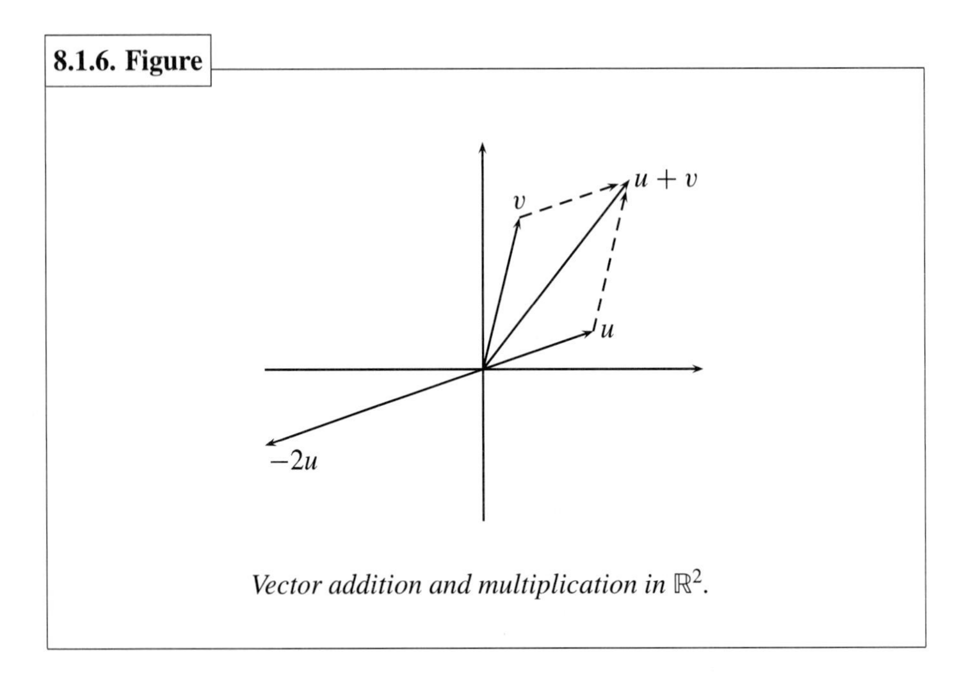

Vector addition and multiplication in $\mathbb{R}^2$.

According to Exercise 9a, $(\mathbb{R}^2, +)$ is an abelian group. To check that the multiplication is a scalar multiplication, check the conditions. We will only do the first: let $\alpha,\ \beta \in \mathbb{R}$ and $(x,\ y) \in \mathbb{R}^2$. Then,

$$\begin{aligned}[\alpha + \beta](x,\ y) &= ([\alpha + \beta]x,\ [\alpha + \beta]y) \\ &= (\alpha x + \beta x,\ \alpha y + \beta y) \\ &= (\alpha x,\ \alpha y) + (\beta x,\ \beta y) \\ &= \alpha(x,\ y) + \beta(x,\ y).\end{aligned}$$

After checking the other conditions we can conclude that $(\mathbb{R}^2, +, \cdot)$ is a vector space.

Example. Although not often thought of as vectors, the field $(\mathbb{R}, +, \cdot)$ can be viewed as a vector space with scalar multiplication defined as regular multiplication of reals. Generalizing, $\mathbb{R}^n$ can be viewed as a vector space over $\mathbb{R}$ for all positive integers n.

Example. To view $C[0,\ 1]$ as a vector space, use function addition and scalar multiplication (page 235). This makes $(C[0,\ 1], +)$ into an abelian group. The check of the conditions is left as an exercise.

We are now ready to define our generalization of length.

8.1.7. Definition

Let V be a vector space over $\mathbb{R}$. Define a function that assigns to every $v \in V$ a unique real number denoted by $\|v\|$. This function is called a ***norm***, and V is called a ***normed space*** if for all $u, v \in V$ and $\alpha \in \mathbb{R}$:

1. $\|v\| \geq 0$,
2. $\|v\| = 0$ if and only if $v = 0$,
3. $\|\alpha v\| = |\alpha| \, \|v\|$, and
4. $\|u + v\| \leq \|u\| + \|v\|$ (Triangle Inequality).

Example. Let $n \in \mathbb{Z}^+$. For $(x_1, \ldots, x_n) \in \mathbb{R}^n$, define

$$\|(x_1, \ldots, x_n)\| = \sqrt{x_1^2 + \cdots + x_n^2}.$$

It is a routine exercise to check that this is a norm on $\mathbb{R}^n$. We will refer to it as the ***standard norm***. Notice that when $n = 1$, $\|x\| = |x|$.

Example. Let $f \in C[0, 1]$. To define a norm for $C[0, 1]$, let

$$\|f\| = \int_0^1 |f(x)| \, dx.$$

For instance, if $f(x) = 2x - 2$, then

$$\|f\| = \int_0^1 |2x - 2| \, dx = \int_0^1 (-2x + 2) \, dx = (-x^2 + 2x)\Big|_0^1 = 1.$$

In the Euclidean plane, we can use the standard norm to calculate distances. Let $X = (x_1, y_1)$ and $Y = (x_2, y_2)$. Then,

$$\begin{aligned} d(X, Y) &= \sqrt{(x_1 - x_2)^2 + (y_1 - y_2)^2} \\ &= \|(x_1 - x_2, y_1 - y_2)\| \\ &= \|(x_1, y_1) - (x_2, y_2)\| \\ &= \|X - Y\|. \end{aligned}$$

(Note: $X - Y = (x_1, y_1) - (x_2, y_2) = (x_1 - x_2, y_1 - y_2)$.) To generalize, given any normed space V, we can view it as a metric space using

$$d(x, y) = \|x - y\|$$

for all $x, y \in V$. To see that this is a metric, check the four conditions. We will show the last one to illustrate a very typical use of the Triangle Inequality.

Take $x, y, z \in V$. Then we have:

$$\begin{aligned} d(x, y) &= \|x - y\| \\ &= \|x - z + z - y\| \\ &\leq \|x - z\| + \|z - y\| \\ &= d(x, z) + d(z, y). \end{aligned}$$

(This metric is called the ***induced metric***.)

Example. The set of complex numbers can be viewed as a normed space by defining the ***absolute value***: if $z = a + bi$ for some $a, b \in \mathbb{R}$,

$$|z| = \sqrt{a^2 + b^2}.$$

This induced metric is confirmation that $\mathbb{C}$ looks like $\mathbb{R}^2$.

EXERCISES

1. Prove that for all $a, b \in \mathbb{R}$, $|a| - |b| \leq |a - b|$.

2. For all $(x_1, y_1), (x_2, y_2) \in \mathbb{R}^2$, prove

$$\sqrt{(x_2 - x_1)^2 + (y_2 - y_1)^2} = 0$$

if and only if $(x_1, y_1) = (x_2, y_2)$.

3. For all $x, y \in \mathbb{R}$, define the metric

$$d(x, y) = |x - y|.$$

Show that $(\mathbb{R}, d)$ is a metric space.

4. If d is the standard metric from $\mathbb{R}^2$, prove $(\mathbb{Z} \times \mathbb{Z}, d)$ is a metric space.

5. Use the metric defined on page 327 to find the distance between the functions $f(x) = x^3$ and $g(x) = 2x + 1$.

6. Let $M \neq \varnothing$ and for all $a, b \in M$ define

$$d(a, b) = \begin{cases} 0 & \text{if } a = b \\ 1 & \text{if } a \neq b. \end{cases}$$

Show that (M, d) is a metric space. (This is called the ***trivial metric***.)

7. Let (M, d) be a metric space. The following function is known as the ***bounded metric***: for all $x, y \in M$, define

$$d'(x, y) = \frac{d(x, y)}{1 + d(x, y)}.$$

Prove that (M, d') is a metric space.

8. Assume that both (M_1, d_1) and (M_2, d_2) are metric spaces. Prove that the following functions are metrics on $M_1 \times M_2$:
   (a) $d((a, b), (a', b')) = \max(d_1(a, a'), d_2(b, b'))$
   (b) $d((a, b), (a', b')) = d_1(a, a') + d_2(b, b')$
9. Prove that the following are abelian groups.
   (a) $(\mathbb{R}^2, +)$
   (b) $(n\mathbb{Z}, +)$ for all $n \in \mathbb{Z}$
   (c) $(\mathrm{M}_{2,2}(\mathbb{R}), +)$
   (d) $(\mathbb{R} \setminus \{0\}, \cdot)$
   (e) $(\{1, -1\}, \cdot)$
   (f) $(\mathbb{Z}_n, +)$ for all $n \in \mathbb{Z}^+$
10. Indicate whether each of the following is a vector space over $\mathbb{R}$. If it is not, then list which conditions fail. Assume the usual definitions of vector addition and scalar multiplication for each set.
   (a) $\{(x, 2x + 1) : x \in \mathbb{R}\}$
   (b) $\{(x, y, 0) : x, y \in \mathbb{R}\}$
   (c) $\{ax^2 + bx + c : b + c = 1 \text{ and } a, b, c \in \mathbb{R}\}$
   (d) $\left\{\begin{bmatrix} a & b \\ c & d \end{bmatrix} : a + b + c + d = 1 \text{ and } a, b, c, d \in \mathbb{R}\right\}$
11. Finish the proof on page 330 that $(\mathbb{R}^2, +, \cdot)$ is a vector space.
12. Show that $(\mathrm{M}_{2,2}(\mathbb{R}), +, \cdot)$ is a vector space over $\mathbb{R}$.
13. Prove $(\mathbb{R}^{\mathbb{R}}, +, \cdot)$ is a vector space over $\mathbb{R}$ where $+$ is function addition and the scalar multiplication is defined as

$$af(x) = a[f(x)]$$

   for every $a \in \mathbb{R}$.
14. Fix $n \in \mathbb{Z}^+$. Prove that

$$\|(x_1, \dots, x_n)\| = \sqrt{x_1^2 + \cdots + x_n^2}$$

   defines a norm on $\mathbb{R}^n$.
15. Define the following for $\mathbb{Z} \times \mathbb{Z}$: for all $m, n \in \mathbb{Z}$,

$$\|(m, n)\| = |m| + |n|.$$

   (a) Prove that this is a norm on $\mathbb{Z} \times \mathbb{Z}$.
   (b) This norm is called the ***taxicab norm*** because of its induced metric. Let $X = (x_1, y_1)$ and $Y = (x_2, y_2)$. Prove that the induced metric for this norm is $d(X, Y) = |x_1 - x_2| + |y_1 - y_2|$.
   (c) Why is this norm called the taxicab norm?
16. Let $\|f\| = \int_0^1 |f(x)|\, dx$. Show that this is a norm on $C[0, 1]$.
17. Prove that the absolute value of complex numbers as defined on page 332 defines a norm for $\mathbb{C}$.

8.2. OPEN SETS

The basic object in topology is the open set. It is needed to define *limit* and *continuous*. Crucial to the definition of an open set is the following:

8.2.1. Definition

Let (M, d) be a metric space and take $x \in M$. For all $\varepsilon > 0$, define

$$D(x, \varepsilon) = \{y \in M : d(x, y) < \varepsilon\}.$$

This set is called an ***open disk*** (about y in M).

If the metric space M is unclear from context, then write $D_M(x, \varepsilon)$ for $D(x, \varepsilon)$.

For the familiar spaces of $\mathbb{R}$ and $\mathbb{R}^2$:

- If $M = \mathbb{R}$ with the standard metric,

$$D(x, \varepsilon) = (x - \varepsilon, x + \varepsilon).$$

- If $M = \mathbb{R}^2$ with the standard metric, $D(X, \varepsilon)$ is the inside of a circle excluding its circumference:

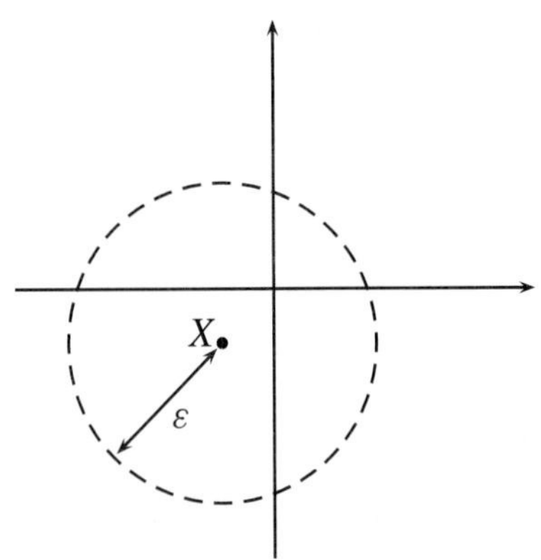

In the picture the dashed lines represent the exclusion of the circumference of the circle, like open circles depict the omission of the endpoints of an interval.

The open disk will allow us to make our main definition.

8.2.2. Definition

Let M be a metric space. A subset U of M is called ***open*** if

for all $x \in U$, there exists $\varepsilon > 0$ so that $D(x, \varepsilon) \subseteq U$.

It is interesting to note that both the metric space M and $\varnothing$ are open. By definition, $D(x,\, \varepsilon)$ is a subset of M for any $x \in M$ and $\varepsilon > 0$. As for the empty set, the proposition

$$(\forall x \in \varnothing)(\exists \varepsilon > 0)[D(x,\, \varepsilon) \subseteq \varnothing]$$

is true because it is equivalent to

$$(\forall x)(x \in \varnothing \Rightarrow (\exists \varepsilon > 0)[D(x,\, \varepsilon) \subseteq \varnothing]).$$

This is true since $x \in \varnothing$ is always false.

Example. Let $(a,\, b)$ be an open interval in $\mathbb{R}$. To prove that it is open and deserving of its name, let $x \in (a,\, b)$. We must isolate x inside an open interval within $(a,\, b)$. To do this, find the distance from x to each endpoint:

$$\vdash x - a \dashv\vdash b - x \dashv$$
$$a \qquad x \qquad b$$

Now let $\varepsilon = \min(x - a,\, b - x)$. If $x - \varepsilon < y < x + \varepsilon$, then

$$a = x - (x - a) \le x - \varepsilon < y < x + \varepsilon \le x + (b - x) = b.$$

Therefore,

$$D(x,\, \varepsilon) = (x - \varepsilon,\, x + \varepsilon) \subseteq (a,\, b).$$

This is illustrated by the next diagram with the assumption that $\varepsilon = x - a$:

$$\vdash x - a \dashv\vdash x - a \dashv$$
$$a \qquad x \qquad b$$

We can use the strategy of the example on other sets. Let A equal the set $(1,\, 2) \cup (4,\, 7)$. To show that A is open, take $x \in A$. In other words, $1 < x < 2$ or $4 < x < 7$. Define:

$$\varepsilon = \begin{cases} \min(x - 1,\, 2 - x) & \text{if } 1 < x < 2 \\ \min(x - 4,\, 7 - x) & \text{if } 4 < x < 7. \end{cases}$$

With this definition, $(x - \varepsilon,\, x + \varepsilon) \subseteq A$.

To show that a set A is not open in a metric space M, we must prove that there exists $x \in U$ so that for all $\varepsilon > 0$, $D(x,\, \varepsilon) \not\subseteq A$.

Example. For this example, see Figure 8.2.3. Let $a,\, b \in \mathbb{R}$ so that $a < b$. The set $A = \{(x,\, y) \in \mathbb{R}^2 : a \le x < b\}$ is not open in $\mathbb{R}^2$. To prove this, we must find a point in A that cannot be isolated in a disk. Any point on the left border will do, so take $(a,\, 3)$. Then for all $\varepsilon > 0$, $D((a,\, 3),\, \varepsilon) \setminus A \neq \varnothing$ because $(a - \varepsilon/2,\, 3)$ is a member of $D((a,\, 3),\, \varepsilon)$ but not of A. There is no trouble with the right border, however, for it is not part of A. In fact, the set $\{(x,\, y) \in \mathbb{R}^2 : a < x < b\}$ is open.

8.2.3. Figure

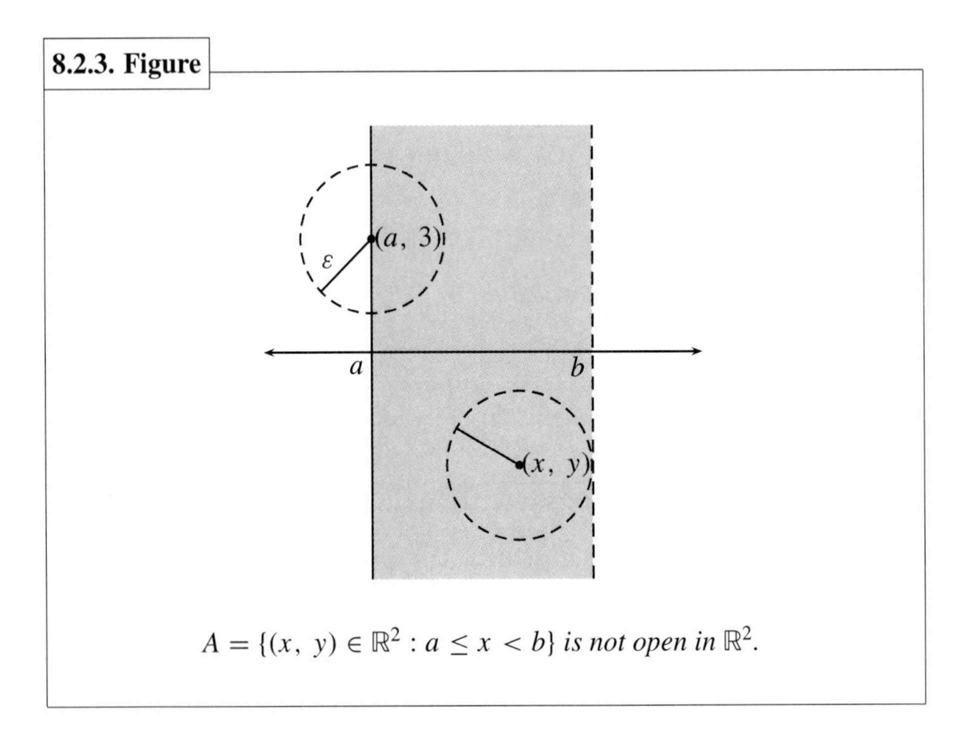

$A = \{(x, y) \in \mathbb{R}^2 : a \leq x < b\}$ *is not open in* $\mathbb{R}^2$.

Often a set will be open in one metric space but not open in another.

Example. Consider the set of points that lie on the real line between 1 and 2. In $\mathbb{R}$ this set is the interval $(0, 1)$, which is open. However, in $\mathbb{R}^2$ this set is $\{(x, 0) : 0 < x < 1, x \in \mathbb{R}\}$ and is not open, for it cannot contain a two-dimensional disk.

Now that we have seen examples of open and non-open sets, we should prove that the disk is actually open and justify its name. The proof is a generalization of the fact that (a, b) is open in $\mathbb{R}$.

8.2.4. Theorem

For any x in a metric space M, $D(x, \varepsilon)$ is open.

Proof. We must show that we can isolate any element of $D(x, \varepsilon)$ inside an open disk. So let $y \in D(x, \varepsilon)$. The open disk about y that we want is

$$D(y, \varepsilon - d(x, y)),$$

as illustrated by the diagram in Figure 8.2.5. To show that this disk is included in $D(x, \varepsilon)$, let $z \in D(y, \varepsilon - d(x, y))$. This means

$$d(y, z) < \varepsilon - d(x, y),$$

8.2.5. Figure

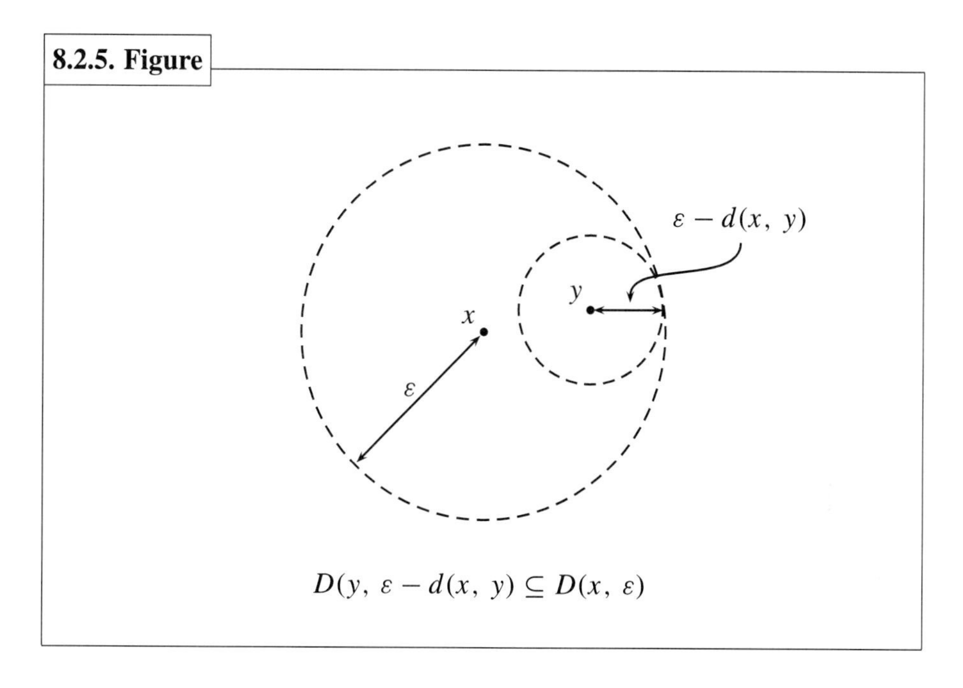

$D(y, \varepsilon - d(x, y) \subseteq D(x, \varepsilon)$

and hence

$$d(x, y) + d(y, z) < \varepsilon.$$

However, the Triangle Inequality states $d(x, z) \leq d(x, y) + d(y, z)$. Thus, $d(x, z) < \varepsilon$ and $z \in D(x, \varepsilon)$. ■

An application of the definition is one way to prove that a set is open. Other ways include the last two theorems of the section.

8.2.6. Theorem

Let M be a metric space.

1. The union of a collection of open sets is open.
2. The intersection of a finite collection of open sets is open.

Proof. Let $\mathscr{F} = \{U_\alpha : \alpha \in \Lambda\}$ be a family of open sets.

1. Take $x \in \bigcup \mathscr{F}$. We must find $\varepsilon > 0$ so that $D(x, \varepsilon) \subseteq \bigcup \mathscr{F}$. Now, there exists $\alpha \in \Lambda$ such that $x \in U_\alpha$. Since U_α is open, we have $\varepsilon > 0$ so that $D(x, \varepsilon) \subseteq U_\alpha$, but this means $D(x, \varepsilon) \subseteq \bigcup \mathscr{F}$.
2. Next, let $x \in \bigcap \mathscr{F}$ and suppose that Λ is finite. Since the names of the elements of the index set are not important, suppose $\Lambda = \{1, 2, \ldots, k\}$ for some $k \in \mathbb{Z}^+$. By definition, $x \in U_\alpha$ for all $\alpha \in \Lambda$.

Since each of these is open, there exists $\varepsilon_\alpha > 0$ so that $D(x, \varepsilon_\alpha)$ is a subset of U_α. Define $\varepsilon = \min(\varepsilon_1, \ldots, \varepsilon_k)$. Then $\varepsilon \le \varepsilon_\alpha$ for all $\alpha \in \Lambda$, and this means by Exercise 2 that

$$D(x, \varepsilon) \subseteq D(x, \varepsilon_\alpha).$$

Hence, $D(x, \varepsilon)$ is a subset of the intersection. ■

Upon first reading it may not be clear where the finiteness condition was used in the last proof. It was crucial in the definition of ε in part 2 of the proof. If there were not only finitely many members of the family, the minimum might not exist.

Example. To see that the finiteness condition is necessary for the intersection of open sets to be open, consider $\{(-1/n, 1/n) : n \in \mathbb{Z}^+\}$. This is a family of open sets in $\mathbb{R}$. However,

$$\bigcap_{i=1}^{\infty}(-1/n, 1/n) = \{0\},$$

which is not open.

The next definition relies on Theorem 8.2.6.

8.2.7. Definition

Take a set A and let

$$\mathcal{F} = \{U : U \text{ is open and } U \subseteq A\}.$$

The ***interior*** of A is $\bigcup \mathcal{F}$ and is denoted by $\operatorname{int}(A)$.

Since it is the union of open sets, the interior of a set A is open. Moreover, it must be the largest open set in A. (See Exercise 17.)

Example. In $\mathbb{R}$:

1. $\operatorname{int}([0, 1]) = (0, 1)$,
2. $\operatorname{int}(\mathbb{Z}) = \varnothing$.

The last equality holds because the empty set is the only open subset of $\mathbb{Z}$.

Example. In $\mathbb{R}^2$:

1. $\operatorname{int}([0, 1] \times [2, 4]) = (0, 1) \times (2, 4)$,
2. $\operatorname{int}(\{(x, y) \in \mathbb{R}^2 : xy \le 1\}) = \{(x, y) \in \mathbb{R}^2 : xy < 1\}$. (See Figure 8.2.8.)

8.2.8. Figure

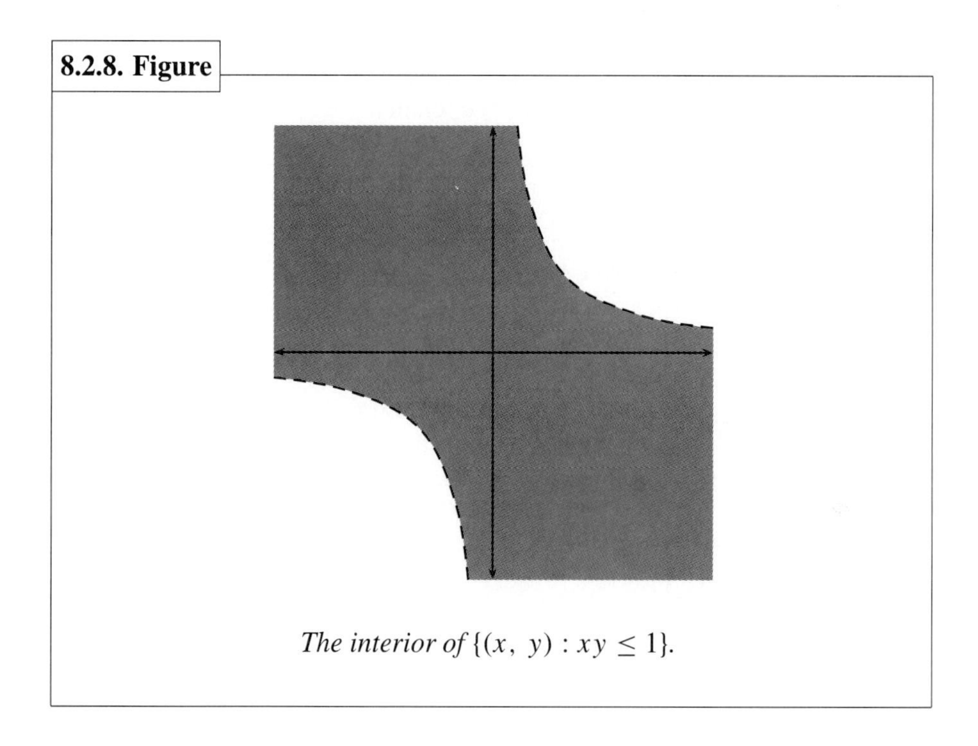

The interior of $\{(x, y) : xy \leq 1\}$.

One way to identify the points in a set's interior is to use the next theorem.

8.2.9. Theorem

If A is a set, then

$$x \in \text{int}(A) \text{ if and only if there exists } \varepsilon > 0 \text{ so that } D(x, \varepsilon) \subseteq A.$$

Proof. Let $\{U_\alpha : \alpha \in \Lambda\}$ be the family of open subsets of A.

($\Rightarrow$) Let $x \in \text{int}(A)$. This means $x \in U_\alpha$ for some $\alpha \in \Lambda$. Since U_α is open, there exists $\varepsilon > 0$ so that $D(x, \varepsilon) \subseteq U_\alpha$. Since $U_\alpha \subseteq A$, $D(x, \varepsilon) \subseteq A$.

($\Leftarrow$) If $D(x, \varepsilon) \subseteq A$ for some $\varepsilon > 0$, x must be an element of the interior, for $D(x, \varepsilon)$ is open. ■

Example. Viewing $\mathbb{Z}$ as a metric space with the standard metric, the interior of $2\mathbb{Z}$ is itself. To see this, let $n \in \mathbb{Z}$. Then,

$$D_{\mathbb{Z}}(2n, 1) = \{m \in \mathbb{Z} : |2n - m| < 1\} = \{2n\}.$$

Therefore, $2\mathbb{Z} \subseteq \text{int}(2\mathbb{Z})$, and since we already have the other inclusion, $\text{int}(2\mathbb{Z}) = 2\mathbb{Z}$. This means that $2\mathbb{Z}$ is open in $\mathbb{Z}$.

EXERCISES

Assume that M denotes a metric space.

1. Prove for all $x, y \in M$ such that $x \neq y$, there exists $\varepsilon > 0$ such that $D(x, \varepsilon) \cap D(y, \varepsilon) = \varnothing$.

2. Suppose $\varepsilon_1 \leq \varepsilon_2$. Prove for any $x \in M$, $D(x, \varepsilon_1) \subseteq D(x, \varepsilon_2)$.

3. Prove that the following are open in the indicated metric spaces. Assume the usual metric for each space.
   (a) $\mathbb{R} \setminus \{0\}$ in $\mathbb{R}$
   (b) $\mathbb{R}^2 \setminus \{(0, 0)\}$ in $\mathbb{R}^2$
   (c) $\mathbb{R}^2 \setminus \{(x, y) \in \mathbb{R}^2 : y = 2x - 1\}$ in $\mathbb{R}^2$
   (d) $\{(x, y) \in \mathbb{R}^2 : 5 < x < 7\}$ in $\mathbb{R}^2$
   (e) $\{(x, y) \in \mathbb{R}^2 : x + y < 1\}$ in $\mathbb{R}^2$
   (f) $\{(x, y, z) \in \mathbb{R}^3 : x^2 + y^2 < 1\}$ in $\mathbb{R}^3$
   (g) $\{f \in C[0, 1] : \|f\| < 1\}$ in $C[0, 1]$

4. Prove every subset of M is open if and only if every singleton is open. (Recall that a singleton is a set with exactly one element.)

5. Prove $A = (1, 2) \cup (4, 7)$ is open by showing $(x - \varepsilon, x + \varepsilon) \subseteq A$ if

$$\varepsilon = \begin{cases} \min(x - 1, 2 - x) & \text{if } 1 < x < 2 \\ \min(x - 4, 7 - x) & \text{if } 4 < x < 7. \end{cases}$$

6. Let (M_1, d_1) and (M_2, d_2) be metric spaces with U_1 open in M_1 and U_2 open in M_2. Determine whether $U_1 \times U_2$ is open in $M_1 \times M_2$ with the following metrics:
   (a) $d((a, b), (a', b')) = \max(d_1(a, a'), d_2(b, b'))$
   (b) $d((a, b), (a', b')) = d_1(a, a') + d_2(b, b')$

7. Assuming that $\mathbb{R}$ is the metric space and the metric is the standard one, verify Theorem 8.2.6 by explicitly proving the following propositions.
   (a) $\bigcup_{n=1}^{\infty} D(0, 1/n)$ is open.
   (b) $\bigcap_{n=1}^{k} D(0, 1/n)$ is open for all $k \in \mathbb{Z}^+$.
   (c) $\bigcap_{n=1}^{\infty} D(0, 1/n)$ is not open.

8. Fix a positive integer n. For all $i \in \mathbb{Z}^+$, let $U_{i,j}$ be an open set in M, $j = 1, \ldots n$. Prove

$$\bigcup_{i=1}^{\infty} \bigcap_{j=1}^{n} U_{i,j}$$

is open in M.

9. Describe the interior of each of the given sets in $\mathbb{R}$:
   (a) $\{0\}$
   (b) $(0, 2]$
   (c) $[0, 2]$
   (d) $(0, 2) \cup \{-1, 3\}$
10. Describe the interior of each of the given sets in $\mathbb{R}^2$:
   (a) $\{(x, y) \in \mathbb{R}^2 : x^2 + y^2 < 1\}$
   (b) $\{(x, y) \in \mathbb{R}^2 : x^2 + y^2 \leq 1\}$
   (c) $\{(x, y) \in \mathbb{R}^2 : x \in \mathbb{Q}\}$
   (d) $\{(x, y) \in \mathbb{R}^2 : y \leq 1/x\}$
11. Use Theorem 8.2.9 to prove the following (assume the standard metric):
   (a) $\text{int}([0, 1]) = (0, 1)$ in $\mathbb{R}$
   (b) $\text{int}(\mathbb{Z}) = \varnothing$ in $\mathbb{R}$
   (c) $\text{int}([0, 1] \times [2, 4]) = (0, 1) \times (2, 4)$ in $\mathbb{R}^2$
   (d) $\text{int}(\{(x, y) : xy \leq 1\}) = \{(x, y) : xy < 1\}$ in $\mathbb{R}^2$
12. If d is the standard metric, then $(\mathbb{R}^2, d)$ is a metric space. Prove that $\text{int}(\mathbb{Z} \times \mathbb{Z}) = \varnothing$ in this metric space.
13. Prove: A is open if and only if $\text{int}(A) = A$.
14. Identify a metric space in which $\text{int}(\mathbb{Z}) \neq \varnothing$.
15. Prove the following about the interior of a set:
   (a) $\text{int}(A) \cup \text{int}(B) \subseteq \text{int}(A \cup B)$
   (b) $\text{int}(A) \cap \text{int}(B) = \text{int}(A \cap B)$
16. Find sets A and B such that $\text{int}(A \cup B) \not\subseteq \text{int}(A) \cup \text{int}(B)$.
17. Let U be an open subset of a set A. Prove $U \subseteq \text{int}(A)$, thus showing that the interior is the largest open subset of a set.

8.3. CLOSED SETS

The difference between an open interval and a closed one lies with the endpoints. Open intervals do not contain the endpoints, but closed intervals do. Similarly, while an open set will not contain its outer edge (whatever that means), closed sets do. We begin by defining what it means to be closed in an arbitrary metric space and then prove some results.

8.3.1. Definition

A set A in a metric space M is ***closed*** if $M \setminus A$ is open.

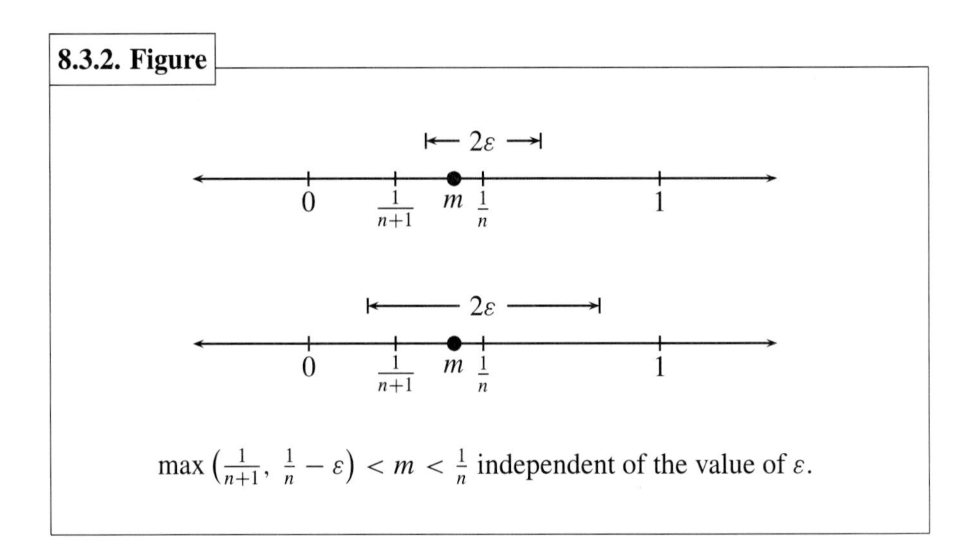

$\max\left(\frac{1}{n+1},\ \frac{1}{n}-\varepsilon\right) < m < \frac{1}{n}$ independent of the value of ε.

According to the definition, both M and $\varnothing$ are closed (see Exercise 1).

For the next examples, recall that if U is the universe, then $\overline{A} = U \setminus A$. When working within a metric space M, it becomes the universe. Hence, here $\overline{A} = M \setminus A$.

Example. Here are some examples of closed sets.

- $[a,\ b]$ is closed in $\mathbb{R}$ for any real numbers $a < b$.
- Since $D(x,\ \varepsilon)$ is open, $\overline{D(x,\ \varepsilon)}$ is closed.
- In $\mathbb{R}^2$, $\{(x,\ y) \in \mathbb{R}^2 : xy \geq 1\}$ is closed. (See Figure 8.2.8.)

There are sets that are neither open nor closed. In $\mathbb{R}$, $(0,\ 1]$ provides a simple example. The set $A = \{(x,\ y) \in \mathbb{R}^2 : a \leq x < b\}$ is such an example in $\mathbb{R}^2$. Here is a more interesting one.

Example. The set $A = \{1/n : n \in \mathbb{Z}^+\}$ is neither open nor closed in $\mathbb{R}$.

- To see that A is not open, let n be a positive integer and consider $D(1/n,\ \varepsilon) = (1/n - \varepsilon,\ 1/n + \varepsilon)$ for any $\varepsilon > 0$. As seen in Figure 8.3.2, there exists $m \in \mathbb{R}$ such that
$$\max\left(\frac{1}{n+1},\ \frac{1}{n} - \varepsilon\right) < m < \frac{1}{n},$$
so $D(1/n,\ \varepsilon) \not\subseteq A$. Hence, A is not open.
- To check that A is not closed, we show that $\overline{A}$ is not open. Notice $0 \in \overline{A}$. Suppose $\varepsilon > 0$. Then there exists $n \in \mathbb{Z}^+$ such that $1/n < \varepsilon$. Thus, $D(0,\ \varepsilon) \cap A \neq \varnothing$, which means $D(0,\ \varepsilon) \not\subseteq \overline{A}$.

We shall see shortly that there is another way to show that $\{1/n : n \in \mathbb{Z}^+\}$ is not closed.

To emphasize the point of the examples, remember that although the complement of an open set is closed and the complement of a closed set is open, the terms are not negations of each other. There are sets that are both open and closed, and there are sets that are neither. Hence, we cannot show that a set is closed by proving that it is not open. Similarly, we cannot show that a set is open by showing that it is not closed. Therefore:

$$\begin{aligned} A \text{ not open} &\not\Rightarrow A \text{ closed} \\ A \text{ not closed} &\not\Rightarrow A \text{ open} \\ A \text{ open} &\not\Rightarrow A \text{ not closed} \\ A \text{ closed} &\not\Rightarrow A \text{ not open} \end{aligned}$$

However, for the closed case the next theorems may help.

8.3.3. Theorem

Let M be a metric space.

1. The intersection of a collection of closed sets is closed.
2. The union of a finite collection of closed sets is closed.

Proof. Let $\{C_\alpha : \alpha \in \Lambda\}$ be a family of closed sets. To prove that the intersection $\bigcap_{\alpha\in\Lambda} C_i$ is closed we must show that its complement is open. By Theorem 3.5.4,

$$\overline{\bigcap_{\alpha\in\Lambda} C_\alpha} = \bigcup_{\alpha\in\Lambda} \overline{C_\alpha}.$$

But since each C_α is closed, $\overline{C_\alpha}$ is open. According to Theorem 8.2.6, we may conclude $\bigcup_{\alpha\in\Lambda} \overline{C_\alpha}$ is open. Similarly, if Λ is a finite index set equal to, say, $\{1, 2, \ldots, k\}$, then

$$C_1 \cup C_2 \cup \cdots \cup C_k$$

is closed because its complement

$$\overline{C_1} \cap \overline{C_2} \cap \cdots \cap \overline{C_k}$$

is open. ■

The largest open set within a set is its interior. Corresponding to this is the smallest closed set that is a superset of a given set. This is our next definition.

8.3.4. Definition

Let A be a set and

$$\mathcal{F} = \{F : F \text{ is closed and } A \subseteq F\}.$$

The ***closure*** of A is the set $\bigcap \mathcal{F}$ and is represented by $\mathrm{cl}(A)$.

Example. In $\mathbb{R}$, $\mathrm{cl}((0,\ 1)) = [0,\ 1]$ and $\mathrm{cl}(\mathbb{Z}) = \mathbb{Z}$.

Example. In $\mathbb{R}^2$,
- $\mathrm{cl}((0,\ 1) \times (2,\ 4)) = [0,\ 1] \times [2,\ 4]$,
- $\mathrm{cl}(\{(x,\ y) \in \mathbb{R}^2 : xy < 1\}) = \{(x,\ y) \in \mathbb{R}^2 : xy \leq 1\}$.

To understand the last equality examine Figure 8.2.8.

Example. It should not be surprising that a set A is closed in a metric space M if and only if $\mathrm{cl}(A) = A$. The proof of this is left as an exercise. We will, however, prove that the closure of a set is closed. To see this, examine the complement of $\mathrm{cl}(A)$:

$$\begin{aligned}\overline{\mathrm{cl}(A)} &= \overline{\bigcap\{F : F \text{ is closed and } A \subseteq F\}} \\ &= \bigcup\{\overline{F} : F \text{ is closed and } A \subseteq F\}.\end{aligned}$$

Since each F is closed, each $\overline{F}$ is open. Therefore, the last union is open.

The way to picture a closed set is to think of a set that includes its outer edge. The problem is that for some sets the "edge" is sometimes hard to imagine. For example, what points are on the "edge" of $\mathbb{Z} \times \mathbb{Z}$ in $\mathbb{R}^2$? We, therefore, introduce the next idea.

8.3.5. Definition

An element $x \in M$ is an ***accumulation point*** of A if for all $\varepsilon > 0$,

$$(D(x,\ \varepsilon) \setminus \{x\}) \cap A \neq \varnothing.$$

Accumulation points are sometimes called ***cluster points***. Basically, an accumulation point of a set A has the property that there are points in A that are arbitrarily close to it. (See Figure 8.3.6.) We exclude the accumulation point from the criterion since it is always very close to itself!

8.3.6. Figure

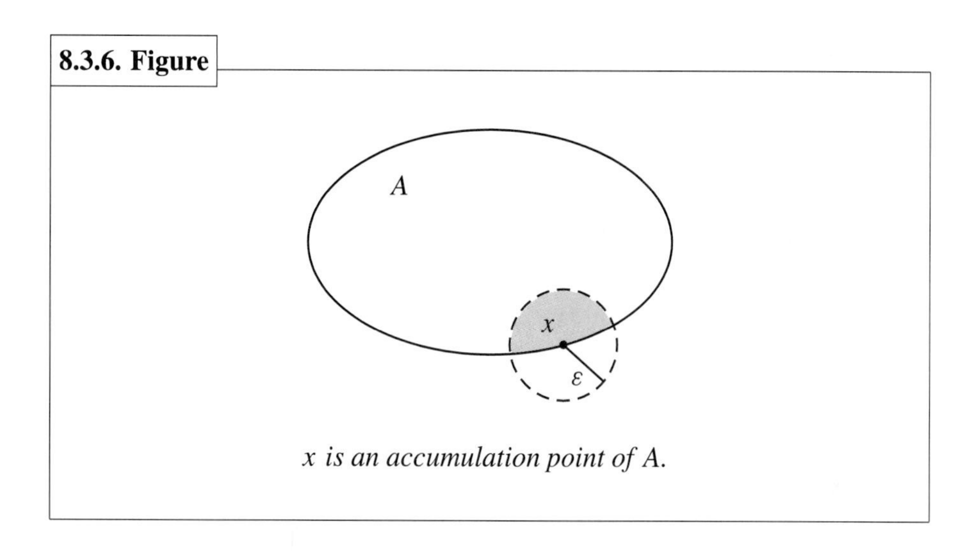

x is an accumulation point of A.

Example. Define $A = \{(x, y) \in \mathbb{R}^2 : 0 \leq x < 2\}$. Let $\varepsilon > 0$. To prove that $(0, 0)$ is an accumulation point, take $a \in \mathbb{R}$ so that $0 < a < \min(1, \varepsilon)$. Then,

$$(a, 0) \in [D((0, 0), \varepsilon) \setminus \{(0, 0)\}] \cap A,$$

which means the desired intersection is nonempty.

In the last example, the interior point $(1, 0)$ is also an accumulation point. However, an accumulation point need not be an element of the set.

Example. In the metric space $\mathbb{R}$, 0 is an accumulation point for both $(0, 1)$ and $[0, 1]$. For $\mathbb{R}^2$, $(1, 0)$ is an accumulation point for

$$\{(x, y) \in \mathbb{R}^2 : x^2 + y^2 < 1\}$$

and

$$\{(x, x - 1) : x \in \mathbb{R}\}.$$

Being in the interior is not a guarantee of being an accumulation point.

Example. Take the metric space $\mathbb{Z} \times \mathbb{Z}$ with the standard metric. Then for all $a, b \in \mathbb{Z}$, (a, b) cannot be accumulation point because

$$D_{\mathbb{Z}\times\mathbb{Z}}((a, b), 1/2) = \{(a, b)\}.$$

Therefore, no subset of $\mathbb{Z} \times \mathbb{Z}$ can have an accumulation point in this metric space.

We are now ready for another test for closure.

8.3.7. Theorem

A set is closed if and only if it contains its accumulation points.

Proof. Let A be a subset of a metric space M.

($\Rightarrow$) Assume A is closed. Let x be an accumulation point of A. To obtain a contradiction, imagine $x \notin A$. This means $x \in \overline{A}$, but this set is open. Hence, there is an $\varepsilon > 0$ with the property $D(x, \varepsilon) \subseteq \overline{A}$. Therefore, $D(x, \varepsilon) \cap A = \varnothing$, contradicting the fact that x is an accumulation point.

($\Leftarrow$) Now suppose A contains all its accumulation points. We must show $\overline{A}$ is open. Take $x \in \overline{A}$. By hypothesis, x cannot be an accumulation point of A. This means there exists $\varepsilon > 0$ so that $D(x, \varepsilon) \cap A = \varnothing$. Therefore, $D(x, \varepsilon)$ is a subset of $\overline{A}$. ■

In the proof we did not need to worry about x not being an element of $D(x, \varepsilon)$. For instance, in the second part x was taken in $\overline{A}$ and, hence, not in A.

Example. An interesting closed set that we must mention is the ***Cantor Set***. It is defined by first setting a sequence of sets. (See Figure 8.3.8.) Let

$$F_1 = [0, 1].$$

The next term is defined by removing the middle third of F_1, namely

$$F_2 = [0, 1/3] \cup [2/3, 1].$$

The next set is obtained by removing the middle third of each of the intervals that make up F_2:

$$F_3 = [0, 1/9] \cup [2/9, 1/3] \cup [2/3, 7/9] \cup [8/9, 1].$$

This process then continues so that the nth term is:

$$F_n = [0, 1/3^{n-1}] \cup [2/3^{n-1}, 1/3^{n-2}] \cup \cdots \cup [(3^{n-1} - 1)/3^{n-1}, 1].$$

The Cantor Set is defined to be

$$\bigcap_{n=1}^{\infty} F_n.$$

Its accumulation points are the endpoints of all the closed intervals of each F_n. Therefore, the Cantor Set is closed because is contains its accumulation points.

8.3.8. Figure

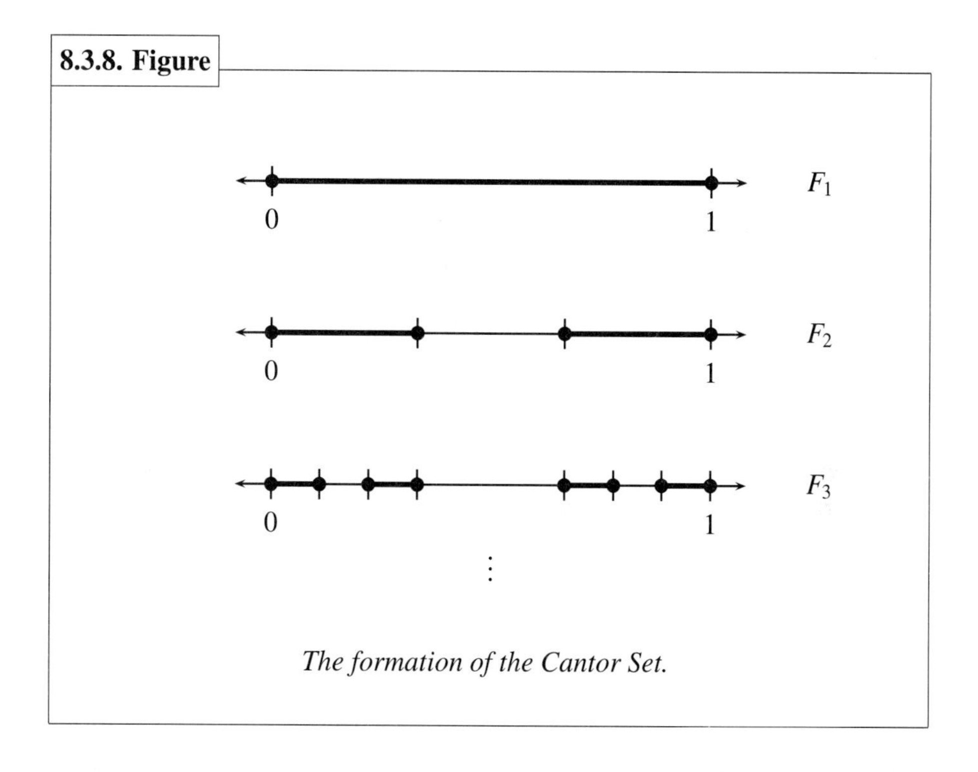

The formation of the Cantor Set.

EXERCISES

Assume that M denotes a metric space.

1. Show that M and $\varnothing$ are closed.
2. Let $x \in M$. Prove $\{x\}$ is closed in M.
3. Show that the following are closed in the given metric spaces.
   (a) $\{0\}$ in $\mathbb{R}$
   (b) $\{(0, 0)\}$ in $\mathbb{R}^2$
   (c) $\{(x, y) \in \mathbb{R}^2 : y = 2x - 1\}$ in $\mathbb{R}^2$
   (d) $\{(x, y) \in \mathbb{R}^2 : 5 \leq x \leq 7\}$ in $\mathbb{R}^2$
   (e) $\{(x, y) \in \mathbb{R}^2 : x + y \leq 1\}$ in $\mathbb{R}^2$
   (f) $\{(x, y, 1) : x, y \in \mathbb{R}\}$ in $\mathbb{R}^3$
   (g) $\{f \in C[0, 1] : \|f\| \leq 1\}$ in $C[0, 1]$
4. Let $A \subseteq M$. Show that $A \setminus \text{int}(A)$ is closed.
5. Take $a \in M$. Prove that $\{b \in M : d(a, b) \leq 1\}$ is closed using both of the following methods:
   (a) the definition of closed. **(b)** Theorem 8.3.7.

6. Fix a positive integer n. For all $i = 1, \dots n$, let $U_{i,\,j}$ be a closed set in M for every $j \in \mathbb{Z}^+$. Prove
$$\bigcup_{i=1}^{n} \bigcap_{j=1}^{\infty} U_{i,\,j}$$
is closed in M.
7. Let $A \subseteq B \subseteq M$. Prove that if a is an accumulation point of A, then it is an accumulation point of B.
8. Suppose that A and B are subsets of M. Prove the following about the closure:
   (a) $A \subseteq \text{cl}(A)$
   (b) if $A \subseteq B$, then $\text{cl}(A) \subseteq \text{cl}(B)$
   (c) $\text{cl}(\varnothing) = \varnothing$
   (d) $\text{cl}(A) = \text{cl}(\text{cl}(A))$
   (e) $\overline{\text{cl}(A)} = \text{int}(\overline{A})$
9. Let $A,\ B \subseteq M$.
   (a) Let $\text{AP}(X)$ denote the set of accumulation points of X. Prove
$$\text{AP}(A \cup B) = \text{AP}(A) \cup \text{AP}(B).$$
   (b) Show $\text{cl}(A) = A \cup \text{AP}(A)$.
   (c) Use these results to prove $\text{cl}(A \cup B) = \text{cl}(A) \cup \text{cl}(B)$
10. Prove A is closed if and only if $\text{cl}(A) = A$.
11. Fix a set A. Take any set C such that $A \subseteq C$ and C is closed. Prove $\text{cl}(A) \subseteq C$. (This shows that the closure is the smallest closed superset of a given set.)
12. Describe the set of accumulation points of each of the following as subsets of the given metric space:
   (a) $(0,\ 2)$ in $\mathbb{R}$
   (b) $(0,\ 2]$ in $\mathbb{R}$
   (c) $(0,\ 2) \cup \{-1,\ 3\}$ in $\mathbb{R}$
   (d) $\{(x,\ y) \in \mathbb{R}^2 : x^2 + y^2 < 1\}$ in $\mathbb{R}^2$
   (e) $\{(x,\ y) \in \mathbb{R}^2 : x^2 + y^2 \leq 1\}$ in $\mathbb{R}^2$
   (f) $\{(x,\ y) \in \mathbb{R}^2 : x \in \mathbb{Q}\}$ in $\mathbb{R}^2$
13. Let $A \subseteq M$.
   (a) Let a be an accumulation point of A. Show that for all $\varepsilon > 0$, $D(a,\ \varepsilon) \cap A$ contains infinitely many points.
   (b) Prove that A is closed if A is finite.
14. Prove that the Cantor Set is closed without appealing to Theorem 8.3.7.
15. Write a formula describing the accumulation points of the Cantor Set.

8.4. ISOMETRIES

Similar to homomorphisms in algebra, there is an important property preserving function in topology. It is the isometry. It preserves distances between elements of metric spaces.

8.4.1. Definition

Let (M, d) and (N, d') be metric spaces. A function $\phi\colon M \to N$ such that for all $x, y \in M$,

$$d(x, y) = d'(\phi(x), \phi(y))$$

is called an ***isometry***.

Example. A ***translation*** is a particular type of isometry. It is a function that maps the Cartesian plane to itself by sliding points. More precisely, given any vector (a, b), the function

$$T_{[a,b]}\colon \mathbb{R}^2 \to \mathbb{R}^2$$

defined by

$$T_{[a,b]}(x, y) = (x + a, y + b)$$

shifts points in the plane. For instance, $T_{[1,3]}(-2, -1) = (-1, 2)$ and $T_{[1,3]}(0, -2) = (1, 1)$. Moreover, the image under $T_{[1,3]}$ of the line segment with endpoints $(-2, -1)$ and $(0, -2)$ is the segment with endpoints $(-1, 2)$ and $(1, 1)$. (See Figure 8.4.2.)

To show that $T_{[a,b]}$ is an isometry, take $(x_1, y_1), (x_2, y_2) \in \mathbb{R}^2$. Using the standard metric, let

$$d_1 = d(T_{[a,b]}(x_1, y_1), T_{[a,b]}(x_2, y_2))$$

and

$$d_2 = d((x_1, y_1), (x_2, y_2)).$$

Now calculate:

$$\begin{aligned} d_1 &= \sqrt{([x_1 + a] - [x_2 + a])^2 + ([y_1 + b] - [y_2 + b])^2} \\ &= \sqrt{(x_1 - x_2)^2 + (y_1 - y_2)^2} \\ &= d_2. \end{aligned}$$

Therefore, the function preserves distances.

8.4.2. Figure

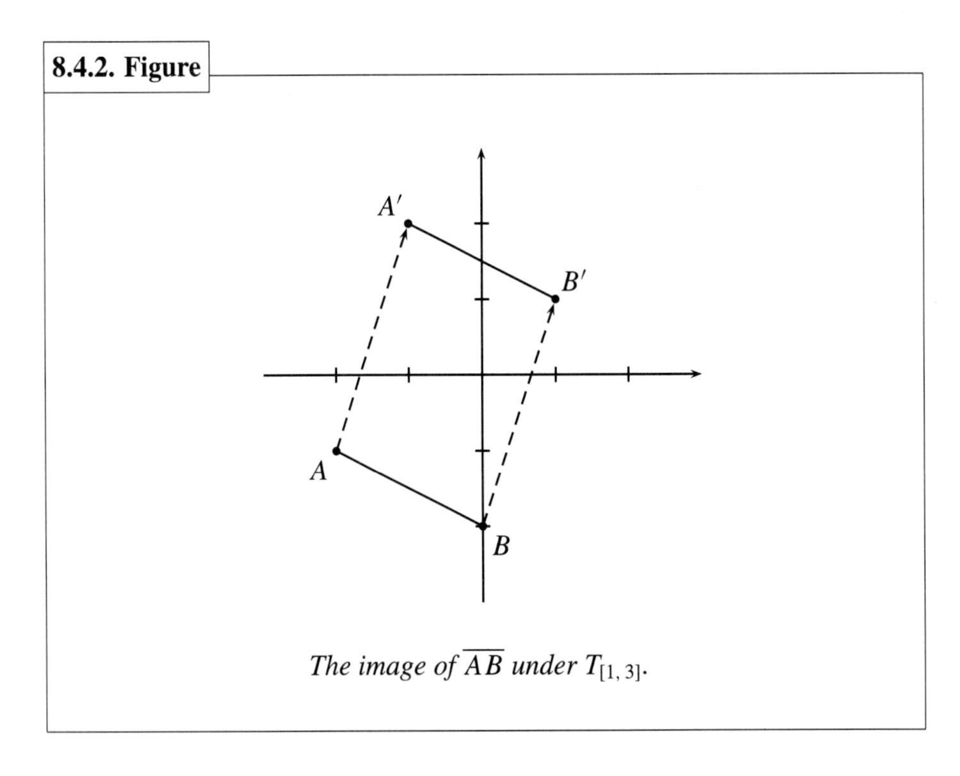

The image of $\overline{AB}$ under $T_{[1,\,3]}$.

The next example requires a little trigonometry.

Example. A ***rotation*** of a point A around a center point C is the motion along the arc of a circle with center C and radius of length AC through a fixed arc length, say α. This is represented in Figure 8.4.3. Algebraically this means the following: Let $(x,\ y)$ be a point in $\mathbb{R}^2$. Let us rotate $(x,\ y)$ α radians around the origin as in Figure 8.4.3. Denote this rotation by R_α. Let r be the radius, so $r = \sqrt{x^2 + y^2}$, and let β be the angle that the line through the origin and $(x,\ y)$ makes with the x-axis. This means

$$x = r\cos\beta,\ y = r\sin\beta$$

and

$$x' = r\cos(\alpha + \beta),\ y' = r\sin(\alpha + \beta).$$

Using the addition formulas from trigonometry and then substituting yields

$$x' = r\cos\alpha\cos\beta - r\sin\alpha\sin\beta = x\cos\alpha - y\sin\alpha$$

and

$$y' = r\sin\alpha\cos\beta + r\cos\alpha\sin\beta = x\sin\alpha + y\cos\alpha.$$

8.4.3. Figure

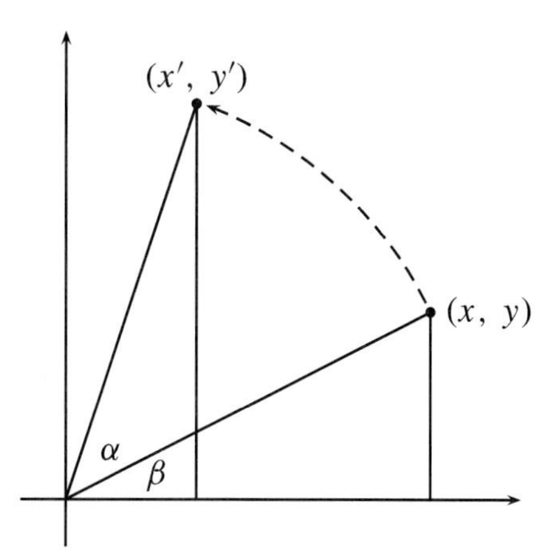

A rotation of (x, y) through α radians about the origin.

Therefore,

$$R_\alpha(x, y) = (x \cos\alpha - y \sin\alpha, \ x \sin\alpha + y \cos\alpha).$$

(Notice that if $\alpha > 0$, the rotation is counterclockwise.)

To see that R_α is an isometry, let $(x_1, y_1), (x_2, y_2) \in \mathbb{R}^2$. As before,

$$d_1 = d(R_\alpha(x_1, y_1), R_\alpha(x_2, y_2))$$

and

$$d_2 = d((x_1, y_1), (x_2, y_2)).$$

If we let $\Delta x = x_1 - x_2$ and $\Delta y = y_1 - y_2$,

$$\begin{aligned} d_1 &= \sqrt{[\Delta x \cos\alpha - \Delta y \sin\alpha]^2 + [\Delta x \sin\alpha + \Delta y \cos\alpha]^2} \\ &= \sqrt{(\Delta x)^2[\cos^2\alpha + \sin^2\alpha] + (\Delta y)^2[\cos^2\alpha + \sin^2\alpha]} \\ &= \sqrt{(x_1 - x_2)^2 + (y_1 - y_2)^2} \\ &= d_2. \end{aligned}$$

In many ways, an isometry is to metric spaces as homomorphisms are to rings. To further see this connection, we will close the section by proving that under a particular assumption, isometries preserve open sets like ring homomorphisms preserve subrings. To prove this, we first need a lemma.

8.4.4. Lemma

Let (M, d) and (N, d') be metric spaces. If $\phi\colon M \to N$ is an isometry, then for all $x \in M$,

$$\phi[D_M(x, \varepsilon)] = D_N(\phi(x), \varepsilon) \cap \phi[M].$$

Proof. We will set up a sequence of equalities:

$$\begin{aligned}
\phi[D_M(x, \varepsilon)] &= \{\phi(a) : a \in D_M(x, \varepsilon)\} \\
&= \{\phi(a) : d(x, a) < \varepsilon \text{ and } a \in M\} \\
&= \{\phi(a) : d'(\phi(x), \phi(a)) < \varepsilon \text{ and } a \in M\} \\
&= \{\phi(a) : \phi(a) \in D_N(\phi(x), \varepsilon) \text{ and } a \in M\}\} \\
&= D_N(\phi(x), \varepsilon) \cap \phi[M].
\end{aligned}$$

The third equality holds because ϕ is an isometry. ■

Notice that if the isometry ϕ is onto, then $\phi[D_M(x, \varepsilon)] = D_N(\phi(x), \varepsilon)$. This is the case with the next example. (See Exercise 16b.)

Example. Let us check the lemma by finding the image of $D((4, 3), 1)$ under $R_{\pi/2}$:

$$\begin{aligned}
R_{\pi/2}[D((4, 3), 1)] &= \{R_{\pi/2}(x, y) : (x, y) \in D((4, 3), 1)\} \\
&= \{(-y, x) : (x, y) \in D((4, 3), 1)\} \\
&= \{(-y, x) : \sqrt{(4 - x)^2 + (3 - y)^2} < 1\} \\
&= D((-3, 4), 1).
\end{aligned}$$

The second equality holds because

$$R_{\pi/2}(x, y) = (x \cos \pi/2 - y \sin \pi/2,\ x \sin \pi/2 + y \cos \pi/2).$$

This work confirms the lemma since $R_{\pi/2}(4, 3) = (-3, 4)$.

Example. Let $f \in C[0, 1]$. Using the usual metric for this set, the open disk around f of radius 1 is

$$D(f, 1) = \{g \in C[0, 1] : \int_0^1 |f(x) - g(x)|\, dx < 1\}.$$

Now fix $g \in C[0, 1]$ and define a function $\phi\colon C[0, 1] \to C[0, 1]$ by

$$\phi(h) = h + g.$$

According to Exercise 10, ϕ is an isometry that is onto, so let us check the image of $D(f, 1)$ under ϕ:

$$\begin{aligned}
h \in \phi[D(f, 1)] &\Leftrightarrow e \in D(f, 1) \text{ and } \phi(e) = h \\
&\Leftrightarrow e \in D(f, 1) \text{ and } e + g = h \\
&\Leftrightarrow \int_0^1 |f(x) - e(x)|\, dx < 1 \text{ and } e + g = h \\
&\Leftrightarrow \int_0^1 |f(x) - [h(x) - g(x)]|\, dx < 1 \\
&\Leftrightarrow \int_0^1 |f(x) + g(x) - h(x)|\, dx < 1 \\
&\Leftrightarrow d(f + g, h) < 1 \\
&\Leftrightarrow h \in D(f + g, 1) \\
&\Leftrightarrow h \in D(\phi(f), 1).
\end{aligned}$$

The fourth biconditional holds since ϕ is onto.

We now prove the theorem.

8.4.5. Theorem

Let (M, d) and (N, d') be metric spaces and $\phi\colon M \to N$ an onto isometry. If U is open in M, then $\phi[U]$ is open in N.

Proof. Assume U is open in M. Let $y \in \phi[U]$. This means $\phi(x) = y$ for some $x \in U$. Since U is open, we have $\varepsilon > 0$ so that $D_M(x, \varepsilon) \subseteq U$. Hence,

$$\phi[D_M(x, \varepsilon)] \subseteq \phi[U].$$

Then using the last lemma and the fact that ϕ is onto,

$$\phi[D_M(x, \varepsilon)] = D_N(y, \varepsilon),$$

and so $\phi[U]$ is open in N. ■

Example. Define $A = (1, 2) \times (-1, 3)$, the open rectangle with vertices at $(1, -1)$, $(2, -1)$, $(1, 3)$, and $(2, 3)$. Therefore,

$$T_{[4, 2]}[A] = (5, 6) \times (1, 5),$$

which is also an open set.

EXERCISES

1. Find the following and graph the result:
   (a) $T_{[1,\,2]}(0,\ 4)$
   (b) $T_{[-1,\,-2]}(0,\ 4)$
   (c) $T_{[0,\,6]}(7,\ -2)$
   (d) $R_{\pi/2}(3, 5)$
   (e) $R_{\pi/4}(3, 5)$
   (f) $R_{\pi}(3, 5)$
2. Find the following images and graph the result:
   (a) $T_{[3,\,-2]}[(-1,\ 5) \times \{0\}]$
   (b) $T_{[3,\,-2]}[(-1,\ 5) \times (-1,\ 5)]$
   (c) $R_{\pi/2}[\mathbb{R} \times \{0\}]$
   (d) $R_{\pi}[(-1,\ 5) \times \{0\}]$
3. Find the following pre-images and graph the result:
   (a) $T^{-1}_{[-1,\,-5]}[\{0\} \times \mathbb{R}]$
   (b) $T^{-1}_{[1,\,5]}[\{0\} \times \mathbb{R}]$
   (c) $R^{-1}_{\pi/2}[\{(1, 0)\}]$
   (d) $R^{-1}_{\pi/2}[\{0\} \times (2,\ 6)]$
4. Let a, b, c, and d be real numbers. Show:
$$T_{[a,\,b]} \circ T_{[c,\,d]} = T_{[c,\,d]} \circ T_{[a,\,b]}$$
5. Prove that the composition of isometries is an isometry.
6. Prove that every isometry is one-to-one.
7. Give an example of an isometry that is not onto.
8. Demonstrate that Theorem 8.4.5 is false if the onto condition is dropped.
9. Show that if $f : \mathbb{R}^2 \to \mathbb{R}^2$ is an isometry, then so is $f^{-1} : \text{ran}(f) \to \mathbb{R}^2$.
10. Let $g \in C[0,\ 1]$ and define $\phi : C[0,\ 1] \to C[0,\ 1]$ by $\phi(h) = h + g$.
    (a) If $g(x) = x^2 + 1$, find $\phi(f)$ when $f(x) = x - 2$.
    (b) Prove that ϕ is an isometry with respect to $C[0,\ 1]$.
    (c) Show that ϕ is onto.
11. Take two metric spaces $(M,\ d)$ and $(N,\ d')$. Let $\phi : M \to N$ be an isometry that is also a surjection. Prove that if U_α is an open set in M for all $\alpha \in \Lambda$, then $\phi[\bigcup_{\alpha\in\Lambda} U_\alpha]$ is open in N.
12. Let $\psi : M \to N$ be an onto isometry between two metric spaces $(M,\ d)$ and $(N,\ d')$. Prove the following:
    (a) If C is closed in M, then $\psi[C]$ is closed in N.
    (b) If C_α $(\alpha \in \Lambda)$ are closed in M, then $\psi[\bigcap_{\alpha\in\Lambda} C_\alpha]$ is closed in N.
13. A ***reflection*** P through the x-axis is a function that maps every element to its mirror image using the x-axis as the mirror. Algebraically this means that for all $(x,\ y) \in \mathbb{R}^2$, $P(x,\ y) = (x,\ -y)$.
    (a) Graph $(3,\ 4)$ and $P(3,\ 4)$.
    (b) Graph $(2,\ -2)$ and $P(2,\ -2)$.
    (c) Graph $(-4,\ 0)$ and $P(-4,\ 0)$.
    (d) Using the previous three graphs as a guide, sketch the reflection of an arbitrary point in $\mathbb{R}^2$ through the x-axis.
    (e) Prove that a reflection through the x-axis is an isometry.

14. A ***glide reflection*** is a composition of a translation and a reflection. Given a vector that is parallel to the x-axis, $(a,\ 0)$, the glide reflection through the x-axis in the direction of $(a,\ 0)$ is

$$G_{[a,\,0]}(x,\ y) = P \circ T_{[a,\,0]}(x,\ y).$$

   (a) Graph $(3,\ 4)$ and $G_{[3,\,0]}(3,\ 4)$.
   (b) Graph $(2,\ -2)$ and $G_{[3,\,0]}(2,\ -2)$.
   (c) Graph $(-4,\ 0)$ and $G_{[3,\,0]}(-4,\ 0)$.
   (d) Let $a \in \mathbb{R}$. Using the previous three graphs as a guide, sketch the glide reflection of an arbitrary point in $\mathbb{R}^2$ through the x-axis in the direction of $(a,\ 0)$ and find an equation.
   (e) Prove that a glide reflection is an isometry.

15. Find the inverse of each of the following:
   (a) $T_{[a,\,b]}$
   (b) R_α
   (c) P
   (d) $G_{[a,\,0]}$

16. Prove that each of the following are surjections:
   (a) $T_{[a,\,b]}$
   (b) R_α
   (c) P
   (d) $G_{[a,\,0]}$

8.5. LIMITS

Remember that we opened the chapter with a mention of limits. Since limits are crucial to the study of topology and analysis, our last section will introduce them. For our purposes, we will focus on sequences of real numbers. Everything that is done here can be translated to an arbitrary metric space.

To formally define convergence, we will use open intervals. We need to write a precise test that will determine if a sequence gets arbitrarily close to some real number.

8.5.1. Definition

Let a_n be a sequence and take $a \in \mathbb{R}$. We say that a_n ***converges*** to a if

$$(\forall \varepsilon > 0)(\exists N \in \mathbb{Z}^+)(\forall n \in \mathbb{Z}^+)(n \geq N \Rightarrow |a_n - a| < \varepsilon).$$

In this case we write $a_n \to a$ or the familiar

$$\lim_{n\to\infty} a_n = a.$$

We say that a is the ***limit*** of a_n. If a sequence does not converge, then it is said to ***diverge***.

8.5.2. Figure

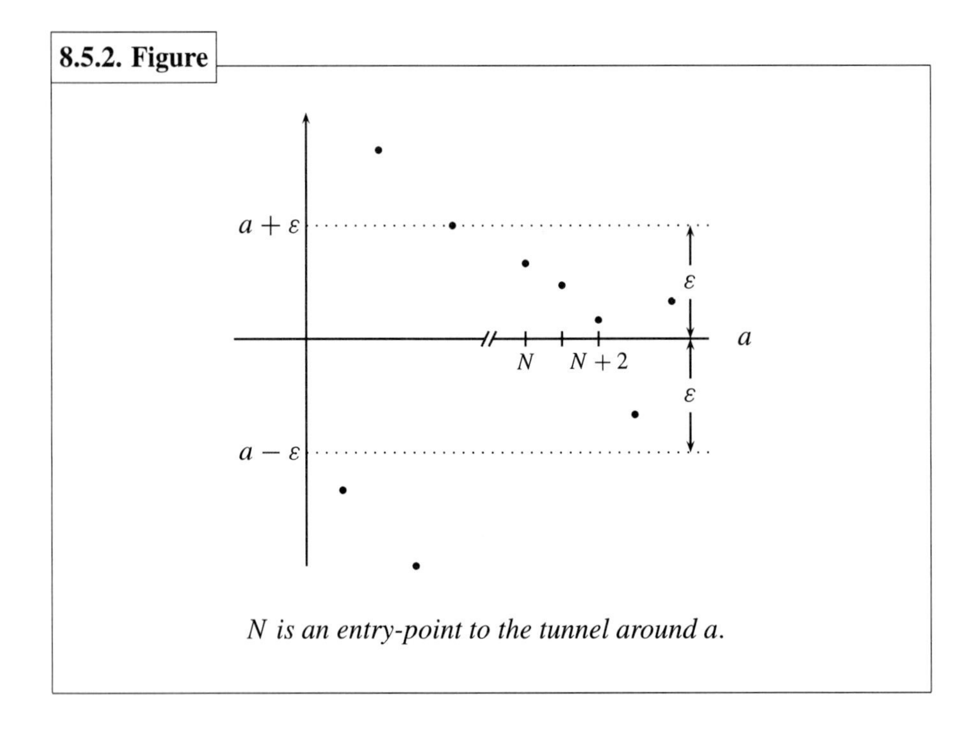

N is an entry-point to the tunnel around a.

The intuition for this definition is known from calculus. The idea is that the distance between a_n and a decreases as n becomes large. This means that for any distance ε around a, there is a subscript beyond which all terms of the sequence are within ε of a. This is viewed as a ***tunnel*** in Figure 8.5.2, and N is the entry point to the tunnel. In summary:

ε	the radius of the tunnel surrounding a
N	the entry point of the tunnel
n	the indices

No matter what radius is chosen, we can always go out far enough in the sequence to find an entry point to the tunnel from which the terms do not escape.

Example. Take the sequence $-1,\ 1/2,\ -1/3,\ 1/4,\ \ldots$ Its formula is

$$a_n = \frac{(-1)^n}{n}.$$

To see that $a_n \to 0$, take $\varepsilon > 0$. We must find a positive integer N so that

$$(*) \qquad n \geq N \Rightarrow \left| \frac{(-1)^n}{n} - 0 \right| < \varepsilon.$$

8.5.3. Figure

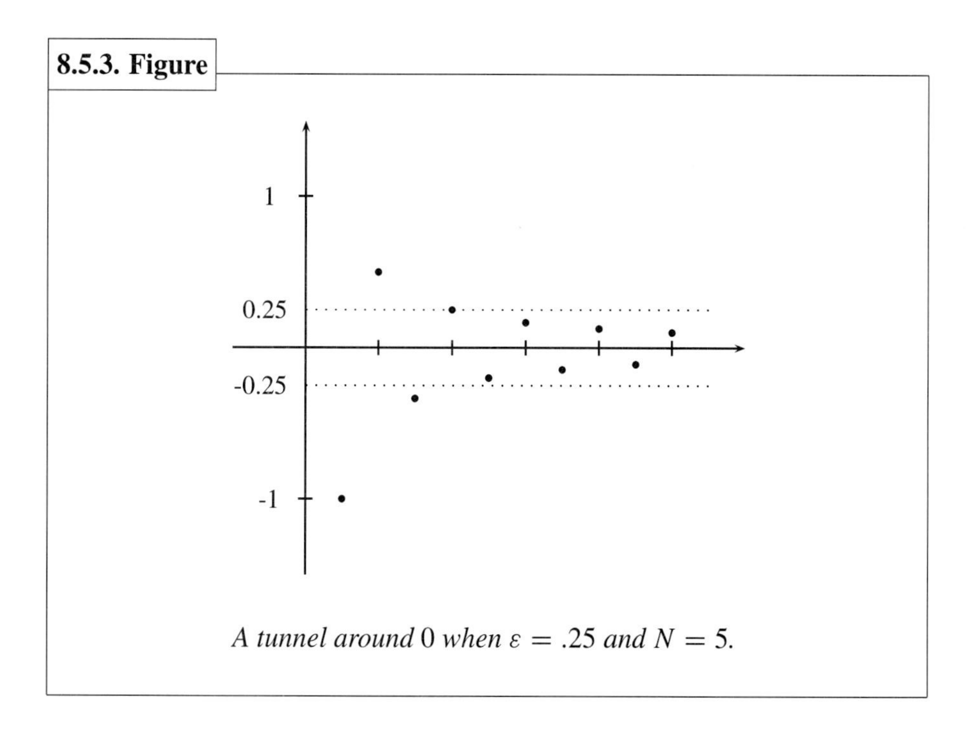

A tunnel around 0 *when* $\varepsilon = .25$ *and* $N = 5$.

(Note: if $\varepsilon = 0.25$, we may take $N = 5$. See Figure 8.5.3.) To do this, we observe

$$\left|\frac{(-1)^n}{n} - 0\right| = \left|\frac{(-1)^n}{n}\right| = \frac{1}{n}.$$

Hence, we need $1/n < \varepsilon$, and this means we want $n > 1/\varepsilon$. Therefore, choose N to be any positive integer greater than $1/\varepsilon$. Now show $(*)$ by Direct Proof: Assume $n \geq N$. Then $n > 1/\varepsilon$. Therefore, $1/n < \varepsilon$ and

$$\left|\frac{(-1)^n}{n} - 0\right| = \left|\frac{(-1)^n}{n}\right| = \frac{1}{n} < \varepsilon.$$

Now to try one without a picture.

Example. To prove $(2^n/n!) + 1 \to 1$, let $\varepsilon > 0$. We must find N so that

$$n \geq N \Rightarrow \left|\left(\frac{2^n}{n!} + 1\right) - 1\right| < \varepsilon.$$

Notice for all $n > 1$,

$$\left|\left(\frac{2^n}{n!} + 1\right) - 1\right| = \frac{2n}{n!} = \frac{2}{(n-1)!} \leq \frac{2}{n-1}.$$

8.5.4. Figure

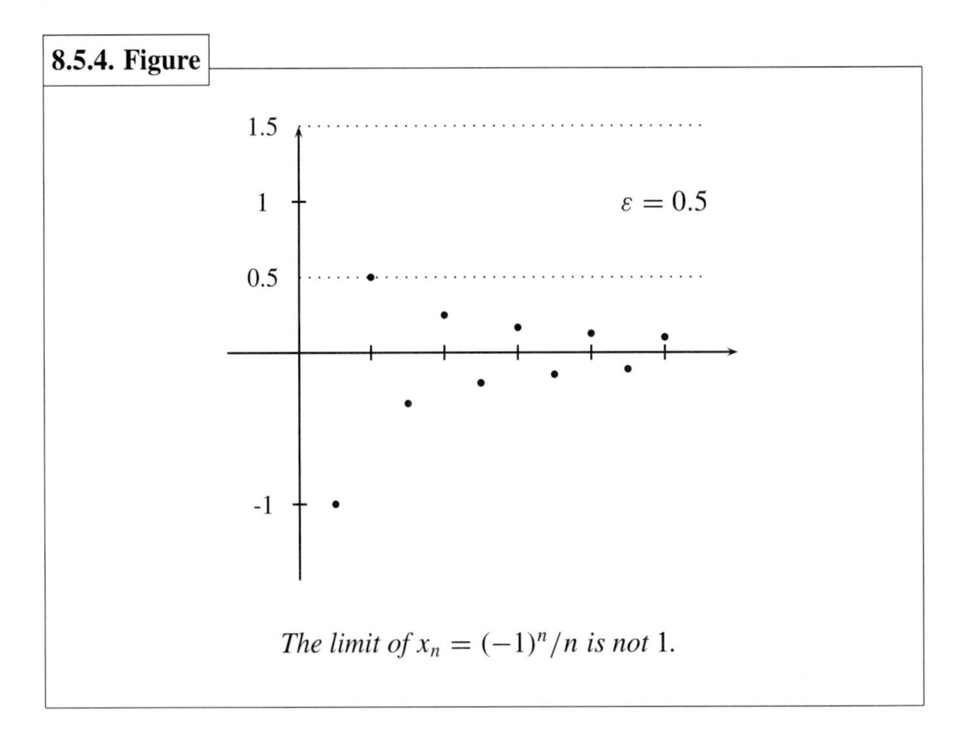

The limit of $x_n = (-1)^n/n$ *is not* 1.

The inequality holds since the denominator of the first fraction is at least as large as the second. Therefore, it suffices to make $2/(n-1) < \varepsilon$ for large n. We claim that any positive integer $N > (2/\varepsilon) + 1$ works. To show this take $n \geq N$. Then

$$\frac{2}{n-1} \leq \frac{2}{N-1} < \frac{2}{(2/\varepsilon + 1) - 1} = \frac{2}{2/\varepsilon} = \varepsilon.$$

If a sequence a_n does not converge to a, this means

$$(\exists \varepsilon > 0)(\forall N \in \mathbb{Z}^+)(\exists n \in \mathbb{Z}^+)(n \geq N \wedge |a_n - a| \geq \varepsilon).$$

This is obtained by negating the definition of convergence. It says that there is a tunnel of radius ε that has the property that no matter how far out in the sequence we go, there is a term beyond that is not in the tunnel. This is illustrated in the next example.

Example. To show that $a_n = (-1)^n/n$ does not converge to 1, take $\varepsilon = 0.5$. Figure 8.5.4 shows that this choice provides the required tunnel.

Example. From calculus we know that the sequence $1, -1, 1, -1, \ldots$ does not converge. To prove this, suppose that it does converge to some $a \in \mathbb{R} \setminus \{1\}$. Set $\varepsilon = |1 - a|$, the distance from 1 to a. Let $N \in \mathbb{Z}^+$. By definition of the sequence, there exists $n \geq N$ so that $a_n = 1$. Let n_0 be such an index. Then, $n_0 \geq N$, and

$$|a_{n_0} - a| = |1 - a| = \varepsilon > 0.$$

Hence, a_n does not converge to a. To show that the sequence does not converge to -1, take $\varepsilon = |-1-a|$.

When we speak of the limit of a sequence, there is an underlying assumption that there is exactly one limit if it exists. The next theorem proves that fact.

8.5.5. Theorem

If the limit of a sequence exists, then it is unique.

Proof. Let a_n be a sequence and suppose $a_n \to a$ and $a_n \to a'$. In order to reach a contradiction, suppose that the limits are not equal. Let

$$\varepsilon = \frac{|a - a'|}{2}.$$

Since $\varepsilon > 0$, there exists $N_1,\ N_2 > 0$ such that if $n \geq N_1$, then

$$|a_n - a| < \varepsilon,$$

and if $n \geq N_2$,

$$|a_n - a'| < \varepsilon.$$

Let $N = \max(N_1,\ N_2)$ and assume $n \geq N$. Calculating with the Triangle Inequality we find:

$$\begin{aligned} |a - a'| &= |a - a_n + a_n - a'| \\ &\leq |a - a_n| + |a_n - a'| \\ &< \varepsilon + \varepsilon \\ &= |a - a'|. \end{aligned}$$

(The strict inequality follows because n is greater than or equal to both N_1 and N_2.) Therefore, $|a - a'| < |a - a'|$, a contradiction. Thus, the limit must be unique. ■

Let a_n be a sequence of real numbers such that $\lim_{n\to\infty} a_n = a$. If this is the case, then $a_n \neq 2^n$, for this sequence continues to grow and approaches no real number. This sequence is not bounded. We say that a_n is ***bounded*** if there exists $M \in \mathbb{R}$ such that $|a_n| \leq M$ for all $n \in \mathbb{Z}^+$. The constant M is called a ***bound***, and it is not unique.

Example. The sequence $a_n = 1 - 1/n$ is bounded since $|1 - 1/n| \leq 1$ for all positive integers n. Thus, 1 is a bound for a_n. Moreover, any real number greater that 1 will be a bound for the sequence. Similarly, a bound of $b_n = \cos n$ is also 1.

Notice that the first sequence in the example converges, but the second one does not. Thus, we may conjecture that convergence is sufficient for a sequence to have a bound but not necessary. To see sufficiency, suppose that a_n is a sequence that converges to a. This yields a positive integer N such that for all $n \geq N$, $|a_n - a| < 1$. Let

$$M = \max\{|a_1|, |a_2|, \ldots, |a_{N-1}|, 1 + |a|\},$$

so M is the maximum of the first $N - 1$ terms of the sequence and $1 + |a|$. Therefore, for $n \geq N$,

$$|a_n| = |a_n - a + a| \leq |a_n - a| + |a| \leq 1 + |a| \leq M,$$

and we may conclude $|a_n| \leq M$ for every $n \in \mathbb{Z}^+$.

We close the section with an application of this result. This theorem will allow us to calculate some limits without using ε in the proof.

8.5.6. Theorem

Take $k \in \mathbb{R}$. If $a_n \to a$ and $b_n \to b$, then

1. $a_n + b_n \to a + b$
2. $a_n - b_n \to a - b$
3. $a_n b_n \to ab$
4. $a_n/b_n \to a/b$ if $b \neq 0$ and $b_n \neq 0$ for all n
5. $ka_n \to ka$

Proof. We will prove the multiplication result and leave the rest to the exercises. Assume $a_n \to a$ and $b_n \to b$. We then know both sequences are bounded, so there exist real numbers M_1 and M_2 such that $|a_n| \leq M_1$ and $|b_n| \leq M_2$ for all $n \in \mathbb{Z}^+$. Define M to be the maximum of M_1 and M_2. By Exercise 13,

$$|a| \leq M_1 \leq M.$$

Now take $\varepsilon > 0$. There exists positive integers N_1 and N_2 such that for every $n \in \mathbb{Z}^+$,

$$\text{if } n \geq N_1, \text{ then } |a_n - a| < \varepsilon/2M,$$

and

$$\text{if } n \geq N_2, \text{ then } |b_n - b| < \varepsilon/2M.$$

Define $N = \max(N_1, N_2)$. Therefore, if $n \geq N$,

$$\begin{aligned} |a_n b_n - ab| &= |a_n b_n - ab_n + ab_n - ab| \\ &\leq |a_n b_n - ab_n| + |ab_n - ab| \\ &= |b_n||a_n - a| + |a||b_n - b| \\ &< M \cdot \varepsilon/2M + M \cdot \varepsilon/2M \\ &= \varepsilon. \end{aligned}$$

Therefore, $a_n b_b \to ab$. ■

EXERCISES

1. Find a bound of each of the following sequences.
   (a) $a_n = 7$
   (b) $a_n = (-1)^n$
   (c) $a_n = 3 - 1/n$
   (d) $a_n = 3\cos(n - 4)$
2. Give an example of a strictly increasing sequence of real numbers that is bounded.
3. Choose $k \in \mathbb{R}$ and define $a_n = k$ for all $n \in \mathbb{Z}^+$. Prove that a_n converges to k.
4. Prove that the following converge to 0:
   (a) $a_n = 1/n$
   (b) $a_n = 1/n^2$
   (c) $a_n = 2^{-n}$
   (d) $a_n = 3 - 3n/n$
5. Prove that the following do not converge to 0:
   (a) $a_n = 2n^2/(6n^2 + n + 1)$
   (b) $a_n = 2^n$
   (c) $a_n = \cos n$
   (d) $a_n = (1/2)^n + 1$
6. Find the limits of the following and prove the results:
   (a) $a_n = 1/n^2$
   (b) $a_n = (n + 5)/(n - 2)$
   (c) $a_n = (n^2 + 2n - 1)/(n^3 - 1)$
   (d) $a_n = 1/\sqrt{n}$
7. Prove that the following do not converge.
   (a) $a_n = n$
   (b) $2n^2/(n + 1)$
   (c) $a_n = 2^n$
   (d) $a_n = n - 1/n$
8. Prove the remaining parts of Theorem 8.5.6.
9. Let $a_n \to a$ where $a \neq 0$ and $a_n \neq 0$ for all $n \in \mathbb{Z}^+$. Prove $1/a_n \to 1/a$.
10. Prove that for all $k \in \mathbb{Z}^+$, $a_n^k \to a^k$.
11. Show if $a_n \to a$, then $|a_n| \to a$. Is the converse true?
12. Prove the ***Sandwich Theorem***: Let a_n, b_n and c_n be sequences of real numbers. If there exists $\ell \in \mathbb{R}$ such that $a_n \to \ell$, $c_n \to \ell$, and $a_n \leq b_n \leq c_n$ for all $n \in \mathbb{Z}^+$, then $b_n \to \ell$.

13. Let a_n be a sequence that converges to some real number ℓ. Suppose there exists $b, c \in \mathbb{R}$ such that $b \le a_n \le c$ for all $n \in \mathbb{Z}^+$. Prove that $b \le \ell \le c$.
14. Let $a_n \to 0$ and b_n bounded. Prove $a_n b_n \to 0$. (Is this true if the bounded condition is removed?)

CHAPTER EXERCISES

1. Find an example of the following:
   (a) a function $M_{2,2}(\mathbb{R}) \times M_{2,2}(\mathbb{R}) \to \mathbb{R}$ that is a non-trivial metric on $M_{2,2}(\mathbb{R})$.
   (b) a function $M_{2,2}(\mathbb{R}) \to \mathbb{R}$ that is a norm on $M_{2,2}(\mathbb{R})$.
2. Show that $(C[0, 1], +)$ is an abelian group, but $(C[0, 1], \circ)$ is not a group.
3. Let V and U be two vector spaces over a field F. A function $T : V \to U$ is called a ***linear transformation*** if for all vectors $a, b \in V$ and scalars $\alpha \in F$,
   - $T(a + b) = T(a) + T(b)$, and
   - $T(\alpha a) = \alpha T(a)$.

   (Notice that a linear transformation preserves the operations of a vector space, so it can be considered a homomorphism for vector spaces.) For each of the following, either prove that the function is a linear transformation or show that it is not by providing a counterexample. View each domain and each codomain as a vector space over $\mathbb{R}$.
   (a) $f : \mathbb{R}^2 \to \mathbb{R}^3$, $f(x, y) = (2x, 3y, x + y)$.
   (b) $g : \mathbb{R}^3 \to \mathbb{R}^2$, $g(x, y, z) = (0, 0)$
   (c) $\psi : \mathbb{R}[X] \to \mathbb{R}$, $\psi(a_0 + a_1X + \cdots + a_{n-1}X^{n-1} + a_nX^n) = a_0$
   (d) $h : \mathbb{R}^2 \to \mathbb{R}^2$, $h(x, y) = (x + y, xy)$
   (e) $\phi : M_{2,2}(\mathbb{R}) \to R$, $\phi\left(\begin{bmatrix} a & b \\ c & d \end{bmatrix}\right) = ad - bc$
4. Let $S, T : U \to V$ be two linear transformations between the vector spaces U and V over a field F. Demonstrate that each of the following are also linear transformations:
   (a) $S + T$
   (b) αS (where α is a scalar)
   (c) $S \circ T$ (provided $V \subseteq U$)
5. Give an example of both of the following:
   (a) a linear transformation $\mathbb{R}^2 \to \mathbb{R}^2$ that is an isometry.
   (b) a linear transformation $\mathbb{R}^2 \to \mathbb{R}^2$ that is not an isometry.

6. Give an example of an uncountable metric space M and a countable $A \subseteq M$ such that A is open.
7. Let A be a subset of a metric space M. We say that A is ***dense*** if $\text{cl}(A) = M$. Prove that the following are equivalent:
   - A is dense.
   - The only closed superset of A is M.
   - The only set B such that A and B are disjoint is $B = \varnothing$.
   - A intersects every nonempty open subset.
8. Answer the following about the Cantor Set:
   (a) Prove that .25 is a member of the Cantor Set.
   (b) Describe the interior of the Cantor Set.
   (c) Prove that the set of endpoints of the Cantor Set is countable.
   (d) Show that the Cantor Set is equinumerous with $[0, 1]$.
9. Let $\phi: (M, d) \to (N, d')$ be an isometry and assume that U is open in N. Show $\phi^{-1}[U]$ is open in M.
10. Take a metric space M and assume $A \subseteq M$. Define the ***boundary*** of A to be

$$\text{bd}(A) = \text{cl}(A) \cap \text{cl}(\overline{A}).$$

Show the following:
   (a) $\text{bd}(A) = \varnothing$ if and only if A is both open and closed
   (b) $\text{bd}(\text{int}(A)) \subseteq \text{bd}(A)$
   (c) Assuming $\text{bd}(A)$ and $\text{bd}(B)$ are disjoint:
      (i) $\text{int}(A \cup B) = \text{int}(A) \cup \text{int}(B)$
      (ii) $\text{bd}(A \cap B) = [\text{cl}(A) \cap \text{bd}(B)] \cup [\text{bd}(A) \cap \text{cl}(B)]$
11. Define the function $R_{\alpha, (a, b)}: \mathbb{R}^2 \to \mathbb{R}^2$ by

$$R_{\alpha, (a, b)} = T_{[a, b]} \circ R_\alpha \circ T_{[-a, -b]}.$$

This is a ***rotation*** α radians around a point $(a, b) \in \mathbb{R}^2$.
   (a) Graph each point and its image under the given rotation.
      (i) $(4, 5)$, $R_{\pi, (1, 3)}$
      (ii) $(-1, -1)$, $R_{\pi, (1, 3)}$
      (iii) $(4, 5)$, $R_{\pi/2, (-2, 1)}$
      (iv) $(-1, -1)$, $R_{\pi/2, (-2, 1)}$
   (b) Find the following images and inverse images.
      (i) $R_{\pi, (-2, 1)}[(2, 4) \times \{0\}]$
      (ii) $R_{\pi, (-2, 1)}[\mathbb{R} \times \{0\}]$
      (iii) $R^{-1}_{\pi/2, (-2, 1)}[(1, 2) \times (1, 2)]$
      (iv) $R^{-1}_{\pi/2, (-2, 1)}[\{0\} \times (2, 4)]$
   (c) Prove that $R_{\alpha, (a, b)}$ is an isometry.

12. Let ℓ be the line in the Cartesian plane given by the equation $y = mx + b$. Assume $m \neq 0$. Let α be the angle ℓ makes with the x-axis. Further, let $(x_0, 0)$ be the x-intercept of the line. Define the ***reflection*** through the line ℓ as

$$P_\ell = T_{[x_0, 0]} \circ R_\alpha \circ P \circ R_{-\alpha} \circ T_{[-x_0, 0]},$$

where P is as defined in Exercise 8.4.13.

(a) Graph each point and its image under P_ℓ where ℓ is describe by the given equation.
   (i) $(1, 1)$, $y = x + 2$
   (ii) $(-3, 3)$, $y = x + 2$
   (iii) $(2, 3)$, $y = -2x + 2$
   (iv) $(-3, -2)$, $y = -2x + 2$

(b) Find the following images and inverse images where ℓ is the line $y = x + 2$.
   (i) $P_\ell[[-1, 1] \times [1, 3]]$
   (ii) $P_\ell[\mathbb{R} \times \{0\}]$
   (iii) $P_\ell^{-1}[[1, 2] \times [1, 2]]$
   (iv) $P_\ell^{-1}[\{0\} \times \mathbb{R}]$

13. Suppose ℓ is the line given by $y = mx + b$. The function $G_{\ell, k} \colon \mathbb{R}^2 \to \mathbb{R}^2$ where $k \in \mathbb{R}$ is called a ***glide reflection*** and is defined as

$$G_{\ell, k} = P_\ell \circ T_{[k, km]}.$$

(a) Graph each point and its image under $G_{\ell, 2}$ where ℓ is the given line.
   (i) $(1, 0)$, $y = x + 2$
   (ii) $(-3, 4)$, $y = x + 2$
   (iii) $(1, 0)$, $y = -2x + 2$
   (iv) $(1, 5)$, $y = -2x + 2$

(b) Let $(x_0, y_0) \in \mathbb{R}^2$. Prove that the line through the points (x_0, y_0) and $T_{[k, km]}(x_0, y_0)$ is parallel or equal to $y = mx + b$.

(c) Describe the geometric purpose of the value of k in $T_{[m, km]}$.

(d) Let ℓ be the line $y = x+2$. Find the following images and inverse images:
   (i) $G_{\ell, 2}[[-2, 0] \times \{0\}]$
   (ii) $G_{\ell, 2}[\{0\} \times \mathbb{R}]$
   (iii) $G_{\ell, 2}^{-1}[\{(0, 0)\}]$
   (iv) $G_{\ell, 2}^{-1}[\{(0, 2)\}]$

14. Take a function $f: M \to N$ where (M, d) and (N, d') are metric spaces. We say that f is ***continuous*** at $x_0 \in M$ if for all $x \in M$,

$$(\forall \varepsilon > 0)(\exists \delta > 0)(d(x, x_0) < \delta \Rightarrow d'(f(x), f(x_0)) < \varepsilon).$$

Prove that every isometry is continuous.

15. Prove that the following functions $\mathbb{R} \to \mathbb{R}$ are continuous at the given x_0 value.
    (a) $f(x) = 4$ at $x_0 = 7$
    (b) $g(x) = x$ at $x_0 = 2$
    (c) $h(x) = 4x + 8$ at $x_0 = 1$
    (d) $f(x) = |x|$ at $x_0 = 0$
16. A sequence a_n is a ***Cauchy sequence**** if for all $\varepsilon > 0$, there exists N such that if $m, n \geq N$, then $|a_m - a_n| < \varepsilon$. It can be proven that every Cauchy sequence converges. Here, prove the converse, namely that every convergent sequence is a Cauchy sequence.
17. Let a_n be a sequence so that for all $n \in \mathbb{Z}^+$,

$$|a_n - a_{n+1}| \leq \frac{1}{2}|a_{n-1} - a_n|.$$

Prove that a_n is a Cauchy Sequence.

18. Let a_n be a sequence of real numbers. A sequence b_i is a ***subsequence*** of a_n if there exists a strictly increasing function $f: \mathbb{Z}^+ \to \mathbb{Z}^+$ such that $b_i = a_{f(i)}$. Often we denote $f(i)$ by n_i so that the subsequence can be written as a_{n_i}. For each of the following, write the first five terms of each subsequence using the given f.
    (a) $a_n = 1/n$, $f(i) = 2i$
    (b) $a_n = (-1)^n$, $f(i) = 4i$
    (c) $a_n = 2n + 1$, $f(i) = i^2$
    (d) $a_n = 2 \cdot 3^n$, $f(i) = i + 2^i$
19. Let a_n be a sequence and a_{n_i} a subsequence. Prove that if there exists $a \in \mathbb{R}$ such that $a_n \to a$, then $a_{n_i} \to a$.
20. Let A be a set. A ***topology*** in A is a family $\mathcal{F}$ of subsets of A such that $\varnothing, A \in \mathcal{F}$ and for all $\mathcal{E} \subseteq \mathcal{F}$:
    - $\bigcup \mathcal{E} \in \mathcal{F}$, and
    - if $|\mathcal{E}| < \aleph_0$, then $\bigcap \mathcal{E} \in \mathcal{F}$.

    The pair $(A, \mathcal{F})$ is called a ***topological space***, and the sets in $\mathcal{F}$ are called ***open sets***. Prove that the following are topological spaces:
    (a) $(A, \{\varnothing, A\})$
    (b) $(\{0, 1\}, \{\varnothing, \{0\}, \{0, 1\}\})$
    (c) $(A, \mathbf{P}(A))$

---

*Augustin-Louis Cauchy (Paris, France, 1798 – Sceaux, France, 1857): Among Cauchy's many contributions to mathematics was his development of complex analysis and the introduction of rigor to the calculus.

(d) $(\mathbb{R}, \mathcal{S})$, where $\mathcal{S} = \{D(x, \varepsilon) : x \in \mathbb{R} \text{ and } \varepsilon \in \mathbb{R}^+\}$.
(e) $(\mathbb{R}^2, \mathcal{T})$, where $\mathcal{T} = \{D(X, \varepsilon) : X \in \mathbb{R}^2 \text{ and } \varepsilon \in \mathbb{R}^+\}$.

(Note: the last two are called ***Euclidean topologies***.)

21. Prove that if $|A| \geq \aleph_0$, then $\mathcal{T} = \{\varnothing\} \cup \{A : \overline{A} \text{ is finite}\}$ is a topology.
22. Let $\{\mathcal{T}_\alpha : \alpha \in \Lambda\}$ be a family of topologies in A.
    (a) Prove $(A, \bigcup_{\alpha\in\Lambda} \mathcal{T}_\alpha)$ is a topological space.
    (b) Show that $(A, \bigcap_{\alpha\in\Lambda} \mathcal{T}_\alpha)$ may not be a topological space.
23. Let $A \subseteq \mathbb{R}$ and let $\mathcal{U}$ be a family of open sets (in $\mathbb{R}$). We say that $\mathcal{U}$ is an ***open cover*** for A if $A \subseteq \bigcup \mathcal{U}$. Find infinite open covers for the following sets.
    (a) $\{0\}$
    (b) $[0, 2]$
    (c) $[0, 2] \cup [3, 5]$
    (d) $\mathbb{R}$
24. Let $A \subseteq \mathbb{R}$. An open cover $\mathcal{U} = \{U_\alpha : \alpha \in \Lambda\}$ of A has a ***finite subcover*** if there exists a finite set $\mathrm{K} \subseteq \Lambda$ such that $\{U_\alpha : \alpha \in \mathrm{K}\}$ is also a cover of A. Find finite subcovers of the open covers found in the first three parts of the previous problem.
25. Prove the ***Heine-Borel Theorem****: Let $A \subseteq \mathbb{R}$. Every open cover of A has a finite subcover of A if and only if A is closed and bounded. (Note: a subset of the real numbers that satisfies the conditions of the Heine-Borel Theorem is called ***compact***.)

---

*Émile Borel (Saint-Affrique, France, 1871 – Paris, France, 1956) and Heinrich Eduard Heine (Berlin, Prussia, 1821 – Halle, Germany, 1881): The theorem that bears their names was initially worked on by Cauchy. Heine stated the theorem in terms of uniformly continuous functions. (A function is ***uniformly continuous*** when for all $\varepsilon > 0$, there exists $\delta > 0$ such that $|x - y| < \delta$ implies $|f(x) - f(y)| < \varepsilon$.) Borel's contribution was to state the result in terms of open covers.

IV Appendices

A Logic Summary

Rules of Inference (1.3.2) and Replacement (1.4.2)

Addition [Add]	$p \vdash p \vee q$
Conjunction [Conj]	$p, q \vdash p \wedge q$
Constructive Dilemma [CD]	$(p \Rightarrow q) \wedge (r \Rightarrow s),\ p \vee r \vdash q \vee s$
Destructive Dilemma [DD]	$(p \Rightarrow q) \wedge (r \Rightarrow s),\ \sim q \vee \sim s \vdash \sim p \vee \sim r$
Disjunctive Syllogism [DS]	$p \vee q,\ \sim p \vdash q$
Modus Ponens [MP]	$p \Rightarrow q,\ p \vdash q$
Modus Tolens [MT]	$p \Rightarrow q,\ \sim q \vdash \sim p$
Simplification [Simp]	$p \wedge q \vdash p$
Transitivity [Trans]	$p \Rightarrow q,\ q \Rightarrow r \vdash p \Rightarrow r$
Associative Laws [Assoc]	$(p \wedge q) \wedge r \equiv p \wedge (q \wedge r)$
	$(p \vee q) \vee r \equiv p \vee (q \vee r)$
Commutative Laws [Com]	$p \wedge q \equiv q \wedge p$
	$p \vee q \equiv q \vee p$
Contrapositive Law [Contra]	$p \Rightarrow q \equiv \sim q \Rightarrow \sim p$
Distributive Laws [Distr]	$p \wedge (q \vee r) \equiv (p \wedge q) \vee (p \wedge r)$
	$p \vee (q \wedge r) \equiv (p \vee q) \wedge (p \vee r)$
De Morgan's Laws [DeM]	$\sim(p \wedge q) \equiv \sim p \vee \sim q$
	$\sim(p \vee q) \equiv \sim p \wedge \sim q$
Double Negation [DN]	$p \equiv \sim\sim p$
Exportation [Exp]	$(p \wedge q) \Rightarrow r \equiv p \Rightarrow (q \Rightarrow r)$
Material Equivalence [Equiv]	$p \Leftrightarrow q \equiv (p \Rightarrow q) \wedge (q \Rightarrow p)$
	$p \Leftrightarrow q \equiv (p \wedge q) \vee (\sim p \wedge \sim q)$
Material Implication [Impl]	$p \Rightarrow q \equiv \sim p \vee q$
Tautology [Taut]	$p \wedge p \equiv p$
	$p \vee p \equiv p$

Domain with the Quantifier (2.2.3)
If D is a domain for a predicate $P(x)$, then:

$$(\forall x \in D)P(x) \equiv (\forall x)[x \in D \Rightarrow P(x)],$$
$$(\exists x \in D)P(x) \equiv (\exists x)[x \in D \wedge P(x)].$$

Universal Instantiation [UI] (2.4.1)
If a is a constant from a nonempty domain D, then

$$(\forall x \in D)P(x) \vdash P(a).$$

Universal Generalization [UG] (2.4.2)
Let a be an arbitrary constant symbol from a nonempty domain D. If $P(a)$ contains no particular constants from D, then

$$P(a) \vdash (\forall x \in D)P(x).$$

Existential Generalization [EG] (2.4.3)
Let D be a nonempty domain. If a is an element from D, then

$$P(a) \vdash (\exists x \in D)P(x).$$

Existential Instantiation [EI] (2.4.4)
If a is a constant symbol from a nonempty domain D that has no prior occurrence in the proof, then

$$(\exists x \in D)P(x) \vdash P(\hat{a}).$$

Direct Proof [DP] (2.5.1)
For propositional forms $h_1, \ldots, h_k, p, q,$

$$h_1, \ldots, h_k \vdash p \Rightarrow q$$
if and only if
$$h_1, \ldots, h_k, p \vdash q$$

Indirect Proof [IP] (2.5.2)
For any propositional forms $h_1, \ldots, h_k, p,$

$$h_1, \ldots, h_k \vdash p$$
if and only if
$$h_1, \ldots, h_k, \sim p \vdash q \wedge \sim q$$

for some propositional form q.

Biconditional Proof [BP] (2.6.1)
Let $h_1, \ldots, h_k, p, q$ be propositional forms. Then,

$$h_1, \ldots, h_k \vdash p \Leftrightarrow q$$
if and only if
$$h_1, \ldots, h_k \vdash p \Rightarrow q \text{ and } h_1, \ldots, h_k \vdash q \Rightarrow p.$$

Shorter Rule of Biconditional Proof (Page 88)
Used when the steps for one part of a biconditional are the steps for the other in reverse order. These proofs are simply a sequence of biconditionals without the reasons. This is a good method when only rules of replacement are required.

Equivalence Rule (2.6.2)
To prove that the propositional forms $p_1, p_2, \ldots, p_k$ are pairwise equivalent, prove:

$$p_1 \Rightarrow p_2,\ p_2 \Rightarrow p_3,\ \ldots,\ p_{k-1} \Rightarrow p_k,\ p_k \Rightarrow p_1.$$

Proof of Disjunctions (2.6.3)
For any propositional forms p and q:

$$\sim p \Rightarrow q \vdash p \vee q.$$

Proof by Cases [CP] (2.6.4)
For any positive integer k, if $p \equiv p_1 \vee \cdots \vee p_k$, then

$$p_1 \Rightarrow q,\ \ldots,\ p_k \Rightarrow q \vdash p \Rightarrow q.$$

B Summation Notation

This appendix is provided as a review of summation notation. Let a_n and b_n be sequences of real numbers, $k \in \mathbb{R}$, and $m, n \in \mathbb{Z}^+$. (The work done here can be generalized to arbitrary rings and polynomial rings.)

1. We begin with the definition of summation notation.
   (a) Finite sum:

$$\sum_{i=1}^{n} a_i = a_1 + a_2 + \cdots + a_n$$

   (b) Infinite sum:

$$\sum_{i=1}^{\infty} a_i = a_1 + a_2 + a_3 + \cdots$$

   If the indices are not a sequence of positive integers but instead come from a set I, we may write

$$\sum_{i \in I} a_i.$$

2. The product of a real number and a sum uses the Distributive Law. The work here and below also applies to sums with other indices.

$$\begin{aligned} k\sum_{i=1}^{n} a_i &= k(a_1 + a_2 + \cdots + a_n) \\ &= (ka_1 + ka_2 + \cdots + ka_n) \\ &= \sum_{i=1}^{n} ka_i. \end{aligned}$$

3. Two sums can be added together as follows using the Associative and Commutative Laws:

$$\begin{aligned}\sum_{i=1}^{n} a_i + \sum_{i=1}^{n} b_i &= (a_1 + a_2 + \cdots + a_n) + (b_1 + b_2 + \cdots + b_n)\\ &= (a_1 + b_1) + (a_2 + b_2) + \cdots + (a_n + b_n)\\ &= \sum_{i=1}^{n}(a_i + b_i).\end{aligned}$$

4. Combining the last two remarks we note:

$$\sum_{i=1}^{n} a_i - \sum_{i=1}^{n} b_i = \sum_{i=1}^{n}(a_i - b_i).$$

5. Multiplication also relies on the Distributive Law:

$$\begin{aligned}\left(\sum_{i=1}^{n} a_i\right)\left(\sum_{j=1}^{m} b_i\right) &= (a_1 + a_2 + \cdots + a_n)(b_1 + b_2 + \cdots + b_m)\\ &= a_1(b_1 + \cdots + b_m) + \cdots + a_n(b_1 + \cdots + b_m)\\ &= a_1b_1 + \cdots + a_1b_m + \cdots + a_nb_1 + \cdots + a_nb_m\\ &= \sum_{i=1}^{n}\left(\sum_{j=1}^{m} a_ib_j\right)\\ &= \sum_{i=1}^{n}\sum_{j=1}^{m} a_ib_j.\end{aligned}$$

6. Lastly, we have this result that allows us to "switch" indices:

$$\begin{aligned}\sum_{i=1}^{n}\sum_{j=1}^{m} a_ib_j &= \left(\sum_{i=1}^{n} a_i\right)\left(\sum_{j=1}^{m} b_i\right)\\ &= \left(\sum_{j=1}^{m} b_i\right)\left(\sum_{i=1}^{n} a_i\right)\\ &= \sum_{j=1}^{m}\sum_{i=1}^{n} b_ja_i\\ &= \sum_{j=1}^{m}\sum_{i=1}^{n} a_ib_j.\end{aligned}$$

C Greek Alphabet

upper	lower	name
Α	α	*alpha*
Β	β	*beta*
Γ	γ	*gamma*
Δ	δ	*delta*
Ε	ε	*epsilon*
Ζ	ζ	*zeta*
Η	η	*eta*
Θ	θ	*theta*
Ι	ι	*iota*
Κ	κ	*kappa*
Λ	λ	*lambda*
Μ	μ	*mu*
Ν	ν	*nu*
Ξ	ξ	*xi*
Ο	o	*omicron*
Π	π	*pi*
Ρ	ρ	*rho*
Σ	σ	*sigma*
Τ	τ	*tau*
Υ	υ	*upsilon*
Φ	ϕ	*phi*
Χ	χ	*chi*
Ψ	ψ	*psi*
Ω	ω	*omega*

Bibliography

Artin, Michael. *Algebra*. Basel: Birkhäuser, 1998.

Bittinger, Marvin L. *Logic, Proofs, and Sets*. Reading, Mass.: Addison-Wesley, 1982.

Brabenec, Robert L. *Introduction to Real Analysis*. Boston: PWS-KENT Publishing Company, 1990.

Burton, David M. *Elementary Number Theory*. 4th ed. Dubuque, Iowa : McGraw Hill, 2001.

———. *The History of Mathematics: An Introduction*. 4th ed. Boston: WCB/McGraw Hill, 1999.

Cain, George L. *Introduction to General Topology*. Reading, Mass.: Addison-Wesley, 1994.

Copi, Irving M. *Symbolic Logic*. 5th ed. Upper Saddle River, NJ : Prentice Hall, 1979.

Dean, Neville. *The Essence of Discrete Mathematics*. London; New York: Prentice Hall, 1997.

Ebbinghaus, Heinz-Dieter, Jörg Flum, and Wolfgang Thomas. *Mathematical Logic*. 2nd ed. New York: Springer, 1996.

Enderton, Herbert B. *Elements of Set Theory*. New York: Academic Press, 1977.

Eves, Howard. *An Introduction to the History of Mathematics*. 6th ed. Philadelphia: Saunders College Publishing, 1990.

Fraleigh, John B. *A First Course in Abstract Algebra*. 6th ed. Reading, Mass.: Addison-Wesley, 1999.

Franceschetti, Donald R., ed. *Biographical Encyclopedia of Mathematicians*. New York: Marshall Cavendish, 1999.

Grimaldi, Ralph P. *Discrete and Combinatorial Mathematics: An Applied Introduction*. 4th ed. Reading, Mass.: Addison-Wesley, 1999.

Halmos, Paul R. *Naive Set Theory*. New York: Springer-Verlag, 1987.

Hungerford, Thomas W. *Abstract Algebra: An Introduction*. 2nd ed. Fort Worth: Saunders College Publishing, 1997.

Kolman, Bernard, Robert C. Busby, and Sharon Cutler Ross. *Discrete Mathematical Structures*. 4th ed. Upper Saddle River, N.J.: Prentice Hall, 2000.

Marsden, Jerrold E. *Elementary Classical Analysis*. 2nd ed. New York: W. H. Freeman and Company, 1993.

Roseman, Dennis. *Elementary Topology*. Upper Saddle River, N.J.: Prentice Hall, 1999.

Rosen, Kenneth H. *Discrete Mathematics and Its Applications*. 4th ed. Boston: WCB/McGraw Hill, 1999.

———. *Elementary Number Theory and its Applications*. 4th ed. Reading, Mass.: Addison-Wesley, 2000.

Ross, Kenneth A. and Charles R. B. Wright. *Discrete Mathematics*, 4th ed. Upper Saddle River, N.J.: Prentice Hall, 1999.

Rotman, Joseph J. *A First Course in Abstract Algebra*. 2nd ed. Upper Saddle River, N.J.: Prentice Hall, 2000.

Simmons, George F. *Introduction to Topology and Modern Analysis*. Malabar, Fla.: R.E. Krieger Publishing Company, 1983, 1963.

Strang, Gilbert. *Linear Algebra and Its Applications*. 2nd ed. Wellesley, Mass.: Wellesley-Cambridge Press, 1998.

Teller, Paul. *A Modern Formal Logic Primer: Predicate Logic & Metatheory*. Englewood Cliffs, N.J.: Prentice Hall, 1989.

van Dalen, Dirk. *Logic and Structure*. 3rd ed. New York: Springer, 1997.

Venit, Stewart and Wayne Bishop. *Elementary Linear Algebra*. 4th ed. Boston: PWS Publishing Company, 1996.

The source for some of the historical notes can be found at the The MacTutor History of Mathematics Archive, *http://www-history.mcs.st-andrews.ac.uk*, hosted by the University of St. Andrews, Scotland.

Selected Solutions

1.1. PROPOSITIONS

2. (a) No
   (b) Yes
   (c) Yes
   (d) No
   (e) No
   (f) Yes
   (g) No
   (h) No
   (i) No

3. (a) False
   (b) False
   (c) False
   (d) True
   (e) True
   (f) False
   (g) True
   (h) True
   (i) True
   (j) True
   (k) False
   (l) False

4. (a) or, if
   (c) if, if and only if

5. (a) Yes
   (c) No
   (e) Yes

6. (a) Hypothesis: *the triangle has two congruent sides*
   Conclusion: *it is isosceles*
   (c) Hypothesis: *the data is widely spread*
   Conclusion: *the standard deviation is large*
   (e) Hypothesis: *it has a solution*
   Conclusion: *the system of equations is consistent*

7. (a) The contrapositive of the first implication in the last problem is *if it is not isosceles, then the triangle does not have two congruent sides*, and its converse is *if it is isosceles, then the triangle has two congruent sides.*
   (b) The inverse of that same implication is *if the triangle does not have two congruent sides, then it is not isosceles.*

8. (b) *If the integer is divisible by four, it is even.*
   The integer is even if it is divisible by four.
   (d) *The integer being even is necessary for the integer to be divisible by four.*

9. (b) *A polygon is a triangle if and only if it has three sides.* (Note: if this sentence is not intended to be a definition, then it would be a conditional.)
   (d) *If it is a multiple of nine, then it is divisible by three.*
   (g) *If a line has positive slope, then it increases from left to right.*

1.2. PROPOSITIONAL FORMS

1. (a) See the solution to Exercise 6h.

2. (c) *The angle sum of a triangle is 180, and* $3 + 7 \neq 10$.
   (e) $3+7 = 10$ *if and only if the sine function is not continuous.*
   (g) *If the angle sum of a triangle is 180 or the sine function is continuous, then* $3 + 7 \neq 10$.
   (k) *Either the angle sum of a triangle is* 180 *or both* $3 + 7 = 10$ *and the sine function is continuous.*

3. (a) $R \wedge Q$
   (c) $Q \Rightarrow \sim R$
   (e) $R \Leftrightarrow (Q \Rightarrow \sim P)$

4. *The determinant being nonzero is sufficient for the matrix being invertible.*
   The matrix being invertible is a necessary condition for the determinant to be nonzero.

6. (c) We give the last column of the truth table:

P	Q	$(P \vee Q) \wedge \sim(P \wedge Q)$
T	T	F
T	F	T
F	T	T
F	F	F

(e) Use the order of operations:

P	Q	R	$P \wedge Q$	$P \wedge Q \vee R$
T	T	T	T	T
T	T	F	T	T
T	F	T	F	T
T	F	F	F	F
F	T	T	F	T
F	T	F	F	F
F	F	T	F	T
F	F	F	F	F

(h) The following is the first eight of sixteen lines and the last column of the truth table:

P	Q	R	S	$P \Rightarrow Q \Leftrightarrow R \Rightarrow S$
T	T	T	T	T
T	T	T	F	F
T	T	F	T	T
T	T	F	F	T
T	F	T	T	F
T	F	T	F	T
T	F	F	T	F
T	F	F	F	F

7. The propositional form for an exclusive or is $(P \vee Q) \wedge \sim(P \wedge Q)$.

8. (b) Define:
P := *some functions have a derivative*,
Q := *a square is round*, and
R := *everyone loves mathematics*.
So, $P \equiv$ T, $Q \equiv$ F, and $R \equiv$ F (we have to be honest!), and the sentence is represented as

$$(P \Rightarrow Q) \Rightarrow R.$$

Now to see that the proposition is true, examine the relevant line of the truth table:

P	Q	R	$P \Rightarrow Q$	$(P \Rightarrow Q) \Rightarrow R$
T	F	F	F	T

1.3. RULES OF INFERENCE

2. (b) Prove $[\sim(P \wedge Q) \wedge P] \Rightarrow \sim Q$ is a tautology.
(d) Prove $[(P \Rightarrow Q) \wedge (Q \Rightarrow R) \wedge P] \Rightarrow R$ is a tautology.

3. (b) Let $P \equiv$ T, $Q \equiv$ F and $R \equiv$ F.
(e) Either let $P \equiv$ F, $Q \equiv$ F, $R \equiv$ T, $S \equiv$ F, or $P \equiv$ T, $Q \equiv$ F, $R \equiv$ F, $S \equiv$ F.

4. (a) Trans
(b) Conj
(c) Add
(d) MP
(e) CD
(f) DS
(g) MT
(h) DD
(i) Simp

5. (a) $P \Rightarrow Q$
(c) $\sim P$
(e) *The number is not divisible by four.*
(g) *The function is decreasing.*

6. (a) Line 4: S does not follow by Simp.
Line 5: Addition requires only line 3.
Line 6: Here DD requires $\sim\sim Q \vee \sim\sim S$, and the conclusion is $\sim P \vee \sim R$.
Line 7: $\sim R$ does not follow by Simp.

7. (b) **Proof.**

1. $P \Rightarrow Q$ — Given
2. $Q \Rightarrow R$ — Given
3. P — Given
⟨Show $R \vee Q$⟩
4. $P \Rightarrow R$ — 1, 2 Trans
5. R — 3, 4 MP
6. $R \vee Q$ — 5 Add ■

8. (a) **Proof.**

1. $P \Rightarrow Q$ — Given
2. $P \vee (R \Rightarrow S)$ — Given
3. $\sim Q$ — Given
⟨Show $R \Rightarrow S$⟩
4. $\sim P$ — 1, 3 MT
5. $R \Rightarrow S$ — 2, 4 DS ■

(f) **Proof.**

1. $P \Rightarrow (Q \wedge R)$ — Given
2. $(Q \vee S) \Rightarrow (T \wedge U)$ — Given
3. P — Given
⟨Show T⟩
4. $Q \wedge R$ — 1, 3 MP
5. Q — 4 Simp
6. $Q \vee S$ — 5 Add
7. $T \wedge U$ — 2, 6 MP
8. T — 7 Simp ■

(j) **Proof.**

1. $P \Rightarrow Q$	Given
2. $Q \Rightarrow R$	Given
3. $R \Rightarrow S$	Given
4. $S \Rightarrow T$	Given
5. $P \vee R$	Given
6. $\sim R$	Given
$\langle$Show $T\rangle$	
7. $P \Rightarrow R$	1, 2 Trans
8. $R \Rightarrow T$	3, 4 Trans
9. $(P \Rightarrow R) \wedge (R \Rightarrow T)$	7, 8 Conj
10. $R \vee T$	5, 9 CD
11. T	6, 10 DS ■

(l) **Proof.**

1. $P \Rightarrow Q$	Given
2. $Q \Rightarrow R$	Given
3. $R \Rightarrow S$	Given
4. $(P \vee Q) \wedge (R \vee S)$	Given
$\langle$Show $Q \vee S\rangle$	
5. $Q \Rightarrow S$	2, 3 Trans
6. $(P \Rightarrow Q) \wedge (Q \Rightarrow S)$	1, 5 Conj
7. $P \vee Q$	4 Simp
8. $Q \vee S$	6, 7 CD ■

1.4. RULES OF REPLACEMENT

2. (b) Both forms are tautologies.
   (c) The relevant columns are:

P	Q	$P \wedge Q$	$(P \Leftrightarrow Q) \wedge (P \vee Q)$
T	T	T	T
T	F	F	F
F	T	F	F
F	F	F	F

   (f) Both forms are false when $P \equiv$ T, $Q \equiv$ T, and $R \equiv$ F.

3. (a) Assoc (b) DN (c) Com (d) DM (e) Contra (f) Impl (g) Equiv (h) Taut (i) Com (j) Dist (k) Contra (l) Exp

4. (d) $\sim P \vee \sim Q$
   (h) $(\sim P \vee Q) \vee P$
   (l) $(\sim Q \Leftrightarrow T) \wedge Q \vee P$

5. (b) Line 3: This is not a correct application of DeM.
   Line 4: This is not Com.
   Line 5: Simp allows us to conclude the first conjunct.

6. (b) **Proof.**

1. $\sim(P \wedge Q) \Rightarrow (R \vee S)$	Given
2. $\sim P$	Given
3. $\sim S$	Given
$\langle$Show $R\rangle$	
4. $\sim P \vee \sim Q$	2 Add
5. $\sim(P \wedge Q)$	4 DeM
6. $R \vee S$	1, 5 MP
7. $S \vee R$	6 Com
8. R	3, 7 DS ■

7. (b) **Proof.**

1. P	Given
$\langle$Show $\sim Q \Rightarrow P\rangle$	
2. $P \vee Q$	1 Add
3. $Q \vee P$	2 Com
4. $\sim\sim Q \vee P$	3 DN
5. $\sim Q \Rightarrow P$	4 Impl ■

(d) **Proof.**

1. $P \Rightarrow Q$	Given
$\langle$Show $(P \wedge R) \Rightarrow Q\rangle$	
2. $\sim P \vee Q$	1 Impl
3. $(\sim P \vee Q) \vee \sim R$	2 Add
4. $\sim R \vee (\sim P \vee Q)$	3 Com
5. $(\sim R \vee \sim P) \vee Q$	4 Assoc
6. $(\sim P \vee \sim R) \vee Q$	5 Com
7. $\sim(P \wedge R) \vee Q$	6 DeM
8. $(P \wedge R) \Rightarrow Q$	7 Impl ■

(k) **Proof.**

1. $P \Rightarrow \sim(Q \Rightarrow R)$	Given
$\langle$Show $P \Rightarrow \sim R\rangle$	
2. $\sim P \vee \sim(\sim Q \vee R)$	1 Impl
3. $\sim P \vee (\sim\sim Q \wedge \sim R)$	2 DeM
4. $\sim P \vee (\sim R \wedge \sim\sim Q)$	3 Com
5. $(\sim P \vee \sim R) \wedge (\sim P \vee \sim\sim Q)$	4 Dist
6. $\sim P \vee \sim R$	5 Simp
7. $P \Rightarrow \sim R$	6 Impl ■

(n) **Proof.**

1. $P \Leftrightarrow (Q \wedge R)$	Given
$\langle$Show $P \Rightarrow Q\rangle$	
2. $[P \Rightarrow (Q \wedge R)] \wedge [(Q \wedge R) \Rightarrow P]$	1 Equiv
3. $P \Rightarrow (Q \wedge R)$	2 Simp
4. $\sim P \vee (Q \wedge R)$	3 Impl
5. $(\sim P \vee Q) \wedge (\sim P \vee R)$	4 Dist
6. $\sim P \vee Q$	5 Simp
7. $P \Rightarrow Q$	6 Impl ■

8. (d) **Proof.**

1. $P \Rightarrow (Q \Rightarrow R)$	Given
2. $R \Rightarrow (S \wedge T)$	Given
$\langle$Show $P \Rightarrow (Q \Rightarrow T)\rangle$	
3. $(P \wedge Q) \Rightarrow R$	1 Exp
4. $(P \wedge Q) \Rightarrow (S \wedge T)$	2, 3 Trans
5. $\sim(P \wedge Q) \vee (S \wedge T)$	4 Impl
6. $[\sim(P \wedge Q) \vee S] \wedge [\sim(P \wedge Q) \vee T]$	5 Dist
7. $[\sim(P \wedge Q) \vee T] \wedge [\sim(P \wedge Q) \vee S]$	6 Com
8. $\sim(P \wedge Q) \vee T$	7 Simp
9. $(P \wedge Q) \Rightarrow T$	8 Impl
10. $P \Rightarrow (Q \Rightarrow T)$	9 Exp ■

(g) **Proof.**

1. $(P \vee Q) \Rightarrow (R \wedge S)$	Given
2. $\sim P \Rightarrow (T \Rightarrow \sim T)$	Given
3. $\sim R$	Given
(Show $\sim T$)	
4. $\sim R \vee \sim S$	3 Add
5. $\sim(R \wedge S)$	4 DeM
6. $\sim(P \vee Q)$	1, 5 MT
7. $\sim P \wedge \sim Q$	6 DeM
8. $\sim P$	7 Simp
9. $T \Rightarrow \sim T$	2, 8 MP
10. $\sim T \vee \sim T$	9 Impl
11. $\sim T$	10 Taut ■

9. (a) $P \wedge Q \equiv \sim(\sim P \vee \sim Q)$, $P \Rightarrow Q \equiv \sim P \vee Q$, and $P \Leftrightarrow Q \equiv \sim(\sim P \vee \sim Q) \vee \sim(P \vee Q)$.

2.1. PREDICATES AND SETS

1. (a) $2^2 - 1 = 0$, false
   (b) 1 *is a rational number*, true
   (c) f *is a linear function*, true
   (d) $-4 \in \mathbb{N}$, false
   (e) *there is a natural number n so that* $6 = 2^n$, false
   (f) *every integer is greater than 0*, false
   (g) $1 + 3 = 4$, true
   (h) $3 = 1$, false
   (i) $f - g = 1$, true
   (j) $\sin x$ *and* e^x *are increasing functions*, false
   (k) $0 \in \mathbb{Z}$ *and* $0 \in \mathbb{N}$, true
   (l) $\cos^2 x + \sin^2 x = 1$, true for all $x \in \mathbb{R}$

2. (a) true (b) false (c) true (d) true (e) true (f) true (g) true (h) false (i) false (j) false (k) true (l) false (m) false (n) false

3. (a) No (c) No (g) No

4. (c) $\{0, 1, 2, 3, \ldots\}$
   (e) $\{.01, .02, .03, \ldots, .98, .99\}$

5. (a) $(4, \infty)$
   (b) $[-6, -5]$

6. (b) $x \in \mathbb{R}$ *and* $-3 < x \leq 3$.
   (d) $x = a/b$ *where* $a, b \in \mathbb{Z}$ *and* $b \neq 0$.
   (g) $x \in \mathbb{Z}$ *and* $x > 0$.
   (i) $x = na$ *where n is a positive even integer.*

2.2. QUANTIFICATION

1. (a) *For every* $x \in \mathbb{Z}$, $x - x = 0$.
   $x - x = 0$ *for all* $x \in \mathbb{Z}$.
   (c) *There exists* $x \in \mathbb{Z}$ *such that for every integer* y, $y + x = x$.
   There exists $x \in \mathbb{Z}$ *so that* $y + x = x$ *holds for all* $y \in \mathbb{Z}$.

2. (a) $(\forall x \in \mathbb{Z})R(x)$
   (c) $(\exists x \in \mathbb{N})P(x)$
   (e) $R(x) \vee (\exists x \in \mathbb{Z}){\sim}R(x)$
   (g) $(\forall x \in \mathbb{Q})(\exists y < x)Q(y)$

3. (c) *For every* $x \in \mathbb{R}$, *there exists* $y \in \mathbb{R}$ *such that* $x > 5$ *and* $y + 5 = 9$.
   (d) *There is a real number x such that* $x^2 \neq 4$, *or for every* $x \in \mathbb{R}$, $x + 5 = 9$ *if and only if* $x \leq 5$.

4. (a) True (b) True (c) False (d) False (e) True (f) False (g) False (h) True (i) True (j) False (k) True

6. (a) False (c) False (e) True

7. (b) True (d) True (f) False

8. (c) $(\forall x > 0)(\exists y \leq -3)R(x, y)$

9. (a) all bound
   (c) last one free

10. $R(1, 2) \Rightarrow (\forall x)[S(x) \wedge (\exists y)T(y, 3)]$

11. (c) $(\forall x)$: $(\exists y)(\exists z)P(x, y, z)$
    $(\exists y)$: $(\exists z)P(x, y, z)$
    $(\exists z)$: $P(x, y, z)$

12. $(\exists k)(\forall x)[f(x) = k]$

16. (b) $(\exists x, y)P(x) \wedge (\forall z)Q(z, y)$
    (e) $(\forall x \in \mathbb{Z})(\exists y \in \mathbb{Q})(\forall z, w \in \mathbb{R})$ $P(x, y, z, w)$

2.3. NEGATING QUANTIFIERS

1. (a) No (b) No (c) No (d) No (e) Yes (f) No

2. (a) $(\forall x)[Q(x) \wedge \sim R(x)]$
   (c) $(\forall x)(\forall y)[\sim P(x) \wedge \sim Q(x, y)]$
   (e) $(\forall x)R(x) \wedge (\exists x)([Q(x) \wedge P(x)] \vee [\sim Q(x) \wedge \sim P(x)])$

3. (a) No:
   There is a real number that does not have a square root.
   There exists a real number that does have a square root.

(c) No:
Some multiple of four is not a multiple of two.
All multiples of two are not multiples of four.

(e) No:
There exists $x \in \mathbb{R}$ such that x is odd but x^2 is even.
For every $x \in \mathbb{R}$, x is odd and x^2 is odd.

(g) Yes

4. (a) *There exists $x \in \mathbb{R}$ such that for all $y \in \mathbb{Z}$, $y/x \neq 9$.*
   (c) *There is no function with y-intercept equal to -3.*

5. (a) *$x = 0$ is not a solution to $x + 1 = 0$.*
   (c) *The product of 1 and 3 is odd.*
   (e) *The function $f(x) = x^2$ has a minimum but not a maximum.*

2.4. PROOFS WITH QUANTIFIERS

1. (a) Line 3: c should be written with a hat.
   Line 4: c is not a new symbol.
   Line 5: Wrong application of DS.
   (c) Line 4: c is not a new symbol.
   Line 5: c lost its hat!
   Line 6: This is an illegal use of UG since $\hat{c}$ is particular in line 5.

2. (b) **Proof.**

1. $(\forall x)P(x)$	Given
2. $(\forall x)[Q(x) \Rightarrow \sim P(x)]$	Given
$\langle$Show $(\forall x)\sim Q(x)\rangle$	
3. $P(a)$	1 UI
4. $Q(a) \Rightarrow \sim P(a)$	2 UI
5. $\sim\sim P(a)$	3 DN
6. $\sim Q(a)$	4, 5 MT
7. $(\forall x)\sim Q(x)$	6 UG ■

(e) **Proof.**

1. $(\exists x)P(x)$	Given
$\langle$Show $(\exists x)[P(x) \vee Q(x)]\rangle$	
2. $P(\hat{c})$	1 EI
3. $P(\hat{c}) \vee Q(\hat{c})$	2 Add
4. $(\exists x)[P(x) \vee Q(x)]$	3 EG ■

(i) **Proof.**

1. $(\exists x)P(x)$	Given
2. $(\forall x)[P(x) \Rightarrow Q(x)]$	Given
$\langle$Show $(\exists x)Q(x)\rangle$	
3. $P(\hat{c})$	1 EI
4. $P(\hat{c}) \Rightarrow Q(\hat{c})$	2 UG
5. $Q(\hat{c})$	3, 4 MP
6. $(\exists x)Q(x)$	5 EG ■

(m) **Proof.**

1. $(\exists x)(\exists y)P(x, y)$	Given
$\langle$Show $(\exists y)(\exists x)P(x, y)\rangle$	
2. $(\exists y)P(\hat{c}, y)$	1 EI
3. $P(\hat{c}, \hat{d})$	2 EI
4. $(\exists x)P(x, \hat{d})$	3 EG
5. $(\exists y)(\exists x)P(x, y)$	4 EG ■

3. (b) **Proof.**

1. $(\forall x)[P(x) \Rightarrow Q(x)]$	Given
2. $(\forall x)[Q(x) \Rightarrow R(x)]$	Given
3. $\sim(\forall x)R(x)$	Given
$\langle$Show $(\exists x)\sim P(x)\rangle$	
4. $(\exists x)\sim R(x)$	3 QN
5. $\sim R(\hat{c})$	4 EI
6. $P(\hat{c}) \Rightarrow Q(\hat{c})$	1 UI
7. $Q(\hat{c}) \Rightarrow R(\hat{c})$	2 UI
8. $P(\hat{c}) \Rightarrow R(\hat{c})$	6, 7 Trans
9. $\sim P(\hat{c})$	5, 8 MT
10. $(\exists x)\sim P(x)$	9 EG ■

5. (b) $-24 = 6(-4)$
   (d) Since

$$6n + 12 = 6(n + 2)$$

and

$$n + 2 \in \mathbb{Z},$$

6 divides $6n + 12$.

   (e) $2^3 \cdot 3^4 \cdot 7^5 = 6(2^2 \cdot 3^3 \cdot 7^5)$

6. (a) Since $a = 1(a)$ and $a \in \mathbb{Z}$, 1 divides a.

7. (c) Let a be an even integer. This means $a = 2n$ for some $n \in \mathbb{Z}$. Then,

$$a^2 = (2n)^2 = 2(2n^2).$$

Thus, a^2 is even.

8. (b) $x = 1$ satisfies the equation since

$$1^2 + 2(1) - 3 = 1 + 2 - 3 \\ = 0.$$

(Note: we do not need to find both solutions to prove this!)

9. (c) Let $a \in \mathbb{R}$. We must find y so that $a - y = 10$, so we should choose y to equal $a - 10$. Check:

$$a - (a - 10) = a - a + 10 = 10.$$

   (e) Let $a, b, c \in \mathbb{R}$. The quadratic formula guarantees a complex solution to $ax^2 + bx + c = 0$. (Remember that a real number is also a complex number.)

2.5. DIRECT AND INDIRECT PROOF

1. (a) Line 4: The reason should be 3, DN. Line 11: The justification for this line is not valid. It refers to a line within the subproof.

2. (d) **Proof.**

Line	Statement	Reason
1.	$P \Rightarrow Q$	Given
2.	$R \Rightarrow Q$	Given
	$\langle$Show $(P \vee R) \Rightarrow Q\rangle$	
3.	→ $P \vee R$	Assumption
	$\langle$Show $Q\rangle$	
4.	$(P \Rightarrow Q) \wedge (R \Rightarrow Q)$	1, 2 Conj
5.	$Q \vee Q$	3, 4 CD
6.	Q	5 Taut
7.	$(P \vee R) \Rightarrow Q$	3-6 DP ■

(g) **Proof.**

Line	Statement	Reason
1.	$R \Rightarrow \sim S$	Given
	$\langle$Show $(P \wedge Q) \Rightarrow (R \Rightarrow \sim S)\rangle$	
2.	→ $P \wedge Q$	Assumption
	$\langle$Show $R \Rightarrow \sim S\rangle$	
3.	→ R	Assumption
	$\langle$Show $\sim S\rangle$	
4.	$\sim S$	1, 3 MP
5.	$R \Rightarrow \sim S$	3-4 DP
6.	$(P \wedge Q) \Rightarrow (R \Rightarrow \sim S)$	2-5 DP ■

(i) **Proof.**

Line	Statement	Reason
1.	$P \Rightarrow (Q \Rightarrow R)$	Given
2.	$R \Rightarrow (S \wedge T)$	Given
	$\langle$Show $P \Rightarrow (Q \Rightarrow T)\rangle$	
3.	→ P	Assumption
	$\langle$Show $Q \Rightarrow T\rangle$	
4.	→ Q	Assumption
	$\langle$Show $T\rangle$	
5.	$Q \Rightarrow R$	1, 3 MP
6.	R	4, 5 MP
7.	$S \wedge T$	2, 6 MP
8.	$T \wedge S$	7 Com
9.	T	8 Simp
10.	$Q \Rightarrow T$	4-9 DP
11.	$P \Rightarrow (Q \Rightarrow T)$	3-10 DP ■

3. (b) **Proof.**

Line	Statement	Reason
1.	$P \vee (Q \wedge R)$	Given
2.	$P \Rightarrow S$	Given
3.	$Q \Rightarrow S$	Given
	$\langle$Show $S\rangle$	
4.	→ $\sim S$	Assumption
5.	$\sim P$	2, 4 MT
6.	$Q \wedge R$	1, 5 DS
7.	Q	6 Simp
8.	S	3, 7 MP
9.	$S \wedge \sim S$	4, 9 Conj
10.	S	4-9 IP ■

(e) **Proof.**

Line	Statement	Reason
1.	$[(P \vee Q) \vee R] \Rightarrow (Q \wedge R)$	Given
	$\langle$Show $\sim P \vee (Q \wedge R)\rangle$	
2.	→ $\sim[\sim P \vee (Q \wedge R)]$	Assumption
3.	$\sim\sim P \wedge \sim(Q \wedge R)$	2 DeM
4.	$P \wedge \sim(Q \wedge R)$	3 DN
5.	P	4 Simp
6.	$P \vee Q$	5 Add
7.	$(P \vee Q) \vee R$	6 Add
8.	$Q \wedge R$	1, 7 MP
9.	$\sim(Q \wedge R) \wedge P$	4 Com
10.	$\sim(Q \wedge R)$	9 Simp
11.	$(Q \wedge R) \wedge \sim(Q \wedge R)$	8, 10 Conj
12.	$\sim P \vee (Q \wedge R)$	2-11 IP ■

4. (a) **Proof.**

Line	Statement	Reason
1.	$P \Rightarrow Q$	Given
2.	$P \vee (R \Rightarrow S)$	Given
3.	$\sim Q$	Given
	$\langle$Show $R \Rightarrow S\rangle$	
4.	→ R	Assumption
	$\langle$Show $S\rangle$	
5.	→ $\sim S$	Assumption
6.	$\sim P$	1, 3 MT
7.	$R \Rightarrow S$	2, 6 DS
8.	$\sim R$	5, 7 MT
9.	$R \wedge \sim R$	4, 8 Conj
10.	S	5-9 IP
11.	$R \Rightarrow S$	4-10 DP ■

5. (b) **Proof.**

Line	Statement	Reason
1.	$(P \wedge Q) \vee (R \wedge S)$	Given
	$\langle$Show $\sim S \Rightarrow (P \wedge Q)\rangle$	
2.	→ $\sim(P \wedge Q)$	Assumption
	$\langle$Show $\sim\sim S\rangle$	
3.	$R \wedge S$	1, 2 DS
4.	$S \wedge R$	3 Com
5.	S	4 Simp
6.	$\sim\sim S$	5 DN
7.	$\sim(P \wedge Q) \Rightarrow \sim\sim S$	2-6 DP
8.	$\sim S \Rightarrow (P \wedge Q)$	7 Contra ■

6. (a) Suppose a divides b. Then $b = ak$ for some $k \in \mathbb{Z}$. Multiplying, $bd = akd$, which means a divides bd.

7. (a) Let a and b be even. This means $a = 2k$ and $b = 2\ell$ where $k, \ell \in \mathbb{Z}$. Thus,

$$a + b = 2k + 2\ell = 2(k + \ell),$$

which is even.

(e) Let $a = 2k + 1$ and $b = 2\ell + 1$ for some $k, \ell \in \mathbb{Z}$. Then $ab = (2k + 1)(2\ell + 1) = 2(2k\ell + k + \ell) + 1$, an odd integer.

8. The first proof goes like this: Take two even integers a and b. Let $k, \ell \in \mathbb{Z}$ such that $a = 2k$ and $b = 2\ell$. Assume $a + b$

is odd. This means $a + b = 2m + 1$ for some integer m. Therefore, $2k + 2\ell = 2m + 1$, which yields $1 = 2(k + \ell - m)$. This is a contradiction, for 1 is not even. Therefore, $a + b$ is even.

10. (a) Take $n \in \mathbb{Z}$ and assume that n is odd. This means there exists an integer k such that $n = 2k + 1$. Calculating we find $n^4 = 16k^4 + 32k^3 + 24k^2 + 8k + 1$, which is odd. Thus, we have proven the contrapositive of the implication.

11. (b) Let a and b be solutions to the equation. Then

$$\sqrt{2a - 5} - 2$$

equals

$$\sqrt{2b - 5} - 2.$$

Therefore,

$$\sqrt{2a - 5} = \sqrt{2b - 5}.$$

Squaring both sides yields $2a - 5 = 2b - 5$. Hence, $a = b$.

2.6. MORE METHODS

1. (a) **Proof.**

1.	$(P \vee Q) \Rightarrow \sim R$	Given
2.	$S \Rightarrow R$	Given
3.	$(\sim P \vee Q) \Rightarrow S$	Given
	$\langle$Show $P \Leftrightarrow \sim S\rangle$	
$(\Leftarrow)$ 4.	P	Assumption
	$\langle$Show $\sim S\rangle$	
5.	$P \vee Q$	4 Add
6.	$\sim R$	1, 5 MP
7.	$\sim S$	2, 6 MT
$(\Rightarrow)$ 8.	$\sim S$	Assumption
	$\langle$Show $P\rangle$	
9.	$\sim(\sim P \vee Q)$	3, 8 MT
10.	$\sim\sim P \wedge \sim Q$	9 DeM
11.	$P \wedge \sim Q$	10 DN
12.	P	11 Simp
13.	$P \Leftrightarrow \sim S$	4-12 BP ■

2. (b) Each line appeals to only one rule of replacement:

$$P \Rightarrow (Q \vee R)$$
$$\Leftrightarrow \sim P \vee (Q \vee R)$$
$$\Leftrightarrow (\sim P \vee Q) \vee R$$
$$\Leftrightarrow (\sim P \vee \sim\sim Q) \vee R$$
$$\Leftrightarrow \sim(P \wedge \sim Q) \vee R$$
$$\Leftrightarrow (P \wedge \sim Q) \Rightarrow R$$

4. (b) Let a be an integer.

   $(\Rightarrow)$ Let a be an odd integer. This means there exists $k \in \mathbb{Z}$ such that $a = 2k + 1$. Then $a + 1 = 2(k + 1)$, an even integer.

   $(\Leftarrow)$ Now suppose that $a + 1$ is even. In other words, $a + 1 = 2k$, some integer k. Therefore, $a = 2k - 1 = 2(k - 1) + 1$, so a is odd.

   (e) Suppose $c \neq 0$. Then:

   a divides b

$$\Leftrightarrow b = ak, \text{ some } k \in \mathbb{Z}$$
$$\Leftrightarrow bc = ack, \text{ some } k \in \mathbb{Z}.$$

   The second biconditional is true because $c \neq 0$.

7. We will prove the three implications.
   - Assume a is divisible by 3. Then $a = 3k$ for some $k \in \mathbb{Z}$. This gives $3a = 9k$, so $3a$ is divisible by 9.
   - Suppose $3a$ is divisible by 9. This means there exists an integer k such that $3a = 9k$. Then, $a = 3k$ and $a + 3 = 3k + 3 = 3(k + 1)$. Thus, $a + 3$ is divisible by 3.
   - Lastly, let $a + 3 = 3\ell$, some $\ell \in \mathbb{Z}$. So, $a = 3\ell - 3 = 3(\ell - 1)$, and this means a is divisible by 3.

8. Take a, $b \in \mathbb{Z}$ and suppose $ab = 2k$ for some $k \in \mathbb{Z}$. Assume a is not even, so $a = 2\ell + 1$ for some integer ℓ. Then $(2\ell + 1)b = 2k$ and $b = 2(k - b\ell)$, so b is even.

9. (a) Let $a \in \mathbb{Z}$. Suppose $a = 0$ or $b = 0$. If $a = 0$, then $ab = 0 \cdot b = 0$, and if $b = 0$, then $ab = a \cdot 0 = 0$.

   (e) Consider two cases:

   (Case 1) The line is vertical. Then it can be described by the equation $x = a$ for some $a \in \mathbb{R}$. This intersects the x-axis at $(a, 0)$.

   (Case 2) The line is not vertical. This means that all points on the line satisfy $y = mx + b$ for some m, $b \in \mathbb{R}$ and $m \neq 0$. This line then intersects the x-axis at $x = -b/m$.

10. (a) We have two cases to consider: First assume $a = 2k$ for some integer k. Then $a(a+1) = 2k(2k+1) = 4k^2 + 2k = 2(2k^2 + k)$, an even integer. Now let $a = 2k + 1$ for some $k \in \mathbb{Z}$. Calculating we again find that a is even since $a(a+1) = (2k+1)(2k+2) = 4k^2+6k+2 = 2(2k^2+3k+1)$.

11. (a) Although absolute value involves two cases, we need three for this one. In all three of them, $|x| = |-x|$:

    (Case 1) If $x = 0$, then $x = -x = 0$.
    (Case 2) Suppose $x > 0$. Then $|x| = x$ and $|-x| = --x = x$.
    (Case 3) If $x < 0$, then $|x| = -x$ and $|-x| = -x$.

3.1. SET BASICS

1. (a) False (b) True (c) True (d) True (e) True (f) True (g) False (h) True (i) True (j) True

2. (a) conjunction
   (b) Venn
   (c) rational
   (d) intersection or union
   (e) the universe
   (f) sometimes
   (g) sometimes
   (h) elements

3. (b) $\{0\}$
   (d) $\{0,\ x^2+x+1,\ 2x^2+2x+2,\ \ldots\}$

4. (b) $\{a/b : a,\ b \in \mathbb{Z}^+\}$
   (e) $\{(x,\ y) \in \mathbb{R}^2 : x > 0 \text{ and } y < 0\}$
   (i) $\{x : x \neq x\} = \varnothing$
   (l) $\{ax^3+bx^2+cx+d : a,\ b,\ c,\ d \in \mathbb{R}\}$

5. (c) $\{0, 2\}$
   (e) $\{1, 3, 5, 7, 8, 9, 10\}$
   (f) $\{(0, 3), (0, 4), (0, 5), \ldots, (6, 6)\}$
   (g) $\{0, 2, 4, 6\}$
   (j) U
   (l) $\{0, 2, 4, 6\}$

6. (a) $\{x \in \mathbb{R} : a \leq x \leq b\}$
   (b) $\{x \in \mathbb{R} : a \leq x\}$

7. (a) $(2.5,\ 3]$ (d) $(4,\ 12]$

8. $\varnothing$

9. (a) $\{17\}$ (d) $[6,\ 32) \setminus \{17\}$

10. (b)

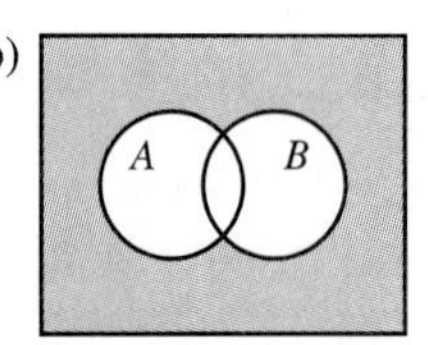

(c)

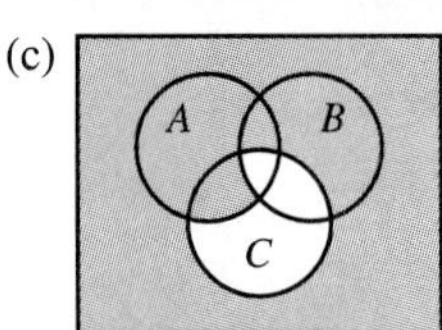

(e)

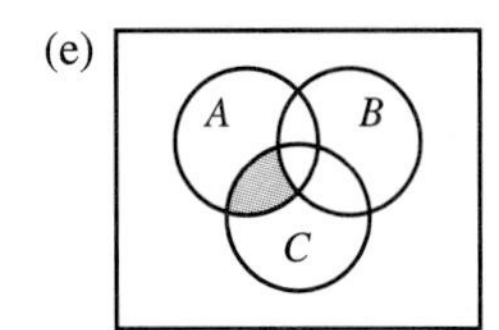

11. (a) F (c) A (e) D (g) C (i) E

3.2. SUBSETS

1. (a) False (b) True (c) True (d) False (e) False (f) False (g) False (h) False (i) True (j) True

2. (a) True (b) False, .5 (c) True (d) False, π (e) True (f) False, $\pi/4$ (g) False, .5 (h) True

3. (a) Let $A \subseteq B$; $x \in A$; $x \in B$; *Modus Ponens*

4. (b) Let $a \in (0,\ 1)$. Then $a \in \mathbb{R}$ and $0 < a < 1$. A little algebra then gives us $0 \leq a \leq 1$. Hence, $a \in [0,\ 1]$.
   (c) $0 \in [0,\ 1]$ but $0 \notin (0,\ 1)$.

5. (c) Let $a \in A$. Then $a \in A$ or $a \in B$ by Addition. Therefore, $a \in A \cup B$.
   (g) Assume $A \subseteq B$ and let $a \in A \cap C$. By definition $a \in A$ and $a \in C$. Thus, $a \in B$, and then $a \in B \cap C$.

(i) Suppose A is nonempty. This means there exists $a \in A$. Hence, $a \notin \overline{A}$.

(k) Let $A \subseteq B$. Suppose $(1, a) \in \{1\} \times A$. By hypothesis, $a \in B$. Therefore, $(1, a) \in \{1\} \times B$.

7. Let $A = \{1\}$ and $B = \{1, 2\}$.

8. (a) $B \subseteq A$ (b) $A = B$

9. Let $a \in \mathbb{R}$. Then, $a = a + 0i$, which means $a \in \mathbb{C}$.

11. (a) Assume $(\forall x)[P(x) \Rightarrow Q(x)]$. Let $a \in \{x : P(x)\}$. This means $P(a)$. Hence, $Q(a)$ follows by *Modus Ponens*, so $a \in \{x : Q(x)\}$.

(d) If a is any element, then the hypothesis gives us that $Q(a)$ is true. Hence, $a \in \{x : Q(x)\}$.

3.3. EQUALITY OF SETS

1. (a) No, $\cap$ is not a connective.
(b) Yes
(c) No, $\wedge$ is not a set operation.
(d) Yes
(e) No, the phrase $x \in \{1, 2\}$ is not a set.
(f) Yes
(g) No, $\vee$ is not a set operation.
(h) Yes

2. (a) $x \in \overline{A}$ and $x \in B$;
$x \in B \setminus A$
(b) $x \in A \setminus (B \setminus C)$;
$x \in A$ and $\sim(x \in B \setminus C)$;
$x \in A$ and $\sim(x \in B$ and $x \notin C)$;
$x \in A$ and $(x \in \overline{B}$ or $x \in C)$;
$x \in A$ and $x \in \overline{B} \cup C$
(c) $A \cup B = B$;
$B \subseteq A \cup B$;
$x \in A$ or $x \in B$;
$x \in B$;
$x \in A$;
$x \in B$;
$A \cup B = B$;
$x \in A \cup B$;
$x \in B$

4. (a) $a \in \overline{\varnothing} \Leftrightarrow a \notin \varnothing \Leftrightarrow a \in U$

(d) Let x be an arbitrary element of the universe. If $x \notin A$, then $x \in \overline{A}$. Hence, $x \in A \cup \overline{A}$. Therefore, since we also have $A \cup \overline{A} \subseteq U$, $A \cup \overline{A}$ is equal to U.

(f) Let $a \in A \setminus A$. Then $a \in A$ and $a \notin A$, a contradiction. So, $A \setminus A = \varnothing$.

(i) For every element a:

$$a \in A \setminus B \Leftrightarrow a \in A \wedge a \notin B$$
$$\Leftrightarrow a \in A \wedge a \in \overline{B}$$
$$\Leftrightarrow a \in A \cap \overline{B}$$

5. (a)

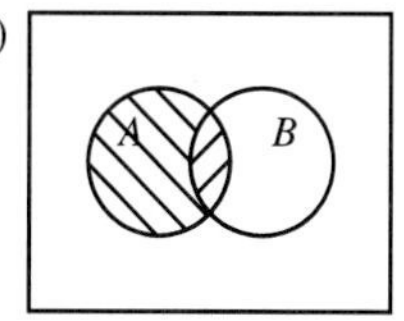

For every element a,

$$a \in (A \cap B) \cup (A \cap \overline{B})$$

is equivalent to

$$(a \in A \wedge a \in B) \vee (a \in A \wedge a \in \overline{B})$$

which is then equivalent to

$$(a \in A \vee a \in A) \wedge (a \in A \vee a \in \overline{B}) \wedge$$
$$(a \in B \vee a \in A) \wedge (a \in B \wedge a \in \overline{B}).$$

This is then equivalent to $a \in A$ because $a \in B \vee a \in \overline{B}$ is always true and $a \in B \wedge a \in \overline{B}$ is always false.

(e)

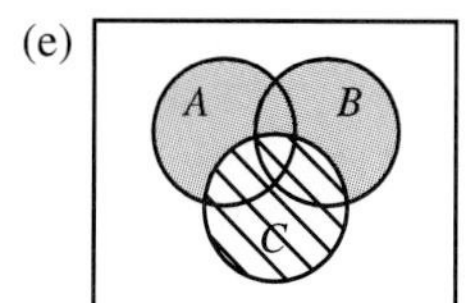

$$a \in (A \cup B) \setminus C$$
$$\Leftrightarrow (a \in A \cup B) \wedge a \notin C$$
$$\Leftrightarrow (a \in A \vee a \in B) \wedge a \notin C$$
$$\Leftrightarrow (a \in A \wedge a \notin C) \vee (a \in B \wedge a \notin C)$$
$$\Leftrightarrow (a \in A \setminus C) \vee (a \in B \setminus C)$$
$$\Leftrightarrow a \in (A \setminus C) \cup (B \setminus C)$$

6. (c) Let $U \subseteq A$. Since we already know that $A \subseteq U$, we have $A = U$.

(e) We will show both directions.

($\Rightarrow$) Assume $A \times B = \varnothing$, and suppose $A \neq \varnothing$. So, there exists $a \in A$. If B is nonempty, then there is a $b \in B$. Thus,

$(a, b) \in A \times B$, a contradiction. Therefore, $B = \varnothing$.

($\Leftarrow$) Let $A = \varnothing$ or $B = \varnothing$. Hence, either $a \in A$ or $b \in B$ is always false. Thus, $A \times B = \varnothing$.

7. Take $a, c, m \in \mathbb{Z}$.
   ($\subseteq$) Let $x \in A$. Then $x = a + mk$ for some $k \in \mathbb{Z}$. Then $a + mk$ equals

   $$a + m(c + [-c + k]).$$

   Thus, $x \in B$.

   ($\supseteq$) Assume $x \in B$. This means there exists $k \in \mathbb{Z}$ such that

   $$x = a + m(c + k).$$

   Since $c + k \in \mathbb{Z}$, $x \in A$.

9. (a) $.5 \in \mathbb{Q}$ but $.5 \notin \mathbb{Z}$
   (f) $2x^3 \in A$ but $2x^3 \notin B$

11. This is the case only when the universe is empty!

12. (a) Let $A = B$ and $B = C$. Then,

    $$x \in A \Leftrightarrow x \in B \Leftrightarrow x \in C$$

    for all x.

13. Since $A \subseteq B \Rightarrow A \cup B = B$ is Exercise 2c, we only need to prove the last three implications.
    - Let $A \cup B = B$ and show $A \setminus B = \varnothing$. Let $a \in A \setminus B$. Then $a \in A$ but $a \notin B$. Since $A \cup B \subseteq B$, $a \in B$, a contradiction! Therefore, $A \setminus B$ is empty.
    - Assume $A \setminus B = \varnothing$. Since we know $A \cap B \subseteq A$, to show $A \cap B = A$, let $a \in A$. If $a \notin B$, then $A \setminus B \neq \varnothing$. Hence, we must have $a \in B$ and we conclude $a \in A \cap B$.
    - Let $A \cap B = A$ and take $a \in A$. Then $a \in A \cap B$. Thus, a is an element of B, and we have shown that $A \subseteq B$.

3.4. FAMILIES OF SETS

1. (a) $\{\{1, 2\}, \{3, 4\}, \{1, 4\}, \{1, 3\}\}$
   (b) $\{\{3, 4\}\}$
   (c) $\{\{1, 2\}\}$
   (d) $\{\{1, 2\}, \{3, 4\}\}$
   (e) $\{\}$
   (f) $\{\{1, 2\}, \{1, 4\}\}$

2. (a) False (b) True (c) False (d) True (e) False (f) False (g) False (h) False (i) True (j) False (k) True (l) True (m) True (n) False (o) True (p) False (q) True (r) True (s) False (t) True (u) False (v) True

3. (a) $\{0, 1, 2, 3\}$
   (b) $\{0, 1, 2, 3, 4, 5, 6, 7, 8\}$
   (c) $\{0, 1, 2\}$
   (d) $\{4, 5, 6, \ldots\}$
   (e) $\{0, 1, 2\}$
   (f) $\{0\}$

6. (a) $\mathscr{E} = \{B_n : n \in \mathbb{N}\}$ where for all $n \in \mathbb{N}$, $B_n = A_{n+1}$.
   (c) $\mathscr{E} = \{B_n : n \in \mathbb{Z} \cap [6, \infty)\}$ where for all integers $n \geq 6$, $B_n = A_{n-5}$.

8. (a) Let $I \subseteq J$ and take $B \in \mathscr{E}$. This means $B = A_i$ for some $i \in I$. Hence, $i \in J$ and we have $B \in \mathscr{F}$.
   (c) Let $B \in \{A_i : i \in I \cap J\}$, so $B = A_i$ for some $i \in I \cap J$. Therefore $i \in I$ and $i \in J$. Hence, $B \in \mathscr{E}$ and $B \in \mathscr{F}$, which means $B \in \mathscr{E} \cap \mathscr{F}$.

9. (a) Take $A_1 = \{1, 2\}$, $A_2 = \{1, 2\}$, $I = \{1\}$, and $J = \{2\}$. Then $\mathscr{E} \cap \mathscr{F} = \{\{1, 2\}\}$ but $\{A_i : i \in I \cap J\} = \varnothing$.

11. (a) 1 (b) 2 (c) 4 (d) 8 (e) 2^n

12. (b) $\{\varnothing, \{\varnothing\}, \{\{1\}\}, \{\{2\}\}, \{\{1, 2\}\}, \{\varnothing, \{1\}\}, \{\varnothing, \{2\}\}, \{\varnothing, \{1, 2\}\}, \{\{1\}, \{2\}\}, \{\{1\}, \{1, 2\}\}, \{\{2\}, \{1, 2\}\}, \{\varnothing, \{1\}, \{2\}\}, \{\varnothing, \{1\}, \{1, 2\}\}, \{\varnothing, \{2\}, \{1, 2\}\}, \{\{1\}, \{2\}, \{1, 2\}\}, \{\varnothing, \{1\}, \{2\}, \{1, 2\}\}\}$
    (f) $\{\varnothing, \{\varnothing\}, \{\{\varnothing\}\}, \{\{\varnothing, \{\varnothing\}\}\}, \{\varnothing, \{\varnothing\}\}, \{\varnothing, \{\varnothing, \{\varnothing\}\}\}, \{\{\varnothing\}, \{\varnothing, \{\varnothing\}\}\}, \{\varnothing, \{\varnothing\}, \{\varnothing, \{\varnothing\}\}\}\}$

15. (a) $\mathbf{P}(A \cup B) \subseteq \mathbf{P}(A) \cup \mathbf{P}(B)$ can be false. For example, let $A = \{1\}$ and $B = \{2\}$. The other inclusion is true. To set this, let $C \in \mathbf{P}(A) \cup \mathbf{P}(B)$. This means $C \in \mathbf{P}(A)$ or $C \in \mathbf{P}(B)$. Then $C \subseteq A$ or $C \subseteq B$. Hence, $C \subseteq A \cup B$, and $C \in \mathbf{P}(A \cup B)$.
    (d) Neither inclusion is true.

3.5. GENERALIZED UNION AND INTERSECTION

1. (a) $\{1, 2, 3, 4\}$
   (b) $\{1\}$
   (c) $\varnothing$
   (d) $\{\varnothing\}$
   (e) $\{1, 2, 3, 4\}$
   (f) $\varnothing$
   (g) $\{1\}$
   (h) $\varnothing$
   (i) $\varnothing$
   (j) U

2. (a) $\{0, 1, 2, 3\}$
   (b) $\{0\}$
   (c) $\mathbb{N}$
   (d) $\{0\}$
   (e) $\mathbb{Z}^+$
   (f) $\{4, 5, 6, \dots\}$

3. (a) $\mathbb{Z}$
   (b) $\{\dots, -12, -6, 0, 6, 12, \dots\}$

4. (a) $\mathbb{Z}^+$
   (b) $\varnothing$

5. (a) Let $a \in \bigcup_{i=0}^{k+1} D_i$. This is equivalent to

   $a \in D_i$ for some i $= 0, \dots, k+1$

   if and only if $a \in D_i$ for some i $= 0, \dots, k$ or $a \in D_{k+1}$ if and only if

   $$a \in \bigcup_{i=0}^{k} D_i \cup D_{k+1}.$$

   (d) Let $a \in \bigcap_{i=0}^{\ell} D_i$. Then $a \in D_i$ for all $i = 0, \dots, \ell$. Therefore, since $k \le \ell$, $a \in D_i$ for all $u = 0, \dots, k$. Hence, $a \in \bigcap_{i=0}^{k} D_i$.

8. (a) $a \in \bigcup\{A, B\} \Leftrightarrow a \in A$ or $a \in B \Leftrightarrow a \in A \cup B$

10. (a) Let $B \subseteq A_i$ for some $i \in I$. Take $a \in B$. Then $a \in A_i$, which yields $a \in \bigcup_{i\in I} A_i$. Thus, $B \subseteq \bigcup_{i\in I} A_i$.

12. We will prove part 1:

$$\begin{aligned} a \in \overline{\bigcap_{i\in I} A_i} &\Leftrightarrow a \notin \bigcap_{i\in I} A_i \\ &\Leftrightarrow \sim(a \in \bigcap_{i\in I} A_i) \\ &\Leftrightarrow \sim(\forall i \in I)(a \in A_i) \\ &\Leftrightarrow (\exists i \in I)(a \notin A_i) \\ &\Leftrightarrow (\exists i \in I)(a \in \overline{A_i}) \\ &\Leftrightarrow a \in \bigcup_{i\in I} \overline{A_i}. \end{aligned}$$

17. (a) Let $\mathscr{F}$ be a family of sets.

$$\begin{aligned} &\bigcup\{\mathscr{F}\} \\ &= \{x : x \in A \text{ for some } A \in \{\mathscr{F}\}\} \\ &= \{x : x \in \mathscr{F}\} \\ &= \mathscr{F} \end{aligned}$$

   (c) Similar to the proof of Theorem 3.5.5.

18. (a)
$$\begin{aligned} &x \in \bigcup(\mathscr{E} \cup \mathscr{F}) \\ &\Leftrightarrow x \in A \text{ for some } A \in \mathscr{E} \cup \mathscr{F} \\ &\Leftrightarrow x \in A \text{ for some } A \in \mathscr{E} \text{ or } x \in A \text{ for some } A \in \mathscr{F} \\ &\Leftrightarrow x \in \bigcup \mathscr{E} \text{ or } x \in \bigcup \mathscr{F} \\ &\Leftrightarrow x \in \bigcup \mathscr{E} \cup \bigcup \mathscr{F}. \end{aligned}$$

19. (a) Let $\mathscr{E} = \{\{1\}, \{1, 2\}\}$ and $\mathscr{F} = \{\{1\}, \{2\}\}$.

20. (a) True
    (b) False
    (c) False
    (d) False

21. (b) Let $m, n \in \mathbb{N}$. We must either show $(2^m)\mathbb{Z} \subseteq (2^n)\mathbb{Z}$ or $(2^n)\mathbb{Z} \subseteq (2^m)\mathbb{Z}$. If $m \le n$, then $2^m \mid 2^n$. In other words, $2^n = 2^m\ell$ for some integer ℓ. Hence, if $x \in (2^n)\mathbb{Z}$, then $x = 2^n k$ for some $k \in \mathbb{Z}$. This yields $x = 2^m \ell k$, which means $x \in (2^m)\mathbb{Z}$. Thus, in this case, $(2^n)\mathbb{Z} \subseteq (2^m)\mathbb{Z}$. On the other hand, if $n < m$, then we have the other inclusion.

22. Yes

4.1. THE FIRST PRINCIPLE

1. (b) $P(1)$: $1 = (1+1)! - 1$
   $P(n+1)$: $1 \cdot 1! + 2 \cdot 2! + \cdots + (n+1) \cdot (n+1)! = (n+2)! - 1$
   (e) $P(1)$: *If A has* 1 *element, then* $\mathbf{P}(A)$ *has* 2 *elements.*
   $P(n+1)$: *If A has* $n+1$ *elements, then* $\mathbf{P}(A)$ *has* 2^{n+1} *elements.*
   (f) $P(1)$: *Every set of size* 1 *contains a.*
   $P(n+1)$: *Every set of size* $n+1$ *contains a.*

2. (a) For the basis case:

$$1 = 1(1+1)/2 = 2/2 = 1.$$

For the induction step, assume that the n case holds:

$$1 + 2 + \cdots + n = \frac{n(n+1)}{2}.$$

To show $1 + 2 + 3 + \cdots + (n+1)$ equals

$$\frac{(n+1)(n+2)}{2},$$

add $n+1$ to both sides of the induction hypothesis. The right side is then equal to

$$\begin{aligned}\frac{n(n+1)}{2} &+ (n+1)\\ &= \frac{n(n+1)}{2} + \frac{2(n+1)}{2}\\ &= \frac{(n+1)(n+2)}{2}.\end{aligned}$$

(d) To show it works for $n = 1$:

$$[1(1+1)/2]^2 = 1 = 1^3$$

Next assume

$$1^3 + \cdots + n^3 = [n(n+1)/2]^2.$$

To show that the $n+1$ case holds, add $(n+1)^3$ to both sides of the induction hypothesis. This gives,

$$\begin{aligned}1^3 &+ 2^3 + \cdots + (n+1)^3\\ &= [n(n+1)/2]^2 + (n+1)^3\\ &= n^2(n+1)^2/4 + 4(n+1)^3/4\\ &= (n+1)^2(n^2+4n+4)/4\\ &= (n+1)^2(n+2)^2/4\\ &= [(n+1)(n+2)/2]^2.\end{aligned}$$

(f) Since $(1+1)!-1 = 2!-1 = 1 = 1\cdot 1!$, the basis case holds. So, assume

$$1 \cdot 1! + \cdots + n \cdot n! = (n+1)! - 1.$$

Adding $(n+1)(n+1)!$ to both sides of this equations yields the following on the right-hand side:

$$\begin{aligned}(n+1)! &- 1 + (n+1)(n+1)!\\ &= (n+2)(n+1)! - 1\\ &= (n+2)! - 1.\end{aligned}$$

3. (a) $1/n$
   (b) $2^n - 1$
   (c) $(n-1)2^n + 1$

4. (a) The basis case holds since $1 < 2^1$. Next assume $n < 2^n$. Then

$$\begin{aligned}n + 1 &\le n + n\\ &= 2n\\ &< 2 \cdot 2^n\\ &= 2^{n+1}.\end{aligned}$$

   (c) For the induction step, assume that the formula holds for n. To show the $n+1$ case, add $1/(n+1)^2$ to both sides of the n case. The right hand side of the inequality becomes

$$\begin{aligned}2 &- 1/n + 1/(n+1)^2\\ &= 2 - \frac{(n+1)^2}{n(n+1)^2} + \frac{n}{n(n+1)^2}\\ &= 2 - \frac{n^2+3n+1}{n(n+1)^2}\\ &< 2 - \frac{n^2+3n}{n(n+1)^2}\\ &= 2 - \frac{n+3}{(n+1)^2}\\ &< 2 - \frac{1}{(n+1)^2}.\end{aligned}$$

6. (c) We will show only the induction step: Let $n \ge 4$ and assume $2^n < n!$. Therefore, $2 \cdot 2^n < 2n!$. Hence,

$$\begin{aligned}2^{n+1} &< 2n!\\ &< (n+1)n!\\ &= (n+1)!.\end{aligned}$$

4.2. COMBINATORICS

1. 1, 7, 21, 35, 35, 21, 7, 1

2. (a) $\binom{n}{0} = \frac{n!}{n!0!} = \frac{n!}{n!} = 1$

3. (a) $x^7 + 7x^6y + 21x^5y^2 + 35x^4y^3 + 35x^3y^4 + 21x^2y^5 + 7xy^6 + y^7$.
   (c) $x^5 - 5x^4 + 10x^3 - 10x^2 + 5x - 1$

4. (a) If $n \in \mathbb{N}$, then $2^n = (1+1)^n$. But this equals

$$\binom{n}{0}1^n \cdot 1^0 + \binom{n}{1}1^{n-1} \cdot 1^1 + \cdots,$$

which is the same as

$$\binom{n}{0} + \binom{n}{1} + \cdots + \binom{n}{n}.$$

5. (a) Let $r \in \mathbb{Z}^+$. By induction on n: We must first check the equation for $n = r$:

$$\binom{r}{r} = 1 = \binom{r+1}{r+1}.$$

For the induction step, assume

$$\binom{r}{r} + \cdots + \binom{n}{r} = \binom{n+1}{r+1}.$$

If we add $\binom{n+1}{r}$ to both sides of this equation, the right-hand side becomes

$$\binom{n+1}{r+1} + \binom{n+1}{r} = \binom{n+2}{r+1}$$

by Pascal's Identity.

6. Use the fact that

$$\begin{aligned} r\binom{n}{r} &= r\frac{n!}{r!(n-r)!} \\ &= \frac{n!}{(r-1)!(n-r)!} \\ &= n\frac{(n-1)!}{(r-1)!(n-r)!} \\ &= n\binom{n-1}{r-1}. \end{aligned}$$

8. Use Pascal's identity.

4.3. THE SECOND PRINCIPLE

2. Define $a_1 = 1$ and $a_n = na_{n-1}$. To show $a_n = n!$ for all $n \in \mathbb{Z}^+$, proceed by induction. The basis step is clear, for $1 = 1!$. Now assume $a_n = n!$. Then

$$\begin{aligned} a_{n+1} &= (n+1)a_n \\ &= (n+1)n! \\ &= (n+1)!. \end{aligned}$$

3. (a) The $n = 1$ case holds since $(-1)^1 = -1$. Next, assume $a_n = (-1)^n$. Therefore,

$$a_{n+1} = -(-1)^n = (-1)^{n+1}.$$

(c) By strong induction on n: The basis case holds since

$$a_1 = 3 \cdot 2^1 - 2 \cdot 3^1 = 0$$

and

$$a_2 = 3 \cdot 2^2 - 2 \cdot 3^2 = -6.$$

Next, assume $a_k = 3 \cdot 2^k - 2 \cdot 3^k$ for all positive integers $k \le n$. We then compute,

$$\begin{aligned} a_{n+1} &= 5a_n - 6a_{n-1} \\ &= 5(3 \cdot 2^n - 2 \cdot 3^n) - \\ &\quad 6(3 \cdot 2^{n-1} - 2 \cdot 3^{n-1}) \\ &= 5 \cdot 3 \cdot 2^n - 5 \cdot 2 \cdot 3^n - \\ &\quad 6 \cdot 3 \cdot 2^{n-1} + 6 \cdot 2 \cdot 3^{n-1} \\ &= 3 \cdot 2^{n-1}(5 \cdot 2 - 6) - \\ &\quad 2 \cdot 3^{n-1}(5 \cdot 3 - 6) \\ &= 3 \cdot 2^{n-1}(4) - 2 \cdot 3^{n-1}(9) \\ &= 3 \cdot 2^{n+1} - 2 \cdot 3^{n+1}. \end{aligned}$$

4. (a) We will use strong induction on n: For the basis case, since $F_3 = 2$ and $\rho \approx 1.618$, $F_3 > \rho^1$. For the induction step, suppose $F_{k+2} > \rho^k$ for all positive integers $k \le n$. We have seen that $\rho^{-1} + \rho^{-2} = 1$. Therefore,

$$\begin{aligned} F_{n+3} &= F_{n+2} + F_{n+1} \\ &> \rho^n + \rho^{n-1} \\ &= \rho^{n+1}(\rho^{-1} + \rho^{-2}) \\ &= \rho^{n+1}. \end{aligned}$$

7. Assume (MI2). Let $P(n)$ be a predicate. To prove (MI), suppose $P(1)$ and for all $n \in \mathbb{Z}^+$,

$$P(n) \Rightarrow P(n+1).$$

We must show $P(n)$ for every positive integer n using strong induction. $P(1)$ is true by assumption. Assume $P(k)$ is true for all $k = 1, \ldots n$. Thus, in particular, $P(n)$ is true. Hence, $P(n+1)$ holds.

4.4. THE WELL-ORDERING PRINCIPLE

1. (a) True
   (b) False
   (c) False
   (d) False
   (e) False
   (f) False

2. $1/2,\ 1/3,\ 1/4,\ \ldots$ will work for all four sets.

3. (a) The set is finite, so it is well-ordered.
   (c) $\pi/2,\ \pi/3,\ \pi/4,\ \ldots$ is a strictly decreasing sequence in the set. Hence, the set cannot be well-ordered.
   (e) It is well-ordered by Corollary 4.4.4.
   (g) It is well-ordered by Corollary 4.4.4.

4. Let $B \subseteq A \subseteq \mathbb{R}$ with $A,\ B \neq \varnothing$. Assume that A is well-ordered. To see that B is well-ordered, let $C \neq \varnothing$ such that $C \subseteq B$. Then $C \subseteq A$. Hence, C has a minimum.

7. Let A be a nonempty subset of $\mathbb{Z}^-$. Define
$$B = \{-n : n \in A\}.$$
Therefore, $B \subseteq \mathbb{Z}^+$, and B is nonempty. So by (WO), B has a minimum. Call it m. We claim that $-m$ is the maximum of A. First, $-m \in A$ since $m \in B$. Second, let $n \in A$. Then $-n \in B$, which yields $m \leq -n$. Hence, $-m \geq n$.

5.1. AXIOMS

1. (a) identity
   (b) nonzero integers
   (c) identity
   (d) Associative Axiom
   (e) Commutative Axiom
   (f) closed
   (g) $m < n$, $m = n$, or $m > n$

2. (a) Commutative
   (b) Additive Inverse
   (c) Closure of $\mathbb{Z}^+$
   (d) Commutative
   (e) Trichotomy
   (f) Multiplicative Identity
   (g) Commutative
   (h) Cancellation
   (i) Closure of $\mathbb{Z}$
   (j) Additive Identity

3. (a) Let $x \in \mathbb{Z}$.
$$\begin{aligned} x+4 &= 10 \\ (x+4)+-4 &= 10+-4 \text{ [Equality]} \\ x+(4+-4) &= 10+-4 \text{ [Assoc]} \\ x+0 &= 10+-4 \text{ [Add Inv]} \\ x &= 10+-4 \text{ [Add Id]} \\ x &= 6 \end{aligned}$$
   (e) For any integer x:
$$\begin{aligned} x^2 - x + 6 &= 12 \\ (x-3)(x+2) &= 12 \end{aligned}$$
   By the Zero Property, $x - 3 = 0$ or $x+2 = 0$. Using the Additive Inverse Law, we find $x = 3$ or $x = -2$.

4. (a) Take $a,\ b,\ c \in \mathbb{Z}$.
$$\begin{aligned} & a+(b+c) \\ &= a+(c+b) \quad \text{[Com]} \\ &= (a+c)+b \quad \text{[Assoc]} \\ &= (c+a)+b \quad \text{[Com]} \end{aligned}$$

5. (a) Let $a,\ b \in \mathbb{Z}$.
$$\begin{aligned} & (a+b)^2 \\ &= (a+b)(a+b) \\ &= (a+b)a + (a+b)b \quad \text{[Dist]} \\ &= a(a+b) + b(a+b) \quad \text{[Com]} \\ &= a^2 + ab + ba + b^2 \quad \text{[Dist]} \\ &= a^2 + ab + ab + b^2 \quad \text{[Com]} \\ &= a^2 + 2ab + b^2. \end{aligned}$$

6. 4 is not the additive inverse of 5 because $4+5 \neq 0$, and 7 is not the multiplicative inverse of 8 since $7 \cdot 8 \neq 1$.

7. (a) $a(-b)+ab = a(-b+b) = a \cdot 0 = 0$
   (c) $(a+b)+(-a)+(-b) = (a-a)+(b-b) = 0+0 = 0$

9. (a) To see that $*$ is an operation, let a, a', b, and b' be integers. Assume $a = a'$ and $b = b'$. Then,
$$\begin{aligned} a * b &= a + b + 2 \\ &= a' + b' + 2 \\ &= a' * b'. \end{aligned}$$

(b) Let $a \in \mathbb{Z}$. Then $a * -2 = a - 2 + 2 = a$ and $-2 * a = -2 + a + 2 = a$.
(c) Let $n \in \mathbb{Z}$. Then

$$n * -n - 4 = n - n - 4 + 2 \\ = -2$$

and

$$-n - 4 * n = -n - 4 + n + 2 \\ = -2.$$

11. (a) $2 < 5$ since $5 - 2 = 3 \in \mathbb{Z}^+$
(d) $6 \not< 6$ because $6 - 6 = 0 \notin \mathbb{Z}^+$

12. (a) Assume $a < b$. This means $b - a \in \mathbb{Z}^+$. Now calculate: $(b - c) - (a - c) = b - a$. Thus, $(b - c) - (a - c) \in \mathbb{Z}^+$.
(e) Let $a < b$ and $b < c$. Then $b - a \in \mathbb{Z}^+$ and $c - b \in \mathbb{Z}^+$. Since $\mathbb{Z}^+$ is closed under addition, $(c - b) + (b - a) = c - a \in \mathbb{Z}^+$. Therefore, $a < c$.

14. (a) Let $a \in \mathbb{Z}$. If $a = 0$, then $a^2 = 0$. If $a > 0$, then $a \in \mathbb{Z}^+$. Since $\mathbb{Z}^+$ is closed under multiplication, $a^2 \in \mathbb{Z}^+$. Lastly, if $a < 0$, then $-a \in \mathbb{Z}^+$. Thus, $(-a)(-a) = a^2 \in \mathbb{Z}^+$. In all cases, $a^2 \geq 0$.

5.2. DIVISIBILITY

1. (a) $0 = 0 \cdot 4 + 0$
(b) $24 = 5 \cdot 4 + 4$
(c) $-5 = 7 \cdot -1 + 2$
(d) $-100 = 14 \cdot -8 + 12$

2. (a) 6 (c) 14 (e) 7
(b) 3 (d) 1 (f) 25

3. (a) $a = \pm b$
(b) $b \nmid c$
(c) $b \mid a$
(d) $n = ua + vb$
(e) $\gcd(b, m) \neq 1$
(f) $a \mid c$
(g) $0 \leq r < b$

6. Let $m, n \in \mathbb{Z}$ with $m \neq 0$ and divide m into n. We have two cases: First suppose $m > 0$. This case follows from the Division Algorithm since $m = |m|$. Next, let $m < 0$ and divide n by $-m$. Then there exists $q, r \in \mathbb{Z}$ such that $n = q(-m) + r$ and $0 \leq r < -m$. Therefore, $n = -qm + r$, and since $|m| = -m$, $0 \leq r < |m|$.

7. (a) Let $a, b \neq 0$ and $b = ak$ for some $k \in \mathbb{Z}$. To show $a \leq |b|$, assume $a > |b|$. We have two cases. First assume $b > 0$. Then $a > ak$. If $a > 0$, then k is positive and $1 > k$, a contradiction. If $a < 0$, then k is negative and $1 < k$, another problem. Therefore, $a \leq |b|$. On the other hand, if $b < 0$, then $-b > 0$ and $a | - b$. Now proceed as before.

8. Let $a \mid b$ and $b \mid c$. Assume $n \in A$. This means $n \mid a$. Therefore, $n \mid b$ and then $n \mid c$. Hence $n \in C$.

11. Let $d \mid a$ and $d \mid b$. Furthermore, there exist $u, v \in \mathbb{Z}$ such that $\gcd(a, b) = ua + vb$. Therefore, $d \mid \gcd(a, b)$.

13. (a) a (c) a
(b) a (d) a

14. (b) $c = 1$

17. Let $\gcd(a, c) = \gcd(b, c) = 1$. Let $d \in \mathbb{Z}^+$ and $d \mid ab$ and $d \mid c$. Since a and c are relatively prime, there exist integers u and v such that $1 = ua + vc$. Thus, $b = uab + vcb$, so $d \mid b$. Since c and b are relatively prime, $d = 1$.

18. (c) Since a and b are relatively prime, we have integers u and v such that $au + bv = 1$. Hence, $2au + 2bv = 2$. Let $d \in \mathbb{Z}^+$ such that d is a common divisor of $a + b$ and $a - b$. This means $a + b = dk$ and $a - b = d\ell$ for some $k, \ell \in \mathbb{Z}$. Hence, $2a = d(k + \ell)$ and $2b = d(k - \ell)$. Substituting into the above equation we find $d(k + \ell)u + d(k - \ell)v = 2$. In other words, d divides 2, which means $d = 1$ or $d = 2$.
(e) Assume $\gcd(a, b) = 1$ and write $a + b = c\ell$ with $\ell \in \mathbb{Z}$. To show that a and c are relatively prime, let $d \in \mathbb{Z}^+$ such that $d \mid a$ and $d \mid c$. This means there are integers k_1 and k_2 such that $a = dk_1$ and $c = dk_2$. Since $b = c\ell - a$,

$$b = dk_2\ell - dk_1,$$

so $d \mid b$. However, a and b are relatively prime. Hence, $d = 1$. (Similarly, b and c are shown to be relatively prime.)

19. $\sqrt{2} \in \mathbb{Q}$; $a, b \in \mathbb{Z}$; $2b^2 = a^2$; a is even; $2b^2 = 4k^2$; $b^2 = 2k^2$; b is even.

5.3. PRIMES

1. (a) $2^0 \cdot 3^0 \cdot 5^0 \cdot 7^1 \cdot 11^0 \cdot 13^0 \cdot 37^0$
   (b) $2^1 \cdot 3^1 \cdot 5^1 \cdot 7^0 \cdot 11^0 \cdot 13^0 \cdot 37^0$
   (c) $2^2 \cdot 3^1 \cdot 5^0 \cdot 7^0 \cdot 11^0 \cdot 13^0 \cdot 37^0$
   (d) $2^2 \cdot 3^0 \cdot 5^2 \cdot 7^0 \cdot 11^0 \cdot 13^0 \cdot 37^0$
   (e) $2^1 \cdot 3^0 \cdot 5^1 \cdot 7^0 \cdot 11^1 \cdot 13^0 \cdot 37^1$
   (f) $2^6 \cdot 3^0 \cdot 5^6 \cdot 7^0 \cdot 11^0 \cdot 13^0 \cdot 37^0$

4. Suppose $p \mid a^n$. By Theorem 5.3.3, $p \mid a$, so there exists $k \in \mathbb{Z}$ such that $a = pk$. Then, $a^n = (pk)^n = p^n k^n$, which means $p^n \mid a^n$.

7. Suppose $e \mid a$ and $e \mid b$. Write
$$e = p_1^{t_1} p_2^{t_2} \cdots p_k^{t_k}$$
for some $t_i \in \mathbb{N}$ $(i = 1, \ldots k)$. This means for each i, $p_i^{t_i} \mid a$. Hence $p_i^{t_i}$ divides $p_i^{r_i}$. From this we conclude $t_i \le r_i$. Similarly, $t_i \le s_i$.

8. Let $b^2 = a^3\ell$ with $\ell \in \mathbb{Z}$. Write
$$a = p_1^{r_1} p_2^{r_2} \cdots p_k^{r_k}$$
and
$$b = p_1^{s_1} p_2^{s_2} \cdots p_k^{s_k}.$$
Then
$$p_1^{2s_1} \cdots p_k^{2s_k} = p_1^{3r_1} \cdots p_k^{3r_k}\ell,$$
and
$$p_1^{2s_1-3r_1} p_2^{2s_2-3r_3} \cdots p_k^{2s_k-3r_k} = \ell.$$
Since ℓ is an integer, $2s_i - 3r_i \ge 0$ for all i. Hence, $s_i \ge 3/2r_i \ge r_i$. From here we conclude $a \mid b$.

11. (a) 8 (b) 18 (c) 12 (d) 120

13. Let $\ell = p_1^{\max(r_1, s_1)} \cdots p_k^{\max(r_k, s_k)}$.
   - Since $r_i, s_i \le \max(r_i, s_i)$ for all $i = 1, \ldots, k$, a and b divide ℓ.
   - Let $m \in \mathbb{Z}^+$, $a \mid m$, and $b \mid m$. Write a prime power decomposition for m:
   $$m = p_1^{m_1} p_2^{m_2} \cdots p_k^{m_k}.$$
   (What if the factorization for m contains primes other than $p_1, \ldots, p_k$?) Since $a \mid m$ and $b \mid m$, $r_i \le m_i$ and $s_i \le m_i$. Therefore $\max(r_i, s_i) \le m_i$, and $\ell \le m$.

5.4. CONGRUENCES

1. (a) True (b) False (c) True (d) True (e) False (f) True

2. (a) 5 (b) 3 (c) 4 (d) 2 (e) 0 (f) 4

3. (a) True (b) True (c) False (d) False (e) False (f) False (g) False (h) True

4. (a) 2

7. Let $m \in \mathbb{Z}^+$.
$$\begin{aligned} m \mid a &\Leftrightarrow a = mk, \text{ some } k \in \mathbb{Z} \\ &\Leftrightarrow a = 0 + mk, \text{ some } k \in \mathbb{Z} \\ &\Leftrightarrow a \equiv 0 \pmod{m}. \end{aligned}$$

9. (a) Let a be even. Then $a = 2k$ for some $k \in \mathbb{Z}$. Hence, $a \equiv 0 \pmod 2$.
   (c) Suppose $a = 2k$ with $k \in \mathbb{Z}$. Then $a^2 = 4k^2$. Hence, $a^2 \equiv 0 \pmod 4$.

11. By induction on n:
   - Since $6 = 1 + 5(1)$, $6^1 \equiv 1 + 5(1) \pmod{25}$. Therefore, the basis case holds for $n = 1$.
   - Assume $6^n \equiv 1 + 5n \pmod{25}$. This means there exists an integer k such that $6^n = 1 + 5n + 25k$. Therefore,
   $$\begin{aligned} &6^{n+1} \\ &= 6 + 6(5n) + 6(25k) \\ &= 1 + 5 + 6(5n) + 25k \\ &= 1 + 5 + 5n + 5(5n) + 25k \\ &= 1 + 5(n+1) + 25(n+k). \end{aligned}$$
   Thus, $6^{n+1} \equiv 1 + 5(n+1) \pmod{25}$.

12. (a) $\ldots, -9, -2, 5, 12, \ldots$
    (d) $\ldots, -5, 0, 5, 10, \ldots$
    (f) $\ldots, -8, 1, 10, 19, \ldots$

13. (a) $\{\ldots, -7, -2, 3, 8, \ldots\}$
    (b) $\{\ldots, -6, 0, 6, 12, \ldots\}$
    (c) $\{\ldots, -8, -3, 2, 7, \ldots\}$
    (d) $\varnothing$
    (e) $\mathbb{Z}$
    (f) $\varnothing$

15. (a) Since c and m are relatively prime, there are integers b and v such that $bc + vm = 1$. Then, $bc = 1 - vm$. Therefore, $bc \equiv 1 \pmod{m}$.

16. (b) Let $a \equiv b \pmod{m}$, so there exists $k \in \mathbb{Z}$ so that $a = b + mk$. To show that $\gcd(a, m) = \gcd(b, m)$, we will show that the common divisors of a and m are the common divisors of b and m.

    - Let $d \mid a$ and $d \mid m$, so $a = d\ell_1$ and $m = d\ell_2$ for some $\ell_1, \ell_2 \in \mathbb{Z}$. Then

    $$\begin{aligned} b &= a - mk \\ &= d\ell_1 - d\ell_2 k \\ &= d(\ell_1 - \ell_2 k). \end{aligned}$$

      Hence, $d \mid b$, and it is a common factor of b and m.
    - Similarly, if d is a common factor of b and m, then it is a common factor of a and m.

18. Let $a^2 \equiv a \pmod{p}$. Then, p divides $a(1-a)$. Because p is prime, $p \mid a$ or $p \mid (1 - a)$. If the first disjunct is true, then $a \equiv 0 \pmod{p}$. If the second disjunct holds, then $a \equiv 1 \pmod{p}$.

6.1. RELATIONS

1. (a) domain: $\{0, 2, 4, 6\}$
   range: $\{1, 3, 5, 7\}$
   (c) domain: $\mathbb{R}$
   range: $\mathbb{Z}$
   (e) domain: $\varnothing$
   range: $\varnothing$
   (g) domain: $\mathbb{R}$
   range: $\mathbb{R}$
   (i) domain: $[-1, 1]$
   range: $[-1, 1]$
   (k) domain: $\{e^x\}$
   range: $\{ax : a \in \mathbb{R}\}$

2. (a) $\{(1, 3), (3, 6)\}$ (c) $\{\}$

3. For (a), $(R \circ S)^{-1} = \{(3, 1), (6, 3)\}$. Also, $S^{-1} = \{(2, 1), (3, 2), (4, 3)\}$ and $R^{-1} = \{(0, 1), (3, 2), (6, 4)\}$, so

$$S^{-1} \circ R^{-1} = \{(3, 1), (6, 3)\}.$$

4. (a) $\{(x, y) \in \mathbb{R}^2 : x^2 + y^2 = 1\}$
   (c) $\{(x, y) \in \mathbb{R}^2 : (x - 2)^2 - y^2 = 0\}$

5. (a) $\varnothing$
   (c) $\{(0, 1), (3, 2), (6, 4)\}$
   (e) $\mathbb{R} \times \mathbb{Z}$
   (g) $\{(x, y) \in \mathbb{R}^2 : x^2 + y^2 = 1\}$

6. (a) To prove $R \circ I_A = R$, we note that:

$$\begin{aligned} &(x, y) \in R \circ I_A \\ &\Leftrightarrow (x, z) \in I_A \text{ and } (z, y) \in R \\ &\Leftrightarrow x = z \text{ and } (z, y) \in R \\ &\Leftrightarrow (x, y) \in R. \end{aligned}$$

   where z, if it exists, is an element of A.

8. Take $(y, y) \in I_{\text{ran}(R)}$. Then $y \in \text{ran}(R)$, and there exists $x \in \text{dom}(R)$ such that $(x, y) \in R$. This implies $(y, x) \in R^{-1}$. Hence, $(y, y) \in R \circ R^{-1}$.

10. (a) Let $(x, y) \in R^{-1}$. This means (y, x) is an element of R. Hence, $y \in A$ and $x \in B$, and $(x, y) \in B \times A$.
    (c) Let $(x, y) \in R^{-1} \circ S^{-1}$, so there exists $z \in B$ such that $(x, z) \in S^{-1}$ and $(z, y) \in R^{-1}$. Thus, $(z, x) \in S$ and $(y, z) \in R$. Therefore, $(y, x) \in S \circ R$. Since $S \circ R \subseteq A \times C$, $y \in A$ and $x \in C$. Thus, (x, y) is an element of $C \times A$.

12. (a) Let $R \subseteq S$ and take $(b, a) \in R^{-1}$. Then, $(a, b) \in R$, so $(a, b) \in S$. Thus, $(b, a) \in S^{-1}$.
    (c)

$$\begin{aligned} &(b, a) \in (R \cap S)^{-1} \\ &\Leftrightarrow (a, b) \in R \cap S \\ &\Leftrightarrow (a, b) \in R \wedge (a, b) \in S \\ &\Leftrightarrow (b, a) \in R^{-1} \wedge (b, a) \in S^{-1} \\ &\Leftrightarrow (b, a) \in R^{-1} \cap S^{-1}. \end{aligned}$$

6.2. EQUIVALENCE RELATIONS

1. | | |
|---|---|
| reflexive | (d), (e) |
| irreflexive | (a), (b), (f) |
| symmetric | (b), (c), (e), (f) |
| asymmetric | (a), (f) |
| antisymmetric | (a), (d), (f) |
| transitive | (a), (d), (e), (f) |

2. (b) $A = \mathbb{Z}$ and $R = \{(n,\ n+1) : n \in \mathbb{Z}\}$

3. (a) Let $a,\ b,\ c \in \mathbb{R} \setminus \{0\}$.
   1. Since $a \neq 0$, $a^2 > 0$. Therefore, we conclude that $a \sim a$.
   2. Assume $a \sim b$. This means $ab > 0$. From this $ba > 0$, so $b \sim a$.
   3. Let $a \sim b$ and $b \sim c$. Then $ab > 0$ and $bc > 0$, which yields $a > 0$ and $b > 0$ or $a < 0$ and $b < 0$. If this first disjunct is true, then $c > 0$, for $bc > 0$. Therefore, $ac > 0$. Similarly, if the second disjunct is true, $ac > 0$.

   (b) $[1] = (0,\ \infty)$
   $[-3] = (-\infty,\ 0)$

5. $c \in A \Leftrightarrow c \in [c]_\sim \Leftrightarrow c \in \bigcup_{a \in A}[a]_\sim$.

8. (a) Let $(a,\ b),\ (c,\ d),\ (e,\ f)$ be ordered pairs in $\mathbb{Z} \times \mathbb{Z}$.
   1. $(a,\ b) \sim (a,\ b)$ because $ab = ab$.
   2. Suppose $(a,\ b) \sim (c,\ d)$. This means $ab = cd$. Therefore, $cd = ab$, and we have $(c,\ d) \sim (a,\ b)$.
   3. Assume we have $(a,\ b) \sim (c,\ d)$ and $(c,\ d) \sim (e,\ f)$. Then $ab = cd$ and $cd = ef$. Thus, $ab = ef$, which means $(a,\ b) \sim (e,\ f)$.

   (b) $\{(a,\ 2a) : a = \pm 1\} \cup$
   $\{(2a,\ a) : a = \pm 1\}$

9. (a) The relation is not reflexive.

10. (b) 6

11. We will show that the family is a partition.
    1. Let $x \in \mathbb{R}$. Then $x \in (\lceil x \rceil - 1,\ \lceil x \rceil]$. Thus, $x \in \bigcup_{n \in \mathbb{Z}}(n,\ n+1]$. Since we know that the other inclusion holds,
    $$\mathbb{R} = \bigcup_{n \in \mathbb{Z}} (n,\ n+1].$$
    2. Next take $m,\ n \in \mathbb{Z}$ such that $m \neq n$. Then $m < n$ or $n > m$. Suppose $m < n$ and let $x \in (m,\ m+1] \cap (n,\ n+1]$. Therefore,
    $$x \leq m + 1 \leq n < x,$$
    a contradiction. This means that the collection is pairwise disjoint.

14. (a) R is reflexive
    $$\Leftrightarrow (a,\ a) \in R \text{ for all } a \in A$$
    $$\Leftrightarrow (a,\ a) \in R^{-1} \text{ for all } a \in A$$
    $$\Leftrightarrow R^{-1} \text{ is reflexive.}$$
    (c) To show sufficiency, let R be symmetric. Then
    $$(a,\ b) \in R \Leftrightarrow (b,\ a) \in R$$
    $$\Leftrightarrow (a,\ b) \in R^{-1}.$$
    For necessity, suppose $R = R^{-1}$ and take $(a,\ b) \in R$. Therefore, $(a,\ b) \in R^{-1}$, which gives $(b,\ a) \in R$.

15. (b) Let R and S be symmetric. To prove $R \cap S$ is symmetric, let $(a,\ b) \in R \cap S$. By hypothesis, $(b,\ a) \in R$ and $(b,\ a) \in S$. Hence $(b,\ a) \in R \cap S$.

17. (a) R is reflexive if and only if $(a,\ a) \in R$ for all $a \in A$ if and only if $a \in R(a)$ for all $a \in A$.

6.3. FUNCTIONS

1. (a) 38 (b) 0

2. (a) $\{\ldots,\ -14,\ -7,\ 0,\ 7,\ 14,\ \ldots\}$
   (c) $\{\ldots,\ -11,\ -4,\ 3,\ 10,\ 17,\ \ldots\}$

3. (a) If $x = 1$, then y is either 4 or 6.
   (c) If $x = 3$, then y is either 1 or -1.

4. (a) Yes
   (b) No, 1
   (c) Yes
   (d) No, 1
   (e) Yes
   (f) No, $[2]_5$
   (g) Yes

5. (a) $f : \mathbb{R} \to \mathbb{R}$
   $f : \mathbb{R} \to [0,\ \infty)$
   (d) $f : (0,\ \infty) \to \mathbb{R}$
   $f : (0,\ \infty) \to \mathbb{C}$

6. (a) $f(x) = x$
   (d) $g(x) = \lceil |x| \rceil$

7. (a) Let $x,\ y \in \mathbb{R}\setminus\{0\}$ and assume $x = y$. Then $1/x = 1/y$ and $f(x) = f(y)$.

9. Let $(a,\ b),\ (c,\ d) \in \mathbb{R}\times\mathbb{R}$. Since $f(a) \in \mathbb{R}$ and $g(b) \in \mathbb{R}$, $(f(a),\ g(b)) \in \mathbb{R}\times\mathbb{R}$. Therefore, the domain and codomain of ϕ is $\mathbb{R}\times\mathbb{R}$. Now assume $(a,\ b) = (c,\ d)$. This means $a = c$ and $b = d$. Because both f and g are functions, $f(a) = f(c)$ and $f(b) = f(d)$. Therefore,

$$\begin{aligned}\phi(a,\ b) &= (f(a),\ f(b))\\ &= (f(c),\ f(d))\\ &= \phi(c,\ d).\end{aligned}$$

13. (b) $\{(0,\ 0),\ (-1,\ 1)\}$

6.4. FUNCTION OPERATIONS

1. (a) Let $x,\ y \in \mathbb{R}$ and suppose $x = y$. Then, since g is a function, $g(x) = g(y)$, and because f is a function,

$$f(g(x)) = f(g(y)).$$

Therefore, $(f \circ g)(x) = (f \circ g)(y)$.

(b) Take two real numbers x and y. Let $x = y$. Therefore, $f(x) = f(y)$ and $g(x) = g(y)$. Thus,

$$\begin{aligned}(f+g)(x) &= f(x) + g(x)\\ &= f(y) + g(y)\\ &= (f+g)(y).\end{aligned}$$

2. (b) Let $(m,\ n) \in \mathbb{Z}\times\mathbb{Z}$. Then, $\phi(m,\ n) = \psi(m,\ n)$ since $m + n = n + m$.

3. We have two cases to consider: In the first case, let $0 \le x$. Then $g(x) = 2x + 1$, and since $|x| = x$, $f(x) = 2x + 1$. In the second case, let $x < 0$. In this case, $g(x) = -2x + 1$, and $f(x) = 2|x| + 1 = -2x + 1$ because $|x| = -x$.

4. (b) Since $f(0) = -3$, $g(0) = 3$, and 0 is an element of the domains of both functions, $f \neq g$.

6. (b) Part (i) is not possible, but for part (ii), $(g \circ f)(4) = 1$.

7. (a) $(f \circ f)(x) = x^4$
   (c) $(\phi \circ \phi)(x,\ y) = (10x - 2y,\ -5x + 11y)$

9. (a) Let $x \in \operatorname{dom}(f)$.

$$\begin{aligned}f(-x) &= \sqrt{(-x)^2 + (-x)^4}\\ &= \sqrt{x^2 + x^4}\\ &= f(x).\end{aligned}$$

   (b) $g(-x) = -|-x| = -|x| = g(x)$.

11. (a) $\{(1,\ 2),\ (3,\ 4)\}$
   (c) $\{(-3.3,\ -5.3),\ (1.2,\ 4.2),\ (7,\ 15)\}$

12. (b) Let $A,\ B \subseteq \operatorname{dom}(f)$.

$$\begin{aligned}&(x,\ y) \in f\restriction(A \cap B)\\ &\Leftrightarrow y = f(x) \text{ and } x \in A \cap B\\ &\Leftrightarrow y = f(x) \text{ and } x \in A \text{ and } x \in B\\ &\Leftrightarrow (x,\ y) \in f\restriction A \text{ and } (x,\ y) \in f\restriction B\\ &\Leftrightarrow (x,\ y) \in f\restriction A \cap f\restriction B.\end{aligned}$$

16. Let g and h be increasing functions. We will show $g + h$ is increasing. Let $x \le y$. We may then conclude $g(x) \le g(y)$ and $h(x) \le h(y)$. Therefore,

$$\begin{aligned}(g+h)(x) &= g(x) + h(x)\\ &\le g(y) + h(y)\\ &= (g+h)(y).\end{aligned}$$

6.5. ONE-TO-ONE AND ONTO

1. (a) We will only show $(f \circ g)(x) = x$.

$$\begin{aligned}(f \circ g)(x) &= f(g(x))\\ &= f(\tfrac{1}{3}x - \tfrac{2}{3})\\ &= 3(\tfrac{1}{3}x - \tfrac{2}{3}) + 2\\ &= x - 2 + 2\\ &= x.\end{aligned}$$

2. (a) $f^{-1}(x) = \frac{1}{7}x - \frac{3}{7}$
   (c) $h^{-1}(x) = -1 + \ln \frac{1}{2}x$

3. Let f be invertible. This means f^{-1} is a function.

   ($\Rightarrow$) Let $f(x) = y$. Since f^{-1} is a function, $f^{-1}(f(x)) = f^{-1}(y)$. Thus, $x = f^{-1}(y)$.

   ($\Leftarrow$) Suppose $x = f^{-1}(y)$. Therefore, $f(x) = f(f^{-1}(y))$ because f is a function. Hence, $f(x) = y$.

7. (b) Let $g(x, y) = g(x', y')$ where (x, y) and (x', y') are elements of $\mathbb{R}^2$. This means $(3y, 2x) = (3y', 2x')$, which gives $3y = 3y'$ and $2x = 2x'$. Thus, $(x, y) = (x', y')$.
   (d) Take $(n, x), (m, y) \in \mathbb{Z} \times \mathbb{R}$. Let $\phi(n, x) = \phi(m, y)$. Hence, $3n = 3m$ and $e^x = e^y$. So, $n = m$ and
   $$x = \ln e^x = \ln e^y = y$$
   since the natural logarithm is a function.

9. (a) Take $x, y \in \text{dom}(g)$ and let $g(x) = g(y)$. Since f is a function,
   $$f(g(x)) = f(g(y)).$$
   Then, $x = y$ because $f \circ g$ is an injection.

11. (a) Since $f(2) = f(-2)$, f cannot be one-to-one.
   (c) $\phi(\{a\}) = \{a\}$ and $\phi(\{a, b\}) = \{a\}$, but $\{a\} \neq \{a, b\}$.

12. (b) Let $y \in (0, \infty)$. If we choose $x = \ln y$, then $g(x) = e^{\ln y} = y$.
   (d) Take $n \in \mathbb{Z}$. Then $\phi(0, n) = 0 + n = n$.

13. (a) -5 does not have a pre-image since $e^x > 0$ for all real numbers x.
   (c) $(0, -1)$ does not have a pre-image, for the second coordinate is always nonnegative.

14. (b) Take $f : \mathbb{R} \to [0, \infty)$ and $g : \mathbb{R} \to \mathbb{R}$ defined by $f(x) = x^2$ and $g(x) = |x|$.

16. Let $\phi : A \to C$ and $\psi : B \to D$ be bijections.
   1. Take $(a, b), (a', b') \in A \times B$. Let
   $$\gamma(a, b) = \gamma(a', b').$$
   This means
   $$(\phi(a), \psi(b)) = (\phi(a'), \psi(b')).$$
   Hence, $\phi(a) = \phi(a')$ and $\psi(b) = \psi(b')$. Since these functions are injections, $a = a'$ and $b = b'$. In other words, $(a, b) = (a', b')$.
   2. Let $(c, d) \in C \times D$. Since ϕ and ψ are surjections, there exists $a \in A$ and $b \in B$ so that $\phi(a) = c$ and $\psi(b) = d$. Therefore, $\gamma(a, b) = (c, d)$.

20. (b) This function is always onto, but may not be one-to-one. For example, define $A = \mathbb{R} \setminus \{1, 2\}$. Let $f \restriction A$ be the identity on A. If $g(x) = x$ with $\text{dom}(g) = \mathbb{R}$, then $\phi(g) = f \restriction A$. But, if h is defined as $h(x) = x$ if $x \neq 1, 2$, $h(1) = 2$ and $h(2) = 1$, then $\phi(h) = f \restriction A$, too.

6.6. IMAGES AND INVERSE IMAGES

1. (a) $(3, 7]$
   (b) $(-\infty, 1)$
   (c) $(-1, 0)$
   (d) $(-.5, .5) \cup (2, 3.5)$

4. (d) (i) $\{2n : n \in \mathbb{Z}\}$
   (ii) $\{2, 4, 6\}$
   (iii) $\mathbb{Z}$
   (iv) $\{1\}$

5. We will check the first one: Let y be an element of $\psi[\varnothing]$. This means $\psi(x) = y$ for some $x \in \varnothing$. This is impossible. Therefore, $\psi[\varnothing]$ is empty.

7. (b) Clearly, $f[A] \subseteq \text{ran}(f)$. So we must show the other inclusion. Let $y \in \text{ran}(f)$. This means there exists $x \in \text{dom}(f)$ such that $f(x) = y$. Since $\text{dom}(f) = A$, $y \in f[A]$.

8. (a) Let $\phi(x) \in \phi[C]$. Then there exists $x' \in C$ so that $\phi(x') = \phi(x)$. Hence, $x = x'$ since ϕ is one-to-one, and then $x \in C$. The converse is clear by the definition of $\phi[C]$.
   (b) Let $f(x) = x^2$. This function is not one-to-one on $\mathbb{R}$. Moreover, $f[[0, 1]]$ equals $[0, 1]$, but $f(-1) = 1$.

10. Let $C \subseteq D \subseteq \text{ran}(f)$. We will show $f^{-1}[C]$ is a subset of $f^{-1}[D]$. Let $x \in f^{-1}[C]$. So, $f(x) \in C$, which implies that $f(x) \in D$. Thus, $x \in f^{-1}[D]$.

11. (a) We will show that (i) is false. Define
   $$f : \mathbb{R} \to \mathbb{R}$$
   by $f(x) = x^2$. Take $U = (1, 2)$ and $V = (-2, -1)$. Then $U \cap V$ is empty,

but $f[U] = f[V] = (1, 4)$, so the images are not disjoint.

12. Let $\text{ran}(g) \subseteq \text{dom}(f)$ and $A \subseteq \text{dom}(g)$.
   ($\Rightarrow$) Take $y \in (f \circ g)[A]$. Therefore, $(f \circ g)(x) = y$ for some $x \in A$. This means that $f(g(x)) = y$. So, $y \in f[g[A]]$ since $g(x) \in g[A]$.
   ($\Leftarrow$) Now take $y \in f[g[A]]$, so there exists $y' \in g[A]$ such that $f(y') = y$. Then there exists $x \in A$ so that $g(x) = y'$. Hence,
   $$y = f(g(x)) = (f \circ g)(x),$$
   which means $y \in (f \circ g)[A]$.

14. (a) Let ϕ be a bijection. To prove the first inclusion, let $y \in \phi[A \setminus C]$. Then $\phi(x) = y$ for some $x \in A \setminus C$. In other words, $x \in A$ but $x \notin C$. Therefore, $y \in \phi[A] \subseteq B$, and since ϕ is one-to-one, Exercise 8 implies that $\phi(x)$ is not an element of $\phi[C]$, for otherwise $x \in C$. The other inclusion is proven by using the onto condition.
   (b) Use $f(x) = x^2$.

6.7. CARDINALITY

1. (a) countable (b) countable (c) countable (d) countable (e) uncountable (f) uncountable (g) countable (h) countable (i) countable (j) uncountable (k) countable

2. (a) 0 (b) 1 (c) 0 (d) 1

3. (b) Since the range of every characteristic function is $\{0, 1\}$, we only need to check when the function returns 1: Let $a \in \text{dom}(\chi)$.
   $$\chi_{A\cap B}(a) = 1$$
   $$\Leftrightarrow a \in A \cap B$$
   $$\Leftrightarrow a \in A \text{ and } a \in B$$
   $$\Leftrightarrow \chi_A(a) = 1 \text{ and } \chi_B(a) = 1$$
   $$\Leftrightarrow \chi_A(a)\chi_B(a) = 1.$$

4. (a) $f(x) = \cos x$ is a bijection on $[0, \pi]$.
   (c) $g\colon \mathbb{R} \to (0, \infty)$ is a bijection when $g(x) = e^x$.
   (e) $f(n) = -n$ is a bijection $\mathbb{Z}^+ \to \mathbb{Z}^-$.
   (g) Let $f\colon \mathbb{Z} \to \mathbb{Z} \times \mathbb{Z}$ be defined to snake through $\mathbb{Z} \times \mathbb{Z}$ similar to the function that shows that $\mathbb{Q}$ is countable. For example,
   $$f(0) = (0, 0) \quad f(5) = (1, 2)$$
   $$f(1) = (1, 0) \quad f(6) = (2, 2)$$
   $$f(2) = (1, 1) \quad f(7) = (2, 1)$$
   $$f(3) = (0, 1) \quad f(8) = (2, 0)$$
   $$f(4) = (0, 2) \quad f(9) = (2, -1).$$
   The negatives will be pre-images for the left halfplane.

5. (a) Use Exercise 6.5.8.

8. Let $a, b \in A$ and assume $\phi(a) = \phi(b)$. We have three cases.
   (Case 1) Let $a = a_n$ and $b = a_m$ for some $n, m \in \mathbb{Z}^+$. Then $\phi(a) = a_{n+1}$ and $\phi(b) = a_{m+1}$. Since the elements of the sequence are distinct, $n+1 = m+1$. Thus, $a = b$.
   (Case 2) Here $a = a_n$ for some n, but b is not a term of the sequence. So, $\phi(a) = a_{n+1}$, which cannot equal b. Thus, this case is impossible.
   (Case 3) Lastly, suppose both a and b are not elements of the sequence. Then $a = b$, for $\phi\restriction A \setminus \{a_1, a_2, \ldots\}$ is the identity map.

9. (a) $f(n) = -n - 1$
   (c) $\phi(a) = (a, b)$ for some fixed $b \in B$
   (e) Use the inclusion map.
   (g) $\phi(a, 0) = (a, 1)$

10. (a) Let $A \preccurlyeq B$ and $B \approx C$. This means that there exists an injection $\phi\colon A \to B$ and a bijection $\psi\colon B \to C$. Therefore, $\psi \circ \phi$ is an injection $A \to C$, so $A \preccurlyeq C$.

11. Let $|A| = n$. Then $|\mathbf{P}(A)| = 2^n$, which is greater than n.

14. Let $\phi\colon \mathbb{N} \to A$ and $\psi\colon A \to B$ be onto. Then $\psi \circ \phi$ is a surjection $\mathbb{N} \to B$.

15. (a) Apply Theorem 6.7.10.
   (b) If $A \cap B = \varnothing$, then it is countable. So suppose there exists $a \in A \cap B$. Since

A is countable, there exists a surjection $f: \mathbb{N} \to A$. Define $h: \mathbb{N} \to A \cap B$ by

$$h(n) = \begin{cases} f(n) & \text{if } f(n) \in A \cap B \\ a & \text{otherwise.} \end{cases}$$

This function is onto, for if $y \in A \cap B$, then there exists $n \in \mathbb{N}$ such that $f(n) = y$. Therefore, $h(n) = f(n) = y$.

16. (a) Apply Theorem 6.7.10.
 (b) Use induction and Exercise 15b.

7.1. TYPES OF RINGS

1. (a) This is not a ring: no additive inverses.
 (c) This is not a ring: no additive identity.
 (e) This is a ring.
 (g) This is not a ring: multiplication is not a binary operation on this set, for $(2i)(2i) = -4$.
 (i) This is not a ring: no additive inverses.

2. We already know that addition and multiplication of integers are associative and commutative. The additive identity is in this set, for $0 = n(0)$. If $a \in \mathbb{Z}$, then $-na = n(-a) \in n\mathbb{Z}$. Therefore, every element in $n\mathbb{Z}$ has an inverse in $n\mathbb{Z}$. Lastly, we know that the integers satisfy the Distributive Law. Thus, this is a commutative ring.

3. There is no additive identity, and inverses are missing.

6. The Distributive Law does not hold.

7. The additive inverse of 1_R.

8. (b) Let $[a]_n \in \mathbb{Z}_n$. Then

$$[a]_n + [0]_n = [a + 0]_n = [a]_n,$$

and

$$[0]_n + [a]_n = [0 + a]_n = [a]_n.$$

 (c) Let $[a]_n \in \mathbb{Z}_n$. Then

$$\begin{aligned}[a]_n + [n - a]_n &= [a + n - a]_n \\ &= [n]_n \\ &= [0]_n.\end{aligned}$$

Similarly, $[n - a]_n + [a]_n = [0]_n$.

9. (a) See the proof of 5.1.3.
 (c) We will prove the first equality:

$$\begin{aligned}(a \otimes b) &\oplus (-a \otimes b) \\ &= (a \oplus -a) \otimes b \\ &= 0_R \otimes b \\ &= 0_R.\end{aligned}$$

 (e) We calculate:

$$\begin{aligned}(a \oplus b) &\oplus (-a \oplus -b) \\ &= a \oplus [b \oplus (-a \oplus -b)] \\ &= a \oplus [b \oplus (-b \oplus -a)] \\ &= a \oplus [(b \oplus -b) \oplus -a] \\ &= a \oplus (0_R \oplus -a) \\ &= a \oplus -a \\ &= 0_R.\end{aligned}$$

10. (a) $\begin{bmatrix} 6 & 1 \\ 8 & -1 \end{bmatrix}$ (b) $\begin{bmatrix} 29 & 4 \\ -24 & -6 \end{bmatrix}$

13. Let $(R, \oplus, \otimes)$ be an integral domain and take $a, b \in R$. Suppose $a \otimes b = 0_R$. If $a \neq 0_R$, then a is not a zero divisor. Hence, $b = 0_R$.

15. (a) [2] and [2]
 (b) [2] and [6], [3] and [4].

17. It is a field. In this case $0_R = 1_R$.

18. (a) Every nonzero real number is a unit.
 (d) [1], [3]

19. (a) Let a be nilpotent and denote 1_R by 1. This means there exists $n \in \mathbb{Z}^+$ such that $a^n = 0_R$. Then, the product

$$(1 \oplus a)(1 \oplus \cdots \oplus (-1)^{n-1}a^{n-1})$$

equals $1 \oplus (-1)^{n-1}a^n$, and this is 1. Thus, $1 \oplus a$ is a unit.

20. We will only show that every element of $R \times R'$ has an additive inverse in that set. Let $(a, b) \in R \times R'$. Since R and R' are rings, $-a \in R$ and $-b \in R'$. Therefore,

$$\begin{aligned}(a, b) &\boxplus (-a, -b) \\ &= (a \oplus -a, b \oplus' -b) \\ &= (0_R, 0_{R'}),\end{aligned}$$

and $(0_R, 0_{R'})$ is the additive identity of $R \times R'$.

7.2. SUBRINGS AND IDEALS

3. (b) This is not a subring: it is not closed under multiplication.
   (d) This is a subring.

6. We will rely on Theorem 7.2.2.

   ($\Rightarrow$) Let S be a subring of R. Therefore, $0 \in S$. Now let $a, b \in S$. By Theorem 7.2.2, $ab \in S$. Furthermore, $-b \in S$, which gives $a - b \in S$. Thus, the three conditions are satisfied.

   ($\Leftarrow$) Suppose S is a subset of R that satisfies the three conditions. Take a and b in S. Then ab and $0 - b = -b$ are elements of S. Thus,

   $$a - (-b) = a + b \in S,$$

   so by Theorem 7.2.2, S is a subring of R.

8. (b) $\{[0]\}, \{[0], [3]\}, \{[0], [2], [4]\}, \mathbb{Z}_6$

10. Let $I = \{(2m, 2n) : m, n \in \mathbb{Z}\}$. We will check the four conditions:

    1. Let $(2a, 2b), (2c, 2d) \in I$. The sum of these ordered pairs is

       $$(2[a + c], 2[b + d]),$$

       and the product is

       $$(2[2ac], 2[2bd]).$$

       These are both elements of I.
    2. $(0, 0) \in I$, since $0 = 2(0)$.
    3. Let $(2a, 2b) \in I$. Its additive inverse is $(-2a, -2b)$. This is in I because $-2a = 2(-a)$ and $-2b = 2(-b)$.
    4. Take $(2a, 2b) \in I$ and $(r, s) \in \mathbb{Z} \times \mathbb{Z}$. Then $(r, s) \cdot (2a, 2b) = (2ar, 2bs)$, and $(2a, 2b) \cdot (r, s) = (2ar, 2bs)$. So, I is an ideal of $\mathbb{Z} \times \mathbb{Z}$.

11. $(0, 1) \cdot (1, 1) = (0, 1)$, which is not an element of the set.

14. We have already seen that the union is a subring. Let $r \in R$ and $a \in \bigcup_{\alpha \in \Lambda} J_\alpha$. This means there exists $\alpha \in \Lambda$ such that $a \in J_\alpha$. Since J_α is an ideal, $ra \in J_\beta$ and $ar \in J_\beta$. Hence, both ra and ar are elements of $\bigcup_{\alpha \in \Lambda} J_\alpha$.

16. (a) $\{[0], [2], [4], [6], [8]\}$
    (d) $\{([0], [0]), ([1], [1]), ([2], [2]), ([3], [3]), ([4], [0]), ([5], [1]), ([0], [2]), ([1], [3]), ([2], [0]), ([3], [1]), ([4], [2]), ([5], [3])\}$

18. A generator is $(1, 3)$.

22. Use Exercise 21. Let F be a field and I an ideal of F. Suppose $I \neq \{0\}$. Then there exists $u \in I \setminus \{0\}$. Hence, u is a unit and $I = F$.

26. We will only show the following: To see that it is a left ideal, let $r, s \in R$. Then $s(ra) = (sr)a$. Since $sr \in R$, $s(ra) \in \langle a \rangle$.

28. (c) Suppose $\gcd(m, n) = 1$. Then there are integers u and v so that $um + vn = 1$. Therefore, $1 \in \langle m, n \rangle$ from which we conclude $\langle m, n \rangle = \mathbb{Z}$.

32. (a) $\{[0]\}$ (c) $\mathbb{Z}_6 \times \{[0]\}$

34. Let p be prime. To show that

    $$I = \{(pa, b) : a, b \in \mathbb{Z}\}$$

    is a maximal ideal, let J be an ideal of $\mathbb{Z} \times \mathbb{Z}$ such that $I \subset J$. Take $(m, n) \in J \setminus I$. This means $p \nmid m$, so $\gcd(m, p) = 1$. Hence, there exist $u, v \in \mathbb{Z}$ such that $um + vp = 1$. Since $(m, n) \in J$,

    $$(u, 1)(m, n) \in J.$$

    Hence, $(um, n) \in J$. Also, $(vp, 1-n) \in J$ because $(vp, 1 - n) \in I$. Thus,

    $$(um, n) + (vp, 1 - n) = (1, 1)$$

    is an element of J. Therefore, $J = \mathbb{Z} \times \mathbb{Z}$.

7.3. FACTOR RINGS

1. (a) $\{\ldots, -4, 0, 4, 8, \ldots\}$
   $\{\ldots, -3, 1, 5, 9, \ldots\}$
   $\{\ldots, -2, 0, 2, 10, \ldots\}$
   $\{\ldots, -1, 3, 7, 11, \ldots\}$
   (c) $\{[0], [5]\}, \{[1], [6]\}, \{[2], [7]\}, \{[3], [8]\}, \{[4], [9]\}$
   (f) $\{([0]_3, [0]_2), ([0]_3, [1]_2)\}, \{([1]_3, [0]_2), ([1]_3, [1]_2)\}, \{([2]_3, [0]_2), ([2]_3, [1]_2)\}$
   (h) $\langle([1]_6, [1]_4)\rangle$,
   $([1], [0]) + \langle([1]_6, [1]_4)\rangle = \{([2], [1]), ([3], [2]), ([4], [3]), ([5], [0]), ([0], [1]), ([1], [2]),$

$([2], [3]), ([3], [0]), ([4], [1]),$
$([5], [2]), ([0], [3]), ([1], [0])\}$

2. (a) $\{\{(a, b)\} : a, b \in \mathbb{Z}\}$
   (c) $\{\{(n+a, m+a) : a \in \mathbb{Z}\} : n, m \in \mathbb{Z}\}$

3. (a) 7 (c) 8

5. Let I be an ideal of R and $a \in R$.
   ($\Rightarrow$) Suppose $a + I$ is the additive identity of R/I. Let $b + I \in R/I$. Then,
   $$(a + I) + (b + I) = b + I.$$
   Hence, $a + b - b \in I$. In other words, $a \in I$.
   ($\Leftarrow$) Let $a \in I$. To show that $a + I$ is the additive identity of R/I, take $b \in R$ and add $a + I$ to $b + I$. This yields $a + b + I$. Since $a + b - b = a$ is an element of I, we have $a + b + I = b + I$.

8. (a) Take $R = \mathbb{Z}$ and $I = 4\mathbb{Z}$.

10. Use Lagrange's Theorem.

7.4. HOMOMORPHISMS

1. (b) We have already seen that this is a function, so let $a, b \in \mathbb{Z}$. To see that ϕ is a ring homomorphism, calculate:
   $$\begin{aligned}\phi([a]_{12} + [b]_{12}) &\\ &= \phi([a + b]_{12})\\ &= [a + b]_6\\ &= [a]_6 + [b]_6\\ &= \phi([a]_{12}) + \phi([b]_{12}),\end{aligned}$$
   and
   $$\begin{aligned}\phi([a]_{12} \cdot [b]_{12}) &\\ &= \phi([ab]_{12})\\ &= [ab]_6\\ &= [a]_6 \cdot [b]_6\\ &= \phi([a]_{12}) \cdot \phi([b]_{12}).\end{aligned}$$

5. (b) To see that the function does not preserve addition, note that $f(1 + 1) = e^2$, but $f(1) + f(1) = 2e \neq e^2$. Neither does the function preserve multiplication.

6. (a) Let $a, b \in R$. Then
   $$\begin{aligned}\phi(a + b) &= a + b + I\\ &= (a + I) + (b + I)\\ &= \phi(a) + \phi(b),\end{aligned}$$
   and
   $$\begin{aligned}\phi(ab) &= ab + I\\ &= (a + I)(b + I)\\ &= \phi(a)\phi(b).\end{aligned}$$
   (b) Exercise 7.3.5 is required for the derivation:
   $$\begin{aligned}\ker(\phi) &= \{a \in R : \phi(a) = 0 + I\}\\ &= \{a \in R : a + I = 0 + I\}\\ &= \{a \in R : a \in I\}\\ &= I.\end{aligned}$$

7. (a) Let $\phi(1) = e$. Take $b \in R'$. Since ϕ is onto, there exists $a \in R$ such that $\phi(a) = b$. Thus,
   $$eb = \phi(1)\phi(a) = \phi(1a) = b.$$
   Similarly, $be = b$. Therefore, e is the unity of R'.
   (b) Use the homomorphism
   $$\phi: \mathbb{Z} \to \mathbb{Z} \times \mathbb{Z}$$
   defined by $\phi(n) = (n, 0)$.

9. We will find the kernel and image of part (b): The kernel of ϕ is $\{[0]_{12}, [6]_{12}\}$, and its image is $\mathbb{Z}_6$, which means that the homomorphism is not one-to-one, but it is onto.

12. (a) Let $a, b \in R$. Then
   $$\begin{aligned}(\phi \circ \psi)(a + b) &\\ &= \phi(\psi(a + b))\\ &= \phi(\psi(a) + \psi(b))\\ &= \phi(\psi(a)) + \phi(\psi(b))\\ &= (\phi \circ \psi)(a) + (\phi \circ \psi)(b).\end{aligned}$$
   That the composition preserves multiplication is proven similarly.

14. We must show that ψ is a homomorphism that is a bijection.

1. Let $(a, b), (c, d) \in R \times R$. We will only check that ψ preserves addition:

$$\psi((a, b) + (c, d))$$
$$= \psi((a + c, b + d))$$
$$= (\phi(a + c), \phi(b + d))$$
$$= (\phi(a) + \phi(c), \phi(b) + \phi(d))$$
$$= (\phi(a), \phi(b)) + (\phi(c) + \phi(d))$$
$$= \psi(a, b) + \psi(c, d).$$

(The multiplication check is similar.)

2. Since we now know that ψ is a homomorphism, use the kernel test to check if it one-to-one. Let $(a, b) \in R \times R$ and assume $\psi(a, b) = (0, 0)$. This means that $\phi(a) = \phi(b) = 0$. Since ϕ is an injective homomorphism, $a = b = 0$.
3. Let $(a', b') \in R' \times R'$. Then there exists $a, b \in R$ such that $\phi(a) = a'$ and $\phi(b) = b'$. Hence,

$$\psi(a, b) = (\phi(a), \phi(b)) = (a', b').$$

16. (a) Use $\phi(r) = r + \{0\}$.

17. (a) Let $a', b' \in R'$ such that $a'b' = 0$. Since ϕ is onto, there exist $a, b \in R$ so that $\phi(a) = a'$ and $\phi(b) = b'$. Since ϕ is a homomorphism,

$$0 = a'b' = \phi(a)\phi(b) = \phi(ab).$$

Since ϕ is one-to-one, $ab = 0$, but R is an integral domain. Hence, $a = 0$ or $b = 0$, which gives $a' = 0$ or $b' = 0$.

7.5. POLYNOMIALS

1. -25

2. (a) 1
(c) 4
(e) $\begin{bmatrix} 5 & 6 \\ 12 & 13 \end{bmatrix}$

3. (a) $f(X) = g(X)[-3 + X] + 0$
(c) $f(X) = g(X)[1 + X + X^2 + X^3 + X^4] + 2$

4. The quotient is $1 + X^2$ and the remainder is X.

7. (b) Assuming that I is an ideal, to prove $I = \langle X \rangle$, let $f(X) \in I$. This means $f(X)$ equals

$$a_1X + a_2X^2 + a_3X^3 + \cdots$$

where $a_i \in \mathbb{Q}$. Therefore, $f(X)$ is the product

$$(a_1 + a_2X + a_3X^2 + \cdots)(X).$$

9. (a) We will only prove the following: Let $f(X) \in (\mathbb{Z} \times \mathbb{Z})[X]$ and $g(X) \in K$. We will show $f(X)g(X) \in K$. Write $g(X)$ as

$$(a_0, 0) + (a_1, 0)X + \cdots$$

with $a_i \in \mathbb{Z}$ and write $f(X)$ as

$$(b_0, c_0) + (b_1, c_1)X + \cdots$$

where $(b_i, c_i) \in \mathbb{Z} \times \mathbb{Z}$. Therefore, $f(X)g(X)$ equals

$$d_0 + d_1X + d_2X^2 + \cdots$$

where

$$d_i = \sum_{j=0}^{i} (b_j, c_j)(a_{i-j}, 0)$$
$$= \sum_{j=0}^{i} (b_j a_{i-j}, 0).$$

Hence, $f(X)g(X) \in K$.

(b) $K = \langle (1, 0) \rangle$

15. (b) Let $f(X), g(X) \in \mathbb{Z}[X]$ and write

$$f(X) = a_0 + a_1X + \cdots$$

and

$$g(X) = b_0 + b_1X + \cdots.$$

Then $\psi(f(X) + g(X))$ equals

$$\psi([a_0 + b_0] + [a_1 + b_1]X + \cdots).$$

This is then equal to $a_0 + b_0$, which is the same as the sum of

$$\psi(a_0 + a_1X + \cdots)$$

and

$$\psi(b_0 + b_1X + \cdots).$$

Similarly, we can show that multiplication is preserved.

16. For part (b), the kernel is

$$\{a_1X + a_2X + \cdots : a_i \in \mathbb{Z}\},$$

and the image is $\mathbb{Z}$.

19. (a) Write $f(X)$ as

$$a_0 + a_1X + \cdots + a_mX^m$$

and $g(X)$ as

$$b_0 + b_1X + \cdots + b_nX^n.$$

Suppose $m \leq n$. In this case, $f(X)+g(X)$ is the polynomial

$$(a_0 + b_0) + \cdots + (a_m + b_m)X^m + \cdots + b_nX^n.$$

Hence,

$$\deg[f(X) + g(X)] = n = \max(m, n).$$

Similarly, if $n < m$, then the degree of $f(X) + g(X)$ is $m = \max(m, n)$.

8.1. SPACES

1. Let $a, b \in \mathbb{R}$. Then

$$|a| = |a - b + b| \leq |a - b| + |b|.$$

Therefore, $|a| - |b| \leq |a - b|$.

3. We will only check that the Triangle Inequality holds: For every $x, y, z \in \mathbb{R}$:

$$\begin{aligned} d(x, y) &= |x - y| \\ &= |x - z + z - y| \\ &\leq |x - z| + |z - y| \\ &= d(x, z) + d(z, y). \end{aligned}$$

5. 1/4

6. Let $a, b, c \in M$. Then $d(a, b) \geq 0$ and

$$d(a, b) = 0 \Leftrightarrow a = b$$

by definition of d. Clearly, $d(a, b) = d(b, a)$. For the Triangle Inequality, we have two cases. If $d(a, b) = 0$, then the inequality holds. If $d(a, b) = 1$, then either $a \neq c$ or $b \neq c$. Hence, $d(a, c) + d(c, b) = 1$, and the inequality holds.

8. (b) We will check the first two properties. Let $(a, b), (a', b') \in M_1 \times M_2$. First,

$$d((a, a'), (b, b')) \geq 0$$

since $d_1(a, a') \geq 0$ and $d_2(b, b') \geq 0$. For the second,

$$\begin{aligned} &d((a, a'), (b, b')) = 0 \\ &\Leftrightarrow d_1(a, a') = 0 \text{ and } d_2(b, b') = 0 \\ &\Leftrightarrow a = a' \text{ and } b = b' \\ &\Leftrightarrow (a, b) = (a', b'). \end{aligned}$$

9. (b) We have four conditions to check.
   1. Since regular addition is associative in general, it is associative on this set.
   2. The additive identity is an element of $n\mathbb{Z}$ since $0 = n \cdot 0$.
   3. Let $a \in n\mathbb{Z}$. This means $a = nk$ for some $k \in \mathbb{Z}$. Then $-a = -nk = n(-k)$. Hence, $-a \in n\mathbb{Z}$.
   4. Since regular addition is commutative, it is commutative on this set.

10. (a) No (b) Yes (c) No (d) No

15. (a) Let $(m, n), (m', n') \in \mathbb{Z} \times \mathbb{Z}$.
   1. $\|(m, n)\| = |m| + |n| \geq 0$
   2. $\|(m, n)\| = 0$
   $$\begin{aligned} &\Leftrightarrow |m| + |n| = 0 \\ &\Leftrightarrow |m| = 0 \text{ and } |n| = 0 \\ &\Leftrightarrow m = 0 \text{ and } n = 0 \\ &\Leftrightarrow (m, n) = (0, 0) \end{aligned}$$
   3. Let $a \in \mathbb{R}$.
   $$\begin{aligned} \|a(m, n)\| &= \|(am, an)\| \\ &= |am| + |an| \\ &= |a|(|m| + |n|) \\ &= |a|\|(m, n)\| \end{aligned}$$
   4. $\|(m, n) + (m', n')\|$
   $$\begin{aligned} &= \|(m + m', n + n')\| \\ &= |m + m'| + |n + n'| \\ &\leq |m| + |n| + |m'| + |n'| \\ &= \|(m, n)\| + \|(m', n')\| \end{aligned}$$

8.2. OPEN SETS

1. Let $x, y \in M$ and $x \neq y$. Let $\varepsilon = d(x, y)/2$. If $z \in D(x, \varepsilon) \cap D(y, \varepsilon)$, then $d(x, z) < d(x, y)/2$ and $d(y, z) < d(x, y)/2$. Hence,

$$\begin{aligned} d(x, y) &\leq d(x, z) + d(y, z) \\ &< d(x, y), \end{aligned}$$

a contradiction.

3. (b) Let $(x, y) \in \mathbb{R}\setminus\{(0, 0)\}$. Let ε be the distance from this point to the origin. In other words,

$$\varepsilon = \sqrt{x^2 + y^2}.$$

Then $D((x, y), \varepsilon) \subseteq \mathbb{R} \setminus \{(0, 0)\}$.

(d) Let $S = \{(x, y) \in \mathbb{R}^2 : 5 < x < 7\}$. Take $(a, b) \in S$. This means $5 < a < 7$. Define $\varepsilon = \min(a - 5, 7 - a)$. We must now show that

$$D((a, b), \varepsilon) \subseteq S.$$

Let $(x, y) \in D((a, b), \varepsilon)$. Hence,

$$\sqrt{(x - a)^2 + (y - b)^2} < \varepsilon.$$

Hence

$$\begin{aligned} (x - a)^2 &\leq (x - a)^2 + (y - b)^2 \\ &< \varepsilon^2. \end{aligned}$$

Therefore, $|x - a| < \varepsilon$, which means $-\varepsilon < x - a < \varepsilon$. Since $\varepsilon = \min(a - 5, 7 - a)$, we have

$$-a + 5 < x - a < 7 - a.$$

Thus, $5 < x < 7$ and $(x, y) \in S$.

(f) Take $(a, b, c) \in \{(x, y, z) \in \mathbb{R}^3 : x^2 + y^2 < 1\}$. Let $\varepsilon = 1 - \sqrt{a^2 + b^2}$ and show $D((a, b, c), \varepsilon)$ is a subset of the given set.

6. (b) Let $(x, y) \in U_1 \times U_2$. This means $x \in U_1$ and $y \in U_2$. Since these two sets are open in their respective metric spaces, there exist $\varepsilon_1 > 0$ and $\varepsilon_2 > 0$ such that

$$D_{M_1}(x, \varepsilon_1) \subseteq U_1$$

and

$$D_{M_2}(x, \varepsilon_2) \subseteq U_2.$$

Define $\varepsilon = \min(\varepsilon_1, \varepsilon_2)$. We must show

$$D_{M_1 \times M_2}((x, y), \varepsilon) \subseteq U_1 \times U_2.$$

Let $d((a, b), (x, y)) < \varepsilon$. Then

$$d_1(a, x) + d_2(b, y) < \varepsilon.$$

So, $d_1(a, x) < \varepsilon_1$ and $d_2(a, x) < \varepsilon_2$. This means $a \in U_1$ and $b \in U_2$. Hence, $(a, b) \in U_1 \times U_2$.

7. (b) This intersection equals $(0, 1/k)$, an open set.

8. Since

$$\bigcap_{j=1}^{n} U_{i,j}$$

is a finite intersection of open sets, it is open for every i. Therefore, the union of this collection is open.

9. (a) $\varnothing$ (c) $(0, 2)$

10. (a) $\{(x, y) \in \mathbb{R}^2 : x^2 + y^2 < 1\}$

(c) $\varnothing$

11. (a) Since $(0, 1)$ is an open subset of the closed interval $[0, 1]$,

$$(0, 1) \subseteq \text{int}([0, 1]).$$

To prove the other inclusion, let $x \in \text{int}([0, 1])$. By the theorem, there exists $\varepsilon > 0$ such that

$$(x - \varepsilon, x + \varepsilon) \subseteq [0, 1].$$

Hence, $x \in (0, 1)$.

15. (a) Let $x \in \text{int}(A) \cup \text{int}(B)$. This means $x \in \text{int}(A)$ or $x \in \text{int}(B)$. If the first disjunct is true, then there exists an open $U \subseteq A$ such that $x \in U$. But, U is also a subset of $A \cup B$. Therefore, $x \in \text{int}(A \cup B)$. Similarly, if the second disjunct is true, then $x \in \text{int}(A \cup B)$.

8.3. CLOSED SETS

2. Fix $x \in M$. To show that $\{x\}$ is closed, we must show $M \setminus \{x\}$ is open. Let $y \in M \setminus \{x\}$. Define $\varepsilon = d(x, y)$. Since $x \notin D(y, \varepsilon)$, $D(y, \varepsilon) \subseteq M \setminus \{x\}$.

3. (c) Let (a, b) be any point in $\mathbb{R}^2$ not on $y = 2x - 1$. Let ℓ be the line containing (a, b) with slope $-1/2$. Then

there exists (c, d) on ℓ so that (c, d) is between (a, b) and $y = 2x - 1$. If we let ε be the distance from (a, b) to (c, d), then $D((a, b), \varepsilon)$ does not intersect $y = 2x - 1$.

5. (b) Fix $a \in M$. Let $S = \{b \in M : d(a, b) \leq 1\}$. Let x be an accumulation point of S. Then for all $\varepsilon > 0$,

$$[D(x, \varepsilon) \setminus \{x\}] \cap S \neq \varnothing,$$

so there exists $w_\varepsilon \in M$ and $w_\varepsilon \neq x$ such that both $d(x, w_\varepsilon) < \varepsilon$ and $d(a, w_\varepsilon) \leq 1$. Therefore, for all $\varepsilon > 0$,

$$\begin{aligned} d(a, x) &\leq d(x, w_\varepsilon) + d(a, w_\varepsilon) \\ &< \varepsilon + 1. \end{aligned}$$

Hence, $d(a, x) \leq 1$, and $x \in S$.

8. (a) Let $x \in A$. Since the closure is the intersection of every closed superset of A, $x \in \text{cl}(A)$.

9. (c) $\text{cl}(A \cup B)$

$$\begin{aligned} &= A \cup B \cup \text{AP}(A \cup B) \\ &= A \cup B \cup \text{AP}(A) \cup \text{AP}(B) \\ &= \text{cl}(A) \cup \text{cl}(B). \end{aligned}$$

10. To prove sufficiency, let A be closed. We know that $A \subseteq \text{cl}(A)$, and since $\text{cl}(A)$ is the intersection of all closed supersets of A, $\text{cl}(A) \subseteq A$. For necessity, let $\text{cl}(A) = A$. Let a be an accumulation point of A. Since $A \subseteq \text{cl}(A)$, a is also an accumulation point of $\text{cl}(A)$. However, $\text{cl}(A)$ is closed, so $a \in \text{cl}(A)$. Hence, $a \in A$, so A is closed.

12. (a) $[0, 2]$
    (c) $[0, 2]$
    (e) $\{(x, y) \in \mathbb{R}^2 : x^2 + y^2 \leq 1\}$

8.4. ISOMETRIES

1. (a) $(1, 6)$ (d) $(-5, 3)$

2. (a) $(2, 3) \times \{0\}$ (c) $\{0\} \times \mathbb{R}$

3. (a) $\{1\} \times \mathbb{R}$ (c) $\{(0, -1)\}$

5. Let $f, g : \mathbb{R}^2 \to \mathbb{R}^2$ be isometries. Take (a, b) and (c, d) in $\mathbb{R}^2$. Then,

$$d([f \circ g](a, b), [f \circ g](a, b))$$

equals

$$d(f(g(a, b)), f(g(a, b))).$$

Since f is an isometry, this equals,

$$d(g(a, b), g(a, b)),$$

which in turn is the same as

$$d((a, b), (a, b))$$

because g is an isometry.

10. (a) $\phi(f) = x^2 + x - 1$
    (b) Let $f, h \in C[0, 1]$. We calculate distances:

$$\int_0^1 |\phi(f) - \phi(h)|\, dx$$

equals

$$\int_0^1 |f(x) + g(x) - h(x) - g(x)|\, dx,$$

and this is equal to

$$\int_0^1 |f(x) - h(x)|\, dx.$$

Therefore, ϕ is an isometry.

12. (a) Let C be closed in M. This means $M \setminus C$ is open. To show $N \setminus \psi[C]$ is open, let y be an element of this set. Since ψ is a bijection, there exists $x \in M \setminus C$ such that $\psi(x) = y$. Because $M \setminus C$ is open, there exists $\varepsilon > 0$ so that

$$D_M(x, \varepsilon) \subseteq M \setminus C.$$

Since $\psi[D_M(x, \varepsilon)] = D_N(y, \varepsilon)$,

$$D_N(y, \varepsilon) \subseteq N \setminus \psi[C].$$

Hence, $\psi[C]$ is closed in N.

13. (e) Let $(a, b), (c, d) \in \mathbb{R}^2$. Then,

$$P(a, b) = (a, -b)$$

and

$$P(c, d) = (c, -d).$$

Therefore, P is an isometry because

$$\sqrt{(a - c)^2 + (b - d)^2}$$

is equal to

$$\sqrt{(a - c)^2 + (-b - -d)^2}.$$

15. (a) $T_{[-a,\,-b]}$ (b) $R_{-\alpha}$

16. (a) Let $(c, d) \in \mathbb{R}^2$. Then

$$T_{[a,\,b]}(c-a,\, d-b)$$

equals

$$(c-a+a,\, d-b+b),$$

which is (c, d).

8.5. LIMITS

1. (a) 7 (b) 1 (c) 3 (d) 3

3. If $\varepsilon > 0$, then for any $n \in \mathbb{Z}^+$,

$$|a_n - k| = |k - k| < \varepsilon.$$

4. (a) Let $\varepsilon > 0$ and choose N to be a positive integer greater than $1/\varepsilon$. Let $n \geq N$. This gives $n > 1/\varepsilon$, and we conclude that $|1/n| < \varepsilon$.

5. (a) By Theorem 8.5.6,

$$\lim_{n\to\infty} \frac{2n^2}{6n^2+n+1}$$

equals

$$\lim_{n\to\infty} \frac{2n^2/n^2}{(6n^2+n+1)\cdot 1/n^2},$$

and this limit equals $1/3$. Since limits are unique, the limit cannot equal 0.

(d) Let $\varepsilon = 1$. Then we have

$$|(1/2)^n + 1 - 0| > 1.$$

6. (a) 0 (b) 1 (c) 0 (d) 0

7. (a) Let $a_n = n$. Suppose there exists $a \in \mathbb{R}$ such that $a_n \to a$. Now for all $n \geq 1 + a$ we have

$$|n - a| \geq 1,$$

a contradiction. Thus, the series does not converge.

(b) Suppose $2a_n = n^2/(n+1)$ converges, so we may conclude that the series is bounded. In other words, there exits $M > 0$ so that for all $n \in \mathbb{Z}^+$,

$$\left|\frac{2n^2}{n+1}\right| \leq M.$$

However, if we choose any positive integer $n > M + 1$, we have a contradiction because

$$\begin{aligned}\frac{2n^2}{n+1} &\geq \frac{n^2}{n+1}\\ &\geq \frac{n^2-1}{n+1}\\ &= \frac{(n+1)(n-1)}{n+1}\\ &= n-1\\ &> M.\end{aligned}$$

Hence, the series cannot converge.

9. Use Theorem 8.5.6.

12. Let $\varepsilon > 0$. Since $a_n \to \ell$ and $c_n \to \ell$, there exist positive integers N_1 and N_2 such that for all n,

$$n \geq N_1 \Rightarrow |a_n - \ell| < \varepsilon$$

and

$$n \geq N_2 \Rightarrow |c_n - \ell| < \varepsilon.$$

Therefore, let $n \geq \max(N_1, N_2)$. Then,

$$-\varepsilon < a_n - \ell < \varepsilon$$

and

$$-\varepsilon < c_n - \ell < \varepsilon.$$

Hence,

$$-\varepsilon < a_n - \ell \leq b_n - \ell \leq c_n - \ell < \varepsilon.$$

This means $|b_n - \ell| < \varepsilon$.

Index